The papers of H.T. De la Beche (1796 - 1855) in the National Museum of Wales

T. SHARPE
Department of Geology, National Museums & Galleries of Wales, Cardiff CF1 3NP

P.J. McCARTNEY
78 Faïracres Road, Oxford OX4 1TG

AMGUEDDFEYDD AC ORIELAU CENEDLAETHOL CYMRU
NATIONAL MUSEUMS & GALLERIES OF WALES
Geological Series No. 17.
Cardiff, August 1998

© *National Museum of Wales 1998*

ISBN 0 7200 0454 3

REFERENCES TO THIS VOLUME

It is recommended that reference to this volume be made in the following form:

SHARPE, T. and McCARTNEY, P.J. 1998. The papers of H.T. De la Beche (1796-1855) in the National Museum of Wales. 257pp. National Museum of Wales, Geological Series No.17, Cardiff.

FOREWORD

An Act of Parliament dated 31 July 1845 was designed *'to facilitate the completion of a Geological Survey of Great Britain and Ireland, under the Director of the First Commissioner for the time being of Her Majesty's Woods and Works'*. On the one hand, this was the culmination of over thirteen years of political manoeuvring, scientific endeavour, and no little self-promoting scheming by Henry Thomas De la Beche. On the other hand, it formed the final stepping stone for De la Beche (he was knighted in 1842) to fulfil his driving vision of bringing geological mapping and surveying into the public domain as a means of establishing a firm base for the assessment and exploitation of natural resources. As first Director of the Survey, and with an initial scientific staff of twenty-four, De la Beche developed and expanded his programmes throughout the British Isles, until his death in 1855. From his earlier interests as a 'gentleman amateur' geologist, and following his appointment in 1832 as the first geologist appointed by the Ordnance Office to the Trigonometrical Survey of Great Britain, his persistent lobbying was the main factor in lifting the profession into a fully accountable national service. Among his other, related achievements were the establishment of a Museum of Economic Geology (successively becoming the Museum of Practical Geology and the Geological Survey Museum; now part of the Natural History Museum, London) and the Royal School of Mines (now part of Imperial College, University of London).

In all his activities De la Beche corresponded regularly with scientific colleagues and many public figures, including politicians, and he compiled journals of his work and travels, often containing sketches in support of his observations, because he was a gifted artist. These various documents, comprising over 2000 items, form the subject matter of this catalogue. They represent one of the most important archival sources in the British Isles documenting the early establishment, history and development of earth sciences, from a period when geology was at the forefront of scientific methodology and philosophy. The archive is consulted regularly by historians of science, but its full scope and content has never been appreciated because it has never been collated comprehensively as it is in this publication.

Combined chances of history and marriage dictated the fact that after his death, the early De la Beche papers remained in his daughter's family in South Wales. They were donated eventually to the National Museum of Wales in the 1930s, when the first Keeper of Geology, Dr F.J. North, began to use them for a series of papers that brought the work of De la Beche and some of his contemporaries to the fore. North's successor, Dr D.A. Bassett, continued this involvement from the 1960s, including the important process of having the original hand-written documents transcribed into type-written copies so that they could be consulted without endangering the originals.

Beginning in 1974, Dr Paul McCartney, then of Wolfson College at Oxford University, began to visit the National Museum to consult the archive in pursuit of his research on the history of science in government; the visits included a number of longer periods of residence, when the opportunity to organise the archive into an indexed catalogue first developed. Employment eventually took Dr McCartney elsewhere and away from the project, but fortunately the work was then maintained by Tom Sharpe following his appointment to the Museum staff in 1978. This publication is a tribute to their joint studies. Notwithstanding the twenty-five years or so that it has taken to bring this task to fruition, its importance cannot by over-estimated as a collation of one of the important documentary sources of British scientific history. It should be added that the compilation also includes the Museum's archival papers of a leading contemporary and collaborator of De la Beche, William Buckland who was Reader in Geology at Oxford and Dean of Westminster; these papers also came to the Museum via South Wales family connections.

Restrictions of space alone prevent the archive contents being published in full. Instead, each item is summarised to give emphasis to the main content, and is then indexed in various ways as explained in the Introduction. References to previously published use of archival items are themselves valuable bibliographical sources of information.

It is wholly appropriate that the National Museum should publish this fundamental contribution to the history of earth sciences. We had hoped to do so in 1996 to recognise the bicentenary of De la Beche's birth, but financial constraints dictated otherwise. Nevertheless, this publication continues a long involvement by the Department of Geology in researching all aspects of the geology of Wales and related regions. The Principality holds a special place in the historical foundations of geology, not only through De la Beche but in his notable contemporaries (and sometime rivals) Adam Sedgwick and Roderick Impey Murchison, and the National Museum is a major repository for the rocks, fossils, minerals and documentation built from the formative studies that were to be applied throughout the world.

The De la Beche (and Buckland) archives are now conserved to modern standards and are housed in purpose-built archival storage in the Department of Geology at the National Museum of Wales, Cardiff. Enquiries for access and consultation should be addressed to the Department.

Some of his contemporaries commented that De la Beche was a 'dirty dog' ..... 'a thorough jobber and a great intriguer' ..... 'an artful dodger, for ever working for his own interest' ..... [who] 'started the survey very much for his own interest, heedless of that of others'. True or not, and for whatever reason, his archive reminds us of his seminal role in taking geology from an amateur pursuit into a governmentally supported science. The 1845 Parliamentary Act envisaged the completion of a geological survey of Great Britain and Ireland. Today the British Geological Survey continues as one of the leading scientific organisations in the U.K., and reflects the far sighted determination set out in the De la Beche archive from the early 19th Century.

Professor Michael G. Bassett
Keeper of Geology

Henry De La Beche in Swansea c. 1853 (**422.4**)

CONTENTS

	Page
Foreword	3
Contents	5
List of illustrations	5
Introduction	7
A chronology of H.T. De la Beche	10
Publications of H.T. De la Beche	11
Acknowledgements	14
Catalogue	
A to Y (catalogue numbers 1 - 2172)	15
Unidentified authors (catalogue numbers 2173 - 2180)	145
Anonymous (catalogue numbers 2181 - 2228)	145
Miscellaneous printed papers (catalogue numbers 2229 - 2237)	147
Addenda (catalogue numbers 2238 - 2278)	147
Miscellaneous items (catalogue numbers 2279 - 2283)	148
Indexes	
Personal Names	150
Place Names	174
Subject	219
Figures	235
References	253

LIST OF ILLUSTRATIONS

[Bold numbers in brackets indicate the relevant catalogue number]

Frontispiece
    Henry De la Beche in Swansea c.1853 (**422.4**)

Figure
1. Fossil trees in Northumberland by William Buckland (**144**)
2. Ichthyosaur skull structure: William Daniel Conybeare to De la Beche 1821 (**297**)
3. Section of Portland Dirt Bed near Lulworth, Dorset by William Buckland (**187**)
4. Ichthyosaur skull structure: William Daniel Conybeare to De la Beche 1821 (**297**)
5. Durdle Door, Dorset (**341**)
6. Branscombe Cliffs, Devon (**341**)
7. Vertical strata of Chalk, Dorset (**341**)
8. The Bill of Portland, Dorset (**341**)
9. Geology of the coast from Beer, Devon to Lyme Regis, Dorset (**341**)
10. Geology of the coast from Charmouth to Burton Bradstock, Dorset (**341**)
11. Fish teeth and spines, Lias, Dorset (**341**)
12. De la Beche's jetty at Lyme (**341**)
13. Pentacrinite, Lias, and Echinite, Greensand, Lyme (**341**)
14. Plan of Cobb walls, Lyme (**341**)
15. Caerphilly Castle (**342**)

List of Illustrations

16	Caerphilly Castle (**342**)	50	*Heat in Mines* (**377**)
17	Section through the Carneddau, Builth, South Wales (**342**)	51	*State of the Survey of England in January 1838* (**384**)
18	Section from Newport to Brecon, South Wales (**342**)	52	*The Irregularities of Sol* (**395**)
19	*Seeing Mont Blanc and the Valley of Chamonix* (**346**)	53	Henry De la Beche, 1848 (**415**)
20	*A Chamonix Char-a-Banc* (**346**)	54	De la Beche and his daughters Rosie (standing) and Bessie (seated) (**422.1**)
21	Section of Jamaica from Kingston to Buff Bay (**357**)	55	De la Beche and his three eldest grandchildren, Mary, Henry, and Elizabeth Amy Dillwyn (**422.2**)
22	Section from Lluidas Vale to the sea near Old Harbour, Jamaica (**355**)	56	Rosie De la Beche and Bessie Dillwyn (**422.5**)
23	Section north across Jamaica from Kingston (**356**)	57	De la Beche, Rosie, and Tartar (**422.3**)
24	Section along the Hope Valley to Catherine's Peak, Jamaica (**356**)	58	Lewis Llewelyn Dillwyn and Badger (**422.7**)
25	Geological map of Tor Bay by De la Beche (**360**)	59	Bessie Dillwyn (**422.9**)
26	*Seeing the Raphaels at the Vatican* (**363**)	60	Lewis Llewelyn Dillwyn and his daughter Amy (**422.6**)
27	*Milanese* (**363**)	61	Bessie Dillwyn and her daughter Amy (**422.10**)
28	*Taking a Cameo-portrait* (**363**)	62	*A Coprolitic Vision* (**432**)
29	*Visiting the Vatican by torchlight* (**363**)	63	Gaff-rigged schooner off Dorset coast (**433**)
30	*Peasant girl of Pignone* (**363**)	64	Hooken Undercliff, Beer Head, Devon (**433**)
31	*Capucines* (**363**)	65	Beer Head and Whitecliff from Dowlands Cliff, Devon (**433**)
32	*Crater of Vesuvius* (**363**)	66	On N. of Snowdon Peak (**436**)
33	*Lazzaroni* (**363**)	67	Fowey Consols Copper Mine, Cornwall (**434**)
34	Transported [erratic] block, Alpi di Pravolta, Lake of Como (**363**)	68	Trilobites from Bala by Richard Gibbs (**609**)
35	*General section of the stratified rocks in the neighbourhood of La Spezia* (**363**)	69	Letter from Robert Jameson on Edinburgh College Museum letterhead (**798**)
36	Section from Bagni di Lucca (**363**)	70	Arrangement of specimens of materials used in iron-making James Nasmyth to De la Beche, 1842 (**1039**)
37	*Portrait à la Rembrandt* (**363**)		
38	Section from the Carrara Marble Quarries to Lavenza (**363**)	71	Section of the Dowlands landslip, Devon: John Phillips to De la Beche, 1840 (**1339**)
39	Sections from near Como to Monte Francesca (**363**)	72	Section through the Radstock coalfield: Samuel Stutchbury to De la Beche, 1843 (**2004**)
40	*Henri and self* (**363**)	73	Section through part of the Lake District: Adam Sedgwick to De la Beche, 1830 (**1870**)
41	*Lerici from Santerenzo, Gulf of Spezia* (**363**)		
42	*Roman peasant and antiquary* (**363**)	74	Piccadilly front of the new Museum of Economic [Practical] Geology (**2230**)
43	Monk and the Devil (**363**)	75	The Great Hall, Museum of Practical Geology (**2231**)
44	Vico (above) and cliffs between Vico and Sorrento (**363**)	76	Map of Bristol annotated to show outbreaks of cholera and fever, c.1843 (**2232**)
45	Coast from Vesuvius to Capri (**363**)	77	Passport used by De la Beche on his 1819-1820 tour of the continent (**2246**)
46	*Awful Changes* (**367**)		
47	Pencil changes on reverse of *Duria Antiquior* watercolour (**368**)	78	Medal struck for De la Beche and presented to James Holmes (**2280**)
48	*Duria Antiquior*: original watercolour (**368**)		
49	*The Mining Chronicle* (**377**)	79	Miniature bust of De la Beche c.1845 (**2283**)

INTRODUCTION

Known as the De la Beche Collection, the major part of the archive collection of the Department of Geology in the National Museum of Wales comprises the papers of Henry Thomas De la Beche (1796-1855). This includes over 2000 items of correspondence received by him, drafts of his own letters, copies of official minutes, journals of his travels, and some personal items such as family photographs. The collection also contains over 40 letters of William Buckland's, in particular to Lady Mary Cole (1776-1855) and from William Daniel Conybeare (1787-1857). Although this Buckland material has no direct link with De la Beche, it has been regarded as part of the same collection for over 50 years, and is therefore included in this catalogue. De la Beche and Buckland were at the forefront of geology in the first half of the 19th century when the foundations of the science were being laid. They corresponded with many of the pioneers of the subject and their papers record aspects of the development of the science and some of its major institutions.

The primary historical sources contained in the collection include the names of prominent contemporary scientists, both British and Continental, as well as those of politicians and other public figures. These include Greenough, Conybeare, Lyell, Sedgwick, Murchison and Darwin; Elie de Beaumont, Brongniart and Cuvier; Peel, Morpeth and Lincoln, to mention but a few.

The collection was put together by Frederick J. North (1889-1968), first Keeper of Geology at the National Museum of Wales. The papers were donated in the 1930s by the descendants of De la Beche's daughter, Elizabeth, who married into the well-known Dillwyn family of Swansea (Bassett, D.A. 1977, Bassett, M.G. 1990). Most of the collection was presented by De la Beche's great-grandson, Colonel John Illtyd Dillwyn Nicholl (1861-1936) of Merthyr Mawr, Porthcawl and Rice Mansel Dillwyn (b.1875) (also a great grandson). Mrs Olga Welbourn (De la Beche's great great grand-daughter) also presented material to the Museum, some in North's time, and again in the 1970s. Through George Edward Blundell (1867-1939), the Buckland letters were given by Dr Mervyn Henry Gordon (1872-1953) and Mr H.S. Gordon, grandsons of Buckland.

At around the same time, North borrowed the letter books of the Ordnance Survey for the period 1830-1841, transcribed them, and returned the originals to Southampton, where they were subsequently destroyed in the bombing of the Survey's headquarters in 1940. These transcripts in the National Museum of Wales have been referred to as the De la Beche collection in some publications (eg Harley 1971, Oliver 1986), but they are not included here.

Henry Thomas De la Beche (10 February 1796 - 13 April 1855)

A full biography of De la Beche has never been published, although one was prepared for publication by North in the 1950s. North's interest in the history of geology and De la Beche in particular stemmed from a conversation he had with H.G. Seeley (1839-1909) who told North of his time as assistant to Adam Sedgwick. Inspired by this link to the pioneers of geology, North studied the biographies of Sedgwick and his contemporaries and noticed that De la Beche had been overlooked. He resolved to fill the gap, but made little progress until his enquiries led to the discovery of a box crammed with letters and manuscripts in the attic of J.I.D. Nicholl's home at Merthyr Mawr (North unpublished manuscript, NMW Geology Archives). The De la Beche papers were first made available to him in 1930, and some were displayed in an exhibition of maps at the Museum in the summer of 1933 (North 1933a). The newly acquired De la Beche papers, combined with the borrowed Ordnance Survey letter books, provided North with the material for a number of publications on the development of the geological map in Wales, and the formation of the Geological Survey. Most notable is the series published in the *Transactions of the Cardiff Naturalists' Society* (North 1933b, 1934a, 1936a) which contain extensive quotes from the correspondence and provide biographical details about De la Beche. Further papers by North brought his work on De la Beche to a wider audience (North 1934b, 1936b, 1939a, 1939b, 1944, 1951), some taking advantage of appropriate centenaries (North 1939b, 1943), and others relating to De la Beche's co-worker, Conybeare (North 1935, 1956).

While North's publications and other biographical sources (Hamilton 1856, Harrison 1908, Eyles and Eyles 1955, Challinor 1969, Eyles 1971) concentrate on De la Beche's career in the Geological Survey, McCartney (1975, 1977) provides details of his formative years. For a contemporary and personal description of De la Beche see Pym (1882).

De la Beche was born in London, but south-west England is usually regarded as his home and he styled himself 'of Lyme Regis' (Lang 1939). Details of his early life are sketchy, but McCartney (1977) and Fowles (1982) record what is known. Reports of his military 'career' are most commonly in error, even in contemporary publications (such as the *London Illustrated News*, 17 May 1851, p.422) and in histories of the Geological Survey (Flett 1937, Bailey 1952), but McCartney (1977) provides the facts of De la Beche's brief spell at military college. How he acquired his interest in geology is not clear but it is likely that it stems from his acquaintance with the fossil collector Mary Anning (1799-1847) when he moved with his mother to Lyme Regis in 1812 (Torrens 1995, Tickell 1996). Initially, private means allowed De la Beche, as it did many others of the period, to pursue geology as a gentlemanly pastime, and to participate in the activities of the Geological Society. His membership of this society, and the standing it afforded him, provided the professional platform on which his subsequent career was built.

During his time in the South West, De la Beche published a 'wonderful series of classic papers on the geology of West Dorset' (Lang 1941) and worked with Conybeare on the first descriptions of ichthyosaurs and plesiosaurs (Howe et al. 1981, Taylor 1997).

De la Beche made three continental tours, on the first of which in 1819-20, he attempted to climb Mont Blanc. This was the first attempt by an Englishman to ascend the mountain by the St Gervais route (de Beer & North 1950). On these tours De la Beche sought out continental geologists and collections, and his observations provided the basis of several publications (McCartney 1977).

De la Beche's family wealth came from a sugar plantation in Jamaica, and he visited the estate in 1823-24. During his time on the island, De la Beche made notes on the geology of eastern Jamaica and published the first geological account of the island and its first geological map, in the *Transactions of the Geological Society of London* in 1827. This was 'the most significant contribution to the geology of the Antilles in this period' (Draper and Dengo, 1990). Draper (1996) and Donovan (1996) analyse De la Beche's paper and relate its stratigraphy, based largely on De la Beche's experience of the British sequence, to a modern interpretation. De la Beche's observations are recorded in his Jamaican diaries and in his letters to Conybeare. Chubb (1958) describes De la Beche's Jamaican visit, and Robinson (1996) some of his localities. Other, brief, details can be found in Matley (1951), Porter (1990) and Sharpe (1997). Following his return to England, enquiries concerning Jamaica were often directed to De la Beche and information was sought by, amongst others, Charles Darwin.

In Britain, De la Beche's most significant achievements were the establishment of the Ordnance Geological Survey (now the British Geological Survey), the Museum of Economic Geology (later the Museum of Practical Geology, the Geological Museum, and now part of the Natural History Museum, London), the Mining Record Office and the Government School of Mines and of Science Applied to the Arts (the Royal School of Mines, now part of the Imperial College of Science and Technology, London). Various histories of these organisations record De la Beche's involvement, eg Geikie (1895), Reeks (1920), Flett (1937), Bailey (1952), Colvin 1973, Wilson (1985), and particularly Secord (1986) and Cook (1987). Herries Davies (1983, 1995) describes the history of the Geological Survey of Ireland, and the role of De la Beche in that organisation. Many of the papers in the De la Beche archive relate to the formation of these organisations and their subsequent activities, and official correspondence concerning the Geological Survey comprises almost half of the collection. This includes correspondence with Survey staff in the field, and in particular with John Phillips (1800-1874), from whom there are over 300 letters.

Finding himself in financial difficulties as a result of the abolition of slavery and depression of the sugar trade, De la Beche requested funding from the Board of Ordnance in 1832 to allow him to complete the preparation of a geological map of Devon, begun at his own expense (McCartney 1977, Butcher 1996). With encouragement and authority derived from his membership of the Geological Society, De la Beche was appointed Geologist to the Trigonometrical Survey of Great Britain, and his pastime became his profession. In 1835 his mapping was extended into the adjacent county, and the Geological Survey of Cornwall was formed. From this grew the Geological Survey of Great Britain, at its earliest a military initiative under Thomas Colby (Davies 1974, Rose 1996) and later manoeuvred into the civil domain by De la Beche (North 1934a, 1936a, 1936b and McCartney 1977). Miller (1996) reviews De la Beche's work in the southwest, in particular his introduction of the term 'head'.

Rudwick (1972, 1985) and McCartney (1977) describe the controversy which developed when De la Beche found Carboniferous plants in the older Transition rocks during his mapping of the South West. The archive shows how seriously De la Beche took the criticism of his work, and the threat he perceived to his reputation and livelihood.

When the Geological Survey moved into South Wales in 1837, De la Beche set up his headquarters in Swansea to begin mapping the South Wales Coalfield. He was already acquainted with the Swansea naturalist and Member of Parliament Lewis Weston Dillwyn (1778-1855). Through Dillwyn, De la Beche became involved in the recently-formed Swansea Philosophical and Literary Institution, one of whose members, William Edmond Logan (1798-1875), later the founder of the Geological Survey of Canada, was already mapping the coalfield. In the summer of 1838, De la Beche's daughter Elizabeth (Bessie) married Lewis Llewelyn Dillwyn, second son of Lewis Weston Dillwyn (Jenkins 1959). As mentioned previously, it was from the descendants of this marriage that North, and later, McCartney acquired the De la Beche papers. Further information on De la Beche's time in Swansea and on his family links with the Dillwyns can be found in Randall and Rees (1963), Thomas (1979), Painting (1987) and Sharpe (1985).

Once established in an official government post, De la Beche found himself appointed to several government inquiries, such as the Metropolitan Sanitary Commission (Health of Towns Commission) (1843-1849), the Admiralty Coal enquiry (1845-1850), and a Commission to select the building stone for the new Houses of Parliament (1838).

Introduction

De la Beche was an accomplished draughtsman, and used his skill not only in the preparation of plates, maps and sections for his publications but also in sketches and cartoons to express his views on subjects as diverse as Lyell's theoretical geology and tourists in the Vatican. The collection, therefore, is unusually rich in graphical comment (McCartney 1977, Thackray 1985, and especially Rudwick 1975, 1976, 1989, 1992).

William Buckland (12 March 1784 - 14 August 1856)
Unlike De la Beche, the life of Buckland was accorded a published account, by his daughter Mrs E.O. (Anna B.) Gordon (Gordon 1894) and has also been the subject of recent studies by Rupke (1983) and Boylan (1967, 1981, 1984). Biographical details can also be found in Hunt (1886) and Cannon (1970).

Although only a small proportion of the Geology Department archive, the Cardiff collection has been described as 'the most interesting group [of Buckland papers]' (Cannon 1970, p.571). They contain some significant material, particularly Buckland's letters to Lady Mary Cole and her daughters of Penrice Castle on the Gower peninsula near Swansea. This correspondence dates from the period when Buckland was preparing his *Reliquiae Diluvianae* (Buckland 1823), and includes details of the discovery of Paviland Cave. From this cave Buckland excavated the Late Palaeolithic human partial skeleton which came to be known as the 'Red Lady of Paviland' (North 1942, 1966). A lithograph of Buckland lecturing in Oxford on his Paviland discoveries has been described by Edmonds and Douglas (1976) and this lecture is also the subject of one of his letters to Lady Mary Cole in this collection.

Structure of the catalogue
The catalogue comprises two main sections: a summary or description of each item and an index subdivided into personal names, placenames and subject. Although some parts of the collection had previously been given various NMW accession numbers, for consistency the whole collection has been re-accessioned under the number NMW 84.20G.D.1 to 2283. Numbers allocated to individual items (1 to 2283) form the final element of the accession number. The main catalogue entries are arranged in numerical order of the individual item number which appears in bold print in the summaries. These numbers are also those listed in the indexes.

The main catalogue listing is arranged alphabetically by the surname of the author of the item, and by date. Where a date has been assumed on the basis of internal evidence, it is given in square brackets. The letters 'FJN' which follow some of these assumed dates indicate that the date is derived from F.J. North's work on the collection. Undated items are listed at the end of each name. Authors' names are given in full, where these are known, adjacent to the catalogue number, and the name of the recipient opposite. Names of authors and recipients and placenames in their addresses are included in the indexes.

Following the main alphabetical listing are letters from authors who are unidentified or whose names are illegible; anonymous handwritten items; printed items; addenda; and objects such as medals, a quadrant, and a bust. In attempting to maintain a combined numerical, alphabetical and chronological order to the main listing, it has not been possible to insert additional entries, such as those letters whose authors, previously unidentified, have since been recognised. These are listed under addenda.

Efforts have been made to identify letter authors and individuals mentioned within the letters, often on the basis of context or other internal evidence. In some cases, for example with a common name, it has not been possible. Any corrections to these identifications, or further identifications are welcomed.

Considerations of space have precluded full quotation of the material in the collection. Instead, each entry comprises a précis which is designed to give a flavour of the content and tone of the letter. Inevitably, these summaries are highly selective and reflect what we see as relevant. The indexes, however, are based on the actual contents of the letters, and in many cases subjects or names which appear in the index will not appear in the summaries. Full transcripts (by North) or copies of most the letters are available from the Department of Geology at the National Museum of Wales to researchers with specific requests.

Many of the letters have been cited or quoted in various publications on the history of geology. Where these are known, references are given, as they may provide additional information on the context of a letter or a fuller quote.

Many of the items, especially De la Beche's diaries, contain sketches, views, sections, and maps. A full list of these illustrations will be found in the subject index, and in the personal and placename indexes where they are indicated by '(illn.)'.

Other abbreviations used in the catalogue are:
Bull. Geol. Soc. Fr. (Bulletin of the Geological Society of France)
Edinb. Phil. Jnl. (Edinburgh Philosophical Journal)
Mem. Geol. Surv. (Memoirs of the Geological Survey)
Phil. Mag. (Philosophical Magazine)
Phil. Trans. (Philosophical Transactions of the Royal Society of London)
Proc. Geol. Soc. (Proceedings of the Geological Society of London)
QJGS (Quarterly Journal of the Geological Society of London)
Trans. Geol. Soc. Lond. (Transactions of the Geological Society of London)
BAAS (British Association for the Advancement of Science)

A Chronology of H.T. De la Beche

1796	10 February	Born at 21 Wimpole Street, London.
1800		Taken to family estate in Jamaica.
1801		Shipwrecked on Great Inagua, West Indies.
1801		Brought back from Jamaica following death of his father.
1802		At school in Hammersmith, London.
1805		At school in Keynsham, Somerset.
1808		At school in Ottery St Mary, Devon.
1810	3 April	Entered Royal Military College, Great Marlow.
1811	8 October	Dismissed from Royal Military College.
1811		Lived in Dawlish, Devon.
1812		Lived in Charmouth, then settled with mother in Lyme Regis, Dorset.
1816		Tour of Scotland and north of England.
1817		Elected to Geological Society of London.
1818		Married Letitia Whyte of Loughbrickland, County Down.
1819	5 March	Read first paper to Geological Society (on geology of the south coast of England from Bridport Harbour to Babbacombe Bay).
1819	July	Set out on tour of continent.
1819	25 November	Elected Honorary Member of Société de Physique at d'Histoire Naturelle de Geneve
1819		First paper published (On the temperature and depth of Lake Geneva)
1819	2 December	Daughter Elizabeth born in Geneva.
1819	23 December	Elected Fellow of Royal Society.
1820	July	Returned to Lyme Regis.
1821		Fieldwork around Bayeaux and mouth of the Seine.
1822	summer	Fieldwork in southern Pembrokeshire.
1823	9 November	Set out to visit his estate in Jamaica.
1824	December	Returned to England.
1825		Separated from wife.
1826		Divorced.
1827		*A Tabular View* published.
1827		Published Remarks on the geology of Jamaica.
1827	autumn	Set out on continental tour.
1828	summer	Returned to England.
1828	autumn	Set out on continental tour.
1829	June	Returned to England.
1830		*Geological Notes* published.
1830		*Sections and Views* published.
1831-32		Served as Secretary, Geological Society.
1831		*A Geological Manual* published.
1831		Elected member, Geological Society of France.
1832	28 March	Submitted proposal for colouring geologically Ordnance map of Devon.
1832	2 May	Proposal accepted by Office of Ordnance; appointed Geologist to the Trigonometrical Survey of Great Britain.
1832	September	Elected Corresponding Member, Academy of Natural Sciences of Philadelphia.
1833	30 August	Elected Honorary Member, Royal Geological Society of Cornwall.
1834	June or July	Daughter, Rosalie Torre born near Taunton.
1834		First four sheets of geological map published.
1834		Discovered fossils in 'Grauwacke' rocks of Devon: beginning of Great Devonian Controversy.
1834		*Researches in Theoretical Geology* published.
1835		*Geology* (*How to observe* series) published.
1835		Geological Survey of Cornwall formed.
1835-47		Foreign Secretary, Geological Society.
1835	10 August	Museum of Economic Geology established at 6 Craig's Court, Charing Cross.
1837	February	Given honorary title of 'Director' of the Museum.
1837	December	Survey moved into South Wales, based in Swansea.
1837		Honorary Member, Swansea Philosophical Society and Institution.
1838		Appointed to Parliamentary Building Stones Commission.
1838	16 August	Marriage of Elizabeth to Lewis Llewelyn Dillwyn.
1839		Elected Honorary Member, Birmingham Philosophical Institution.
1839		Mining Records Office established.
1839		First report of Geological Survey published.
1840		Appointed to British Association Committee on Railway Sections.
1840		Membre Titulaire, Institut d'Afrique.
1842		Knighted.
1843		Appointed to Royal Commission on the Health of Towns.
1845		Geological Survey Act: Survey transferred to Office of Woods, Forests, etc. Geological Survey of Ireland established as part of Geological Survey of the United Kingdom.
1845		Appointed to Admiralty Coal Enquiry
1846		First Memoir of the Geological Survey published.
1846		Elected Vice President, Geological Society of Ireland.
1847-55		President of the Palaeontographical Society.
1847-49		President of the Geological Society of London.
1848		Companion of the Order of the Bath.
1848		President, Section C (Geology and Physical Geography), British Association (at Swansea).
1848	26 November	Elected Corresponding Associate, Turin Academy.
1849		Launched the series *Figures and descriptions illustrative of British Organic Remains (Decades)*.
1851	14 May	New Museum of Practical Geology opened in Jermyn Street by Prince Albert.
1851		*The geological observer* published.
1851		Royal School of Mines established.
1851		Juror for the Great Exhibition.
1852		Elected member, Imperial Society of Naturalists, Moscow.
1852	8 December	Elected member, Swedish Academy of Science.
1852	28 December	Elected member, Société Météorologique de France.
1853	1 April	School of Mines and Geological Survey and Museum transferred to Department of Science and Art.
1853	29 April	Elected Corresponding Member, Académie de France.
1855	16 February	Awarded Wollaston Medal of Geological Society.
1855	13 April	Died, London, aged 59.
	19 April	Buried in Kensal Green Cemetery, London (Grave no. 12387, Square 28).
1858	17 October	Rosalie died, aged 24; buried with her father.
1866		Elizabeth died, aged 47.

PUBLICATIONS OF H.T. DE LA BECHE

1819 On the depth and temperature, etc. Sur la profondeur et la température du lac de Genève. Lettre adressée au Prof. Pictet par Mr H.T. De la Beche. *Bibliothèque Universelle*, Vol.12, pp.118-126.

1820 On the depth and temperature of the Lake of Geneva. Communicated by Professor Pictet. With a map. *Edinburgh Philosophical Journal*, Vol.2 no.3, 1820, pp.106-110.

1820 Températur de quelques lacs de la Suisse. *Annalen der Physik. L.W. Gilbert*, Vol.66, pp.146-151.

1821 Sur la température des lacs de Thun at de Zug, en Suisse. *Bibliothèque Universelle*, Vol.14, pp.144-145.

1821 Température de quelques lacs da la Suisse. *Annals de Chemie*, Vol.19, pp.77-83.

1821 (with W.D. Conybeare) Notice of the discovery of a new fossil animal, forming a link between the Ichthyosaurus and crocodile; together with general remarks on the osteology of the Ichthyosaurus [read 6 Apr 1821]. *Transactions of the Geological Society of London*, First Series, Vol.5, part 2, pp.559-594.

1822 Remarks on the geology of the south coast of England, from Bridport Harbour, Dorset, to Babbacombe Bay, Devon [read 5 Mar 1819]. *Transactions of the Geological Society of London*, Second Series, Vol.1, part 1, pp.40-47. *Isis*, Vol.12 (1823), pp.180-181.

1822 Notice respecting fossil plants found at Col de Balme, near Chamouny in Savoy [read 3 Nov 1820]. *Transactions of the Geological Society of London*, Second Series, Vol.1, part 1, pp.162-163.

1822 On the geology of the coast of France, and of the inland country adjoining; from Fécamp, Department de la Seine Inferieure, to St Vaast, Department de la Manche. [read 2 Nov 1821]. *Transactions of the Geological Society of London*, Second Series, Vol.1, part 1, pp.73-89.

1822 Versuche über das Frieren mit Oel bedeckten Wassers. *Annalen der Physik. L.W. Gilbert*, Vol.71, pp.435-436.

1823 *Map of 24 miles around the city of Bath...by C. Harcourt Masters, coloured geologically by Rev. W.T. Conybeare and H.T. De la Beche.* Bath.

1824 Notice of the discovery of a large fossil elephant's tusk, near Charmouth, Dorset.[read 2 May 1823] *Transactions of the Geological Society of London*, Second Series, Vol.1, part 2, pp.421-422.

1824 *A selection of the geological memoirs contained in the Annales des Mines.* London, 335pp.

1824 Catalogue of the birds, and of terrestrial and fluviatile molluscae, found in the vicinity of Geneva. *Zoological Journal*, Vol.1, no.1, pp.89-93.

1825 Notice on the temperature of the surface water of the Atlantic, observed during a voyage to and from Jamaica. *Annals of Philosophy*, New Series, Vol.10, pp.333-335.

1825 *Notes on the present conditions of the negroes in Jamaica.* Cadell, London, 64pp.

1825 Notice on the Diluvium of Jamaica. *Annals of Philosophy*, New Series, Vol.10, pp.54-58.

1826 On the Lias of the coast in the vicinity of Lyme Regis, Dorset [read 21 Nov 1823]. *Transactions of the Geological Society of London*, Second Series, Vol.2, part 1, pp.21-30.

1826 On the geology of southern Pembrokeshire [read 16 May 1823]. *Transactions of the Geological Society of London*, Second Series, Vol.2, part 1, pp.1-20.

1826 On the Chalk and sands beneath it (usually termed Green-sand) in the vicinity of Lyme Regis, Dorset and Beer, Devon [read 16 December 1825]. *Transactions of the Geological Society of London*, Second Series, Vol.2, part 1, pp.109-118.

1826 Notice of traces of a submarine forest at Charmouth, Dorset. *Annals of Philosophy, New Series*, Vol.11, p.143.

1827 *Carte des principales Sondes du lac Léman.* Genève: Briquet et Dubois.

1827 Remarks on the geology of Jamaica [read 2 Dec 1825 and 6 Jan 1826]. *Transactions of the Geological Society of London*, Second Series, Vol.2, part 2, pp.143-194.

1827 *A tabular and proportional view of the Superior, Supermedial and Medial Rocks; (Tertiary and Secondary Rocks.)* London: W. Phillips, 64pp. (second edition 1828).

1827 Nummulites in the Greensand Formation. *Philosophical Magazine*, New Series, Vol.4, p.235.

1828 On the geology of Tor and Babbacombe Bays, Devon [abstract] *Proceedings of the Geological Society of London*, Vol.1, no. 4 (1827-28), pp.31-33.

1828 Notes on the habits of a Caryophyllia from Tor Bay, Devon. *Zoological Journal*, Vol.3, no.12, pp.481-486. *Isis*, Vol.23 (1830), pp.1238-1239. *Heusinger's Zeitschrift für die organische Physik*, Vol.3, pp.90-93.

1829 On the geology of Nice [abstract]. *Proceedings of the Geological Society of London*, Vol.1, no.9 (1828-29), pp.87-89.

1829 On the geology of Tor and Babbacombe Bays, Devon [read 16 Nov 1827]. *Transactions of the Geological Society of London*, Second Series, Vol.3, part 1, pp.161-170.

1829 On the geology of the shores of the Gulf of La Spezia [abstract] *Proceedings of the Geological Society of London*, Vol.1, no.14 (1829-39), pp.164-167

1829 On the geology of the environs of Nice and the coast thence to Vintimiglia [read 21 Nov 1828]. *Transactions of the Geological Society of London*, Second Series, Vol.3, part 1, pp.171-185.

1829 Note on the differences, either original or consequent on distrubance, which are observable in the Secondary Stratified rocks. *Philosophical Magazine, New Series*, Vol.6, pp.213-215. *Annales des Sciences Naturelle*, Vol.17, pp.426-445.

1829 Notice on the excavation of valleys. *Philosophical Magazine and Annals of Philosophy*, New Series, Vol.6, pp.241-248.

1829 Sketch of a classification of the European rocks. *Philosophical Magazine, New Series*, Vol.6, p.440. *American Journal of Science*, 1830, Vol.18, p.26. *Bulletin Universel des Sciences et d'Industrie (2e section). Bulletin des Sciences Naturelles et de Géologie*, 1830, Vol.20, pp.17-19.

1830 On the geographical distribution of organic remains contained in the Oolitic series of the great London and Paris Basins, and in the same series of the south of France. *Philosophical Magazine, New Series*, Vol.7, pp.81-95, 202-205, 250-268, 334-351; Vol.8, pp.35-44, 208-213. *Bulletin Universel des Sciences et d'Industrie (2e section). Bulletin des Sciences Naturelles et de Géologie*, Vol.23, pp.191-200.

1830 On the formation of extensive conglomerate and gravel deposits. *Philosophical Magazine, New Series*, Vol.7, pp.161-171. *Bulletin Universel des Sciences et d'Industrie (2e section). Bulletin des Sciences Naturelles et de Géologie*, Vol.22, pp.164-165.

1830 (with W. Buckland) On the geology of Weymouth and the adjacent parts of the coast of Dorsetshire [abstract] *Proceedings of the Geological Society of London*, Vol.1, no.16 (1829-30), pp.217-221.

1830 *Sections and views illustrative of geological phaenomena.* London: Treuttel & Wurtz, viii + 71pp., 40 plates.

1830 *Geological notes.* London: Treuttel & Wurtz, xliii + 70pp.

1831 *A geological manual.* London: Treuttel & Wurtz, 535pp.

1832 *A geological manual.* 2nd edition London: Treuttel and Wurtz, Treutel Jun and Richter; Paris and Strasburgh: Treutel and Wurtz, 564pp.

1832 *Handbuch der Geognosie.* Translated by Heinrich von Dechen. Berlin: Duncker und Humblot, 612pp.

1833 *A geological manual.* 3rd edition London: Charles Knight, vi + 629pp.

1833 *Manuel géologique.* Translated by André-Jean-Francois-Marie Brochant de Villiers. Paris.

1833 Sur les environs de La Spezia. *Mémoires de la Société Géologique de Paris*, Vol.1, pp.23-36.

1834 *Researches in theoretical geology.* London: Charles Knight, xvi + 408pp.

1834 On the anthracite found near Bideford in north Devon. *Proceedings of the Geological Society of London*, Vol.2, no.37, pp.106-107.

1835 Lettre sur la découverte d'empreintes de plantes dans les Schists Subordonés à la formation de le Grauwacke. *Bulletin de la Société géologique de France*, Vol.6, p.90.

1835 Note on the Trappean rocks associated with the (New) Red Sandstone of Devonshire *Proceedings of the Geological Society of London*, Vol.2, no.41, pp.196-198.

1835 *How to observe. Geology.* London: Charles Knight, viii + 312pp.

1835 (with W. Buckland) On the geology of the neighbourhood of Weymouth and the adjacent parts of the coast of Dorset. [read Apr 2 and 16, 1830]. *Transactions of the Geological Society of London*, Second Series, Vol 4, part 1, pp.1-46

1836 [Notice of fossils from the schistose rocks of North Cornwall] *Proceedings of the Geological Society of London*, Vol 2, no 43 (1835-36), pp.225-226

1836 *A selection of geological memoirs contained in the Annales des Mines, written by Brongniart, Humboldt, von Buch and others.* 2nd edition. London: Phillips, xxii + 336pp.

1836 *How to observe. Geology.* 2nd edition, London: Charles Knight, viii + 312pp.

1836 *Untersuchungen über theoretische Geologie.* Translated by C. Hartmann.

1836 *Anteilung zum wissenschaftl. Beobachten. I. Geologie.* Berlin.

1837 *Manuel géologique.* 2nd edition. Translated into French by A.J.M. Brochant de Villiers. Brussels: Meline, xxxvi + 506pp.

1838 *L'Art d'observer en géologie.* Translated by H. de Collegno. Paris: Levrault, viii + 232pp.

1838	*Recherches sur la partie théorique de la Géologie.* Translated by H. de Collegno.
1839	*Coupes et vues pour servir à l'explication des phénomènes géologiques.* Translated by H. de Collegno.
1839	On the influence of atmospheric pressure on the tidal waters of Cornwall and Devon. *Edinburgh New Philosophical Journal*, Vol.26, pp.415-419.
1839	*Report on the geology of Cornwall, Devon and West Somerset.* London: Longman, Orme, Brown, Green and Longmans, xxviii + 648pp.
1839	(with C. Barry, W. Smith and C.H. Smith) Report with reference to the selection of stone for building the new Houses of Parliament. In *Parliamentary Papers for 1839*, Vol.30. London.
1842	On the connection between geology and agriculture in Cornwall, Devon and West Somerset. *Journal of the Agricultural Society of England*, Vol.3, pp.21-35.
1844	On estuaries and their tides. *Philosophical Magazine*, Vol.24, pp.485-491. *Transactions of the Royal Geological Society of Cornwall*, Vol.6, 1846, pp.211-218.
1845	*Report on the state of Bristol and other large towns.* London: Health of Towns Commission, 102pp.
1846	On the formation of the rocks of South Wales and South Western England. *Memoirs of the Geological Survey*, Vol.1, pp.1-296.
1847	(with Lyon Playfair) Gases and explosions in collieries. *Journal of the Franklin Institute of the State of Pennsylvania*, Vol.13, pp.347-353, 427-433.
1848	(with Lyon Playfair) *First Report on the coals suited to the steam navy.* Museum of Practical Geology, HMSO, 67pp. Also *Memoirs of the Geological Survey of Great Britain* Vol.2 part 2, pp.539-630.
1848	Anniversary address of the President. *Proceedings of the Geological Society*, Vol.4, pp.xxi-cxx.
1849	Anniversary address of the President. *Proceedings of the Geological Society*, Vol.5, pp.xix - cxvi.
1849	Geological changes from the Earth's axis of rotation. *Edinburgh New Philosophical Journal*, Vol.47, pp.98-104.
1849	Flora of the Silurian System. Plants of the anthracite formation. Geological climate. Co-existence of Saurian and Molluscous forms. Phosphate of lime in the mineral kingdom. *Edinburgh New Philosophical Journal*, Vol.47, pp.122-132.
1849	Mineralogy. In J.F.W. Herschel (ed.) *A manual of scientific enquiry; prepared for the use of Her Majesty's Navy: and adapted for travellers in general*. London: John Murray, xi + 488pp.
1849	(with Lyon Playfair) *Second report on the coals suited to the steam navy.* Museum of Practical Geology, H.M.S.O., 57pp.
1849	Minutes of evidence in *A Select Committee to enquire into the best means of preventing the occurrence of dangerous accidents in mines.* London, pp.3-24.
1851	(with Lyon Playfair) *Third report on the coals suited to the steam navy.* Museum of Practical Geology, H.M.S.O., 55pp.
1851	*The geological observer.* London: Longman, Brown, Green and Longmans, xviii + 846pp.
1851	Analysis of coal from Van Diemen's Land. *Papers and Proceedings of the Royal Society of Van Diemen's Land*, Vol.1, pp.265-267.
1851	Inaugural discourse, delivered at the opening of the School of Mines and of Science Applied to the Arts, 6 Nov 1851 *Records of the School of Mines and of Science Applied to the Arts*, Vol.1, part 1 (1852), pp.1-22.
1851	Mining, quarrying and metallurgical processes and products. Lecture 1, December 2nd 1851 [Lecture to Society of Arts, Manufactures etc.] pp.17-34
1853	*The geological observer.* 2nd edition. London: Longman, Brown, Green and Longmans, xxviii + 740pp.
1853	*Vorschule der geologie...frei mit Zusätzen bearbeitat von Dr Ernst Dieffenbach.* Brunswick: Vieweg, xviii + 624pp.
1853	Note on the Stigmaria beds of the South Staffordshire coal field. *Records of the School on Mines*, Vol.1, pp.336-339.
1855	(with Trenham Reeks) *Catalogue of specimens illustrative of the composition and manufacture of British pottery and porcelain from the occupation of Britain by the Romans to the present time.* London: Museum of Practical Geology, xxiii + 179pp.

ACKNOWLEDGEMENTS

We would like to thank the following people for their assistance in the compilation of this catalogue: Mrs Olga Welbourn for donating material to the Museum both in the 1930s and 1970s; John Andrews (Chepstow), Philip Attwood (British Museum), Edward Besly (National Museum of Wales), Roger Clark (Bristol City Museum), Ron Cleevely (Natural History Museum), John Fowles (Lyme Regis), Andrew Grout (University of Edinburgh), Stephen Howe (National Museum of Wales), Frank James (Royal Institution), Bill Jones (University of Wales, Cardiff), Simon Knell (University of Leicester), Bob Owens (National Museum of Wales), Eric Robinson (University College, London), Brian Taylor (British Geological Survey), John Tipton (Tenby), and Christopher Titterington (Victoria and Albert Museum) for providing information; John Kenyon, Louise Carey, and Pat Havill of the National Museum of Wales Library and the staff of the Department of Library and Information Services at the Natural History Museum; the Photography Department of the National Museum of Wales; John Elliott (University of Wales, Cardiff) for assistance with data transfer; Beryl Chant, Cindy Howells, Jean Parsons, Margaret Pearce and Jean Powell, current and former members of the Department of Geology at the National Museum of Wales, who have worked on the collection; Douglas Bassett (Cardiff), Michael Bassett (National Museum of Wales), Jennifer Stewart (Cardiff), John Thackray (Natural History Museum) and especially Hugh Torrens (University of Keele) for their comments and suggestions; and Arwel Hughes and particularly Valerie Deisler for their help in preparing the catalogue for publication.

SUMMARIES

1 **Acland**, Thomas Dyke
Holnicote.
9 April 1834
To: H. T. De la Beche
Minehead.

I heard yesterday of your arrival at Minehead, and hasten to invite you to join me on a circuit ride to Exford and Winsford. You will be able to see about 20 miles of the country which you are about to observe, and we can arrange your accepting further hospitality. I understand your daughter is with you.

2 **Acland**, Thomas Dyke
Holnicote.
3 July [1834]
To: H. T. De la Beche
Bishops Lydeard.

I have just returned home after a circuit round the County of Devon. Thank you for the book and for the deposit made by Gardner, on your command. I snatched a peep at the Greyological [sic] Trigonometrical Survey [maps], and am glad this part of the kingdom will be favoured by your skill and industry.

3 **Acland**, Thomas Dyke
Pelham Place,
Hastings.
17 December 1849
To: H. T. De la Beche
Athenaeum Club,
London.

Can you help me to a tolerably accurate statement of the number of acres in Somersetshire? Also a rough approximation of the particular part of the county specified?

4 **Acland**, Thomas Dyke
Pelham Place,
Hastings.
17 December 1849
To: H. T. De la Beche
Athenaeum Club,
London.

I request an unofficial answer on the geology of Somersetshire. I am engaged in a report on Somerset farming. Please introduce me to the gentleman in your office who has got up the geology of Somersetshire as I want to ask a few questions. My ignorance of geology is so great that I need the aid of a living interpreter.

5 **Adare**, Edward Richard Windam
Wyndham-Quin, Viscount
1 October [?1843]
To: [H. T. De la Beche]

Should this weather continue, I presume you will not go to Maesteg. Will you kindly write to me when you and Hichens fix another day?

6 **Adare**, Edward Richard Windam
Wyndham-Quin, Viscount
Dunraven Castle.
28 October [?1843]
To: [H. T. De la Beche]

I am glad you are going to be our guest. What wretched weather for going about. Write a day or two before coming. Bring accompanying gentleman or assistant, as we have plenty of room.
[Cited: North 1936a, p.81]

7 **Adare**, Edward Richard Windam
Wyndham-Quin, Viscount
Dunraven Castle.
24 October [1845 FJN]
To: [H. T. De la Beche]

We are making a large map of my father's estate. Could you get one of your men to make a tracing from the Ordnance map. Has Caerbryn or Brynchwith the better mineral under it?

8 **Agassiz**, Jean Louis Rodolphe
Boston.
28 April 1847
To: [R. I. Murchison]

I should be happy to see your new specimens of the oldest fishes. As for your Silurian system everyone must see the justice of your claims to its full establishment as one great system. Geologists here are of the same opinion. I hope my work on glaciers has reached England, and that impartial readers will find that Forbes could only gain temporary advantage over me. I made a lucky hit by tracing the close analogy between the flora of the European Miocene and the actual flora of the U.S. I cannot publish a paper now, but please make these remarks known.
[Cited : Secord 1986a, pp.197, 201]

9 **Albert**, H.R.H. Prince
Windsor Castle.
31 October 1851
To: H. T. De la Beche

As President of the Royal Commission for the Exhibition of 1851 I transmit a commemorative medal to you for your valuable services as Juror. Cordial thanks for assistance.

10 **Albert**, H.R.H. Prince
Society of Arts, Manufactures, and Commerce,
Adelphi, London.
5 July 1853
To: H. T. De la Beche

Council of the Society express thanks for lecture on some of the results of the Great Exhibition, and as acknowledgement request your acceptance of volumes containing lectures delivered. As President I convey their thanks for aid given in elucidating practical influences of the Exhibition.

11 **Anning**, Charles Churchill
Lyme Regis.
July 1847
To: H. T. De la Beche

At the request of Roberts I send an account of my late aunt, Mary Anning, the first seller of fossils at Lyme, 1799-1847. In August 1800 she survived after being struck by lightning. Her nurse, Elizabeth Haskings was killed. After the death of his father, Richard Anning, her brother began fossil collecting. Large head of supposed crocodile found. Mary Anning excavated the remainder 12 months later. It was purchased by H. H. Henley for about £23.
[Enclosure] Account of incident resulting in his wife's death, by John Haskings. It may be said the death of the nurse was the life of the child. Dated 27 August 1821.
[Cited: Lang 1958, pp.91-93; Howe et al. 1981, p.12; Herries Davies 1982, pp.104-105; Torrens 1995a, p.270; Tickell 1996, p.10]

12 **Arago**, Dominique François Jean
Permanent Secretary,
Institut de France,
Académie Royale des Sciences,
Paris.
18 October 1830
To: H. T. De la Beche
Post Office,
Plymouth,
Devon.

Official thanks for works sent to the Library of the Institute: *Geological Notes*, 1830, and *Sections and Views*, 1830.

13 **Arbuthnot**, George
Treasury.
5 January 1852
To: [H. T. De la Beche]

I am anxious for information about quicksilver mines in California. It is said that the discovery of these mines so close to the silver mines of South America would so facilitate the production of silver as to make it keep pace with the increased supplies of gold. I suggest this is erroneous. I am very ignorant of mining operations and as they are frequently referred to with reference to currency questions, it would be very useful to me to gain a little sense on the subject.

14 **Argyll**, George Douglas Campbell
8th Duke of
Castle Howard,
York.
16 December 1851
To: H. T. De la Beche
Geological Museum,
Jermyn Street,
Piccadilly,
London.

Perhaps Dr Playfair may have shown you the sketch and description I sent him some time ago of the bed of nickel ore at Inveraray. Since I wrote to him we have found the bed in a perfectly straight line along the strike of the strata at the distance of 100 feet farther eastward, preserving the same character and appearance. I am very anxious now to have the best advice as to the best way of working.

15   **Argyll**, George Douglas Campbell,   To: H. T. De la Beche
        8th Duke of
        6 February 1853

You were kind enough to say that you would procure advice for me on the terms which might be offered by Mr Phillips for a lease of one of my nickel veins. Now it appears to me that whilst the sum they offer to pay down for the old workings is rather larger than I expected, the lordship is very much smaller and as the latter is by far the most important in the event of the vein being found to be productive, I am very little disposed to be bought off.

16   **Argyll**, George Douglas Campbell,   To: H. T. De la Beche
        8th Duke of
        Inveraray,
        Argyllshire.
        21 September 1853.

You were kind enough to say that you would procure for me any advice I might require from the Government Mining Engineer, whose name I forget. I want to know what are the usual stipulations in mining leases (not referring to coal or ironstone, but to mineral veins) to secure adequate working, and prevent dilatory and superficial working, whether this may arise from design, or from want of capital or knowledge on the part of the Lessees. I believe it is usual to stipulate for a certain number of miners being employed. Any information you can give or get for me would oblige.

17   **Arnoux**, G.   To: H. T. De la Beche
        Stoke on Trent.
        10 September 1852

I have sent you a small sample of feldspar which comes from Arendal (Norway). Could Dr Percy analyse it? Sorry you could not send the bricks, but we must submit to the rules. Re your specimen from an Indian temple, I am convinced it is of Persian manufacture. In fact this is the very beginning of pâte tendre.   [Letter possibly incomplete.]

18   **Artis**, Elizabeth   To: [H. T. De la Beche]
        Woodcroft,
        near Peterborough.
        25 January [1848]

In reply to your letter, herewith details concerning my late husband Edmund Tyrell Artis (1789-1847), collector and antiquarian. Enclosed catalogue of part of his collection. He was the owner of the Club House, Doncaster, where he died after a short illness.
[Cited: Crowther 1984, pp.14-15]

19   **Atherstone**, W. Guybone   To: H. T. De la Beche
        Graham's Town.
        4 August 1849

I have sent you two boxes of fossils, and more will follow in a few weeks. I have kept the list of localities, which I will send with next lot. I take a great interest in geology, though, like my friend Bain, I am self-taught. I will also make a map and sections. The want of a good map is a great drawback to my providing accurate geological information of the district.

20   **Auckland**, George Eden, Earl of   To: H. T. De la Beche
        Admiralty.
        29 August 1847

It seems to me that there are some remarks in the enclosed paper worthy of your attention in your perquisitions on the properties of coal.

21   **Auckland**, George Eden, Earl of   To: H. T. De la Beche
        Admiralty.
        18 October 1847

I should like to hear what you are doing at Putney, first as regards artificial fuels, and secondly whether the coals from Labuan or Borneo have been submitted to you. The Commander of the Nemesis prefers the latter to English coal.

22   **Auckland**, George Eden, Earl of   To: H.T. De la Beche
        Admiralty.
        24 December 1847

Thanks for the report on Warlich's patent coal. It proves I am not wrong in having entered into a contract with him. Experiments ought to be tried to ascertain not only what coal makes the best patent fuel, but also what will be the properties of inferior coal after the process. Coals of Burdwan, Borneo, and Australia might be rendered useful. Please report in January.

23   **Auckland**, George Eden, Earl of   To: H.T. De la Beche
        Admiralty.
        15 January 1848

I would like first to see the report in manuscript. It will be well to separate it from the general report and lay it before Parliament.

24   **Austen**, Robert Alfred Cloyne   To: H. T. De la Beche
        Teignmouth.   Tiverton,
        5 May 1835   Devon.

Shipments of clay from this port are from 18,000 to 20,000 tons annually. Granite exports negligible. I send you an account of the raised beach. I accumulate instances of faults in red sandstone. Scarcity of animal remains make me despair of determining the age of the Bovey deposit. I have found new cases of faults in the limestone, and offer a solution to the problem. We leave for my father's tomorrow where I shall be happy to receive you.
[Attached] An account of a raised beach on the eastern coast of Devon, Tor Bay.

25   **Austen**, Robert Alfred Cloyne   To: H. T. De la Beche
        Ogwell House.   Tavistock,
        23 August 1837   Devon.

Sorry that another engagement prevents my joining you to work out the eastern boundary of culms. You must come to stay with us at Ogwell. Sedgwick was here. When you come please bring a coloured copy of the Ordnance map. I should much like to see the French accounts to which you allude.
[Cited: Rudwick 1985, p.216]

26   **Austen**, Robert Alfred Cloyne   To: H. T. De la Beche
        London.   Post Office,
        15 December 1837   Swansea.

My paper came off last night. My account of disputed ground concerns only my own neighbourhood. I have both you and Sedgwick and Murchison against me on certain points. My conclusion was that it is incorrect to separate the Carboniferous strata from the Transition, that it is only the upper portion of a thick series. Want of conformity an exception, not the rule.
[Cited: Rudwick 1985, pp. 227, 228]

27   **Austen**, Robert Alfred Cloyne   To: H. T. De la Beche
        Ogwell.   Swansea,
        29 May 1838   Glamorganshire.

Re my speculations on the Bovey valley, what is securely established is the location of Greensand below and horizontal beds of "Head" above. Coal, clay, and certain sands are of intermediate age. Herewith a list of organic remains to establish the point as to the Greensand which you were anxious about. What genus does Phillips refer the small bivalve to? If you send me a map, I will mark the area covered by the Bovey clay. I have looked over some interesting early notes I made on Cornwall.

28   **Austen**, Robert Alfred Cloyne   To: H. T. De la Beche
        Ogwell House.   Swansea,
        May 1838   South Wales.

Herewith all I know of Mr Floud's search for coal near Bovey. Also, I was at the pits trying to determine age from fossils. There are so many points of resemblance that we must suppose the same causes to have produced them, and at the same times though levels are now different. With sketches I illustrate the order of events. Half England and half Europe has been beneath the present seas with their present shells. I have heard nothing from Phillips. Sedgwick intends to come to Devon. I don't care about Darwin's objections to my coral reef notions. You must come and visit us.
[Cited: Rudwick 1985, p.238]

29   **Austen**, Robert Alfred Cloyne   To: H. T. De la Beche
        Ogwell.   Swansea,
        22 July 1838   South Wales.

Sedgwick has been with me about a week and we have gone over the country. I've looked carefully at the culm question. They do not go under the slate of S. Devon, and I should be sorry if you had printed it so. Sedgwick has had your letter in which you rely on the Hembury Hill section. My sketch shows the slate, not culm, dips under Chudleigh rock. It appears Sedgwick and Murchison have a paper in preparation. Should the Geological Society publish my paper, I have declined having a map. All I know is in yours. I understand Phillips has no opinion of the Silurian System.
[Cited: Rudwick 1985, pp.247, 257, 288]

30	**Austen**, Robert Alfred Cloyne	To: H. T. De la Beche	
	Ogwell.	Swansea,	
	29 July 1838	Glamorganshire.	

Herewith by return of post all I can give you respecting the Chudleigh limestone, including sketches. Old slates pass under limestone. Unconformity exists. Phillips' note relates only to fossils of S. Devon. I am glad to hear he is to go in. I am rather astonished that so good a geologist should not have looked at Devon with his own eyes before this. Sedgwick is in N. Devon. Half the globe will become Silurian. The mischief is that it is taken as a type.
[Cited: Rudwick 1985, pp. 238, 252]

31 **Austen**, Robert Alfred Cloyne   To: H. T. De la Beche
   Ogwell.   Swansea,
   8 August 1838   Glamorganshire.

I cannot be at Newcastle. I will send my paper with Ibbetson. Carbonaceous grauwacke puzzles me much. I will try my hand at another theory. Suppose a submarine ridge produces a land-locked channel on the N and open sea to the S. You thus have contemporaneous deposits differing in mineral composition. What do you think? I look to you for an impartial account of Newcastle.
[Cited: Rudwick 1985, pp.251, 252, 253; Secord 1986a, p.120]

32 **Austen**, Robert Alfred Cloyne   To: H. T. De la Beche
   Ogwell.   Geological Society,
   10 November 1838   Somerset House,
                                       London.

Newspapers have announced your return to London after search for stone for Houses of Parliament. Let me have maps to finish Goodrington country. Sedgwick informs me some for, some against my paper at Newcastle meeting. I have been to the Cotentin to do some geology and will return there next year. I'm inspecting my limestone fossils. I have found a beautiful junction of granite with carbonaceous grauwacke.
[Cited: Rudwick 1985, pp.252, 257]

33 **Austen**, Robert Alfred Cloyne   To: H. T. De la Beche
   Ogwell.   Swansea,
   8 December 1838   South Wales.

Shipments of Bovey clay from Teignmouth are between 23,000 and 28,000 tons. I could not obtain the price per ton. Nothing doing in Bovey-coal way. I have powerful notion that "carbonaceous grauwacke" is about the age of the "coal measures" from fossil evidence. At Newton culm slabs and conglomerates go under the limestone, and they all rest unconformably on slates and older limestone. In his lectures, Conybeare talked on all kinds of subjects, and very little on geology.

34 **Austen**, Robert Alfred Cloyne   To: H. T. De la Beche
   Ogwell.   Swansea,
   19 March 1839   South Wales.

Your Report is the best description of a district yet made. The "Silurian System" contains too much mineralogical detail. As to a "System" it's "all my eye", in which elegant predicament I also place the "Cambrian". I have worked on my fossils during winter. I am now satisfied that limestone of Newton is in the Carb. beds above conglomerate. I have thought of pleasant days together over ground spoken of in chapter on grauwacke.
[Cited: North 1939b, p.1052; Bassett 1977, p.vi; Rudwick 1985, pp.272, 289; Secord 1986b, p.230]

35 **Austen**, Robert Alfred Cloyne   To: H. T. De la Beche
   Ogwell.   Athenaeum Club House,
   7 April 1839   London.

I wish I could join the fray over Devon. Your local knowledge thumps that of Sedgwick and Murchison. I told Sedgwick you owed them nothing. Be prepared to answer a statement in your notes on Williams' claims. When I read their new view on the age of S. Devon I could not but smile. It prejudices the 2nd volume of the "Silurian System". I have read the whole Report and much like it.
[Cited: Rudwick 1985, p.289]

36 **Austen**, Robert Alfred Cloyne   To: H. T. De la Beche
   Ogwell.   Swansea,
   20 May 1839   South Wales.

I leave Devonshire for good in June. Phillips is not here. I will take him over the Ugbrooke ground. I have long held the older rocks of Devon to be of the age of the Old Red. Murchison and Sedgwick do not acknowledge my authorship. Our society seems to be conducted for a clique. I shall have very little more to do with it. I feel sure Chudleigh limestone is unconformable beneath the Carb. rocks.
[Cited: Rudwick 1985, p.300]

37 **Austen**, Robert Alfred Cloyne   To: H. T. De la Beche
   Gosden House,
   near Guildford.
   27 January 1841

My newspaper gives an account of your last meeting at Somerset House. I send my "account" for the copy of the Report you promised me. I gave my copy to Lovelace. The British Association meet in Devon next summer. I will meet you there. The proposal of the glacial theory is "seasonable".
[Cited: Rudwick 1985, p.370]

38 **Austen**, Robert Alfred Cloyne   To: [H. T. De la Beche]
   [?April or May 1841]

"Devonian" is a foolish name. "Silurian" just as bad. The Silurian period belongs to history not geology. If we adopt such names, we must reform old ones. Why do not you writers on general systematic geology propose a nomenclature like Lyell's? The review of Murchison's work is mortal heavy and out of place in our Whig Review. I hope to attend the meeting of the British Association, this year at Plymouth.   [Part of letter only]
[Cited: Rudwick 1985, p.370]

39 **Austen**, Robert Alfred Cloyne   To: H. T. De la Beche
   Gosden House,   Athenaeum Club,
   Guildford.   London.
   [1841 FJN]

Will you transmit my regrets to Phillips. On his return from the west I hope he appoints a day for coming here. De Verneuil sent me a collection of fossils with "Silurian" added to each. In no single instance is he right, nor did the collection contain a single Silurian fossil.

40 **Austen**, Robert Alfred Cloyne   To: [H. T. De la Beche]
   Chilworth Manor,   President,
   Guildford.   [Geological Society.]
   [1848 FJN]

I have been asked many questions on the subject of beds of phosphate of lime mentioned in the Gardeners Chronicle. I have looked at the base of the North Downs and offer a short paper. It is a long time since we've had anything on economic geology.

41 **Austen**, Robert Alfred Cloyne   To: H. T. De la Beche
   [no date]

I had proposed to be at the Geological to hear your exposition of the aims of the Jermyn St. Institution, but I have accepted an invitation to stay at Sandhurst, where my boy passes for his Commission.

   Aveline, ?Elizabeth   To: H.T. De la Beche
   [?1848]
   [see **2238**]

42 **Aveline**, Henry Thomas   To: H. T. De la Beche
   Oatlands,
   Wrington.
   12 February 1843.

As you have given your permission to correspond, herewith an account of the discovery by myself and my brother at Burrington cave of fossil bones under stalagmite and red clay floor. Sketches show site and samples. The floor never seems to have been broken, so we are the first to find the organic remains here.

43 **Aveline**, William Huddle   To: H. T. De la Beche
   Axminster.   Helston,
   23 June 1836   Cornwall.

Your immediate reply to my letter gave me great pleasure. It seems almost yesterday that we were rambling on the coast together, and it is now twenty years ago. In asking you to recommend some town in Cornwall, I would add it must contain a skilful surgeon. Excuse my morbid letter. I am so stricken that any temporary occupation would kill me in six months.

44 **Aveline**, William Talbot   To: [H. T. De la Beche]
   Welshpool.
   7 October 1850

I did not get Cox's evaluation of the damage done to the plantation of larches until I left Church Stretton. Herewith the facts of the case: young

trees were cut to facilitate obtaining angle of slope. I do not believe the number damaged to be that stated by Cox. All is an exaggeration.
[Enclosure] Estimate by Mr Cox. 61 trees, total damage £7-2-4. 20 September 1850.

45   **Aveline**, William Talbot   To: [H. T. De la Beche]
     Llanrhaeadr,
     Oswestry.
     22 October 1850

Herewith George's evaluation of damage to trees. He tells me it will not be perceptible in 2 years.
[Enclosure] W. George evaluation 16 October 1850. Damage to sum of £1-5-0.

46   **Babbage**, Charles   To: H. T. De la Beche
     Dorset Street.
     22 June 1849

I have a few engineers and others coming on Sunday to look at my drawings and the language by which they are explained. Do if at leisure join us.

47   **Babington**, William   To: H. T. De la Beche
     Sherborne,
     Dorset.
     9 July 1847

Here for your inspection is a geological sketch of the country round Sherborne with a short memoir. Since my removal to Sherborne I have resumed geological researches with much interest. As I have a large family and limited income I wish employment on the Geological Survey.

48   **Bainbridge**, William   To: [H. T. De la Beche]
     Newcastle on Tyne.
     23 June 1844

On the subject of the Mines Registration project, I presume the interview with Peel was unfavourable. I am still working on the draft of the Bill. We think it ought to be introduced into Parliament this session. Would you let us know your opinion? The only new aspect is that expenses are to be drawn from Consolidated fund. The whole annual cost is estimated at below £7000.

49   **Baker**, John   To: H. T. De la Beche
     Warminster.
     20 April 1848

My box of greensand fossils will be sent on the 21st. There are a great number of Greensand fossils from the Junction bed at Maiden Bradley, Wilts. Price of lot, £3.10.0.

50   **Ball**, Robert   To: H. T. De la Beche
     3 Granby Row,
     Dublin.
     12 December 1845

I am much obliged for your intentions to promote the science of zoology and myself. Economic zoology is insufficiently worked. I have been asked to write a primer on fishes for the schools in fishing districts. I look for £500 as a return.

51   **Ball**, Robert   To: H. T. De la Beche
     3 Granby Row,
     Dublin.
     15 February 1846

Thank you for your assistance in obtaining the grant of birds to the Dublin University Museum. Lord Enniskillen is to give us a collection of fossils. Gulpeth has given us all his minerals and promises fossils as well. There seems to be something happening about fisheries and I suspect you are at the bottom of it. We have honoured ourselves by making you a Vice-President of the Geological Society of Ireland.

52   **Ball**, Robert   To: H. T. De la Beche
     3 Granby Row,
     Dublin.
     23 February 1847

Can you find out for me what is to be done about fisheries? The principles I have urged are the right ones: educate the fisherman and leave it to his own intelligence. Pains must be taken to stop the old system. Many are being duped by promotors of plausible but misguided schemes.

53   **Bancroft**, George   To: H. T. De la Beche
     90 Euston Square.
     8 February 1848

I regret that Mrs. Bancroft & I cannot accept your invitation to dinner on the 18th.

54   **Bancroft**, George   To: H. T. De la Beche
     London.
     12 November 1848

I regret I cannot dine at the Geological Society club next Wednesday, as I have another engagement.

55   **Bancroft**, George   To: H. T. De la Beche
     1 Upper Belgrave Street.
     18 December 1848

Your parcel will be forwarded to Washington today. I desire an interchange of national works, but have none of our own at my disposal. I will write home about it today.

56   **Bardwell**, William   To: H. T. De la Beche
     4 Great Queen Street,
     St. James's Park.
     1 November 1849

Being assured that the true principles of metropolitan drainage are contained in my description, which you have already received, I earnestly beg your attentive examination of the plan now in the Court house.

57   **Barff**, John   To: N. F. Lascelles
     Wakefield.   Campden Hall.
     22 April 1847

This letter will introduce William Ramsden, one of your constituents. He has investigated the ventilation of mines & buildings. He is anxious to lay his plans before the enquiry at present going on, and is a respectable scientific man. Family bereavement has prevented my laying before you my own plans for the expansion of our local railways.

58   **Baring**, Francis Thornhill   To: H. T. De la Beche
     London.
     16 August 1841

Lord Melbourne has told me that he will recommend the Queen to confer a Knighthood on you. But it cannot be done in person as she will not hold a levee or appear in public. Consequently it will have to be done by patent which will be more expensive. Will you let me know that I am to say?

59   **Baring**, Francis Thornhill   To: H. T. De la Beche
     Home Office.
     6 May 1851

Sir George Grey asks the return of the papers respecting the vacant appointment of Inspector of Mines in S. Wales with any remarks you have made on the qualifications of the candidates.

60   **Baring**, Francis Thornhill   To: H. T. De la Beche
     Home Office.
     3 February 1852

Sir George Grey has desired me to send you the enclosed papers and ask you whether the Newcastle College of Practical Science can in any way be assisted. He has asked the Board of Trade, and heard that they have no funds applicable to such purposes.

61   **Baring**, Francis Thornhill   To: H. T. De la Beche
     Home Office.
     10 February 1852

I have mentioned your opinion on Mining Schools to Sir George Grey who seems favourably impressed. I am to write to the Newcastle School of Science to express his opinion of the value of Mining Schools and to ask further information without making any promise.

62   **Barker**, James   To: H. T. De la Beche
     Bakewell.
     12 October 1849

Enclosed Rosser's receipts for his weekly salary. I have been to Newcastle, and also visited Phillips at York. We discussed ventilation of mines. The Government and owners are not indifferent to loss of life. I have seen the map of the limestone district in Derbyshire. Phillips says the section is with engraver. Can I have a first copy for use in my lectures to the Bakewell Institution?

63 Barker, James   To: H. T. De la Beche
Bakewell.
14 December 1851

Thank you for the packet of Inaugural Discourses. I am glad to see that the School of Mines has made such a good start. I will send John Percy some ores for assay and analysis, to ascertain losses in smelting. Your school, for the first time, offers facilities for experiments.

64 Barlow, John   To: H. T. De la Beche
Royal Institution.
4 February 1851

Lord Liverpool says thank you for the message respecting barley mills.

    Barr   To: H.T. De la Beche
    [no date]
    [see **2239**]

65 Barry, Charles   To: H. T. De la Beche
Foley Place.
29 October 1838

I hope to hear from you soon as to the appointment of Experimentalists who will do the work on the different building stones. If Faraday has offered to do it gratis, we should accept, as we need cash in hand for further quarry visits. I have a new sample in from Swansea. Please visit the quarry; its name is "Sutton".

66 Barry, Charles   To: H. T. De la Beche
Foley Place.
27 April 1839

Staunton is making a great fuss about his black marble for the new Houses of Parliament. I wish to have your opinion before I defend my opposition to it before the Bar of the House. He is quoting you as an authority. Please give me an answer to my previous letter also.

67 Baylis, B.   To: H. T. De la Beche
Merthyr Tydfil,
South Wales.
15 September 1846

I have to carry out plans for the improvement of the sanitary conditions of a large town, and cannot obtain a copy of the Report of the Health of Towns Commission, except through a Member of that Commission. Please favour me with one.

68 Baylis, B.   To: H. T. De la Beche
Borough Engineer,
Chester.
2 July 1849

I learn you are active on the Metropolitan Sanitary Commission. Will you tell me where I can obtain information about the use of sewage manure for agricultural purposes? The experiment could be tried here.

69 Beaufort, Francis   To: H. T. De la Beche
Admiralty.   3 Alpha Place.
19 April 1832

Please oblige me by answering the questions of my First Lieutenant, Becher, or have them answered.

70 Beaufort, Francis   To: H. T. De la Beche
Admiralty.
2 November 1852

Herewith another drawing of the raised beaches on Loch Tarbert, Jura, made by Comr. E. Bedford.

71 Becher, A.B.   To: [H. T. De la Beche]
September 1831
Counter Current off the Cape of Good Hope

We are inclined to think that the counter current could be taken advantage of by ships going to the eastward against the South East Trade. Its existence is proved by the facts surrounding the wreck of a ship on Wolfe rock, and the drifting of a beacon from False Bay.

72 Becher, A.B.   To: H. T. De la Beche
Admiralty.   3 Alpha Place,
12 May 1832   Alpha Road.

I send you the French work we were speaking of last evening and if you meet with anything useful to me, please mark it in pencil. You will also receive the chart of part of the river Thames.

73 Beddes, Joseph   To: H. T. De la Beche
Office of Woods.
2 December 1847

I have prepared a statement of Votes, Expenditure and Balances upon Votes for your Museum. I would like to show it to you. There has been a payment made for some services connected with the new building in Piccadilly. At present the funds voted are absorbed, and a surplus expenditure of more than £90 shows.

74 Bedlington, William   To: M. Scott
Ovingham,   1 Chapel Place,
Near Gateshead.   Duke Street,
2 May 1848   Westminster.
       London.

Thanks for yours of 28th. I might be able to manage all you require. I don't care much about the place. The climate is bad for Europeans. What sort of workmen will be employed? They will need a deal of drilling and Englishmen as leaders in the different branches of the trade. I should hesitate to undertake the work, if I did not have proper subordinates. Your advice please.

75 Belcher, Edward   To: H. T. De la Beche
Thurloe Square.
25 March 1851

Sorry to have missed you on Sunday. On Saturday night I prepared the Texas collection for your inspection. I hope to be more fortunate next time.

76 Bell, George Gray   To: [H. T. De la Beche]
7 Pit Street,   [Mining Records Office.]
Newcastle upon Tyne.
5 February 1849

Surveys of mines are plotted erroneously on working plans. I have often fruitlessly urged the adoption of a better system. Money and lives are being lost. In plotting, no regard is taken for daily variation of the Meridian. Bearings must be reduced to the Surface Meridian. I offer my services in your Office for which a moderate salary would suffice.

77 Bell, Thomas   To: H. T. De la Beche
New Bond Street.
24 March [1851 FJN]

My sincere thanks for sending me your new book. Though at present a mere tyro in Geology, I mean to make myself better acquainted with your favourite science.

78 Belshaw, Thomas   To: H. T. De la Beche
Royal Institution,
Manchester.
23 April 1850

Playfair informed me that the promised specimens have not been received. I have written, and they will be ready for the Great Exhibition. You can then make a selection for the Museum. Some manufacturers promised to send up specimens now. If they have not done so yet, please tell me and I will remind them.

79 Benett, Charles Cowper   To: H. T. De la Beche
Lyme Regis.
8 April 1847

Waring requests me to forward the accompanying note. He tells me you are coming here on a Government survey. Mrs. Benett and self would be pleased were you to stay with us. Please do not mention Waring's object to anyone here.
[See Waring, H.F. to De la Beche, H.T., 6 April 1847, **2278**].
[See De la Beche, H.T. to Waring, H.F., 9 April 1847, **410** for reply].

80 Benett, Sarah   To: H. T. De la Beche
Lyme Regis.   Bridgend,
25 April 1840   South Wales.

The inhabitants of Lyme Regis are anxious about damage done to the cliffs by the sea and the threat to the church and churchyard here in Lyme Regis. I have been asked to request that you would inspect the damage before we take any steps to remedy it. Your expenses will be paid.

81 Benett, Sarah   To: H. T. De la Beche
Lyme Regis.
29 November [1849 FJN]

Has the Professor who was here last spring made any report on the state of

Church Cliffs? If so will you write a line on the subject. Do you still recommend Putney College for a boy intending to be a civil engineer? I wish you could visit Lyme again.

82 Bianconi, Giacomo — To: [H. T. De la Beche]
Bologna.
12 December 1846

Because I have published today a paper on a very singular geological problem it is my duty and pleasure to send it to you, taking advantage of the presence in this town of Mrs. Somerville. I have surmised that the waters of the Mediterranean were once much higher than the modern level. I have presented two other papers to the Academy of Science, the second of these being on the modernity of the Egyptian delta.

83 Bianconi, Giacomo — To: H. T. De la Beche
Bologna. Geological Survey Office,
3 November 1849 London.

The political situation in Italy has retarded the progress of scientific work. Thank you for your letter of 20th Sept 1847. I have continued slowly my research on the former level of the Mediterranean and published a 3rd memoir which I will send to you when the 4th has been printed. But the question is complicated further by similarities with places other than the Mediterranean.

84 Bianconi, Giacomo — To: H. T. De la Beche
Bologna. Geological Survey Office,
3 July 1853 London.

I recommend to you our young professor of astronomy who is coming to London to visit the Greenwich Observatory. Professor Respighi met Mr. Pentland at Bologna and will call on him if he is in London. I am glad you were pleased with the collection of sulphur rocks which I sent to you. You can count on me for anything you may want either from Italy or Dalmatia where I have connections. We would like for our Museum specimens of genuine Cambrian, Devonian and Silurian rocks.

85 Biram, Ben — To: H. T. De la Beche
Wentworth House.
20 March 1837

Since at your request I inspected the Ardsley Main Colliery after the disaster there, I have considered improvements to the Davy Safety Lamp, and enclose my designs for its improvement. I intend to register them under the designs act if your opinion favourable.

86 Birch, Samuel — To: H. T. De la Beche
British Museum.
3 March 1854

Herewith Egyptian manner of writing the names of the Gods you ask for.

87 Blackwell, J. Kenyon — To: H. T. De la Beche
Abercarn,
Near Newport,
Monmouth.
29 June 1849

Yours of 27th was forwarded here. I regret I cannot investigate the Dudley Mine explosion immediately. My recovery is slow. I hope to see you in London next week.

88 Blackwell, J. Kenyon — To: H. T. De la Beche
Abercarn,
Monmouth.
4 November 1849

I have finished my Survey here and will reach Dudley on my way to Lancashire. I would like to see you in London on Wednesday or Thursday next concerning the extent and contents of reports to be given at the next session.

89 Blackwell, J. Kenyon — To: H. T. De la Beche
Portway Hall, Jermyn Street,
Dudley. London.
3 October 1850

What arrangements will be made under the Mines Inspections Act? Will I be allocated the District of South Wales and Somerset? I need information to make residence plans etc. Would an interview be possible? I have another Report available embracing districts & subjects not covered in my first.

90 Blackwell, Thomas E. — To: H. T. De la Beche
Foxhangers,
Chippenham.
10 May 1850

Herewith specimens of plates engraved for Meteorological purposes, also applicable to other uses.

91 Blayds, Elizabeth — To: [H. T. De la Beche]
South Stoke Vicarage,
Near Bath.
4 March [1845 FJN]

Mr. Blayds informs me you are interested in the collection of the late Miss Benett. The price would be £185 with cabinets. A catalogue is available. You are invited to call.

92 Bontell, Charles — To: H. T. De la Beche
Sandridge Vicarage,
St. Albans.
29 March 1847

I hope to obtain permission to take rubbings from the brass of Louis de Corteville in the Museum of Economic Geology. I wish to present you with a small volume on the subject of monumental brasses.

Booker, Thomas W.
[c.1837]
[see **2240**]

Booker, Thomas W.
24 June 1839
[see **2241-2243**]

93 Boué, Ami. — To: H. T. De la Beche
Société géologique de France, Athenaeum Club,
Paris. London.
28 March 1832

My thanks to you in the name of the Geological Society of France for the 2nd edition of your Manual. I have translated your paper on La Spezia, which most probably will be inserted in our transactions if not in the bulletin now in the press. I hope you will be able to sell copies of my own work in Britain, and beg you instantly to make it known. What do they think in London of Deshayes' ideas on successive tertiary deposits?

94 Bowerbank, James Scott — To: H. T. De la Beche
3 Highbury Grove.
7 June 1848

Great and important events for the Palaeontographical Society are upon the carpet. Owen is aggrieved at Bell's proposed monograph on Reptilia. After talking with Bell he proposed to abandon his work. I proposed to Owen that we should take it up, and a Council was called. Lyell, Bell and myself have been named to bring to maturity the scheme to publish all 110 of his plates. What is your opinion?

95 Bowerbank, James Scott — To: H. T. De la Beche
35 George Street,
Hastings.
14 October 1848

Herewith news of the Palaeontographical Society affairs. The Council have agreed that King should have £5.0.0 per plate. Dinkel and Sowerby are also at work. Phillips is cooperating, as well as Lord Enniskillen. The monograph on reptiles will be a good thing when done. I am trying to make Howse and King of Newcastle cooperate to our benefit. Our reports have done us good service and three good men have offered themselves as local secretaries: Bree for Stowmarket, Saxby for the Isle of Wight & Dickie for Aberdeen.

96 Bowerbank, James Scott — To: H. T. De la Beche
6 Devonshire Terrace,
Ventnor.
1 October 1849

Although out of town, I am working on proofs. Milne Edwards has taken six sheets of his work back to Paris. King's work is in press. It is not as good as Edwards. Jones' work is now printing off. He has also described Cytherinae in Kings's work. New members keep dropping in.

97 **Bradley**, W.H.   To: [H. T. De la Beche]
     Aurangabad,
     Bombay.
     21 September 1847

My friend Capt. Ward has obtained your leave for me to correspond. I am sending you boxes of fossils from India, and describe for you now the geology of the site. I believe Murchison is not long back, and remarked on the utter absence of all vestiges in the Gawil Range of Secondary or Tertiary fossils.

98 **Brand**, H.   To: H. T. De la Beche
     Home Office.
     7 June 1850

There is no intention of presenting the Aberdare Report. If Blackwell applies to Sir G. Grey he will receive copies of his Report. Phillips has received 50 copies.

99 **Brand**, H.   To: H. T. De la Beche
     Home Office.
     12 June 1850

Please read the enclosed from one of the Inspectors of Factories. Is there any such opening as that stated within for his son?
[See Howell, T. to Brand, H., 11 June 1850, **732**]

100 **Brand**, H.   To: H. T. De la Beche
     Home Office.
     15 June [c.1850]

Sir George would be obliged to you to look at the enclosed Bill & to let him have your opinion upon it.

101 **Brande**, William Thomas   To: H. T. De la Beche
     Royal Mint.
     1 April 1840

Longmans inform me you had not time to do the Dictionary article on geology, but would revise and correct one done by me. The article is now written. Can I send it to you for correction? Please abridge and add to it as necessary. I must tell you that I am one of Lyell's faction.

102 **Brande**, William Thomas   To: H. T. De la Beche
     Royal Mint.   Bridgend,
            Glamorganshire.

Thank you for the time and trouble taken with the proof sheets of my article. I am glad to find Phillips so well installed at the Museum.

103 **Brayley**, Edward William   To: H. T. De la Beche
     London Institution.
     21 December 1840

I have applied to the Council of University College for the chair of Mineralogy. Mr. R. Phillips has sent a testimonial. Would you also give a reference? I enclose the syllabi of lectures recently given. Greenough is on the Council. Prof. Graham supports me in Senate.

104 **Brewster**, David   To: H. T. De la Beche
     St. Leonard's College,
     St. Andrews.
     19 December 1845

I observe that in the Explanations of the Vestiges the author refers to my ignorance in quoting your Manual respecting the earliest fossils. Are his observations justified?

105 **Brickenden**, Richard Thomas William Lambart
     Rockingham Hall,   To: H. T. De la Beche
     Cuckfield,   President,
     Sussex.   Geological Survey of
     2 February 1849   the United Kingdom.

I am intensely interested in the Geological Survey of the United Kingdom. I request the favour of an appointment to that Corps of Gentlemen.

106 **Brickwell**, William   To: H. T. De la Beche
     Battersea.
     8 January 1848

Please read the enclosed at your leisure. I trust it contributes a mite to sanitary improvement.

107 **Brickwell**, William   To: W. Buckland
     Artificial Granite Works,
     Albert Road,
     Battersea.
     16 February 1849

I have been engaged on experiments with coke and other fuels. The best manufacturer of coke in London is Cory's. Herewith is a sample to which I will refer in a paper on the subject.

108 [?]**Briggs**, George D.   To: Capt. E. Harvey
     Royal Military College.   14 Light Dragoons,
     27 August 1840   Cardiff,
            Wales.

Herewith the specifications of the theodolite designed to withstand a fall and other damage. It has been used for many years both on a large scale for content, and for more general ones. One was made for Sir Edward Paget.

109 **Bright**, Benjamin Heywood   To: H. T. De la Beche
     59 Pulteny Street,
     Bath.
     4 November 1841

You have essentially helped me in my mining and agricultural interest. Do you think the Everton veins could lie at Newgate Bridge less than 100 fathoms deep? I am watching with great interest all cheap and easy modes of making defences against the surf. A perfect saurian has been found at Whitby. I have sent specimens of galena and strontium to Craig's Court.

110 **Bright**, Benjamin Heywood   To: H. T. De la Beche
     59 Pulteny Street,
     Bath.
     24 August 1842

The collection at Brand Lodge is completely at the disposal of yourself and Phillips. I send you the Journal of Coal Works at Bogie, Scotland 1719-54, and present it to the Museum of Economic Geology. My coalbeds at Newgate promise very well under Walters' management. Has Phillips found any new crustaceans? I must ask him to examine the structure of my trilobites' eyes.

111 **Bright**, Benjamin Heywood   To: H. T. De la Beche
     12 October 1842

The "Mendip Customs" are ready. Shall I send it to you at Ross? What is this capital discovery of our amiable friend of the Link Farm [Phillips]? The Professor tells my sister I shall have a paper upon it when printed, but this does not satisfy my curiosity. Sedgwick had a field day over it. Walters has just opened a vein of coal 30 feet thick. On Monday I go to some warm baths for a week.

112 **Bright**, Benjamin Heywood   To: H. T. De la Beche
     Bishopston,
     Near Stratford on Avon.
     24 October 1842

Thank you for your letter explaining Miss Phillips' discovery. If your sketch be taken literally, no more need be said, but if it is theoretical, there may be questions to ask. Does her discovery confirm Phillips' determination of the place of the Caradoc fossils? I am considering the electrotype for your benefit. I am too much an invalid now to engage in personal research.

113 **Bright**, Henry   To: H. T. De la Beche
     London.   Mall,
     16 April 1825   Clifton,
            Bristol.

Hume has put off his motion respecting Sierra Leone till 22 April. I hope you will before then furnish the information on ships trading in slaves from that Colony. We are to have a committee to consider colonization and emigration. I wish you would favour me with your opinion of the probable success of a white settlement in Jamaica bearing in mind the climate, soil etc. I shall forward your letter to Wilberforce today.

114 **Bright**, Henry   To: H. T. De la Beche
     Winchester.   24 Regent Street,
     24 September 1828   London.

Please give me a list of the persons who can be called together in November. The whole thing is damned if Lord Leaford is at the head of it. Jamaica will have nothing to do with the Colonial Club. The thing may possibly lead to the establishment of horticultural efforts in Jamaica. Do they expect Parliament to meet in November?

115 **Bright**, Henry   To: H. T. De la Beche
     London.
     30 July 1829

Your letter gives rise to your friend's asking what you have seen, where you have been, and what have you brought home. The Jamaica scheme fell dead as a stone, though I think some Jamaica improvements are talked of. I have handed your letter to Broderip, from whom you will hear in a day or two.

116   Bright, Henry   To: H. T. De la Beche
2 Paper Buildings,
Temple.
22 June 1848

As you encouraged me to ask you about the water consumption of a steam engine, I now trouble you with this. From the enclosed analyses of two waters, can you estimate how much coal of merchantable value ought the engine consume in a 10 hour working day? We West Indians must for the future live on water and be obliged to eat crusts.

117   Bright, Richard   To: H. T. De la Beche
11 Saville Row.
3 February 1852

Your present indisposition is due to overwork. You must get away from labour and all anxiety or you will get worse.

118   Bristow, Henry William   To: H. T. De la Beche
Presteigne.
6 September 1844

When you visit me here I would like you to see if my opinion on Brampton Bryan Park is not the right one. Murchison has not noticed a coarse quartzose conglomerate. Herewith sketch of a section. What is its meaning? Please send the balance of my pay for the last quarter.

119   Bristow, Henry William   To: H. T. De la Beche
Knighton,
Radnorshire.
10 November 1844

I am running the Clun outline of the Old Red Sandstone. If fossils are to be the criterion for classifying the strata, then Pentamerus knightii will decide the question, in the teeth of all the evidence. Pray let me know your opinion. Send £10.

120   Bristow, Henry William   To: H. T. De la Beche
Knighton,
Radnorshire.
9 April 1845

I have paid Gibbs the half £10 note, and also advanced him £3. Please send the other half to Knightsford Bridge. I will be delayed as a result of Gibbs' leaving so soon.

121   Bristow, Henry William   To: H. T. De la Beche
Knighton,
Radnorshire.
13 April 1845

I am glad my statement has cleared up the difficulty in the accounts. The other two are easily dealt with. The discrepancy is due to your having deducted travelling allowance after I had already done so.
[4 additional folios of Bristow's Geological Survey accounts preserved with this letter, from 1 July 1844 to 31 March 1845].

122   Bristow, Henry William   To: H. T. De la Beche
Bagber Farm,
Sturminster Newton,
Dorset.
20 December 1846

Thanks for the increase in my pay. I hope to finish the district soon, but frost has put me back a few days.

123   Bristow, Henry William   To: H. T. De la Beche
Bridport,
Dorset.
14 May 1848

Litigation with Harris. My complaint refers to the wilful obstruction offered to the Survey after notice to enter upon his lands had been served. No man has a right to assault another for trespass. He eventually did not give permission to enter. I always find the farmers civil, but this man is a ferocious animal. Your advice is requested.

124   Bristow, Henry William   To: H. T. De la Beche
Upwey,
Dorchester.
28 September 1849

Thanks for your letter of yesterday. I will consult my friends and let you know the result. I value your confidence in recommending me at the Colonial Office for the Surveyorship [of New South Wales].

125   Bristow, Henry William   To: H. T. De la Beche
Upwey,
Dorchester.
7 October 1849

Before deciding finally on the New South Wales Surveyorship I ask confidentially what are the pay prospects on the Survey of G.B.? In case of positive acceptance of the Survey of N.S. Wales, your cooperation is needed to obtain liberal travelling allowances and ample assistance.

126   Bristow, Henry William   To: H. T. De la Beche
Upwey,
Dorchester.
10 October 1849

I will accept the Surveyorship of New South Wales. I regret quitting the survey of the U.K., as the happiest years of my life have been spent in its service.

127   Bristow, Henry William   To: H. T. De la Beche
Upwey,
Dorchester.
2 November 1849

I have no objection to the arrangement for the horse allowance in New South Wales proposed by Hawes, it being understood that further allowances will be arranged by the local authorities.

128   Bristow, Henry William   To: A.C. Ramsay
The King's Arms,
Piddletown,
Dorset.
18 October 1850

I am nearly at the end of this section's work. The year has been hard on my pocket as a result of inn and hotel expenses. Could you ask Sir Henry how the matter of £50 extra salary stands? Apart from the above expenses, other difficulties leave me in a precarious financial state.

129   Bristow, Whiston   To: H. T. De la Beche
Hitchin,   Athenaeum Club,
Hertfordshire.   London.
16 February 1840

Thank you for your letter. I am disappointed you will not pass this way. The offer of hospitality still stands, hoping you'll visit before the end of summer. I must be my own biographer, whereas all know you. Thank you for the kind offer of service, but I have no wish to trespass on your goodness.
[Cited: North 1936a, p.72]

130   Broderip, William John   To: H. T. De la Beche
Somerset House.   Lyme Regis,
17 August 1829   Dorset.

The best way of terminating our official correspondence will be for you to consult Buckland and let me have the Tor & Babbacombe paper with your joint emendations. I will pay every attention to the alterations of the illustrations of Nice. There are two Jamaica land crabs in the Zoological Garden, in good care.

131   Brongniart, Adolphe Theodore   To: H. T. De la Beche
8 February 1830   58 Jermyn Street,
                                                                 London.

Thank you for the interesting specimens which you sent me and which arrived in good condition. I have nothing like them from the Lias, and these ones are so well preserved that they will enable me to give a good description of these remarkable plants. The branch with leaves appears to me to be a branch of a conifer related to the Cypressus. To determine the genus it would be desirable to find the fruits borne by the branches.

132   Brongniart, Adolph Theodore   To: [H. T. De la Beche]
31 December 1847

Some time ago you asked me for information on the works of my father which you could use in the Anniversary Address to the Geological Society. The best I can do is refer you to the speeches made at his funeral by colleagues and pupils. Elie de Beaumont's is the most complete.

133 Brongniart, Alexandre To: H. T. De la Beche
Paris. Paris.
9 June 1828

I have learned with great pleasure from your note that you are in Paris. Now I can thank you for your gifts, especially the 'Tabular View of English formations'. I spend most of my time at Sèvres, and propose you pay me a visit there and dine with me next Thursday. You may perhaps find something of interest to see in the manufacture of porcelain.

134 Bronn, Heinrich Georg To: H. T. De la Beche
Heidelberg,
Germany.
1 January 1850

Herwith is a copy of my Index palaeontologicus. It is a report on the actual state (1847 - it has been with the printers since then) of science, which seizes the moment with all its faults, omissions and errors, without being able to do much to remedy it. I arrive at the same result as M. Agassiz regarding the development of the organised world towards an ever greater perfection, but I believe this law subordinate to another more powerful, that is to say, the law of accommodation of the animal and vegetable organisations to the exterior conditions of their existence.

135 Brown, John To: H. T. De la Beche
Stanway, Ordnance Map Office,
Near Colchester. Tower,
2 June 1837 London.

I have been carrying out experiments on carbonizing bones as described in your book "How to Observe". I commenced the process on 6th April last after delay. Specimens of granite, porphyry etc. were also placed in carbonized water. Have you any suggestions?

136 Brown, John To: H. T. De la Beche
Stanway, Ordnance Map Office,
Near Colchester. Tower,
7 May 1838 London.

Twelve months have elapsed since the experiments on carbonizing bones etc. were commenced, but I am delaying inspection of the result until the time lost due to severe frost has been made up. I am impresssed by the large quantity of matter held in solution after 12 months.

137 Brown, Robert To: H. T. De la Beche
17 Dean Street,
Soho.
4 March [1848 FJN]

The Polar Dinner (barring pemmican) is fixed for 22nd at 7.

138 Brown, Robert To: H. T. De la Beche
17 Dean Street,
Soho.
28 February 1849

I return with thanks the 5 specimens you permitted me to examine.

139 Bryce, James Jr. To: H. T. De la Beche
Belfast.
12 December 1845

I am applying for the Professorship of Geology at our new college. Please favour me with a reference.

140 Buccleuch, Duke of To: H. T. De la Beche
Montagu House.
6 June 1843

I have received the Ordnance Survey Dublin Map. It is well worthy of attention by the Commission.

141 Buch, Baron Christian Leopold von To: H. T. De la Beche
Berlin. Foreign Secretary of
5 March 1842 the Geological Society,
London.

My thanks to you for the offer of the Wollaston Medal. I owe it to you that I have become known favourably in your country. It is true that Fitton has said that all my views on the Tyrol could have been read in Hutton's work, but I did not do so. I have studied the book of nature. Hutton's claims of priority are false. If Hutton has said the same thing as I, then well and good. I have committed no theft and have a right to gratitude from contemporaries. Bunsen will accept the medal on my behalf.

142 Buckland, Mary To: H. T. De la Beche
Islip.
15 March [1850]

How many gas shares has the Dean? Last year he received £30 I am sure. He is much worse. He puzzles everybody, so much reason and so much non-reason. God in his infinite mercy may restore him, but I see no hope.

143 Buckland, Mary To: H. T. De la Beche
Deanery.
20 September [c.1850]

Were you in town I would ask you to call on me. My husband is in same state as when we left home. He is now kept alive by artificial means, and has become very morose. He was sadly distressed at seeing the children. I am told he does not feel. The change is so awful that to dwell on the subject almost shakes my own reason.

144 Buckland, William
May 1816

[Sketch of fossil trees] copied from a drawing by Trevelyan in 1803. Location, Bog Hall near Newbiggin on the coast of Northumberland. 20 such were counted in 1815. All but 4 or 5 were upright. Beneath was a horizontal bed of coal. These plants seem to have been gigantic reeds, floated from a distant point.
[Enclosure in Buckland, W. to Cole, Lady Mary, 3 June 1816, **145**].

145 Buckland, William To: Lady Mary Cole
2 Parliament Street.
3 June 1816

I will be leaving tomorrow, and by October the information on the geology of Glamorganshire which I hope to receive from you and the Misses Talbot will be wanted for insertion in the Geological Society's map of England. Enclosed is a drawing and description of fossil trees described by Trevelyan. Geological subjects were proposed for the first time to a student in logical questions in the schools at Oxford last Thursday.
[See Buckland, W. to Cole, Lady Mary, May 1816, **144** for enclosure].
[Cited: North 1928, p.34; Rupke 1983, p.120]

146 Buckland, William To: Lady Mary Cole
Corpus Christi College, Penrice Castle,
Oxford. Swansea,
3 April 1817 Glamorganshire.

Thank you for the present of the map of Glamorganshire which I found on my return to England last November. A parcel has been forwarded to you containing seeds collected last summer on the tour with Conybeare, when we visited eastern Europe. The journey occupied 5 months of intense labour in seeing every collection and professor that could be heard of and purchasing every map, book and print that has been published relating to geology or to the political economy of the countries we passed through.

147 Buckland, William To: Lady Mary Cole
Corpus Christi College, Penrice Castle,
Oxford. Swansea,
14 July 1817 Glamorganshire.

The enclosed silver bowl was sent 3 weeks since. I have taken it to London and the Antiquarian Society cannot decipher the inscription. It is Roman, crude workmanship, worth only a little more than its weight as old silver. Enclosed are also some seeds, a piece of fungus, and a specimen of a newly discovered tourmaline. It is unlikely I will come to Wales this summer, as I plan to make a geological map of Devon.

148 Buckland, William To: Lady Mary Cole
London.
24 December 1818

Enclosed are some Alpine seeds. I hope soon to have a proper room for specimens in Oxford, the least the University can do to meet the grant I have obtained from the Crown for the establishment of a Professorship in Geology. The philosophical world is here much occupied with Capt. Ross' polar expedition, the problem of red snow, meteoric iron used for harpoons by natives, and the dangers of the ice flows.
[Enclosure:] Lines on the commencement of term.

149 Buckland, William To: Lady Mary Cole
London.
29 December 1818

Here are the seeds from the Alps I promised you. Wollaston has examined the water from red snow, and separated out the colouring globules. I have laid some on a bit of paper for you. The water possesses an offensive odour. Strangways is preparing a memoir on organic remains collected in Russia last summer to be presented to the Geological Society.

150 Buckland, William To: Lady Mary Cole
    Oxford. Penrice Castle,
    13 July 1819 Swansea.

I enclose my long promised copy of King Coal's Levee. The author is a Northumberland gentleman [Mr. Scafe] lately in the Army. A third edition will soon be published by Longman. I went to Kent to witness Miss Lewin's curious experiments with a divining rod. It is all a trick. I hope to be at Penrice in the autumn.
[Cited: Rupke 1983, p.225]

151 Buckland, William To: Lady Mary Cole
    Axminster. Penrice Castle,
    29 October 1819 Swansea.

My plan to come to Glamorganshire has been deranged, as I have had an accident. A piece of iron from my hammer lodged in my eye, and I have had to have 5 or 6 operations. I am recovered now. Lord Bathurst has undertaken to promote my collections by enlisting help in the colonies. At Weymouth I met Mr. Lemon and Lady Charlotte.
[Cited: North 1934a, p.63; North 1938, p.1040 Rupke 1983, pp.121, 225]

152 Buckland, William To: Lady Mary Cole
    Oxford. Penrice Castle,
    5 January 1820 Swansea.

Please thank Sir Christopher for his suite of specimens of the organic remains of the present woods of Gower. Your brother will save his studentship if he resides in Oxford for 3 weeks of next term. I have been lately occupied on a paper about the proofs of the Mosaic Deluge afforded by the neighbourhood of Oxford & London, in answer to Whittacre. Conybeare and I will be making a tour of the Severn district. I mean to propose William Strangways as Vice-President of the Geological Society.
[Cited: Rupke 1983, p.121]

153 Buckland, William To: Lady Mary Cole
    Oxford. Penrice Castle.
    18 May 1820

Please accept this copy of a lecture of mine defending geology against the charges that it leads to opinions inconsistent with the Mosaic chronology and account of the Creation. I am obliged again to postpone my visit to Penrice, as I am joining Greenough and Breunner on a tour of Europe. I have a larger class than ever in spite of all that Greenough has done in his book to discourage geology.
[Cited: Rupke 1983, p.122]

154 Buckland, William To: Lady Mary Cole
    Oxford. Penrice Castle,
    13 June 1820 Swansea.

I am glad you approve of my performance in the way of authorship. The Bishop of London and Mr. Sumner also approve. Sumner liked my interpretation of the first verse of Genesis and the statements I have advanced with respect to the Mosaic Deluge and the cause of the errors of Linnaeus. Lord Grenville also has commended my lecture.

155 Buckland, William To: Miss Jane Talbot
    Oxford.
    11 December 1820

I take advantage of your brother's offer to send a box enclosing the volume of plants collected in 1816, together with a small collection I made last Summer. My journey on the Continent was less adventurous than the last one. Whately composed the enclosed elegy in contemplation of my remaining underground on the Continent.
[Cited: North 1938, p.1040]

156 Buckland, William To: H. T. De la Beche
    Oxford. Mall Buildings,
    18 February 1821 Clifton.
      Bristol.

Herewith a lithographic copy of my epitaph. Thanks for your offer to add your drawings of Alpine fossils to my proposed memoir. I am preparing an account of the Weymouth district, S of the chalk.

157 Buckland, William To: Lady Mary Cole
    Oxford. Bath.
    31 March 1821

I beg to send you by Mr. Talbot some seeds of a Passiflora producing edible fruit. I am promised a large cargo of seeds from New South Wales and New Zealand. This week I received specimens from the Great Lakes of N. America. They contain fossils of the Dudley Limestone. From near Rugby I have just got some splendid bones of rhinoceros, elephants etc.

158 Buckland, William To: Lady Mary Cole
    Oxford. Penrice Castle,
    25 July 1821 Swansea.

Enclosed are some seeds brought back from New Zealand and New South Wales. I received a letter from your brother in Poland on his way to Odessa. We have been busy preparing for the election of Sir. J. Nicholl. I have accepted an invitation to visit the Highlands, for the sake of McCulloch's map now in preparation, and intend to be back in six or eight weeks. I enclose you copies of poems by Mr Duncan and Mr Conybeare.
[Cited: North 1938, p.1040 - 1041]

159 Buckland, William To: Lady Mary Cole
    Bakewell.
    18 September 1821

My design to visit Scotland and the Western Isles has been thwarted by my companion being obliged to return to York. In Trevelyan's family at Wallington I fancied I saw a model of Penrice. I am now working southwards through Derbyshire. I am sorry I cannot congratulate you on Sir. J. Nicholl's success, but his minority was a glorious one. No man in England except Heber would have carried it against him.
[Cited: Rupke 1983, p.225]

160 Buckland, William To: Lady Mary Cole
    Oxford.
    11 October 1821

I met with a formidable accident, having been thrown from a coach coming from Northampton to London. I have been resting my back for a few days. Your brother figures largely in the volume of the Geological Transactions just published. I have also a very nice book on the botany of three northern counties. At the end of it is a scale of temperatures in Newcastle mines.

161 Buckland, William To: H. T. De la Beche
    London. Lyme Regis,
    21 November 1821 Dorset.

Pentland has sent a letter from Paris saying that Cuvier wants you to send drawings of the 17 vertebrae of the fossil crocodile you bought at Honfleur, if you do not intend to publish, as they probably belong to the same species as a specimen which he collected from the same place.
[Cited: McCartney 1977, p.20]

162 Buckland, William To: Miss Talbot
    Oxford.
    26 November 1821

If the spot where you found the bones is within walking distance of Penrice, would you examine carefully if there be any traces of caverns or fissures still remaining which may be connected with that in which the bones were discovered. A similar discovery has been made at Kirkby Moorside. I hope you will get some converts to my theory or supply a better. Conybeare and myself have recommended Miller as successor to Leach at the British Museum.
[Cited: North 1942, pp.96-97; North 1966, pp. 126-127; Rupke 1983, p.33]

163 Buckland, William To: H. T. De la Beche
    Brislington. Lyme Regis,
    20 April 1822 Dorset.

I have never been delighted by a fossil more than by your glorious jaw of Plesiosaurus which deserves to be cast in gold and circulated over the Universe. With Conybeare's concurrence, I have presumed to arrange for Chantry to make a cast of it. In connection with your Lyme paper, I have sent to the Geological Society an account of the valleys of denudation which intersect the coast.
[Cited: McCartney 1977, p.20; Rupke 1983, p.134]

164 Buckland, William To: Lady Mary Cole
    Axminster.
    5 November 1822[?]

Thank you for the present of the plants. I am much flattered by the attention many of my importations of 1816 have received at your hands. Conybeare has informed me of his visit with Sedgwick to your new caves.

His description is imperfect. Could you send me specimens? I will return them immediately with a cast of a coal plant.
[Cited: North 1942, p.124; North 1966, pp.135-136]

165 **Buckland**, William   To: Mary Cole
Oxford.
24 December 1822
Has there really been a discovery of a cave full of bones in the neighbourhood? Dillwyn talks in his letters as if I were coming immediately to Glamorgan to examine it. I asked Talbot to send me a box of specimens. I would have come with Davy and Wollaston had I not been preparing an account of German caves for my book. Cuvier has changed his mind, and the Royal Society has awarded me the Copley Medal, so I am not afraid of further opposition to my hyena story.
[Cited: North 1942, pp.102-103; North 1966, pp.127-128; Edmonds & Douglas 1976, p.147]

166 **Buckland**, William   To: Miss Mary [Theresa] Talbot
31 December 1822
I am sorry I could not assist at the opening of the Paviland Cave. Was the bottom of it flat or inclined? Had it side vaults or communication to the surface? Is it in a cliff or inland? What was the position and depth of mud? Was there stalagmite above the mud? Were the broken bones old or new? Was the skull an elephant's? I am impatient for further accounts. Please have the mouth closed to prevent total destruction.
[Cited: North 1942, pp. 103-104; North 1966, pp.128-129; Rupke 1983, p.68; Howes 1988, pp.301-302]

167 **Buckland**, William   To: Lady Mary Cole
Oxford.   c/o Sir Chr. Cole MP,
15 February 1823   Penrice Castle,
         Swansea.
Readdressed to: United Services Club, London.
I failed in my candidacy for the Presidency of College, but am likely to derive very superior advantages. Today I lectured to an overflowing class, including the Bishop of Oxford. I spoke on the phenomena of the Paviland Cave. The Red Lady is not Antediluvian. Why do you not get up a romance about her, a witch etc? If you have not sent away her bones, pray keep them for me. I should like to have the large thigh bone.
[Cited: North 1942, pp. 107-110, 112, 116; Edmonds 1956, p.291; North 1966, pp.130-134; Boylan 1967, pp.242-243; Edmonds & Douglas 1976, pp.149-152; Rupke 1983, p.64]

168 **Buckland**, William   To: Lady Mary Cole
3 April 1823
The Red Woman's bones arrived safely five days ago. Extreme pressure comes from the printers and I hope to publish by the 1st May, with 25 plates of caves and animals. My lecture room is to be put to rights and filled with £300 worth of cabinets. Your brother writes from Naples that he is to come home via Florence and Paris immediately. Our commemoration is in the middle of June and the prize poem is on geology.
[Cited: Gordon 1894, p. 51; North 1942, p.112; North 1966, p.131; Edmonds & Douglas 1976, pp.153-154]

169 **Buckland**, William   To: H.T. De la Beche
Oxford.   Lyme Regis,
9 September 1823   Dorset.
I have heard nothing of the fossil crustacea expected from Col. Birch and Mr. Woodcock. The Plesiosaurus jaw was accidentally fractured, but has been repaired without trace of damage. I will ask Miller to send you any rare Lias fossils we may have. I hope you still intend to donate the Plesiosaurus jaw to our Museum.
[Cited: Rupke 1983, p.134]

170 **Buckland**, William   To: H.T. De la Beche
Oxford.   23 Mall,
28 September [1823]   Clifton,
         Bristol.
I will send you the Plesiosaurus jaw as requested, but ask you to donate it to our Museum. Names of donors are now affixed to important presents. I conjure you in the name of all future students and educated gentlemen that you will give it to us. The Oxford Museum has been visited with approbation by the Prime Minister & the Bishop of Durham. Mr. Gells' rhinoceros from the Wirksworth cave has just arrived.
[Cited: McCartney 1977, p.20]

171 **Buckland**, William   To: Lady Mary Cole
Oxford.   Penrice Castle.
3 December 1823
Thank you for the specimen of the coal plant. Does it come from yourself or Mr. Lucas? Mr. Strangways is now returning via the Hague and not Naples. I fear sickness in your family and absence have prevented any more researches in the vicinity of Paviland. I am still without any satisfactory theory about the red ochre surrounding the old woman's bones. Granville Penn opposes my views. He finds the same faults with Cuvier, Wollaston and Newton, so I am in good company.
[Cited: North 1942, pp.112-113; North 1966, p.131; Boylan 1967, p.243]

172 **Buckland**, William   To: Lady Mary Cole
Oxford.
22 December 1823
I have returned from London (on a Committee of the Royal Society for selecting granite for London Bridge) and Dropmore. My congratulations to you on the honours Mr. Talbot has obtained at his public examinations. I am at a loss to find a theory which explains the rats' bones in the Paviland Cave. I wish we could prevail on Dillwyn to publish on the fossil plants of the S. Wales coal field.
[Cited: North 1942, p.117]

173 **Buckland**, William   To: H.T. De la Beche
Oxford.   Chepstow,
7 July 1825   Monmouthshire.
Thank you for your admirable sketches for our Society's seal. They were most decidedly the best sent in, and have brought you immortal honour. Thanks also for expediting your paper in the Annals on Jamaican Diluvium. I have just received the King's appointment to the vacant Canonry of Christ Church, with a good income and a good house.
[Cited: Rupke 1983, p.70]

174 **Buckland**, William   To: [H.T. De la Beche]
Oxford.
31 January 1826
As you are deep in the Greensand at the moment I enclose the section made by a well digger sent to me by Mrs. Short of Bickam. The lowest bed is evidently New Red Sandstone. I shall go to the meeting in Bedford St. on Friday and return here Saturday.

175 **Buckland**, William   To: H.T. De la Beche
20 Bedford Street.   White Lion,
4 February 1826   Bristol.
The Council of the Geological Society have proposed you as Secretary. Scrope has been wholly non-resident and inefficient. Murchison and Webster are to be appointed conjointly with yourself. Please send your reply as soon as possible. It may influence your decision to know that Greenough has positively declined the Presidency.

176 **Buckland**, William   To: R.I. Murchison
10 March 1828   Bryanston Place,
         London.
My apologies for delay in sending the Lias specimens, but I had forgotten it was the oolitic you wished for. Was there anything more in the stone line I promised to send you? I could not leave my lectures to come up on Friday last. When do you depart for the continent? Before you go get some information from Pentland on new and good caves with bones near Bordeaux. Mrs. Buckland is ill.

177 **Buckland**, William   To: H.T. De la Beche
Oxford.   Lyme Regis,
26 July 1829   Dorset.
I read your letter on my return last night, and have asked Lonsdale to send you my paper on the rocks between Nice and Col di Tende. It supports your views. I was not aware that you had changed your views on diluvium, and am glad you are here to reject any change in your paper that I may have suggested. I have always considered the term as applicable only to deposits resulting from violent and sudden irruption of water.
[See De la Beche, H.T. to Buckland, W., 23 July 1829, **364**]
[Cited: McCartney 1977, p.27]

178 **Buckland**, William   To: H.T. De la Beche
Oxford.   Jermyn Street,
5 July 1830   London.
I cannot attend the Geological Society tomorrow for the reading of Scrope's

paper. Two canons have died and I must attend funerals. I trust nothing will prevent you from attending the meeting and pointing out to the unlearned the futility of such doctrines as contained in the paper. Let me know how it is received.

179 Buckland, William   To: H.T. De la Beche
   Lyme.   Dartmouth.
   15 September 1830

Many thanks and much praise for your caricature of Actual Causes and the Huttonian theory rediviva. I wish Conybeare would take the book [Lyell's *Principles*] in hand and be stirred up to a reply. I have been here a month with my wife and children, but have found no great novelty in the subterranean world. I wish you had been here to fill up the details of the N.W. corner of our Weymouth map.
[Cited: Rudwick 1975, p.540]

180 Buckland, William   To: H.T. De la Beche
   1 May 1831   Craven Street,
                                                                          Strand.

The two plates of the Weymouth paper are safe. I will send them to Gardener tomorrow. Your report on the difficulties threaten a costly operation in the reproduction of the map. What about the new Prussian method? I have a capital class which I am sure is 30 per cent better off for your Duria Antiquior by way of a syllabus. Mary Anning has found another Ichthyosaurus, for which she asks £40.
[Cited: North 1934a, p.77; McCartney 1977, p.44; Rupke 1983, p.146]

181 Buckland, William   To: H.T. De la Beche
   Oxford.   Secretary, Geological Society,
   18 May 1831   London.

Trimmer has sent me a 40 page paper on the diluvial deposits of Caernarvonshire and the discovery of marine shells in the diluvium of Moel Tryfan. Please enter it in the list for reading at the meeting of the 8th June, in which case I will come.

182 Buckland, William   To: H.T. De la Beche
   25 May [c.1831]

I enclose a copy of my "Reliquiae". When you have finished with it, please present it to the Athenaeum in my name. The Plate will answer all your questions. On Thursday next I lecture on faults and shall need Buddle's section of the Newcastle faults. I have sent M. Anning £5 for copies sold to Stokes, Lonsdale and Broderip. I hope your [Manual] will be ready before my lectures. I have a capital class and your Duria has contributed to its numbers and my entertainment of them.

183 Buckland, William   To: H.T. De la Beche
   Oxford.   Post Office,
   14 October 1831   Exeter.

I have just received a German parody of your Duria Antiquior, and am anxious you should forestall them by bringing out 2 or 3 restorations of scenes from the ancient world, one on the Period before the formation of diluvium, another a lake scene from the freshwater period, and a third a sea scene from the Carboniferous and transition period. Franky was breeched today.
[Cited: McCartney 1977, p.44; Rudwick 1989, pp.241-243; Rudwick 1992, p.54]

184 Buckland, William   To: H.T. De la Beche
   9 [?May] 1832

Do look at this book, and if worthwhile, write a notice of it as having come out while our Weymouth paper was in the press. I send a woodcut of the Lamia bed which I have had done here, that should be inserted in the description.

185 Buckland, William   To: R.I. Murchison
   17 July 1835   Dowager Lady Cawdor's,
                                                                        Pembroke.

Thank you for your account of the Silurian System. The term suits admirably. With respect to the series below, the term Snowdonian System seems best. I hope Sedgwick adopts it, or rather the latinized form of it. You must quickly assert your priority, or Elie de Beaumont will anticipate you with his name Hercynian. Settle with Sedgwick forthwith.
[Cited: Rudwick 1985, p.135; Secord 1986a, pp.99-100]

186 Buckland, William   To: H.T. De la Beche
   28 January [?1836]   Ivy Cottage,
                                                                       Lodge Road,
                                                                      Regents Park.

Please call at Taylor's and with a penknife cut away some of the lines on the block at p.14 of our Weymouth paper. They will want to print off from this block in 2 or 3 days.

187 Buckland, William
   [c.?1836]

Section midway down the cliff 1 furlong E. of Lulworth Cove. There are tree stumps in the actual place on which they grew while the subjacent bed of Portland Stone was horizontal and before the lowest members of the Purbeck Series were deposited.

188 Buckland, William   To: H.T. De la Beche
   [?1836]   Ivy Cottage,
                                                                       Lodge Road,
                                                                      Regents Park.

When I called to check the text [of our Weymouth paper] in press, I found that 2 inches have been cut off the west end of the map. From Lonsdale I learn of no interference from Greenough or the Council. It must be a mistake of Gardeners. I have asked him to suspend the work for a day until he hears from us. Could I see your Devon maps tomorrow morning?

189 Buckland, William   To: H.T. De la Beche
   Geological Society.   Ordnance Survey,
   19 November 1840   Cardiff,
                                                                         S. Wales.

Since only Northampton and Murchison were with me at the meeting of the Railway Section Committee today [of the British Association] we have made little progress other than adding names of further members. Please give your opinion on Sopwith's paper. The business of the Geological Society is mostly on glacial theory. Pray come to the next meeting or you will be behind hand with the march of glaciers.
[Cited: McCartney 1977, p.60]

190 Buckland, William   To: H.T. De la Beche
   14 December [1840 FJN]

The enclosed relates to a box from India, now at the Geological Society. Clarke does not acknowledge it. Will you postpone the contents to the next meeting after May.

191 Buckland, William   To: H.T. De la Beche
   London.
   8 July 1843

I have found the map and sections of the Bristol paper I promised to lend you, which should be with you by the end of the week. At the Geological Society meeting last evening, I spoke on the subject of the tracks of ambulatory fishes. I congratulate you on your grandson's arrival at Sketty. Where shall you go after your headquarters move from Wotton-under-Edge?

192 Buckland, William   To: H.T. De la Beche
   Oxon.
   12 January [1844 FJN]

The enclosed is a capital invention. We need have no more landslips from railway banks, tunnels, cellars, or underground stores. I have sent to Peel your last letter about a hall for granite specimens [in the Museum].

193 Buckland, William   To: H.T. De la Beche
   19 January 1844

Will a specimen of Buckingham marble do for [your Museum]? On how many sides do you wish it polished? I have sent the plates, and named £200 for the Railway Section to Sir Robert Kane.

194 Buckland, William   To: H.T. De la Beche
   24 March [1844 FJN]

Peel tells me he has sent to your Museum some guano from West Africa. Can you get your chemist to analyse the fragments? Sir R. Peel has asked me to Drayton to meet Smith of Deanston. I am not sure I shall be able to go.

195 Buckland, William   To: H.T. De la Beche
   Oxon.
   19 December [1844 FJN]

Sir R. Peel has sent Peach another £20 and wishes him to continue his marine researches in Cornwall. I have written to Lemon, and Sir R. Peel to the Customs Office in London, to find out if a place can be found for him. Last eve I returned from a weekend at Drayton, where 40 farmers were invited to lunch to hear a talk on agriculture.

196   Buckland, William   To: H.T. De la Beche
     10 [?March] [?1847]
I have 2 bronze escutcheons stolen some time ago from the tomb of Edward III. Will it not be desirable to make an analysis of the metals used? It is a case for your Practical Geology shop. My coprolite paper is to go to the Agriculture Journal.
[Cited: McCartney 1977, p.48]

197   Buckland, William   To: H.T. De la Beche
     [1848 FJN]
Sir Robert is charmed with the glass and begs to keep it for a day or two, so I have left it. Will you send the key to the rooms over my hayloft, as your man locked up a dog in them yesterday.

198   Buckland, William   To: H.T. De la Beche
     15 February [No year]
Peach is to have the place at Fowey in a week. Say nothing till it is announced. What say you to the enclosed letter from Dawes. I do not remember him. Hooker may tell us who he is. I should add the contour lines if it can be done without much additional work.

199   Buckland, William   To: H.T. De la Beche
     [undated]
Sir Robert Peel has started the subject, and we want you at the British Museum. The Archbishop has been speaking of it. We have long wished to have you, and I hope it will not be long before it comes to pass. There seems, therefore, not much further need for pressure from without.

200   [Buckland, William]
     [undated]
[Map of Glamorganshire (plain) with light blue shadings on parts of Gower Peninsula.]

     Bunbury, E.H.   To: [?] J.Lindley
     Hampstead.
     28 January 1833
Here are the answers to your questions about hours of daylight at high latitudes.
[See Lindley, J., 1832-33, **862**]

201   Bunsen, Baron Christian Karl Josias von
     9 Carlton Terrace.   To: H.T. De la Beche
     2 March 1850   Geological Survey Office,
                                               Jermyn Street.
The parcel and letters entrusted to my care have been forwarded to Berlin. Von Humboldt will be greatly delighted. Thank you also for the added gift of the maps and sections of the Geological Survey.
[Cited: North 1936a, p.70]

202   Bunsen, Baron Christian Karl Josias von
     Carlton Terrace.   To: H.T. De la Beche
     13 March 1851
Blauner the Swiss collector is going to the Canary Islands, Puerto Rico and the Antilles. Can you procure him introductions of a scientific character?

203   Bunsen, Baron Christian Karl Josias von
     Carlton Terrace.   To: H.T. De la Beche
     Saturday [morning, No date]
My friend and guest Scharnhorst, who possesses the greatest collection of maps in Germany, above 40,000, wishes to be presented to you and inspect the Museum. Would any hour today be convenient to you?

204   Burgess, Thomas   To: [William Buckland]
     Bishop of St Davids,
     32 Park Street.
     24 November 1820
I have been much pleased by your inaugural lecture on geology. The account you give of its own particular objects, its relation to other branches of science, and consistency with Mosaic records must recommend the study of it. I shall be glad to hear what you think of my suggestion about the unity of design in creation.

205   Burgoyne, John Fox   To: H.T. De la Beche
     Ordnance Office.
     17 July 1854
The subject of a datum level for the great Survey is most interesting, and I shall immediately refer it to the Survey Officers.

206   Burgoyne, John Fox   To: H.T. De la Beche
     Ordnance Office,
     Pall Mall.
     2 August 1854
The Military Departments have acquired a tract of land at Aldershot, on which barracks etc. are to be erected. We are doing a detailed survey of the surface, but wish to have a geological survey also. Can you give the information required, or what measures would you recommend for obtaining it?

207   Butler, G.   To: Inspector General of Ordnance
     Office of Fortifications.
     27 January 1837
Copy
The Board acknowledge Colby's letter of the 13th, and your report on it of the 18th, from which it appears that a total sum of £2,200 will be required for the completion of the Geological Survey of Cornwall and Devon, and its extension into South Wales. The Treasury have been informed and approve this estimate.
[See Colby, T.F. to Inspector General of Fortifications, 13 January 1837, **269**]

208   Byham, R.   To: H.T. De la Beche
     Office of Ordnance.   Geological Society,
     2 May 1832   Somerset House.
I acknowledge yours of the 28th March last, submitting proposals to colour geologically 8 sheets of the Ordnance Map of Devonshire etc. The Master General and Board accept your proposals provided the Index of Colours is referred to the Geological Society Council, & you publish at your own risk and cost the indexes, sections etc., the price to be arranged by the map seller, and £300 to be paid to you in 8 instalments of £37.10.0 each. Minor details are to be arranged with Colby.

209   Byham, R.   To: H.T. De la Beche
     Office of Ordnance.   Tiverton,
     18 November 1834   Devon.
Your letter of 8th inst. has been laid before the Board of Ordnance. The statement of expenses incurred (£700) is so far allowed as to give you £87.10.0 for the 8th sheet and £50 more for the 7th when completed. Thus £100 more for the whole work. A further £50 is granted to you in consideration of the testimonials borne as to manner in which you have executed the work.

210   Byham, R.   To: Lt. Col. Colby
     Office of Ordnance.   Map Office,
     17 July 1835   Tower.
Copy
Your letter of the 14th inst. requesting £100 imprest for De la Beche's Cornwall Survey which he has been directed to commence without delay has been submitted to the Board and authorized, and De la Beche is to forward quarterly accounts to the Surveyor General.
[See Colby, T.F. to Byham, R., 14 July 1835, **261**]

211   Byham, R.   To: Inspector General of
     Office of Ordnance.   Fortifications
     3 July 1837
Copy
Colby's letter of 26th ult and De la Beche's representation as to the inadequacy of £500 salary with a request for £25 per quarter for travelling has been laid before the Board. Before sanctioning the allowance, the Board wish a report from you on De la Beche's progress. Does it warrant the allowance requested?

     Byham, R.   To: Inspector General of
     Office of Ordnance.   Fortifications
     16 August 1837
     [see **276e**]

212   Byham, R.   To: Col. Colby
     Office of Ordnance.
     3 January 1838
Copy
Re De la Beche's claim for £25 per quarter as travelling allowance. The

213 **Byham**, R.   To: Inspector General of
Office of Ordnance.   Fortifications
12 January 1838
Copy

Re your memo of the 10th inst on Colby's letter transmitting De la Beche's Estimate for the present year, £1000 has been inserted in the Estimate for 1838-9 for carrying on the Geological Survey.

214 **Byham**, R.   To: H.T. De la Beche
Office of Ordnance.   Ordnance Map Office,
6 July 1838   Tower.

Having considered your letter representing the difficulties attending the preparation of Travelling Vouchers, and taking into account your special circumstances, the Master General and Board grant a fixed allowance of £25 per quarter to cover all travel expenses while engaged on the Geological Survey.

215 **Byham**, R.   To: Inspector General of
Office of Ordnance.   Fortifications
15 March 1839
Copy

The Master General and Board of Ordnance have considered the subject of De la Beche's request for permission to hire Geological Assistants. They consider that application so far satisfactory that it limits his labours to areas covered by Treasury authority.
Forwarded to Colby by order of Edw. Fanshawe 19 March 1839.
[Copy letter also Colby to Inspector General of Fortifications, 21 March 1839, in reply to above letter:] Please grant authority to allow De la Beche to hire assistants.

216 **Callender**, W.R.   To: [H.T. De la Beche]
Manchester.
16 October 1849

I am making a collection of autographs, and although I can readily obtain those of statesmen and theologians, yet from living in the provinces I am unable to procure those of scientists. Please favour me with yours.

217 **Candolle**, Augustin Pyrame de   To: H.T. De la Beche
Geneva.   Messrs. G.W.S. Hibbert & Co.,
5 June 1828   Billiter Court,
                                                                          London.
[Readdressed to:] Hotel [?],
Rue St Honoré,
Paris.

I have the honour to send you the letter for M. Capodistrias which you asked for, and take the opportunity of wishing you a safe and useful journey. I have seen your agent and given him all the information that time would permit on how to dry plants.

218 **Candolle**, Augustin Pyrame de   To: [H.T. De la Beche]
[No date]

Herewith I send you the two notes for which you asked me. I have reduced them to the strictly necessary, especially the one for you. Be so good as to assure your correspondent that I will send him the names of his plants, and that if I need to mention them I will always acknowledge the collector. Thank you for the pains you have taken over the matter.

219 **Canning**, Charles John Canning, Earl   To: H.T. De la Beche
Windsor.
22 August 1843

I acknowledge receipt of the analysis of the water from the well on the north side St. George's Chapel, to be laid before the next General Chapter.

    Canning, Charles John Canning, Earl   To: T.W. Philipps
    [see De la Beche, H.T. to Philipps, T.W., 21 May 1846, **405**]

220 **Canning**, Charles John Canning, Earl   To: H.T. De la Beche
Grosvenor Square.
7 March [No year]

I shall not be able to keep my engagement at the Museum tomorrow. Can I come on some succeeding Monday?

221 **Cardwell**, Edward   To: H.T. De la Beche
Board of Trade.
27 May 1853

If the Treasury recommend anything on your relations with Lord Stanley and myself, they have gone beyond their province. The [Geological Survey's] position will remain the same.

222 **Cardwell**, Edward   To: H.T. De la Beche
Board of Trade.
9 August 1853
Private

There is no necessity for passing the Transfer Bill (Geological Survey) this year. Re the Outfit Question, when Playfair returns, I will direct him to prepare an official letter, though I hate asking for money above my estimate.

223 **Cardwell**, Edward   To: H.T. De la Beche
Board of Trade.
11 August 1854

I have had the Minute re Jukes' appointment redrawn. He will receive £400 from 1 April to this date, and £500 durante fine placito, and upon a separation of the two offices revert to his £400 [as Local Director in Ireland].

224 **Carlisle**, George William Frederick,   To: H.T. De la Beche
7th Earl of
14 November 1848

My thanks for the Geological Survey Maps and Sections.

225 [**Carlisle**], George William Frederick, 7th Earl of
[c. 1848]
Copy

I concur in the spirit of De la Beche's letter. Though Parliament thought a comprehensive system of administration was needed, I feel De la Beche should have the sole superintendence of stocking the New Building and handing it over to the public. He also should have a very principal part in any new administration.

226 **Carlisle**, George William Frederick,   To: [H.T. De la Beche]
7th Earl of
8 October 1849

A New Commission of Sewers is being constituted. The present Commission will be diminished in number. My thanks for your dedicated service on the previous Commission. It will have laid the foundations for your successors.

227 **Carlisle**, George William Frederick,   To: H.T. De la Beche
7th Earl of
26 March 1850

I have made an appointment with Lord Lansdowne to show him the Museum, and thought you might like to meet him.

228 **Carrington**, Frederick, A.   To: Earl of Carlisle
10 Henrietta Street,
Covent Garden.
20 October 1849
Copy

Herewith are the impressions of my two New Zealand maps, and I would also like to thank you for asking Sir H. De la Beche to view my models of English towns. He does not approve of them entirely, as the vertical and horizontal scales are not the same. The House of Lords Committee accepted my explanation that the difference was necessary. I now present my case to you.

229 **Carrington**, Frederick, A.   To: H.T. De la Beche
10 Henrietta Street,
Covent Garden.
5 January 1850

I enclose a copy of my letter to Carlisle which I mentioned to you yesterday. It will clarify my meaning about true scales. I am much grieved you do not countenance my works, when I have spent years on them. The model I asked you to purchase yesterday cost me more than £1000. I would part with it for 200 guineas.
[See Carrington, F.A. to Carlisle, Earl of, 20 October 1849, **228**].

230 **Cawdor**, John Frederick Campbell, To: H.T. De la Beche
     1st Earl of
     75 South Audley Street.
     29 June 1849
Thanks for the trouble you have taken about my black band. I have seen your new Museum which will be very handsome and convenient.

231 **Chadwick**, Edwin To: H.T. De la Beche
     10 October [1842 FJN]
I forward a copy of the Report on the Sanitary condition of the working population, because you looked over the proofs. Additions have been made since then. I will be glad to forward copies to anyone you suggest.
[Cited: Secord 1986b, p.230]

232 **Chadwick**, Edwin To: H.T. De la Beche
     26 October 1842
Thank you for both your letters. Our thoughts jump together. The Sanitary Report has set people thinking in many directions. Leeds has asked for assistance and I have recommended Vetch & Roe. They intend to get Smith of Deanston if they can. I am drawing up a draught copy of instructions which I would like you to look at.

233 **Chadwick**, Edwin To: H.T. De la Beche
     10 November 1842
I am glad to hear you have found a chemist for the Museum of Economic Geology. Madden has confirmed Liebig's view that the analysis of rain water indicates the state of the atmosphere. Could Phillips analyse several samples of London rain water? I have also been asked to report on burials in towns etc. Is Phillips exclusively occupied for you?

234 **Chadwick**, Edwin To: H.T. De la Beche
     28 June 1843
The Notes of the examinations appear to me satisfactory as verifying the Sanitary report. But new ground might be broken. Surveyors of sewers ought to be sifted. On the whole, good is in prospect. I am glad Lincoln's absurd building bill has been held up. When are you returning?

235 **Chadwick**, Edwin To: H.T. De la Beche
     Metropolitan Enquiry.
     21 September 1847
Don't be alarmed at the work required of you. You will only be asked to look over drainage plans, express an opinion on works that obviously should not be proceeded with, and also show some scientific force in the field. You or Ramsay are the only competent ones.

236 **Chadwick**, Edwin To: H.T. De la Beche
     Metropolitan Sanitary Commission,
     Gwydyr House,
     Whitehall.
     14 December 1847
Please look over the enclosed paper of a former Sergeant of Sappers and Miners. Would he not be worth thinking of? I have been looking over the Treasury terms for expenditure of the Ordnance vote of £40,000 per annum. Since there are no conditions as to how it is spent they may direct some to be expended on the Metropolis.

237 **Chadwick**, Edwin To: H.T. De la Beche
     10 June 1848
The Chancellor of the Exchequer has given his consent to the payment. Burgoyne is out of town. I have written to Hall to ask when we might expect observers in town. The Report is to be read on Thursday. We may have to call on you to advise on the question of granites for pavements.

238 **Chadwick**, Edwin To: H.T. De la Beche
     29 July 1848
We have just had a gunpowder plot amongst the sewer men. A man named McManus planned to blow up the Government Offices and the Houses of Parliament. On the advice of Mayne he has been discharged. The only remedy is the abolition of large sewers and general substitution of tubular drainage. The City people have consented to pay their portion of the Survey.

239 **Chadwick**, Edwin To: H.T. De la Beche
     18 August 1848
Austin has made an examination under the Houses of Parliament, and his report shows the state of things to be worse than expected. We must prevent it, as it is totally at variance with our principles. Large reputations must be prevented from doing mischief by supporting error. Yolland has shown me four sheets of the Westminster survey, and they look well. The Metropolitan sewers bill is now safe.

240 **Chadwick**, Edwin To: H.T. De la Beche
     Gwydyr House.
     5 September 1848
Please look at the enclosed, as we have to give an immediate answer. The sewers have to be secured for H.M.'s visit to Parliament. This has been done today and all is well. Barry agrees that the old system must be removed. The new Health Board has not yet been appointed. We have to consider its new boundaries and the extension of the survey.

241 **Chadwick**, Edwin To: H.T. De la Beche
     4 Stanhope Street,
     Victoria Gate,
     Hyde Park.
     [1848]
I have seen Sir James Clark and rejoiced his heart by telling him of your prospects at Windsor. He wishes to be introduced to you. Some friends, including him, are coming Wednesday evening. Would you join us?

242 **Chadwick**, Edwin To: Viscount Ebington
     18 December 1849
The Metropolitan Sanitary Commission is pressed for an early report on the water supply of the Metropolis. We must settle the point of what additional water is needed for an improved system of sewerage, and details of companies capacity for supplying it. We wish to have their works examined. Vetch and Stephenson should give evidence as to this, having studied the problem.
[Enclosed with Ebington, Viscount to De la Beche, H.T., 23 December 1849, **483**].

243 **Chadwick**, Edwin To: H.T. De la Beche
     24 June 1850
Private
I saw Vetch on Saturday. It is only too true that Westminster is committed to a system of sewerage requiring flushing. I understood from him that the Surrey system was also to have an inclination of only four feet in the mile. Read what the witnesses have said in any case.

244 **Chandelon**, Y. To: H.T. De la Beche
     London.
     9 June 1851
Your interest in education and the good wishes you have expressed in favour of our School of Mines in Liège, and the numerous students of geology we have there, lead me to ask you to donate to the institution a specimen of your geological map of Great Britain. I make this official request on behalf of the Belgian members of the International Jury and academic authorities.

245 **Chantrey**, Francis Legatt To: H.T. De la Beche
     Belgrove Place. 43 Pall Mall.
     15 January 1831
I will meet you at the Geological Society Wednesday next at 4 p.m.

246 **Chapman**, S.J. To: H.T. De la Beche
     J. Wiltshire's,
     Shockerwick,
     Bath.
     24 June 1848
I hope the parcel of papers I sent can be recovered. Would you take the matter up? They relate to the history of Bath waters, and could be made into a compact history, if you undertake to put whole matter in order. I shall be happy to have supplied you with all the material.

247 **Chapman**, S.J. To: H.T. De la Beche
     [1848]
Sorry to have bothered you at a busy time. Please return my documents. Can I see any geological map of Scotland. I am trying to trace the boundary of the Carboniferous Limestone. I am also interested in artesian springs as measures of heat.

248 Charles, G.S. To: [H.T. De la Beche]
13 St. James's Place,
St. James's.
18 October 1848

I offer a house as an Office for Clerks, required by the Commission for Sewers in Lambeth.

249 Charlton, Isaac To: H.T. De la Beche
Geological Society,
Somerset House.
21 June 1848

Leave of absence for a fortnight is requested for myself and wife.

250 Charters, S. To: H.T. De la Beche
Spa.
23 May 1840

I am here, so far on my German tour. Can you give me any sort of Woods and Forests commission to facilitate my task? Why were you not at the Geological to hear the joy of the Devonians at last getting into their proper places?

251 Clark, J.A. To: H.T. De la Beche
Brook Street.
[1850 FJN]

I have been prevented from calling at the Museum by a severe cold. Tomorrow afternoon I trust I will be there. I am sending a geological map of Deeside for you to look at, as you may help to improve it.

252 Clark, J.A. To: H.T. De la Beche
Brook Street.
10 April [1852 FJN]

The Prince will be glad to see you on Tuesday at 2.45 p.m. I will call for you at 2.30.

253 Clark, J.A. To: H.T. De la Beche
Brook Street.
[1853 FJN]

The Prince will see us tomorrow at 2.40 p.m. I will call for you at the Museum a little before.

254 Clawes, Thomas To: H.T. De la Beche
9 Pall Mall East.
16 November 1846

You are indebted to me for the sum of £8.1.9 for wine supplied in 1833-34. Your change of address from the Tower is the reason for my not applying earlier.

255 Clibbono, Edward To: T. Oldham
1 April 1850

Enclosed is a letter and specimen which may interest De la Beche. Please take charge of it for me.

256 Clutterbuck, J.W. To: W. Buckland
Nice, Dean of Westminster,
Italy. Deanery,
7 January 1846 Westminster,
 England.

Your note was sent on to me after I left Watford for the continent. I have the opportunity here to collect facts, should you wish to give me instructions.

257 Coates, Thomas To: H.T. De la Beche
59 Lincolns Inn Fields. Tiverton.
17 December 1834

Fitton's engagement with the Society for the Diffusion of Useful Knowledge has been terminated. Will you undertake to write a series of articles on geology? The renumeration is £25 a number, or £12.10 a sheet.

258 Coffin, Walter To: H.T. De la Beche
Llandaff.
25 June 1841

My friend Montague Hunt is desirous of breathing Welsh air instead of that in Craigs Court and it has occurred to me that he may be better prepared for acquiring a knowledge of this county by a little previous acquaintance with geological pursuits.

Coffin, Walter To: H.T. De la Beche
Llandaff.
25 September 1844
[see **2244**]

259 Colby, Thomas Frederick To: H.T. De la Beche
Ordnance Survey Office,
Dublin.
9 April 1832

Thanks for your letter of the 4th. When your proposal came to me, I suggested that the index of colours should be submitted to the Geological Society. I also recommended payment to you of £35.10 per sheet, and that you should undertake publication of memoirs etc. at your own expense. A small work with sections etc. to illustrate the geology of Devon would sell well.
[Cited: Rudwick 1985, p.90]

260 Colby, Thomas Frederick To: H.T. De la Beche
Ordnance Map Office,
Tower.
5 November 1834

Since I received your letter in Dublin, I have been considering how your object could be best carried into effect. You wish a modification of the original proposition. Put the facts of your expenses so far, the failure of your West India funds etc. to the Master General and Board, as your original one was, and it will probably be favourably received. I will do what I can. Do not lose time.

261 Colby, Thomas Frederick To: R. Byham
Ordnance Map Office,
Tower.
14 July 1835
Copy

With reference to the Board's letter of 11th, directing De la Beche to commence the Geological Survey of Cornwall without delay, I request an imprest of £100 to be granted him to defray contingent expenses. Quarterly accounts are to be sent to the Surveyor General.
[See Byham, R. to Colby, T.F., 17 July 1835, **210** for reply].

262 Colby, Thomas Frederick To: H.T. De la Beche
Ordnance Map O[ffice].
28 September 1835

In reply to your letter of the 18th, in which you state that time and expense will be saved by the employment of an intelligent miner, I beg to observe that the survey must proceed with the least possible expense but in a perfect manner. Thus if it accords with this principle your wish to employ a miner is granted. You must pay him out of contingent expenses.
[Cited: North 1936a, p.64; McCartney 1977, p.41]

263 Colby, Thomas Frederick To: F. Mulcaster
Ordnance Map Office,
Tower.
29 April 1836
Copy

Herewith De la Beche's report on the progress of the Geological Survey of Cornwall. Considerable progress has been made, but considerable geographical corrections are required. It is difficult to estimate the time spent on these corrections. The amount spent up to March 1836 has been £578.0.3.

264 Colby, Thomas Frederick To: H.T. De la Beche
25 June 1836

I remember the Dolcoath Mine being long celebrated as the deepest in Cornwall. If plans and sections are complete, ten guineas is not too high a price. The charges must be put down on your contingencies account.

265 Colby, Thomas Frederick To: H.T. De la Beche
Ordnance Survey Office.
14 July 1836

I like your proposal of letting McLauchlan do the coast of sheet 29, and using Still's energies with your own on the Penzance sheet. We cannot be content with a low degree of accuracy on geographical maps. You deserve credit for the way you have kept down the contingent expenses. All I can do for you is pay Mr. Still on the geographical correction account.
[Cited: Harley 1971b, p.118]

266   **Colby**, Thomas Frederick   To: H.T. De la Beche
Mansfield.
15 September 1836

I have been thinking about what can be done to expedite the geographical corrections in Cornwall. I learned of the deplorable state of the Ordnance map from McLauchlan. I think that it should be done sheet by sheet, coast line first. I regret you have had to commence on so defective a map. I have seen your letter about the Bristol meeting. Your opponents motives were to show their own geological knowledge and hasten your memoir. The difference of opinion will be looked on as trivial by geologists. The public will recognize the utility of your own work.
[Cited: Harley 1971b, pp.119,120; Rudwick 1985, p.176]

267   **Colby**, Thomas Frederick   To: H.T. De la Beche
29 November 1836

The Geological Survey estimates must be given to the Clerk of Ordnance to place before Parliament. You know the House have as yet voted nothing for the Geological Survey, and it has been going on at Ordnance Survey expense; though in principle a separate grant is recognised by the Chancellor and Ordnance. Please prepare an abstract of expenses.

268   **Colby**, Thomas Frederick   To: H.T. De la Beche
29 November 1836

I am in the chair for this year at the Geological Society [of Dublin] and must give an address, according to custom. Could you prepare an account of the advances on the Geological Survey and in the Museum of Economic Geology which I can use at the close of the year?

269   **Colby**, Thomas Frederick   To: Inspector General of
Ordnance Map Office,   Fortifications
Tower.
13 January 1837

Copy

In obedience to the Board's order of 2 December 1836 herewith De la Beche's statement of expenses of the Geological Survey of Cornwall. Also enclosed is the estimate by De la Beche for 1837-38. I trust the Board will obtain the consent of the Treasury for the continuance of the Survey. I propose it should proceed in the mineral districts of South Wales, essentially connected with the smelting of ores.
[See Butler, G. to Inspector General of Fortifications, 27 January 1837, **207**].
[Cited: North 1936a, p.35]

270   **Colby**, Thomas Frederick   To: H.T. De la Beche
Ordnance Map Office.
3 April 1837

Reports are in circulation that your map of Devon and Cornwall is inaccurate. To rebut the criticisms of the authors (whom I do not know) I see no alternative except by publishing the work. The Government are anxious, and their anxiety can be allayed by publication. Your accounts have been sent forward. Still and McLauchlan will soon have finished the geographical corrections.
[See De la Beche, H.T. to Colby, T.F., 8 April 1837, **378** for reply].
[Cited: North 1939b, p.1053; Rudwick 1985, p.203]

271   **Colby**, Thomas Frederick   To: [H.T. De la Beche]
17 April 1837

I do not think you have any personal feelings to contend with [concerning the controversy over your Devon map]. It is necessary to solve the difficulties by publishing the Memoir and Sections as well as the map of Devon and Cornwall. A separate account of mines will not meet the question. Publish in progressive parts or in one volume as you wish. How best can we aid you?
[Cited: Rudwick 1985, p.204]

272   **Colby**, Thomas Frederick   To: H.T. De la Beche
Ordnance Survey Office.
18 April 1837

In reply to your letter of the 16th., I gather you consider a single memoir will do for Cornwall and Devon. I therefore suggest it should be printed at public expense, but before I propose this to the Board I would require to know your feelings and wishes, and the state of the work.
[Cited: Rudwick 1985, p.204]

273   **Colby**, Thomas Frederick   To: [H.T. De la Beche]
Ordnance Survey Office.
25 April 1837

I have your letter telling me you have revisited the disputed spot and found evidence in opposition to Sedgwick. I would say nothing about this until you have examined the whole country. The Geological Survey's aim is not to establish theories, but to collect and arrange facts. In your letter to Robe you talk of part publishing. I disagree. No part should be given to the printer until all is prepared. I will do all I can to assist you.
[Cited: McCartney 1977, pp.30, 39; Rudwick 1985, p.206]

274   **Colby**, Thomas Frederick   To: H.T. De la Beche
12 May 1837

With reference to your letter to Robe about the reduced map, do you require all the coast-line, rivers and streams? Hills? Towns and villages? Roads? I want your opinions definitely on all such matters before I apply to the Board.
[Cited: Oliver 1986, pp.9-10]

275   **Colby**, Thomas Frederick   To: H.T. De la Beche
Ordnance Map Office.
23 June 1837

Standidge has been here and asks me to tell you that he has authority from the Stationery Office to proceed with any lithographs for your Memoir.

276   **Colby**, Thomas Frederick
[One folio comprising five copy letters:]

276a   **Colby**, Thomas Frederick   To: Inspector General of
Ordnance Map Office,   Fortifications
Tower.
26 June 1837

I recommend to the Board that De la Beche should not commence any extension of the Geological Survey until he has finished his final field work and publish a Memoir with no arrear of imperfect work. I also recommend that he be allowed the £25 per quarter travelling allowance he request.

276b   **Byham**, R.   To: Inspector General of
Office of Ordnance.   Fortifications
3 July 1837
[See **211** for summary].

276c   **Colby**, Thomas Frederick   To: Inspector General of
Ordnance Map Office,   Fortifications
Tower.
11 July 1837

At the Board's command I enclose a report by De la Beche on his progress so far on the Geological Survey. With the state of the geographical maps holding him up so much, despite my having assigned [McLauchlan and Still] to correct the more than 30 year old Cornish Maps, it is impossible for me to frame a correct opinion on De la Beche's progress so far.

276d   **Colby**, Thomas Frederick   To: Inspector General of
Ordnance Map Office,   Fortifications
Tower.
15 July 1837

My recommendation for allowing De la Beche £25 per quarter for travel was based on my opinion that he needed to travel extensively to complete his work satisfactorily - after which an opinion on its quality could be satisfactorily formed.

276e   **Byham**, R   To: Inspector General of
Office of Ordnance.   Fortifications
16 August 1837

The Board insists on a fuller explanation before granting £25 per quarter travelling allowance to De la Beche.

277   **Colby**, Thomas Frederick   To: [H.T. De la Beche]
6 July 1837

I have the Board's answer to the Inspector General about your application for £25 per quarter travel allowance, but they ask for a report of progress as a preliminary. I therefore have judged it better not to stir the matter further at present.

278   **Colby**, Thomas Frederick   To: [H.T. De la Beche]
Ordnance Map Office,
Tower.
15 July 1837

I have considered Hopkins' proposals and think it desirable to use his services. But our progress would be retarded by combining geographical and geological labours. So I propose giving him copies of Derbyshire quarter sheets as they are completed, and he can lay down geological lines. Your own experience will have convinced you that the combination of geographical and geological work produces delays.

279   **Colby**, Thomas Frederick   To: H.T. De la Beche
Ordnance Survey Office.
25 August 1837

Here is a copy of the orders I have received in reference to your application of 25 June last. I considered the allowance for travelling as a help towards the completion of the Devon and Cornwall Survey, and not as dependent on progress. The Inspector General has requested elaborate information.

280   **Colby**, Thomas Frederick   To: Inspector General
Ordnance Survey Office.   of Fortifications
15 September 1837

[Copy]
Having communicated the Board's order of 22 Aug., to De la Beche, I now forward a letter from him in which he expresses his willingness to afford any specific information requested. I recommend the travel allowance because it became obvious that much travelling would be needed to obtain speedy publication. Also enclosed is another letter of De la Beche of 19th Aug. I presume the Board will not object to the extension of time.

281   **Colby**, Thomas Frederick   To: H.T. De la Beche
18 October 1837

I am glad the Stationery Office have got the order for printing your Memoir. Your travelling allowance demand is again before the Board, and the Inspector General's opposition to it withdrawn. There seems to have been a complete misapprehension about it. I know geological surveys if well done take time. Because of the quality of the work and your reputation, I felt no hesitation in recommending the £25 per quarter travelling allowance.

282   **Colby**, Thomas Frederick   To: H.T. De la Beche
Ordnance Survey Office.
2 November 1837

I transmit a copy of the Board's order of 25 October, 1837, in reply to your application for £25 per quarter for travel, by which you will perceive that the correspondence on the subject has been terminated.

283   **Colby**, Thomas Frederick   To: H.T. De la Beche
Ordnance Survey Office.
16 November 1837

I have read your letter of the 12th. I was not aware that the geological examination of Glamorganshire was so backward as to render it expedient that you commence there. I wished you to commence in Pembrokeshire so that McLauchlan and Still could make corrections there while you published the Devon and Cornwall work. However, let them go on with those corrections, while you commence where it is most convenient for geological examination. Thus there will be no delays.
[Cited: North 1936a, p.45; Winder 1965, p.112]

284   **Colby**, Thomas Frederick   To: H.T. De la Beche
Ordnance Survey Office.
4 December 1837

It would be desirable to have a plate of sections, provided time allows. The funds of the Survey will not allow me to do it if it is elaborate. Let us see the plate at the Tower. Have you examined the line levelled by the British Association in Somerset? I have requested the marks to be left along it, so that future land movements can be verified.

285   **Colby**, Thomas Frederick   To: Mr Lees
Ordnance Survey Office.
14 December 1837

Copy
In reply to your Memo of 6th. inst., on De la Beche's request for £25 per quarter for travel, I think he should merely provide a certificate that his travel has actually exceeded that sum, as an enumeration of distances etc. would occupy much time.

286   **Colby**, Thomas Frederick   To: Inspector General
Ordnance Map Office,   of Fortifications
Tower.
20 July 1838

Copy
I acknowledge the Board's letter of 6th. informing me that the £25 per quarter has been allowed to De la Beche. Yet I feel I must explain why I recommended this. Travelling is of two classes into the second of which his falls. In this case the keeping of accounts can only lead to trouble.

287   **Colby**, Thomas Frederick   To: H.T. De la Beche
2 August 1838

Ferrier informs me you wish to have some sheets of the map. There are objections against sending out unfinished sheets. Yolland has sent you copies of those [of Wales] so far published.

288   **Colby**, Thomas Frederick   To: H.T. De la Beche
Ordnance Map Office,
Tower.
20 November 1838

Herewith is a copy of the Board's letter concerning Fox's Deflector Compass. Please send a receipt for it to this office.

   Colby, Thomas Frederick   To: Inspector General of
21 March 1839   Fortifications

[See Byham, R. to Inspector General of Fortifications, 15 March 1839, 215].

289   **Colby**, Thomas Frederick   To: H.T. De la Beche
11 April 1839

I am glad I have by my last letter to the Inspector General got the answer to our question of the 16th February. Your geological progress should not now be impeded. I will see about your wish to get rid of McLauchlan when I return to London next week.

290   **Colby**, Thomas Frederick   To: H.T. De la Beche
Ordnance Map Office.
13 November 1844

Your report of the 3rd requires some addition before I can submit it to the Board. You do not say where maps are to be coloured or give any other information for the Board's consideration to prevent the evils you spoke of. Please make an additional report after reading the correspondence again.

291   **Colding**, A.   To: H.T. De la Beche
Nottingham.
10 March 1850

Thank you for your letter of introduction to Dr. Percy here in Birmingham, and our kindness to me in London. Please thank Playfair also for his letter of introduction to Hawksley. By examining the gas and water works in Birmingham and Nottingham I have collected useful information for the new water works in Copenhagen.

292   **Cole**, Henry   To: H.T. De la Beche
Carlton Ride Branch of the Public Record Office,
Carlton Terrace,
London.
1 April 1847

Received £150 check for Mr. Wood. Sorry you have been so troubled.

293   **Collingwood**, G.V. Graham   To: [H.T. De la Beche]
University Club.
26 October [No year]

Thank you for recommending my interests concerning chimney pieces to Guidoni. The delays in building are greater than anticipated. The house is in a very unfinished state. When finished will you come to meet Sir Ralph Price. The cliffs at Hastings are very interesting.

294   **Colquhoun**, [?]J.L.   To: H.T. De la Beche
Royal Arsenal,   Woolwich.
28 March 1848

Can you spare me three copies of your Report on Fuel? I should like to present one to the Royal Artillery Library, another to the Artillery Institution, and to obtain one myself. We are making use of sawdust mixed with cinders and dust to raise steam.

295   **Condamine**, Rev. Henry Malcolm de la   To: H.T. De la Beche
6 Dacre Terrace,
Blackheath.
[1849]

When can I lay my map of the Blackheath district before the Geological Society? The London Clay has been bored into, confirming my theoretical section. I am interested in the Tertiary beds and their faults.

296 **Conybeare**, William Daniel To: H.T. De la Beche
Bristol. Lyme.
14 March 1821
Since writing yesterday evening I laid open my small specimen which exhibited beautifully the course of many of the lower maxillary bones, and today I have reinspected Mr. Braikenridge's large skull and thence obtained good sections in the parts which mine did not exhibit. Have an outline sent me of the Ichthyosaurus plate as you mean to arrange it.
[Cited: North 1935, p.27]

297 **[Conybeare]**, William Daniel To: H.T. De la Beche
Bristol. Lyme Regis,
16 December 1821 Dorsetshire.
I agree with you entirely as far as Ichthyosaurus communis and platyodon are concerned. The Oxford specimen belongs to the former. I have made sketches for you. I wish you would try and clear out the part of the head of I. communis which you lithographised. I congratulate you on the discovery of the Plesiosaurus jaw.
[Letter incomplete].

298 **Conybeare**, William Daniel To: H.T. De la Beche
[1821] at Mr.Auriol's, Clifton.
I think I have made out a close correspondence between Ichthyosaurian lower jaw and crocodile. I send you a sketch of Bullock's head hypothetically filled up with these views, that you may confirm, confute or correct them as circumstances serve.
[Cited: North 1935, p.26; McCartney 1975, p.19; McCartney 1977, p.19; Howe et al. 1981, p.19]

299 **Conybeare**, William Daniel To: H.T. De la Beche
Bristol. Lyme Regis,
[1821] Dorsetshire.
In the prospect of our meeting in the winter and doing some more good work as to the old crocodiles I have begun again to turn my attention to the subject and communicate the results. The large Oxford skull which I borrowed has afforded me more perfect details of the lower jaw than I have yet possessed. I hope you will bring with you as many of such specimens as are portable as you can. Try and get a few instructive and cheap specimens for me.

300 **Conybeare**, William Daniel
[1821]
"Notice of the discovery of a new fossil animal forming a link between the Ichthyosaurus and crocodile, together with general remarks on the anatomy of the Ichthyosaurus, from the observations of H.T. De la Beche etc. and W.D. Conybeare etc., communicated by the latter".
[Draft MS as published pages 559 to 566 of *Transactions of the Geological Society*, Series 1, volume 5, pp 559-594; sketches for plate 40 Figures 2, 4, 5, 6, 7, 8, and 12; and notes on *Plesiosaurus* dated 8 March 1821].

301 **Conybeare**, William Daniel To: H.T. De la Beche
Bristol. Halse Hall,
Begun 19 December 1823 Clarendon,
continued 8 January 1824 Jamaica.
In the midst of scenes of exasperation over the slavery question you may find it refreshing to receive news of discussions so harmless as those of science. I spent a very pleasant week in Oxford. There are now glass cases round two sides of Buckland's lecture room. In respect of diluvian bones, the collection is quite unrivalled. I have promised to write an article for Buckland in answer to G. Penn. Royal Society news, Bristol Institution, and Philosophical Society affairs.
[See De la Beche, H.T. to Conybeare, W.D., 6 March 1824, **355** for reply].

302 **Conybeare**, William Daniel To: H.T. De la Beche
Bristol. Halse Hall,
4 March 1824 Clarendon,
 Jamaica.
The Annings have found a complete skeleton of a Plesiosaurus, on which I have lectured in London. It afforded me satisfaction to have my discoveries which had been questioned confirmed. I made my beast roar almost as loud as Buckland's hyenas. Botanists at the Geological Society have been accused by Buckland of great inattention to the interests of science by neglecting fossil plants. The Bristol Institution can now be regarded as firmly established. With reference to the consolidated Slave Act, I fearlessly pronounce the system detestable and intolerable.
[See De la Beche, H.T. to Conybeare, W.D., 13 May 1824, **357** for reply].
[Cited: North 1935, pp.27-29, 32; Lang 1939, pp.152-153; North 1956, pp.137-138; Boylan 1967, p.250; McCartney 1977, pp.24, 44; Rudwick 1992, p.44; Taylor & Torrens 1987, pp.139, 145; Taylor 1994, p.182; Taylor 1997, p.xxiii-xxiv]

303 **Conybeare**, William Daniel To: H.T. De la Beche
Brislington. Halse Hall,
3 April 1824 Clarendon,
 Jamaica.
I hope my letter brings refreshment to your land of physical and moral pestilence. I expect you will do your best to improve matters there, especially with regard to the education and good treatment of the slaves. I have received a very friendly letter from Cuvier. He is preparing a paper on the Ichthyosaurus. Please bring home with you as many different specimens of indigenous flora and fauna as you can.

304 **Conybeare**, William Daniel To: H.T. De la Beche
Brislington. Halse Hall,
3 June 1824 Clarendon,
 Jamaica.
I contrived through forgetfulness to miss the last packet. I am now preparing for a visit to Oxford where I always lay in a fresh stock of intelligence. Touching your translations from the [Annales des Mines] I imagined you had left instructions for proofs to be sent to me, but the first notice I received was a presentation copy after the publication. There is a very favourable review in the Annals of Philosophy. At the Institution Edmund Davy has just finished an excellent course of twelve lectures on chemistry applied to the arts, but it was ill attended. Our last lecturer, Downes, was certainly a very indifferent hand.

305 **Conybeare**, William Daniel To: H.T. De la Beche
Brislington. Halse Hall,
5 July 1824 Clarendon,
 Jamaica.
Intending to preach at Oxford on Trinity Sunday, I went there in high spirits, expecting to meet my brother on his return from an excursion, but remained only one hour, for I learned alarming things about his health. Though I hurried to Mr. Groombridge's house where he was, I never saw him alive. I have no heart for correspondence now, and have also withdrawn from the activities at the Institution.
[Cited: North 1935, pp.34-35; North 1956, p.139

Conybeare, William Daniel
[c.1824]
[see **2245**]

306 **C[onybeare]**, William Daniel To: W. Buckland
Sully. Christ Church,
28 January 1828 Oxford.
The pictures of Scrope's book afford me great satisfaction. I have been distressed by assisting at Charles' death throes. In travelling through Ireland I thought the central plain of that island would be a fine field for you to challenge Fleming on. I stumbled on Griffith in Dublin. He is engaged in conjunction with the Trigonometrical folks. What shall I do about Brochant's nomination of myself to the Academie de France?

307 **Conybeare**, William Daniel To: W. Buckland
Sully. Christ Church,
29 January 1830 Oxford.
I have just received your ugly geological letter. I don't believe a diluvial current crossing strata of unequal hardness would excavate a straight channel. Scrope is a goose. I shall send the Plesiosaurus back on Saturday. I am pleased Broderip trims the Scotch Donkey, and you are recording your old observations on Dorset.
[Cited: Rupke 1983, pp.86-87]

308 **Conybeare**, William Daniel To: W. Buckland
Sully. Salopian Coffee House,
1 February 1830 Charing Cross,
 London.
The valley of the Thames presents an instance just like those of Wye and Avon. I pray you to refer to my paper on the subject. My only theory is that there was once a ridge of plastic clay from Bagshot Heath in this direction which turned the current. I think this neutralizes Scrope's argument, for surely if anything geological is proved , I have demonstrated

that the existing river could not have excavated the valley.
[Cited: Rupke 1983, p.87]

309 Conybeare, William Daniel To: W. Buckland
Sully. Christ Church,
4 August 1834 Oxford.

Hopkins' Derbyshire contains faulty innovations in nomenclature. He ought to have been more at home as to matters of lines. As to Greenough's speech, I was much disgusted with it. It will retard the progress of our science. If the worst treatise in geology were to be crowned by the Royal Society's Gold Medal, it would be his. Lyell has no competitor. Hawkins' book is capital fun.
[Cited: Howe et al. 1981, p.23; Rupke 1983, p.47; Rudwick 1992, p.65]

310 C[onybeare], William Daniel To: W. Buckland
Lyme.
3 November [c. 1840]

As to Wright I have no distinct recollection. If he were the one who did the maps of the Forest of Dean he must be a good hand. As to Glacial Theory I am a sceptic if not an infidel. The British Association could help in mapping drift. At present Murchison and Co. leap to conclusions for premisses that are most vague and partial. The Glacial Theory has always appeared to me a glorious example of hasty unphilosophical and entirely insufficient induction.
[Cited: North 1935, p.61; North 1942, p.125; North 1943, p.4; McCartney 1977, p.60; Rupke 1983, p.102]

311 Conybeare, William Daniel To: W. Buckland
Axminster. Salopian Coffee House,
15 December 1840 Charing Cross,
London.

Though sadly frost bitten at the moment I don't believe in the former geological supremacy of the Frost King. You see a few scratches on a rock and a heap of gravel at its base and then by an argument per saltem get at your Q.E.D. I should like to see Agassiz's book which is to prove all these wonderful wonders.
[Cited: North 1935, pp.60-61; North 1943, p.4; North 1956 pp.143-144; McCartney 1977, p.60; Rupke 1983, p.102]

312 Conybeare, William Daniel To: H.T. De la Beche
Axminster.
[1845]

You promised me reports on the progress of the Survey. None have arrived at the Deanery of Westminster. Please forward them to Axminster.

313 Conybeare, William Daniel To: H.T. De la Beche
Kew Green.
3 January 1853[?]

Here I am on my way to Southampton to sail this day week to rejoin my sick son at Madeira till the summer recalls me to residence at Llandaff. I have just received the enclosed letter on a district you know better than anyone else. Have you any suggestions?

314 Conybeare, William Daniel To: H.T. De la Beche
4 March [No year]
[Copy]

Last week I was so fully taken up with my second and concluding chemical lecture to my good farmers that I was prevented from answering you immediately. I entirely approve of your proposals and am delighted to hope that the matter will be once again in real progress. I do not apprehend that Sedgwick really ever would have found leisure to give much assistance, but I am sure whatever he'd do he would willingly contribute to our present plan. Please think of the means by which the original calcareous matter of shells in limestone at Porth Kerry and Dunraven has been replaced by silica.

315 Conybeare, William Daniel To: H.T. De la Beche
[No date] Museum of Economic Geology,
London.

An attempt has been made (into the importance of which you from your knowledge of the district will fully enter) to ascertain by boring, whether the coal measures will be found beneath the red marl on the south of the Mendips near Bridgewater, as they are in the collieries on the north of these hills. After boring 81 fathoms the Company has given up in despair. I have advised them to consult Buckland and yourself, as the most competent authorities, on the question as to whether or not it should continue.

316 Conybeare, William Daniel
[No date]

Notes written in preparation for 2nd Edition of *Outlines of the Geology of England and Wales*.

317 Cordier, Pierre Louis Antoine To: H.T. De la Beche
Jardin du Roi, Post Office,
Paris. Lyme Regis,
8 July 1829 Dorset.

Here are the results of the analyses of magnesian limestone from La Spezia made by Laugier, also those of Klaproth on the dolomite of St. Gothard. Please tell the Secretary of the Geological Society that I have not received the Proceedings of 1827-28.

318 Coxworthy, Franklin To: H.T. De la Beche
84 Pall Mall.
30 December 1847

I have left a paper at the Geological Society on Electrical Condition applied to geology. I hope it will be read to the Members, though at variance with present notions. If not against the rules, I would like to reply to any remarks made about it.

319 Croker, John Wilson To: H.T. De la Beche
7 Duke Street,
Adelphi.
2 July 1846

I wish to know if turf can be compressed and used as a substitute for charcoal in the smelting of iron ore. I am anxious to develop the abundant minerals on my estate in Kerry, and there is also an abundant supply of this potential fuel.

320 Cumberland, George Jr. To: [H.T. De la Beche]
Culver Street,
Bristol.
June 1847

I have a collection of Oolite fossils for sale valued at £20. I have added to them the best out of the late Mr. Johnson's collection.

321 Cumming, Rev. Joseph George To: H.T. De la Beche
The College,
Castletown,
Isle of Man.
24 February 1847

I am preparing a popular account of the geology of the island. I have 140 subscribers already. Can your name be placed on the list? We regretted Forbes' hurried visit, and hope he will have a longer furlough in the summer.
[Letter written on printed circular for Cumming's *The Physical History of the Isle of Man*.]

322 Currie, Robert To: [H.T. De la Beche]
Sandhill,
Newcastle.
6 February 1851

I am requested to address this appeal to you to subscribe to a fund for Mr. Hutton, who due to speculation in paper manufacture has been entirely received. Prof. Phillips provides details of his merits.

323 Cuvier, Georges To: V. de Rigny
Jardin du Roi,
Paris.
4 October 1828

The bearer of this, M. De la Beche, a most distinguished English naturalist, has asked me to give him a letter of introduction to you. He wishes to visit the country which you have geologically outlined. Should he need any assistance, please give it to him.

324 Cuvier, Georges To: H.T. De la Beche
[1828]

M. Cuvier has the honour to present his compliments to M. De la Beche, and wishing him a safe journey, he sends him the two letters of introduction which he wanted for Marseilles.

325 Cuvier, Georges et al. To: H.T. De la Beche
Administration du Museum d'Histoire Naturelle,
Jardin du Roi,
Paris.
24 November 1829

The Assembly thanks M. De la Beche for the gift of the head of an

Ichthyosaurus, the merit of which is of great importance to science and for the Museum.

326 Cuvier, Georges To: H.T. De la Beche
Académie Royale des Sciences,
Institut de France.
2 April 1832

The Academy thanks you for the 2nd edition of your Geological Manual which has been deposited in the library of the Institute.

327 Danvers, F. Dawes To: H.T. De la Beche
Duchy of Lancaster Office.
17 January 1840

Lord Holland approves of your suggestion respecting the Ogmore minerals and will speak to the Chancellor of the Exchequer.

328 Darlington, John To: H.T. De la Beche
Alison Hall, Commissioner on "the
Near Chorley, comparative values of
Lancashire. British Coals for
15 June 1850 Steam Purposes",
 Putney,
 Near London.

I wrote to you on the 1st June last concerning Lancashire coal, and the Enquiry into the comparative values of steam coals. I have had no answer. Is Lancashire coal being tested, and have you any objection to testing my coal?

329 Darwin, Charles To: [H.T. De la Beche]
12 Upper Gower Street.
7 February 1842

What colour are the horses bred for some generations in Jamaica, without crossing with any foreign blood? The same re cattle? Please give a description of wild cattle. Have you any information on the above points concerning dogs, cats, poultry, pigs, goats etc? Is there any deterioration in the wool of imported sheep?
[Cited: McCartney 1975, p.21; McCartney 1977, p.25; Burkhardt & Smith 1986, p.310]

330 Darwin, Charles To: H.T. De la Beche
Down,
Farnborough,
Kent.
19 August [1847 FJN]

Studer has been here and unfortunately is too much pressed for time to visit your working party. He is obliged for your kind offer. He is going to meet Forbes in Edinburgh. I am glad to hear about the very old rocks under the Silurian. There is something so grand and mysterious at these depths.
[Cited: Burkhardt & Smith 1988, pp.61-62]

331 [Darwin, Charles] To: H.T. De la Beche
Down, Museum of Economic Geology,
Farnborough, Craig's Court,
Kent. Charing Cross,
7 April [1848] London.

At p.49 of my volume on coral reefs you will see why I do not think it probable that these reefs grew on the submarine margin of the denuded coast-line. I assure you I did not spare time or labour in examining thousands of charts and all voyages.
[letter damaged.]
[Cited: Burkhardt & Smith 1988, pp.129-130]

332 Daubeny, Charles Giles Bridle To: H.T. De la Beche
Oxford. Mall,
28 April [?1825 FJN] Clifton,
 Bristol.

I have examined 3 specimens from Jamaica and found only traces of magnesia. I shall be anxious to see your account of Jamaica. I have sent my Memoir on Sicily to Jameson for the next No. of the Geological Society's Transactions.

333 Daubeny, Charles Giles Bridle To: [H.T. De la Beche]
Broad Street,
Oxford.
30 November 1829

Your specimen contains abundance of Magnesia, and is true Dolomite.

40% or more of carbonate of Magnesia present. More accurate analysis can be given if required.

334 Dawes, John J. To: H.T. De la Beche
Southwick House,
Nr. Birmingham.
6 November 1852

Respecting occurrence of Stigmaria in South Staffordshire, these remains are to be found underneath every bed of coal in district. They exhibit evidence of having been produced in the strata which contain them, not drifted in.

335 Day, M.J. To: H.T. De la Beche
[c.1851]

Mem. Geological Observer. Slips 11 to 16 not returned, although 17 et seq. to hand.

336 Day, William To: H.T. De la Beche
Carmarthen.
11 June 1842

I forgot yesterday to draw your attention to the distorted trilobites in the Llandilo flags. What causes this distortion? Most distorted specimens are ferruginous. Is the oxidization and distortion due to a slip?

337 Dechen, Ernst Heinrich Karl von To: H.T. De la Beche
Bonn.
2 December 1850

Thanks for the donation of memoirs of the Geological Survey, and volumes on fossils. In return I offer a selection of minerals from our mines.

338 De la Beche, Elizabeth (Bessie)
[Birth Certificate of Elizabeth De la Beche]
2 Dec. 1819 at Geneva, daughter of Henry Thomas De la Beche and Letitia Whyte, his wife.
[Cited: McCartney 1977, p.12]

339 De la Beche, Elizabeth (Bessie) To: H.T. De la Beche
Lyme. Alpha Place North,
15 March 1832 Regents Park,
 London.

Thank you for your present of the Geographical Annual. I have been reading travel books lately. I like my new music mistress. I wish to do botany this spring with Misses Aveline and Savage. I will try to have my garden ready for your visit this year.

340 De la Beche, Elizabeth (Bessie) [et al] To: H.T. De la Beche
Falmouth. St. Austell,
1 February 1837 Cornwall.

The Falmouth Foolscap
[A parody newspaper which contains various items of contemporary interest, including De la Beche's activities at St. Austell. Entries are signed C.F., Bessie, A.M., and C.]

341 De la Beche, Henry Thomas
20 November 1818 to 11 March 1819
[Pocket Journal]
[128 pages. Record of geological and meteorological observations at Clifton, Wells, Glastonbury, Exeter, Newton, Teignmouth, Haldon Hill, Dawlish, Lyme, Weymouth etc. Descriptions of fossils. 23 sketches, sections, etc.].

342 De la Beche, Henry Thomas
12 March to 5 June 1819
[Pocket Journal]
[110 pages. Bristol, Weymouth, Dorchester, Newport, Brecon, Merthyr Tydfil, Ross, Bath, Oxford, London. Remarks on geology, fossils, etc. 11 sketches, sections etc.].
[Cited: North 1934a, pp.71-76, plates 4, 5; North 1959, pp.62, 63; Thackray 1985, p.182]

De la Beche, Henry Thomas
15 July 1819
[see **2246**]

343 De la Beche, Henry Thomas
19 June to 10 August 1819
[Pocket Journal]
[124 pages, including entries on Kent Coast, and De la Beche's journey

through France, Abbeville, Paris, Avallon, Dijon, Macon. This "Journal" is one of those from which De la Beche assembled his "Diary"(See **346**.) The sketches (9) are mostly pencilled, and evidently the basis of sketches drawn and painted in the diary].
[Cited: McCartney 1977, pp.13, 15]

344 De la Beche, Henry Thomas
11 August - 12 September 1819
[Pocket Journal]
[172 pages, Lyons, Maximieux, Cerdon, Nantua, Bellegarde, Collonge, Geneva, Mt. Blanc, Valley of Chamonix, Martigny, etc. 11 sketches. Basis of "Diary" (**346**)].
[Cited: de Beer & North 1950, pp.495-500; McCartney 1977, pp.13, 15, 16]

345 De la Beche, Henry Thomas
15 September 1819 - 27 February 1820
[Pocket Journal]
[109 pages, Lake of Geneva, Hermance, Geneva, Mt. Salève, Nyon, Thonon, Morges, Ouchey, Vevay, Geneva (daughter Bessie born 2 Dec. 1819). 2 sketches. Basis of "Diary", (**346**)].
[Cited: North 1943, p.5; McCartney 1977, p.13]

346 De la Beche, Henry Thomas
21 July 1819 to 27 September 1819
Diary while travelling in France and Switzerland.
[This copy was evidently composed by De la Beche at a later date (paper carries an 1825 watermark), and was assembled from the pocket journals which correspond to the above period (**343**, **344**, **345**). Coloured illustrations have been more carefully worked up. 80 pages, 13 sketches].
[Cited: de Beer & North 1950, pp.495-500; McCartney 1977, pp.14, 16]

347 De la Beche, Henry Thomas
31 January to 15 March 1820
[Pocket Journal]
[130 pages, Neuchâtel, Geneva, Chaumont, Turin, Genoa, Marengo, Pisa, Florence. 6 sketches].
[Cited: North 1943, p.6; McCartney 1977, p.16]

348 De la Beche, Henry Thomas
16 March to 3 June 1820
[Pocket Journal]
[120 pages, Florence, Bologna, Milan, Turin, Geneva, Chamonix, Lausanne, Lucerne, Payerne, Fribourg, Berne, Thun, Lauterbrunnen, Zofingen, Zug, Zurich, Basle, etc. Various observations on geological features of country traversed, as well as on modes of dress, religious customs etc. 8 sketches and sections].
[Cited: McCartney 1977, p.16]

349 De la Beche, Henry Thomas
4 June to 9 July, and September 1820
[Pocket Journal]
[124 pages, Strasbourg, Karlsruhe, Heidelberg, Darmstadt, Francfort, Hochheim, Maynz, Bingen, Coblenz, Cologne, Aix-la-Chapelle, Maestricht, Louvain, Antwerp, Utrecht, Amsterdam, Haarlem, Leyden, Catwyk, The Hague. Lyme (Sept. 1820). 5 sketches].
[Cited: McCartney 1977, pp.16-17]

350 De la Beche, Henry Thomas
1 April - 15 May 1821
[Pocket Journal]
[107 pages, Le Havre, Harfleur, Montvilliers, Fécamp, Honfleur, Pont l'Eveque, Bayeux etc. 9 sections].

351 De la Beche, Henry Thomas
[Summer of 1822 FJN]
[Geological notes on Pembrokeshire]
[73 pages. Notebook containing descriptions of rock formations of Pembrokeshire , one section of culm-mine, notes on the Elegug, etc. No sketches. No dates given for any of the entries].
[Cited: Oldroyd 1991, p.412]

352 De la Beche, Henry Thomas
9 November 1823 - 28 June 1824
[Diary of Jamaica visit, in one pocket journal]
[174 pages, account of departure from Bristol, Atlantic voyage, stay on the island, including various expeditions in the mountains, remarks on slaves, etc. 5 sketches].
[Cited: Chubb 1958, pp.6-17; McCartney 1977, pp.23, 24-25; Robinson 1996, p.27]

353 De la Beche, Henry Thomas
9 November - 22 December 1823
[Extended Jamaica Diary]
[20 pages, an account of the Atlantic crossing in the 'Kingston', sailing from Bristol to Kingston, Jamaica].
[Cited: Chubb 1958, p.6]

354 De la Beche, Henry Thomas   To: W.D. Conybeare
Halse Hall,   Brislington,
Clarendon,   Bristol.
Jamaica.
8 January 1824
Here are some notes on the geology of the island. White limestone has been examined. Vast quantity of magnetic sand is found on my property. I am informed that there are rocks like the Giants Causeway on the North side of island. I have started collection of land shells and crustacea. My negroes are contented, despite trouble on other parts of the island.
[Cited: North 1935, p.33; Chubb 1958, pp.9-10; McCartney 1977, p.24]

355 De la Beche, Henry Thomas   To: W.D. Conybeare
Halse Hall,   Brislington,
Jamaica.   Bristol.
6 March 1824
I have little doubt that the white compact limestone of Jamaica is the equivalent of the magnesian limestone of England, or the Zechstein. I send you in this letter a section from Luidas Vale to the sea near Old Harbour. Neither Colonists and Abolitionists nor the English Government will long have anything to do with Jamaica. I regret having translated Annales des Mines. I hear a skeleton of a Plesiosaurus has been found, and suppose you have heard all about it.
[Reply to Conybeare, W.D. to De la Beche, H.T.,19 December 1823, **301**].
[Cited: North 1935, p.33; Chubb 1958, p.12; McCartney 1977, pp.22, 23-24; Robinson 1996, p.27]

356 De la Beche, Henry Thomas   To: W.D. Conybeare
Kingston,   Brislington,
Jamaica.   Bristol.
6 April 1824
Since my last letter I have made many geological discoveries. With sketches of three sections, herewith details of the geology of the island. The rocks have much the same positions as in Europe, showing that the same causes have been in action in these distant parts of the globe. Please give data to Wollaston and ask him to calculate the height of the Peak.
[Cited: Chubb 1958, pp.14-15]

357 De la Beche, Henry Thomas   To: W.D. Conybeare
Halse Hall,   Brislington,
Jamaica.   Bristol.
13 May 1824
I have chosen the section from Kingston to Buff Bay over St. George's Gap to describe to you. Is it not interesting to find so general a resemblance between the rocks in this part of the globe and in Europe? Four earthquakes occurred here lately. May not earthquakes be electrical phenomena? I highly approve of Canning's recommendations for slaves, but a proprietor could be saddled with worthless ones unable to cultivate the property.
[Reply to Conybeare, W.D. to De la Beche, H.T., 4 March 1824, **302**].
[Cited: Chubb 1958, p.16; McCartney 1977, pp.22, 26; Sharpe 1997, p.33]

358 De la Beche, Henry Thomas   To: W.D. Conybeare
Halse Hall,   Brislington,
Clarendon,   Bristol.
Jamaica.
29 July 1824
There is very great difficulty in tracing rocks in this country in consequence of dense vegetation. I have not as yet observed any rocks which would be called tertiary. The cavernous rocks of St. Anns call into question the theory that valleys are excavated by the waters which run in them. Present economic state of affairs prevents me from saying when I will be able to return to England.
[Cited: Chubb 1958, p.17; North 1959, p.61; McCartney 1977, p.24]

359 De la Beche, Henry Thomas To: J. Williams
Halse Hall, Belle Vue.
[Jamaica].
25 August 1824
[Draft/Copy]
I acknowledge yours of 20th. Is there any law which requires proprietors to obtain the consent of the Quarter Sessions to his procuring religious instruction for his negroes, when such is to given by persons not of established Church? I do not see either, how my driver's not inflicting punishment in the field will lead to general destruction of property.
[See also Shipman, J, to De la Beche, 10 December 1824, **1905**].
[Cited: Chubb 1958, p.17]

360 De la Beche, Henry Thomas
October 1827
On the Geology of Tor and Babbacombe Bays, Devon.
[Ms. received 23 October, read 2 and 16 November 1827; abridged by G.B. Greenough. Published in *Transactions of the Geological Society*, Second Series, vol 3, Part 1 (1835), pp.161-170, + plates XVIII-XX. See also *Proceedings of the Geological Society*, Vol 1, No.4 (1827-28), pp.31-33].

361 De la Beche, Henry Thomas
[c.1828]
On the geology of Nice
[Ms. (No.1) of paper published in *Transactions of the Geological Society*, Second Series, vol.3, Part 1 (1835), pp.171-185. Certainly the first copy, heavily marked and cut, either by De la Beche or referees assigned by the Geological Society].
[Cited: McCartney 1977, p.27]

362 De la Beche, Henry Thomas
17 December 1828
On the geology of the environs of Nice & the coast thence to Vintimiglia
[Ms. No.2 of paper pages 1, 2, 3, 4, 11 and 12 only. Back leaf annotated (semi-erased:] To be sent to Dr. Buckland. Map to be furnished by Dr. Fitton. Recd. 17 December 1828 by S.T., for Dr. Buckland. Map at Mr. Gardner's.
[Front leaf annotated:] The referee has made so many alterations in the paper referred to him that he thinks it unnecessary to prepare a corrected copy for the press till he knows whether the Council do or do not sanction the extent of the alterations. [Greenough's hand].
I recommend the alterations and new arrangement made by Mr. Greenough to be adopted by the Council. W. Buckland"].
[See also Greenough, G.B., [November 1828], **619**]
[Cited: McCartney 1977, p.27]

363 De la Beche, Henry Thomas
2 November 1828 to 1 June 1829
[Diary when travelling in Italy]
[182 pages, ascent of Vesuvius, Naples, Pompeii, Sorrento, Castello a Mare, Herculaneum, Pozzuoli, Paestum and Salerno, Rome, Tivoli, Siena, Florence, Lucca, Massa, La Spezia, Genoa, Como, 75 sketches, sections].
[Cited: McCartney 1975, pp.16, 18, 20; McCartney 1977, pp.16, 17, 19, 27]

364 De la Beche, Henry Thomas To: W. Buckland
Lyme Regis. Christ Church,
23 July 1829 Oxford.
The parcel and your letter have been forwarded to me from the Geological Society. I approve of Greenough's condensation of my article on the geology of Nice, but I am not content with your theoretical alterations. My ideas about diluvium and sudden and violent causes have changed of late. Your views will therefore be removed. You had better add them in an appendix as objections to my position. You should have been acquainted with Von Buch's and Elie de Beaumont's views before printing your appendix.
[See Buckland, W. to De la Beche, H.T., 26 July 1829, **177** for reply]

365 De la Beche, Henry Thomas
[December 1829]
On the Gulf of La Spezia, and neighbouring coasts
[Ms. No. 1 of "On the Geology of the shores of the Gulf of La Spezia", read 1 January 1830: *Proceedings of the Geological Society of London*, Vol.1 (1826-33), No.14, pp.164-167].

366 De la Beche, Henry Thomas
[December 1829]
On the Shores of the Gulf of Spezia, and neighbouring Coasts
[Ms. No.2 of "On the Geology of the shores of the Gulf of La Spezia", read 1 January 1830: *Proceedings of the Geological Society of London*, Vol.1 (1826-33), No.14, pp.164-167].

367.1-25 De la Beche, Henry Thomas
1830
Awful Changes
[Lithograph, 25 copies].
[Cited: North 1942, p.111; McCartney 1977, p.51; Taylor 1997, p.xxxii]

368 De la Beche, Henry Thomas
[c.1830]
Duria Antiquior
[Watercolour].
[Cited: Howe et al. 1981, cover; Rudwick 1992, dustjacket, p.44; Cartmill 1992, p14; Torrens 1995b, p.266; Tickell 1996, p.13; Chang 1997, pp.100-101; Sharpe 1997, p.31; Taylor 1997, p.xxxiii]

369 De la Beche, Henry Thomas
July 1832
Plan for emancipating the negroes. As the West Indians contend for compensation in the event of the negroes being emancipated let this be granted them through compulsory manumission.

370.1-5 De la Beche, Henry Thomas
[1832]
This is to certify that H. De la Beche Esq FRS etc employed on the Ordnance Survey of Great Britain, is charged with the Geological Examination of the Districts comprised within the Map of Devonshire, containing portions of the adjoining Counties.
[Signed:] Richard Z. Mudge.... for Lt. Colonel Colby.

371.1-2 De la Beche, Henry Thomas
1832
[Lithograph of an island].

372 De la Beche, Henry Thomas
[1832 FJN]
Note on the economical advantages of Geological Maps
Some of the European Governments have been so well aware that by encouraging, or even supporting all the expenses of geological maps.... solely for the purposes of science, other and economical advantages must flow from them, that the question of cui bono, as it refers only to pecuniary advantages, has never been raised. France has supported all the expenses of a geological map, solely with a view to science. Such maps, when the scale is sufficiently large, are of the highest value to the agriculturalist, the miner, the builder, and the road-maker. Other advantages could be adduced, but these are suffient to show the question of cui bono, as it respects pecuniary advantages, can be successfully met.
[Cited: North 1936a, pp.47-48]

373 De la Beche, Henry Thomas To: Bessie De la Beche
Tavistock, W. H. Aveline's Esq,
Devon. Lyme Regis,
23 June 1833 Dorset.
I have been here since last Tuesday, and I now give you a detailed account of the week's activity, day by day. It will show you how I get on. Visits to Dartmoor Prison, mines, geology of Dartmoor, etc. Tell Grandmama not to worry about Jamaica affairs.

374 De la Beche, Henry Thomas To: Messrs. Dolland
15 November 1835
What is the price of gold spectacles with screws of some other metal than steel or iron, as I wish to wear them when using prismatic compass?
[Answer on same note:] £4.14.6.

375 De la Beche, Henry Thomas
[c.1835]
On the geology of the environs of Nice and the coast thence to Vintimiglia
[Ms No.3 of paper published in *Transactions of the Geological Society*, Second Series, vol.3, Part 1 (1835), pp.171-185. Probably a 'clean' copy prepared from previous mss. Some minor differences between this ms. and the published article.]

376 De la Beche, Henry Thomas　　　　　　　　To: T. Colby
Helston.
5 December 1836
Abstract of Expenditure for Geological Survey of Cornwall, according to Board's Order of 11 July 1835. Average for four quarters mentioned £210.14.2.

　　De la Beche, Henry Thomas
　　[post-1836]
　　[see **2255-2256**]

377 De la Beche, Henry Thomas [et al.]
White Hart,
St. Austell,
Cornwall.
2 February 1837
The Mining Chronicle
[A parody newspaper, with caricatures of R.W. Fox's experiments on magnetism, and temperature in mines. None of the entries is signed or initialled].
[Cited: McCartney 1977, p.41]

378 De la Beche, Henry Thomas　　　　　　　　To: T. Colby
Falmouth.
8 April 1837
I have received your letter informing me of reports on the inaccuracy of my map. They originate with Sedgwick & Murchison who obtained the unpublished map from Gardner. It is clear I have personal enemies. My feelings tell me I should resign, but I would prefer an inquiry. At least I will have finished the Cornwall map!
[See Colby, T.F. to De la Beche, H.T., 3 April 1837, **270**].
[Cited: North 1939b, p.1053; McCartney 1977, p.30; Rudwick 1985, p.203]

379 De la Beche, Henry Thomas and others of the Ordnance
23 June 1837 to 30 September 1837
[Copy of correspondence concerning De la Beche's application for £25 per quarter travel allowance, comprising H.T. De la Beche to T. Colby 23 June 1837; T. Colby to Inspector General of Fortifications, 26 June 1837; R. Byham to Inspector General of Fortifications, 3 July 1837; H.T. De la Beche to T. Colby 8 July 1837; T. Colby to Inspector General of Fortifications, 11 July 1837; T. Colby to Inspector General of Fortifications, 15 July 1837; R. Byham to Inspector General of Fortifications, 16 August 1837; Minute 22 August 1837; Minute 19 July 1837; H.T. De la Beche to T. Colby 19 August 1837; H.T. De la Beche to T. Colby 2 September 1837; T. Colby to Inspector General of Fortifications, 15 September 1837; Minute 21 September 1837; G. Butler (for R. Byham) to Inspector General of Fortifications, 27 September 1837; H.T. De la Beche to T. Colby 7 October 1837; R. Byham to Inspector General of Fortifications, 25 October 1837; H.T. De la Beche to Board of Ordnance 30 September 1837].
[Cited: North 1936a, pp.54-56; Harley 1971b, p.119]

380 De la Beche, Henry Thomas　　　　　　　　To: T. Colby
Falmouth.
13 July 1837
With this I forward letter from Hopkins on the Derbyshire Survey. Also Sedgwick's covering letter. If it could be arranged that Hopkins was attached nominally to the Survey, he could map and report on Derbyshire, and we would have every assistance from the Cambridge men.

381 [De la Beche, Henry Thomas]
25 July to 2 August 1837
[5 magnetic bearings from different stations in Cornwall observed with Fox's Dipping Needle Deflector].

382 De la Beche, Henry Thomas
25 July to 4 August 1837
Magnetic dip observed in West Cornwall with Fox's Dipping Needle Deflector.
[9 separate observations].

383 De la Beche, Henry Thomas　　　　　　　　To: A.W. Robe
Falmouth.
9 August 1837
Could you send me the true bearing of the following points on each other, in order that I may calculate the observed magnetic variation at some of them on given days of observation? [Lands End].
[Extract only]

384 De la Beche, Henry Thomas
26 January 1838
State of the Survey of England in January 1838
Printed map: [office copy of the Index to the Ordnance Survey of Great Britain, updated by colours and hatching, showing the state of both the Ordnance Survey and the Geological Survey].

385 De la Beche, Henry Thomas
20 February 1838
Suggestions for a College of Mines
Central Mining College in London with branches in principal mining districts. Professorships in Chemistry, including Mineralogy and Metallurgy, Mechanics and Civil Engineering, Geology. Lectures to be delivered by them both in London and District Colleges. President to be Chief Commissioner of Woods and Forests. Professorship in Practical Mining in each College, etc.
[Cited: McCartney 1977, p.40]

386 De la Beche, Henry Thomas
[1838]
Suggestions for forming a School or College of Mines, with Branch Establishments in the principal Mining Districts. Central school in London, and branches in Cornwall, Derbyshire and Newcastle. Central college to have a Museum. A general school to be attached to the District Colleges, to which district professors should be attached.

387 De la Beche, Henry Thomas　　　　　　　　To: F. Baring
Ordnance Map Office.　　　　　　　　　　　　　　Treasury.
17 March 1838
[Draft/copy]
Organic remains have been collected during the Survey of Cornwall and Devon. They should be figured and described. I request £250 for this purpose, and ask that Phillips be permitted to execute the work. References from Buckland, Sedgwick, and Whewell included.

　　De la Beche, Henry Thomas
　　Liliput.
　　March 1838
　　[see **2257**]

388 De la Beche, Henry Thomas　　　　　　　　To: T. Colby
Ordnance Geological Survey,
Swansea.
1 November 1838
[Draft/copy]
In my letter of yesterday respecting Fox's Dipping Needle Deflector, I omitted the information that the increased cost was due to adapting the instrument to ascertain local variations of the magnetic needle, for comparison with the true bearing as ascertained by the Trigonometrical Survey.

389 De la Beche, Henry Thomas　　　　　　　　To: C. Lemon
Swansea.
25 December 1838
Copy
The prospectus of your mining school reached me safely. There should be either a central school in London with branches in the country, or a completely independent school for Cornwall. Nothing forthcoming from the new Universities. You have a good man in Prideaux.
[Cited: McCartney 1977, p.40]

390 [De la Beche, Henry Thomas]
[?1838]
[Ms. Part of *Report on the geology of Cornwall, Devon and West Somerset*, 1839, 8 pages (see *Report*, pp.283-291) and 6 pages (see *Report*, pp.349-350].

391 [De la Beche, Henry Thomas]
[?1838]
[Ms. of Appendix B. of *Report on the Geology of Cornwall, Devon and West Somerset*, 1839, pp.629-634. Principal item is translation of letter from William de Wrotham and others to Chancellor of England, provided by T. D. Hardy of H.M. Records Office].

392 [De la Beche, Henry Thomas]
[?1838]
[Ms. copy of Charter granted to the Tinners of Cornwall and Devon. Rot. Chart. Antiq. K. No.5, 29 October, 3rd year reign King John [1201], by H. Petrie, Keeper of H.M. Records, Tower. Printed by De la Beche in *Report on the Geology of Cornwall, Devon and West Somerset*, 1839, Appendix A, p.625].

393 [De la Beche, Henry Thomas]
[?1838]
[Ms. copy of Charter granted to the Tinners of Cornwall and Devon. Rot. Chart. 33 Edw I, Nos. 41 and 40, 10 April, 33rd year of reign Edward I [1305], by H. Petrie, Keeper of the Records, Tower. Printed by De la Beche in *Report on the Geology of Cornwall, Devon and West Somerset*, 1839, Appendix A, pp.626-627].

394 [De la Beche, Henry Thomas]
[?1838]
[Ms. copy of Petitions to Parliament, 1st year of reign of Edward III, No.64. (Complaints about destruction caused by Tinners) by T.D. Hardy of H.M. Records Office. Printed, in part, by De la Beche in *Report on the Geology of Cornwall, Devon and West Somerset*, 1839, Appendix A, pp.628-629].

395 De la Beche, Henry Thomas
15 March 1841
The Irregularities of Sol visited upon his System
[Lithograph of cartoon on Glacial Theory].
[See De la Beche, H.T. to Buckland, W., 8 April 1841, **396**].
[Cited: McCartney 1977, pp.60-63; Sharpe 1997, p.32]

396 De la Beche, Henry Thomas   To: W. Buckland
Ordnance Geological Survey,
Tenby,
South Wales.
8 April 1841
While the other day instructing the daughter of our friend Capt. Smyth, now at Cardiff, in the mode of working upon transfer lithographic paper I knocked off the accompanying trifle illustrative of the sudden changes that may, according to someone, I forget who, be brought about in climate from spots on the sun. I was too much pushed for time to finish the affair, but if you will put your hand upon the lower parts of the elephants, you will see more of the effect intended, tropical vegetables and animals bothered by snow suddenly falling because of Mr. Sol's irregularities. If one had time something might be made of this subject. Only 20 or 25 copies of the accompanying are printed. If the Captain has already sent you one, give this to Greenough for me.
[See De la Beche, H.T., 15 March 1841, **395**]

397 De la Beche, Henry Thomas
[November 1841]
[Draft report on state of Geological Survey and Museum of Economic Geology, with reference to the possibility of transferring the former from the Ordnance to the Department of Woods and Forests, the place of organic chemistry in the laboratory of the Museum, the need for lectures on that subject for Agriculturalists, and generally, proposals to maintain the effectiveness of the two services].

398 De la Beche, Henry Thomas   To: [Chief Commissioner
Museum.   of Woods and Forests]
19 March 1842
[Draft/Copy of Letter and two copies of Memorandum]
In transmitting the memorandum concerning the Chemist at the Museum, I would like to point out that Phillips has laboratory facilities in his own home. His income could be suitably adjusted.
[Memo, Copy No.1] Two chemists needed, one for Government Departments etc., and the other to cope with extensive work from the English Agricultural Society.
[Memo, Copy No.2: ditto, with minor changes].

399 De la Beche, Henry Thomas   To: Capt. Boldero
London.
11 April 1842
[Draft/Copy]
In the last papers submitted to Sir George Murray, I neglected to emphasise the independence of Geological Survey from the Trigonometrical Survey. This I do now, and ask the Master General to consider the inconvenience to the Geological Survey of any move to Southampton, and the suitablity of using the office at the Department of Woods.

400 De la Beche, Henry Thomas   To: J. Phillips
London.
27 May 1844
[Draft/Copy]
Reply to yours of 24th. Your Professorship in Dublin seems to all to preclude you from continuing on the Geological Survey of Great Britain. But if the Irish matter goes well, there is opportunity for you there. So if you do not push the G.B. question for this year, all should go well.
[Cited: Davies 1969, p.32; Herries Davies 1983, p.114]

401 De la Beche, Henry Thomas   To: J. Hamel
Ordnance Geological Survey   Imperial Academy of Sciences,
of Great Britain,   St. Petersburg.
Builth,
Brecknockshire.
16 October 1844
[Draft/Copy]
Thanks for yours of 5th. The Master General and Board of Ordnance have desired copies of Ordnance Geological maps to be sent to the Academy.
[See Hamel, J. to De la Beche, H.T., 5 October 1844, **651**].

402 De la Beche, Henry Thomas   To: T.W. Philipps
Geological Survey Office.
12 March 1846
I send you James' letter, as it appears he has mistaken his duties as local director in Ireland. He must spend nine months in the field, and has nothing to do with the Museum in Dublin except to aid it. It will never do to have him dashing down only from time to time to visit Assistants at work. Please explain this to Canning and get his views.

403 De la Beche, Henry Thomas   To: H. James
Geological Survey Office.
13 March 1846
[Draft/Copy]
Confidential
Your fixed residence in Dublin will be opposed. You are expected to spend 9 months in the field. The mere inspection of parties is insufficient service. Your duties are to the Survey only, and all matters of collections etc., pertain to Kane in charge of the Museum in Dublin. Our force is already too small for Ireland.

404 De la Beche, Henry Thomas   To: T.E. James
Geological Survey Office,
London.
23 April 1846
[Draft/Copy]
I regret that your services will not be required further on the Survey. We are compelled to adjust our service to a variety of conditions. You would have until September to consult friends etc., as to alternative employment.
[Cited: Herries Davies 1983, p.131].

405 De la Beche, Henry Thomas   To: T.W. Philipps
Geological Survey Office.
21 May 1846
You will recollect it was proposed to offer Owen the Honorary Appointment of Consulting Palaeontologist to the Survey. His aid would be most important to us. Would you mention this to Canning for his consideration.
[Canning memo on reverse:] I see no objection, except one against Honorary appointments. What was the proposal?

406 De la Beche, Henry Thomas   To: T.E. James
Geological Survey Office,
Custom House,
Dublin.
10 July 1846
Re your letter of 28th. I gave you extra time to make arrangements by keeping you on the Survey for another quarter. Arrangements for the Survey cannot be altered, so I strongly advise you to exert yourself in seeking a fitting situation.

407 **De la Beche**, Henry Thomas To: [H.] James
Fethard.
18 July 1846
[Draft/Copy]
Yours of 11th., arrived yesterday. I never doubted your claim to four days pay, but it must be paid the by Cashier of Woods and Forests, since salaries above 10/- per day are dealt with there. Portlock's documents are public papers, the same as Phillip's, mine, Bone's and Baily's.
[See James, H. to De la Beche, H.T., 11 July 1846, **786**]

408 **De la Beche**, Henry Thomas
[c.1846]
On the formation of the rocks of South Wales and South Western England.
[Ms of *Memoirs of the Geological Survey of Great Britain*, Vol 1, 1846, pp.1-20].

409 **De la Beche**, Henry Thomas
[c.1846]
Silurian rocks.
[Ms of *Memoirs of the Geological Survey of Great Britain*, Vol 1, 1846, pp.20-40].

410 **De la Beche**, Henry Thomas To: H.F. Waring
Geological Survey Office,
London.
9 April 1847
[Draft/Copy]
In reply to yours of 6th inst., I am about to go to Lyme Regis on account of the Crown to report upon the foreshore between Charmouth and Axmouth. Thank you for offer of hospitality, but I should forego such even from old friends should they be directly or indirectly involved in this enquiry.
[See Waring, H.F. to De la Beche, H.T., 6 April 1847, **2278**].

411 **De la Beche**, Henry Thomas To: R. Phillips
Barlow,
Derbyshire.
19 September 1847
Copy
In reply to yours of 17th. You have my assent to your absence, if you obtain permission from Viscount Morpeth.
[See Phillips, R. to De la Beche, H.T., 17 September 1847, **1670**].

412 **De la Beche**, Henry Thomas To: C. Wild
Geological Survey Office. Royal Society.
16 December 1847
[Two Drafts/Copies]
If you still cannot find a Chairman for the meeting, I will attend.
[Letter in format of poem].

413 **De la Beche**, Henry Thomas To: W.H. Fitton
Geological Survey Office.
28 December 1847
[Draft/Copy]
I hold, and still hold, the rules touching receipt of papers to be good. Neither Morris nor Sowerby can say I made any agreement with you about the matter. I did all I could by having the paper read, and this in fact gives to it priority of publication.

414 **De la Beche**, Henry Thomas To: C.E. Trevelyan
Geological Survey Office.
30 June 1848
[Copy]
Re your note of yesterday and query on Mr. Smyth's charges. They were agreed to by Sir George Grey at the Home Office, and given the danger involved in investigating colliery explosions, the talent of Smyth, and the low pay of the Survey, the Treasury will probably view the account as fair and proper.
[See Trevelyan, C.E. to De la Beche, H.T., 29 June 1848, **2037**]

415 **De la Beche**, Henry Thomas
1848
[Portrait] Sir Henry Thomas De la Beche VPRS
Director General of the Geological Survey of the United Kingdom
President of the Geological Society
Painted by H.P. Bone in Enamel from Life. Engraved by W. Walker
London: January 1st 1848, W. Walker, Excudit. 64, Margaret St., Cavendish Square
Proof   Private Plate
To the Fellows of the Geological Society this Engraving is respectfully inscribed by their Obedient Humble Servant, William Walker.

De la Beche, Henry Thomas
[1848]
[see **2258-2259**]

416 **De la Beche**, Henry Thomas To: [Directors,
Geological Survey Office. East India Company]
14 March 1849
[Draft/Copy]
Herewith the memorial from the widow of the late David Hiram Williams, who went to India surveying for coal. I hope the Directors will consider the case of his widow, seeing that he zealously discharged his duties.

417 **De la Beche**, Henry Thomas To: T.W. Philipps
Museum of Practical Geology.
3 May 1849
Mr Smith, the bearer of this letter has applied on the part of the Grand Junction Canal Company for copies of our Railway sections. Would Lord Carlisle sanction this, as we have no provision for arrangements for such cases as yet.

418 **De la Beche**, Henry Thomas To: A.C. Ramsay
Youghal,
Cork.
28 August 1849
[Draft/Copy]
Trimmer should cease from communicating in a roundabout way concerning his resignation. Government will not provide finance for Agricultural Geology either for Ireland or our own Survey. But if he in the meantime does not find other opportunities, I would like to have a confidential conversation with him when I come to London in a month or so.

419 **De la Beche**, Henry Thomas To: his children and
24 December 1849 grandchildren
Though I will not be with you this Christmas, I wish to greet all with a 'Merry Christmas and a Happy New Year'.

420 **De la Beche**, Henry Thomas To: T.W. Philipps
Museum.
18 January 1853
Copy
Any reference from the Treasury connected with Warington Smyth and the Commissioner of Woods is unintelligible to me. Smyth is now in Ireland on Survey duties. Before he left he indicated that he had abandoned contracting himself with Dean Forest. I told Smyth that I highly disapproved of the whole notion both on public and private grounds. Smyth may be expected from Ireland on Thursday or Friday.

421 **De la Beche**, Henry Thomas To: J. Percy
Jermyn Street.
2 July 1853
I have not received a subpoena at the trial of Moore & Fressange. I therefore wish to make you acquainted with the circumstances of my introducing them as metal casters to Dept. Woods. Having had them do some castings for the Museum on the recommendation of Baily. I found their work skilful. Thus while I cannot speak for their honesty, the former cannot be doubted.

422 **De la Beche**, Henry Thomas
[1853-1854]
[13 photographs mostly of H.T. De la Beche and members of Dillwyn family].
422.1 H.T. De la Beche, Rosie De la Beche, Bessie De la Beche photographed by Mary Dillwyn, 1853
[Figured: Sharpe 1985, p.7; Sharpe 1997, p.33]
422.2 H.T. De la Beche and his three eldest grandchildren (Mary, Henry, and Elizabeth Amy) photographed by Mary Dillwyn, 1853
422.3 H.T. De la Beche, Rosie De la Beche and Tartar photographed by Mary Dillwyn, 1853
422.4 H.T. De la Beche

422.5 Rosie De la Beche and Bessie De la Beche photographed by Mary Dillwyn, 1853
422.6 L.Ll. Dillwyn and Amy Dillwyn photographed by Peter Wickens Fry, 1853
422.7 L.Ll. Dillwyn and Badger photographed by Mary Dillwyn, 1853
[Figured: Sharpe 1985, p.7]
422.8 as **422.7**
422.9 Bessie De la Beche photographed by Peter Wickens Fry, 1853
[Figured: Sharpe 1985, p.9]
422.10 Bessie De la Beche and Amy Dillwyn
422.11 Unidentified group of 3 women and 1 man
422.12 Welsh woman and baby photographed by ?Vivian Webber, 1854
422.13 as **422.1**

423 **De la Beche**, Henry Thomas
[No date]
[Ms. translation of] Memoir on the geology of Germany, by A. Boué. Extracted from the *Journal de Physique*, May 1822.

424 [**De la Beche**, Henry Thomas]
[No date, watermarked 1830]
[Ms notes:] Humbug. Importance of our earth. Vast importance of man. Man's creeds.

425 [**De la Beche**, Henry Thomas]
[No date, watermarked 1830]
[Ms notes:] Man's superstitions. Every nation the best. Every church the right one. Superstitions.

426 **De la Beche**, Henry Thomas
[No date]
[Pocket Journal] Notes for an introduction to geology [Contains small amount of text and headings also pencil sketches and other notes concerning De la Beche's 1828 Italian tour using pages in reverse order].

De la Beche, Henry Thomas
[no date, watermarked 1828]
[see **2248**]

427 [**De la Beche**, Henry Thomas]
[No date, watermarked 1831]
Contributions towards a theory of the earth's structure. [Outline of 7 chapters]. Land, Water, Shells, recent and fossil.

428.1-17 [**De la Beche**, Henry Thomas]
[No date, watermarked 1822, 1830, 1831, 1832]
Various mss: Notes to the printer re 2nd and 3rd editions of *Geological Manual*. Contributions towards a theory of the earth. Essay on a theory of the earth, and other notes apparently relating to *Geological Manual* and *Researches in Theoretical Geology*.

429 [**De la Beche**, Henry Thomas]
[No date, watermarked 1825]
[Ms notes on river valleys and River Rhine]

De la Beche, Henry Thomas
[watermark 1829]
[see **2249**]

De la Beche, Henry Thomas
[watermarks 1830-31]
[see **2250-2253**]

De la Beche, Henry Thomas
[no date, watermarked ?1833]
[see **2254**]

430 [**De la Beche**, Henry Thomas]
[No date, watermarked 1841]
Specimens of the more celebrated British Building Materials with reference to Report [plus notes concerning Hartlepool New Colliery winding engine].

431.1-3 [**De la Beche**, Henry Thomas]
[Lithograph of trilobites]
Fig. 1. Trilobite from Builth, Brecknockshire
2. Undescribed fossil from the same place of the natural size
3. The same magnified.
In the Collection of H.T. De la Beche Esquire
C. Hullmandel's Lithogy.
[See Thackray 1985, p.186]

De la Beche, Henry Thomas
[no date, watermarked 1820]
[see **2247**]

432 **De la Beche**, Henry Thomas
[No date, c. 1829]
[Lithograph:] *A Coprolitic Vision*.
[Cited: McCartney 1977, pp.48-49]

De la Beche, Henry Thomas
[no date, watermarked 1849]
[see **2260**]

433 [**De la Beche**, Henry Thomas]
[No date]
[Sketch book containing a few light pencil sketches and watercolours of a sailing boat; Hooken Undercliff, Beer Head, Devon, Beer Head and Whitecliff from Downlands Cliff, Devon; and some coloured sections of Dorset coast. Mainly blank pages].

434 [**De la Beche**, Henry Thomas]
[No date]
Part of the interior, Fowey Consols Copper Mine.
[Sketch for Plate 6, *Report on the Geology of Cornwall, Devon & West Somerset*, 1839]

435.1-2 [**De la Beche**, Henry Thomas]
[No date]
[Sketches for *The Geological Observer*, 1851
Fig 71 p.196 A coral atoll (**435.1**)
Fig 83 p.232 Coral reef (**435.2**)

436 [**De la Beche**, Henry Thomas]
[No date]
[Pencil sketch] "On...Snowdon Peak".

437 [**De la Beche**, Henry Thomas]
[No date]
[Coloured geological sketch map of Gulf of Spezia].

438 [**De la Beche**, Henry Thomas]
[No date]
[Coloured sketch plans and notes for a horse engine].

439 [**De la Beche**, Henry Thomas]
[No date]
[Leaf rubbings and pencil sketch of man's head].

440.1-2 [**De la Beche**, Henry Thomas]
[No date]
[Sketch of ?Jamaican landscape (**440.1**) and sketch of ?Jamaican ?plant (**440.2**)].

De la Beche, Henry Thomas
[no date]
[see **2261-2265**]

441 **De la Beche**, Rosalie Torre   To: H.T. De la Beche
10 November 1849
I envy the places you visit. I went home last Tuesday week as it was the monthly holiday. I have sent the purse to Miss Dillwyn.

442 **Delesse**, Achille-Ernest-Oscar-Joseph   To: H.T. De la Beche
Service des Mines,
Paris.
10 April 1851
Following the advice of my friend M. De Verneuil who has kindly undertaken to deliver this letter to you, I take the liberty of turning to you to ask if it would be possible to have twenty samples of igneous rocks

which play the most important part in the geology of England. I have been working on igneous rocks for many years and published my results in the Annales des Mines, and the Bulletin de la Société Géologique.

443 De Lor, Philippe Francois To: [H.T. De la Beche]
Registrar,
Canton of Geneva.
4 September 1827
[Extract from register of births pertaining to entry concerning Elizabeth De la Beche, 2 December 1819].

444 Denison, William Thomas To: H.T. De la Beche
Hobart Town,
Van Diemen's Land.
1 March 1849
I have forwarded the coal specimens to the Colonial office, with Dr Milligan's Report on four of the coalfields. I ask your assistance in furthering our aims here. Coal will be our main source of wealth. Can you furnish us with specimens illustrative of various formations. We are woefully deficient in men of science. I am trying to instil a better system by using my position of Governor of the convict settlement.

445 Dickinson, J.D. To: H.T. De la Beche
India House.
8 June 1846
Thank you for the information on the progress of Williams' work in India. It would be a blessing were she made independent as regards coal. We look forward to his Official reports from India as regards the successful working of coal.

446 Dickinson, J.D. To: H.T. De la Beche
East India House.
2 March 1850
The death of Williams renders it necessary for us to find someone to continue his work in India. You kindly secured his services for us and we now ask you to provide us with his successor. I enclose a memo of the terms upon which Williams was employed.
[See East India Company to De la Beche, H.T., [1850], **479**]

447 Dieterle To: [H.T. De la Beche]
[No date]
I come to take a letter from M. De la Beche addressed to M. Ebelman, Director of the Sèvres porcelain factory to ask him for some samples of porcelain colours. This letter must have been ready 15 days ago, but a journey prevented me from coming to get it then.

448 Dillwyn, Elizabeth (Bessie) To: H.T. De la Beche
18 April [c.1844]
I am disappointed you will not come here at Easter, but glad you and Rosy will go to the sea for the holiday. The children have not been quite well, and I have had trouble with my nurse here.

449 Dillwyn, Elizabeth (Bessie) To: H.T. De la Beche
Parkwern.
17 November [1846 FJN]
I enclose a fern specimen for Ramsay. Moggridge has promised another next Spring. I sent some shells to Rose. I hope you have received our "Petition". All well here. Sketty folks as usual. Weather wretched.

450 Dillwyn, Elizabeth (Bessie) To: H.T. De la Beche
Parkwern.
15 December 1846
We request that you pay us a visit, as a year has gone by since your last one. The house has been enlarged.
[Letter in form of "Formal Petition"].

451 Dillwyn, Elizabeth (Bessie) To: H.T. De la Beche
[April or May 1849 FJN]
I have despatched my final agreement to take Miss Vickerman as governess here. When will you be able to decide about your coming here? We had an accident with the Brougham last Friday. The children all well, as also Lewis.

452 Dillwyn, Lewis Llewelyn To: H.T. De la Beche
Parkwern.
8 September 1846
I enclosed Roberts' account for copper sales data. He will be glad of the money. He is prepared to give details of mine localities.

453 Dillwyn, Lewis Llewelyn To: D. Nicol
Swansea.
4 July 1848
My apologies for the delay answering yours of 28th. Write to De la Beche upon the subject, using my name as a reference.

454 Dillwyn, Lewis Llewelyn To: H.T. De la Beche
Parkwern.
14 March 1850
Marryat mentioned that Page is desirous of parting with Bandiner's collection of China, and asked me to inform you. When shall we see you and Rosie down here?

455 Dillwyn, Lewis Llewelyn To: H.T. De la Beche
Parkwern.
21 March 1850
Bessie is away for couple of days. We are glad you are coming to visit us. You did not say what information you wanted as regards Buckland's gas shares.

456 Dillwyn, Lewis Weston To: H.T. De la Beche
Sketty Hall.
23 August 1840
All copies of my Bagatelle are sold. Lewis has a strong circle of friends at Tenby. Gutch, Smith & Eden are coming home by land after wrecking their vessel. Logan has embarked suddenly for Quebec.

457 Dillwyn, Lewis Weston To: H.T. De la Beche
Sketty Hall.
19 October 1840
Thanks for the Buckland and Stokes pamphlets. I presume my Conferva nivea is alluded to. When I was in Parliament I received great kindness from Sir James.

458 Dillwyn, Lewis Weston To: H.T. De la Beche
Sketty Hall.
12 November 1844
My attention was given to carnivorous univalves and their bivalve hosts. I never found either in any other bed of the Greensand or in any other secondary strata. The upper Oolite beds at Farleigh provided an exceptional case. Persuade Forbes instead of Phillips to come and look over my collection.

459 Dillwyn, Lewis Weston To: H.T. De la Beche
Sketty Hall.
16 February 1846
Illingworth's translation of the Charter is erroneous. My recovery from my fall is slow. A vast congregation of rats is undermining our prosperity.

460 Dillwyn, Lewis Weston To: H.T. De la Beche
Sketty Hall.
[1846]
Byllywasta not Ballywasta is the correct spelling in De Breos's Charter. Lewis has started for London on some Dock fancy.

461 Dillwyn, Lewis Weston To: H.T. De la Beche
16 May [1847 FJN]
Vivian and John wish to add their names to the Palaeo Subscription. We have problems in the pottery business as a result of the loss of Spanish Trade.

462 Dillwyn, Lewis Weston To: H.T. De la Beche
Sketty Hall.
14 December 1847
We are glad you are coming to Parkwern. The Flora Antarctica is now published, and will be left at the Museum. Arrangements for the Wisdom Meeting advance well. All well at Parkwern.

463 Dillwyn, Lewis Weston To: H.T. De la Beche
Tenby.
11 June 1848
I am at Tenby for health reasons. In the title page of my Bagatelle should I include Vice President of the British Association? I could not procure the hammer headed shark for our Institution. It was purchased by a man and taken by steamer to Bristol.

464 Dillwyn, Lewis Weston   To: H.T. De la Beche
       Sketty Hall.
       8 November 1848

Thanks for the trouble you have taken about my watch. In preparing for the Wisdom Meeting at Birmingham, I hope you did not forget the Continental practice of furnishing accounts of the local natural history with reference to whole Kingdom. The list of rare plants I circulated long ago needs revision. I wish you would get Henfrey to do it.

465 Dillwyn, Mary (Minnie) De la Beche   To: H.T. De la Beche
       Parkwern.
       12 May 1849

We were all sorry you and Rosie were not here at Easter. Mamma has a parrakeet. Papa, Mamma, Harry and Amy send their love.

466 Dixon, Mrs. A.M.   To: H.T. De la Beche
       [1849]

I introduce my son Francis Dixon. West India prospects reduce our circumstances. Please grant employment as Secretary, or in any capacity.

467 Dixon, Francis Graham   To: H.T. De la Beche
       W. Sparling's Esq,
       19 Old Square,
       Lincoln's Inn.
       15 October 1849

Being in a neccessitous position, I request employment of any kind on the New Commission of Sewers.

468 Dixon, Francis Graham   To: H.T. De la Beche
       W. Sparling's Esq,
       19 Old Square,
       Lincoln's Inn.
       15 October 1849

A few words in explanation of my letter of the same date. My mother is spending her last twenty or thirty pounds. My sister will share her fate. Please push my application for an appointment, I am cousin to Colonel Matthew Dixon of the Royal Engineers.

469 Donaldson, Thomas L.   To: H.T. De la Beche
       Bolton Gardens,
       Russell Square.
       22 January 1852

Thank you for compliance with my request and promise for the future. Please accept the accompanying volume as mark of my gratitude and respect.

470 Donkin, Bryan   To: H.T. De la Beche
       Brighton.
       19 March 1832

I have been studying your "Geological Manual" and have concluded there is need for a Geological Dictionary. The beginner should be able to find out the meanings of "Oolite" etc. It ought to be done by a profound Geologist.

471 Duffield, R. Dawson   To: H.T. De la Beche
       Lamarsh Rectory,
       Bures,
       Colchester.
       27 April 1847

I have lately acquired some interesting fossils of elephant teeth and tusks, and am anxious to sell them. Would you pay carriage to Town?

472 Duffield, R. Dawson   To: H.T. De la Beche
       Lamarsh Rectory,
       Bures,
       Colchester.
       3 May 1847

Thank you for your letter. I will send the fossils at the end of this week.

473 Dufrenoy, Ours Pierre Armand Petit-   To: [H.T. De la Beche]
       Paris.
       25 January 1843

I am profoundly touched by the distinction which the Geological Society of London has just accorded Elie de Beaumont and myself, and I thank you for your kind remarks about the Geological Map of France which we have completed. But allow me to return some of the honour. Your fine work on Normandy was the foundation stone of the study of geology in France. It has been a valuable guide to us. Your description of the coast of Calvados appeared to me to be the mark of guidance and comparison.

474 Dufrenoy, Ours Pierre Armand Petit-   To: H.T. De la Beche
       Ecole Nationale des Mines,   London.
       Paris.
       9 August 1848

Thank you very much for receiving and undertaking the distribution of the Annales des Mines which I have sent to London. I have asked the administration for them and am sending them through the kindness of M.Busry.

475 Dufrenoy, Ours Pierre Armand Petit-   To: [H.T. De la Beche]
       Paris.
       25 August 1850

You are doubtlessly already aware that the Section of Mineralogy and Geology has unanimously elected you to the position of corresponding member of the Academy. The Academy has confirmed our decision. We hope that one of these days you will come to one of our meetings and will visit our collections.

476 Dufrenoy, Ours Pierre Armand Petit-   To: [H.T. De la Beche]
       Paris.
       1 June 1851

Thank you for the details of news which you have given me on the works of the Jury of our section. My duties at the Ecole des Mines keep me in Paris, and only after they are carried out can I leave.

477 Dundas, James Whitley Deans   To: H.T. De la Beche
       A[dmiralty] O[ffice]
       2 July 1849

My friend Mr Rigby is anxious for your report on his coal. Can you give it to me tomorrow?
[See Rigby, W. to Dundas, J.W.D., 2 July 1849, **1820**]

478 Dundonald, Lord   To: H.T. De la Beche
       2 Belgrave Road.
       3 June 1852

According to your permission, I leave specimens of bitumen for experimental trials.

479 [East India Company]
       [1850]

Memorandum of the terms upon which the late Mr. Williams was engaged by the Court of Directors to proceed to India as geologist to survey and examine the coal-fields and mines there. To serve for 5 years. £200 allowed for outward passage. £800 per annum salary from date of arrival. Travelling charges to be paid by the Government of India. On the termination of service £200 for passage home.
[See Dickinson, J.D. to De la Beche, H.T., 2 March 1850, **446**].

480 Eastlake, Charles Lock   To: H.T. De la Beche
       7 Fitzroy Square.
       25 December 1851

Belated thanks for the copies of the lectures. I value these contributions to science.

481 Eastlake, Charles Lock   To: H.T. De la Beche
       7 Fitzroy Square.
       11 December 1852

Thank you for sending me a copy of Playfair's lecture on Industrial Instruction on the Continent.

482 Ebelmen, J.   To: H.T. De la Beche
       Office for the Executive Committee,
       Great Exhibition 1851,
       Exhibition Building,
       Kensington Road,
       London.
       [c.1850]

Would you obtain for the Ceramic Museum the following specimens of Indian pottery from the following factories, Almedabad, Lahore, Bhangulpore, Azimgurlh etc. Finally it would be of great interest for the Museum to obtain a piece of Indian enamel on chambered silver.

483 Ebington, Viscount	To: H.T. De la Beche
Southmolton.	
23 December 1849	

I ought to have forwarded you Chadwick's note earlier, but preparation for the Protectionist meeting at Exeter intervened. It was a confused meeting. Acland was the only one heard. I hope you can muster a quorum for the 28th. If not, I will come up.
[See Chadwick, E. to Ebington, Viscount, 18 December 1849, **242**].

484 Edwards, Henri Milne	To: [H.T. De la Beche]
Jardin des Plantes,	
Paris.	
11 February 1848	

In reply to your letter asking for details for the death of the Marquis de Dree, I have to say that this mineralogist is still alive. Being ruined, he has had to give up his collections, but he appears to be in good health. Please remember me to Phillips and Forbes.

485 Edwards, Henri Milne	To: H.T. De la Beche
London.	
10 May 1849	

I have received the series of fossils, 56 specimens representing 33 species or varieties, and I hope to be able to use the greater part of them for the Memoires de la Societe Palaeontologique.

486 Edwards, Henri Milne	To: [H.T. De la Beche]
Académie deple.de la Seine,	
Paris.	
4 February 1853	

My monograph on the fossil polyps of England is being prepared and the Council of the Palaeontographical Society have decided that only 8 plates will be published. I would like to obtain from the Council the simultaneous publication of 10 different plates in the chapter dealing with the Devonian epoch, and I ask you to back me in this request.

487 Egerton, Philip de Malpas Grey	To: H.T. De la Beche
Oulton Park,	
Tarporley.	
21 March 1848	

I am ready with a paper on Pterichthys and an amusing letter on the subject from Hugh Miller.

488 Eichthall, Charles d'	To: H.T. De la Beche
Ostende.	
5 July 1851	

I have read your three reports on coal for the steam navy, and realise I should have sent you greater quantities of coals for experiment and analysis. I have 6 to 10 cwt. available at Catherine Docks, and enclose a letter to Heath & Co. for their delivery to you. This great source of national prosperity is not yet known in the south of Bavaria.

489 Elie de Beaumont, Jean Baptiste	To: H.T. De la Beche
Armand Louis Léonce	
Paris.	
2 December 1828	

Thank you for the copies of the Tabular View. I have already distributed 10 copies to people interested in geology. Many others have asked where they might obtain copies. You have done a great service to science in placing so many facts on one sheet of paper. My thanks too for submitting my name to the Geological Society of London. I hope your journey goes well.
[Cited: McCartney 1977, p.54]

490 Elie de Beaumont, Jean Baptiste	To: H.T. De la Beche
Amand Louis Léonce	
Paris.	
28 May 1829	

I have received your letter of 12 May from Genoa, and look forward to your arrival in Paris. I see that you no longer believe that the characteristics of the alpine strata are those of antiquity. Your memoirs on La Spezia and Nice will be an important step forward for geology. There are some places you should visit as you cross France. I hope you received the letter I sent to you at Corfu.

491 Elie de Beaumont, Jean Baptiste	To: H.T. De la Beche
Armand Louis Léonce	
Rouen.	
12 November 1829	

I have not forgotten the interest you have shown in my travels since leaving you. Von Buch and I have visited many of the places you also have seen. I have also found many new and important facts to add to my researches on the revolutions of the globe, and have documented four different periods in the formation of the Alps. You will not find Von Buch in Paris. He has just written to me from Berlin, but will be back in May.

492 Elie de Beaumont, Jean Baptiste	To: H.T. De la Beche
Armand Louis Léonce	58 Jermyn Street,
Paris.	London.
10 May 1830	

For some months I have been at work engraving the geological map of France. The scale is about three quarters of that of Greenough's map. My memoir is completed. I would have sent it to you if you were not a subscriber to the Annals of Science in which it appears. All your observations on the nature of the stratifaction of deltas appears to me very justified, especially if one understands your limited meaning about the defective horizontality of the beds.

493 Elie de Beaumont, Jean Baptiste	To: H.T. De la Beche
Armand Louis Léonce	Athenaeum Club,
Paris.	London.
21 April 1831	

I will send to you by post the notes you requested. I will copy them on fine paper, and will show the way in which I see the phenomena of the successive upliftings resulting from the gradual cooling of the globe. I am grateful for the place you have given me in your Manual. We have had the pleasure of inscribing you among the members of the Geological Society of France.
[See Elie de Beaumont, J.B. to De la Beche, H.T., 17 May 1831, **494**].
[Cited: McCartney 1977, p.54]

494 Elie de Beaumont, Jean Baptiste	To: H.T. De la Beche
Armand Louis Léonce	
17 May 1831	

[Ms] Researches on some of the revolutions of the surface of the globe, showing different examples of coincidence between the raising of layers in some mountain systems and the sudden changes which have produced lines of demarcation observed between certain consecutive stages of sedimentary terrain.
[See Elie de Beaumont, J.B. to De la Beche, H.T., 21 April 1831, **493**].

495 Elie de Beaumont, Jean Baptiste	To: H.T. De la Beche
Armand Louis Léonce	
Paris.	
4 January 1832	

I returned to Paris several days ago and found there the copy of your Geological Manual which you have had the goodness to send for me, as well as the October number of the Annals of Philosophy. I have read in the first the summary of my ideas which you have inserted there and in the second the complete translation of the work which I had addressed to you. My ideas are so well reproduced that it is evident you must have taken the trouble yourself to translate them. Your manual should be translated into French.

496 Elie de Beaumont, Jean Baptiste	To: H.T. De la Beche
Armand Louis Léonce	Athenaeum Club,
Paris.	London.
5 February 1832	

I have noticed with great pleasure the rapid success of your Geological Manual. The translation will commence as soon as the second edition arrives in Paris. Levrault will publish it. It will give me great pleasure to contribute to the publication of such a useful work, but I can see there is scarcely anything essential I could add to it. Thank you for the 25 copies of the extract of mine. Please give copies to your learned friends Jameson, Sedgwick, Buckland, Daubeny, Greenough, Lyell, Murchison, Conybeare etc. with the author's compliments.

497 Elie de Beaumont, Jean Baptiste	To: H.T. De la Beche
Armand Louis Léonce	Athenaeum Club.
Paris.	London.
8 April 1832	

Thank you for the second edition of your manual. I take exception to

p.267 on formation of the cretaceous rocks. I have not held such an absurd idea. I think that blocks and other large fragments must have been transported by the waters following each "revolution", but I have never gone further than that. Brochant de Villiers will continue the translation and I will supervise. I have read the first volume of Lyell with much pleasure, in spite of the large number of physical errors which occur in numerous places. Thank you for your map of Devonshire. It is greatly to be hoped you will continue this fine enterprise on all the sheets of this map which will be published in the future.
[Cited: McCartney 1977, pp.54, 55; Rudwick 1985, p.91]

498   Elie de Beaumont, Jean Baptiste   To: H.T. De la Beche
Armand Louis Léonce   Totnes.
Paris.
15 June 1832

Cholera in Paris has delayed the translation of your Manual. Brochant was seriously ill for a month. Work has now started again. I will continue to do everything in my power to ensure the execution of this work, so important for our students. I am pleased to learn the English government are favouring your undertaking of a geological map of Devonshire and the adjacent counties. It will be of the very highest importance to science. It seems to me eminently useful if all geologists were to agree amongst themselves to make maps on an uniform projection so that they could be superimposed.

499   Elie de Beaumont, Jean Baptiste   To: H.T. De la Beche
Armand Louis Léonce   Athenaeum Club,
Paris.   London.
4 January 1834

Thank you for the third edition of your Manual. I will, to the best of my ability, make my audience take advantage of all the excellent things contained in it. It will really be epoch-making in the science by making all the positive data of geology more accessible to the public. Brochant's translation is finished and had just been put on sale the very day before your third edition arrived. I am extremely grateful to you for so kindly sending me a copy of your Devonshire map.

500   Elie de Beaumont, Jean Baptiste   To: H.T. De la Beche
Armand Louis Léonce
Paris.
20 December 1834

A few days ago I received a copy of your Researches in theoretical geology which you were so good as to send me. It forms a worthy sequel to your Manual. Never have I read anything more capable of linking geologists' observations to sound ideas of physics, chemistry and astronomy. I have immediately arranged for it to be translated into French by Boulanger. Brochant and I will revise the translation where necessary and it will be printed by the same bookseller who printed the translation of the Manual.
[Cited: Rudwick 1985, p.114]

501   Elie de Beaumont, Jean Baptiste   To: H.T. De la Beche
Armand Louis Léonce   London.
Paris.
22 April 1835

Boulanger, the translator of your Researches in theoretical geology, has almost arrived at the end of his task without any serious difficulty. Brochant has revised only the first few champters, and is taking the same pains with it as he took with your Manual. I have read with great pleasure the precis which you sent me on the craters of elevation. I found it very clear. Von Buch is coming to Paris soon. We are engaged on the translation into French of his work on the Canary Isles, which will be enriched by a large number of new facts from observations made in different countries during several years.

502   Elie de Beaumont, Jean Baptiste   To: H.T. De la Beche
Armand Louis Léonce   Ordnance Map Office,
Paris.   London.
15 February 1836

The French translation of your Researches in theoretical geology is still not yet completed, and I have too much on my hands to speed it up. The translation of the Manual has had so much success that although 1200 copies were printed, it will soon be out of print. The publisher is thinking of reprinting it based on your 3rd edition. Do you plan a 4th? Thank you for your charming work How to observe, geology. It will be a handbook for all travellers. The same publisher wants to translate it.

503   Elie de Beaumont, Jean Baptiste   To: H.T. De la Beche
Armand Louis Léonce   London.
Paris.
25 November 1836

I hope at last to be able to arrange a complete publication of the last editions of the three principal works with which you enriched the teaching of geology. I have been very vexed at the way the matter has dragged on. De Collegno would with pleasure take over the work. I have spent the summer travelling in Switzerland and Austria. I would need a book to tell you all they contain. Our geological map is still in the hands of the engravers who creep like tortoises, and I think they shall be a year on it.

504   Elie de Beaumont, Jean Baptiste   To: H.T. De la Beche
Armand Louis Léonce
Paris.
15 February 1842

Dufrenoy and myself send to you a copy of the work we are publishing on the geological structure of France. There is a map on a reduced scale at the end. We have had to leave certain gaps in our work, as it has been impossible, for example to indicate with certainty to what stage of the Jurassic certain parts of the Alpine Jura belong. In all cases of this kind we have used one colour. We are inclined to adopt colours as you have chosen for your Devon work on the Carbonaceous system. Von Dechen's map of Europe incorporates our work, but in a very peculiar way. I am glad you are now preparing a map of South Wales.

505   Elie de Beaumont, Jean Baptiste   To: H.T. De la Beche
Armand Louis Léonce   Ordnance Map Office,
Paris.   London.
24 January 1843

Thank you for your letter as Secretary to the Geological Society announcing that Dufrenoy and myself have been honoured with the Wollaston Medal for our geological map of France and the memoir which accompanied it. Please convey to the Council our profound thanks. Our full names for the medal are Pierre Armand Dufrenoy and Léonce Elie de Beaumont.

506   Elie de Beaumont, Jean Baptiste   To: H.T. De la Beche
Armand Louis Léonce   Museum of Economic Geology,
26 March 1843   London.

Thank you for accepting in our names the Wollaston medals, and for finding the means of delivering them. Pentland will bring them to Paris, as well as the maps and sections, I hope. I learn with great pleasure that you are working on a publication of fossils in support of the geological map of the Ordnance. This will produce a work still more classical. Like you I believe that the Daguerrotype and similar processes will be called upon to render great service to this branch of the science.

507   Elie de Beaumont, Jean Baptiste   To: H.T. De la Beche
Armand Louis Léonce   Museum of Economic Geology,
Paris.   Craig's Court,
12 June 1847   London.

I beg permission to introduce to you M. Paul L. Luuyt, a pupil engineer who is travelling in England to perfect his instruction. He has passed two years at the Ecole Polytechnique and has proved himself an outstanding student. I trust you will give him every help. Thank you for your gifts of maps and other works to the Ecole.

508   Elie de Beaumont, Jean Baptiste   To: H.T. De la Beche
Armand Louis Léonce
Paris.
30 March 1850

Thank you for your letter of 28 February in which you inform me that you intend to put a summary of my ideas in your forthcoming work The geological observer. My most recent ideas are to be found in an article in Dictionnaire universel d'histoire naturelle T XII, p.167. Thank you for the assistance you gave to Charles Deville last year. He is still very active in publishing his observations in mineralogy and geology.

509   Elie de Beaumont, Jean Baptiste   To: H.T. De la Beche
Armand Louis Léonce
Paris.
8 May 1850

I thank you for the attention you have given to my article in the Dictionnaire of d'Orbigny. As you will have noticed it has been abridged as the original was found to be too long. Many of my ideas have been omitted and the article represents only a third of the original outline. I am working on my principal publication Notice sur les systems de montagnes.

510 **Elie de Beaumont**, Jean Baptiste To: [?]
Armand Louis Léonce
Paris.
[1851]

Dufrenoy has just forwarded to me the greater part of the documents relative to the Ecole des mines which you wanted. They indicate the five courses taught. He has forgotten to send the two maps you asked for, and I will send them later. My little book is not yet finished. Please thank De la Beche for me if you see him.

511 **Enys**, John Samuel To: H.T. De la Beche
Eastbourne. Ordnance Map Office,
29 May 1837 Tower,
London.

The engine is proceeding well though slowly. I have tried the effect of idea of pumping water up a cliff 5 times as high as Beachy Head, and eyes are opened wider than ever before known. It will do. The windmills have been visited and I find 30 square feet of canvas nearer a bushel of wheat ground on horse power than 25. The power varies of course exceedingly with the wind.

512 **Enys**, John Samuel To: H.T. De la Beche
Eastbourne. Ordnance Map Office,
25 June 1837 Tower,
London.

The above is the result of a second visit to Beachy Head, where there was no fog. The first time it was so thick I could only examine the beach joints and could not distinguish the strongest marked joints when horizontal. My first opinions are now quite changed. Davies Gilbert wished me to show it to Mantell which I did at the Chalk Pit during his geological excursion. At Shoreham bridge he gave us a lecture on granite.

513 **Enys**, John Samuel To: H.T. De la Beche
Enys. Post Office,
20 December 1837 Swansea.

The whole answer to your present letter may be short. I will see Jordan today, being about to visit Falmouth, and in way to Truro enquire for Arthur and send you an opinion respecting the Hammerers, and write to Richards the engineer about the model in the only mine with which I am aquainted. I interpret your silence to mean that the observations on engines were not adapted to your report. Robert Fox is very sanguine about his laminated clay effecting an important revolution in geological ideas and proving the direction of former electric currents.

514 **Enys**, John Samuel To: H.T. De la Beche
Enys. Post Office,
22 April 1838 Swansea.

As you will probably alter the wording of the contents of my last letter to you, I conceive I had better see the proofs, as the meaning if injured can often be restored by the slightest alteration. Enclosed are details of engines. So much the worse for Robert Fox. His rough statements respecting the Durham coal-beds are at variance with those referred to and caught my attention.

515 **Enys**, John Samuel To: H.T. De la Beche
Enys. Post Office,
23 May 1838 Swansea.

Not having heard of the decision respecting what part you intend to put in your appendix, I have sent you the corrected figures respecting the pitwork. Poor Henwood is in a great fright at Duchy prospects and with justice. I should be sorry to see him thrown on the world again, though as far as Cornwall is concerned perhaps his presence in another mining district, would be beneficial.

516 **Enys**, John Samuel To: H.T. De la Beche
Enys. Post Office,
24 July 1838 Swansea.

Here are some of R.W. Fox's latest results. The question is a simple one as to Newcastle. I must join you, and that either at Swansea or London. If at the former place I should like to be there a day or two before you start, to see the country and a Welsh coal mine, and the copper works. Then the mode of progression to the north may be arranged at leisure. I have no very definite object in view where all is new.

517 **Enys**, John Samuel To: H.T. De la Beche
October 1838 Post Office,
Swansea.

I have sent you the accounts for granite as I have received them. Paul Williams I find applied to the same contractor as I did and the terms are such as the latter would instantly supply any quantity that might be ordered. I saw Williams this morning and find there will be great difficulty in getting figures for the quantity of granite exported for the years other than those mentioned.

518 **Enys**, John Samuel To: H.T. De la Beche
[c.1848]

I do not suppose the Cornish engineering character for exaggeration will suffer much by 14lbs of water evaporated by 1lb of coal burned in Cornish boilers instead of the true reading of 14 cubic feet of water evaporated by 1 Bush. of coal. You would I think be puzzled to produce the where and when the statement was made, except at page 15 of the blue book you gave me yesterday. I am glad to say this is the only fault I can find.

519 **Enys**, John Samuel To: H.T. De la Beche
Enys.
20 November 1849

Certain gentlemen are desirous of supplying a large quantity of greenstone for the repair of the roads and streets of London in accordance with the recommendation in your geological report of Cornwall and Devon, p.483. If you have not changed your opinion since 1839 would you give them any assistance in your power in obtaining a fair trial of the particular greenstone? I hear they are ready to supply it on reasonable terms.

520 **Enys**, John Samuel To: H.T. De la Beche
Enys.
10 December 1849

Herewith you will receive a specimen of stone said to be superior in toughness and durability to that used in the trial at Penryn, which has been forwarded for your inspection by Mr. Paul Williams. It comes from Flushing and can be shipped with great facility.

521 **Enys**, John Samuel To: H.T. De la Beche
[No date] Falmouth.

I have drawn up a disclaimer of too much dependence on the table which if appended, or something similar has changed my impression of propriety of its insertion, a doubt of which was the sole cause of disinclination to finish it properly. Have you a table of logarithms, as the rule is to be worked by them, and I cannot put my hand on them even if they exist as I believe they do in the room over the kitchen.

522 **Enys**, John Samuel To: H.T. De la Beche
Enys. Falmouth.
[No date]

Last night I completed 3 sets of figures. The original is returned not for itself, but an account of your notes on it. My present steam notions are added. I forgot to mention the question of velocity of steam.
[Enclosure]: Steam notions applicable to geology.

523 **Enys**, John Samuel To: H.T. De la Beche
[No date] Falmouth.

The temperature given - 50° the mean surface temperature x 50ft. for each degree gives the depth in feet below the surface, and consequently the force of steam due to the temperature and the resistance of the column of water in lbs due to the depth is readily obtained. In reference to R.W. Fox's water circulation, I have always considered it impossible except through the medium of an interposed stratum of steam.

524 **Evans**, D. Th. To: H.T. De la Beche
4 Brick Court,
Temple.
26 April 1851

Herewith a copy of the Morning Chronicle containing the first of my Letter Reports on lead mining and smelting. I have suggested a "School of Mines" and would be encouraged by men of science and influence. I have also written articles for same paper on explosions in mines. Thank you for your note of thanks to my presentation of a specimen of brown coal to your Museum.

525 Faraday, Michael — To: H.T. De la Beche
Wimbledon.
28 September 1848
Can you lend me the fine separate crystals of Titanium you have at the Museum. I will not hurt them.
[Cited: James 1996, no.2107]

526 Faraday, Michael — To: H.T. De la Beche
Royal Institution.
29 June 1849
Lyell said he would move you to add our name to the list of those receiving 40 copies of a work (Forbes', I think). We need your help for our Library.

527 Faraday, Michael — To: H.T. De la Beche
Royal Institution.
29 June [1849]
Thanks. You are ever very kind to me and to us.

528 Fawcett, Joshua — To: [H.T. De la Beche]
Low Moor,
Bradford,
Yorkshire.
4 November 1842
Thank you for the Report on the Selection of Stone. It is more copious than I expected. It will assist me in describing the stone of Yorkshire churches. My work on them is already out of print.

529 Ferrier, J. — To: H.T. De la Beche
Ordnance Map Office.
1 December 1836
I send you a rough statement of the expenditure of the Geological Survey of Cornwall, from its commencement, to enable you to make up the abstract called for by Colby.

530 Fischer, J.C. — To: H.T. De la Beche
Schaffhausen,
Switzerland.
4 December 1851
I am disappointed to find not a syllable of my two articles in the Exhibition Catalogue. Nor my invention of Cast Steel produced from coke iron in one direct fusion. Please send my illustrations to Mr Kreeft, that he might keep them for me. Please keep the exemplar as a gift from me.

531 Fitton, William Henry — To: H.T. De la Beche
Clifton. — Lyme Regis.
11 November 1825
Your purpose of delivering your paper on Jamaica to the Geological Society at the next meeting, on 18th inst., was so acceptable to Lyell the Secretary, that I write again on the subject. Lyell will reckon on your paper and no other subject will be provided that evening. I hope you will deliver it personally and assist discussion afterwards. This practice has been introduced to solicit observations from the members. With illustrations, drawings, discussion and the tea-party afterwards, I really do not know any Society of which the meetings, hitherto, have been more popular and agreeable.

532 Fitton, William Henry — To: [H.T. De la Beche]
53 Upper Harley Street.
22 August 1828
Though I have not had an answer from Mr. Conybeare, to whom I wrote about his proposed name derived from the "Silva-Anderida", I will tell you what has occurred to me, in answer to your enquiry respecting a name for the beds between the green-sand and the Portland. For on reflection nothing better has offered itself than a choice among the names already in use. I wish much to look over your map of Normandy with yourself, as I mean to insert a reduction of it, in my own general map.

533 Fitton, William Henry — To: H.T. De la Beche
Brasted, — Devon.
Sevenoaks,
Kent.
15 July 1830
Having returned from London this evening and brought down with me your volume of Notes, I cannot deny myself the pleasure of congratulating you upon its completion, as a contribution of the greatest interest and importance to geology. The success with which you have given uniformity and cooperation to originals, so multifarious and so different in scale and in relation of heights to horizontal spaces is admirable. Your drawing lives and your mangrove thickets, and the movement of the Niagara waters - are excellent. The work can be appealed to by writers of books on geology whether elementary or otherwise. I greatly envy your power of lithographying so freely and with such clearness.
[Cited: North 1936a, pp.76-77]

534 Fitton, William Henry — To: H.T. De la Beche
Highwood Hill.
20 December 1831
As you have asked my opinion on a point in the history of your "Cretaceous group" and "Wealden rocks" I shall give you in writing what has occurred to me on the subject. If Mantell had not been misled by erroneous identifications it is impossible that a man of his knowledge and sagacity could have come so near the truth and missed it. My inferences have all been confirmed by more extended observations. The credit then which is due to Mantell is that of having in the first instance collected and investigated the fossil remains of his vicinity with a skill and zeal above all praise, and of having gone as far, in his geology of the strata, as the then existing state of the subject would allow.

535 Fitton, William Henry — To: H.T. De la Beche
Athenaeum.
14 June 1842
You will be glad to hear that, after our conversation, a meeting here today (of Buckland, Murchison, P.Duncan, Lonsdale's Secretary and myself) it has been resolved to raise a subscription for Lonsdale of not less than £5 from each person. It is also proposed to address a short memorial to Sir. R. Peel requesting to grant a pension to Lonsdale.

536 Fitton, William Henry — To: H.T. De la Beche
Athenaeum.
21 June 1842
Thank you for your donation of £10 for the Lonsdale subscription. We will reach £500 or at least £400. As for pension I much doubt. It is certain however that Buckland is our channel of communication with Sir Robert Peel, and I was glad to read what you say on this point as it has caused me to write back to Buckland to know the best mode of presentation.

537 Fitton, William Henry — To: H.T. De la Beche
Putney.
18 July 1842
The subscription for Lonsdale has done well. We have £548 and want only £42 to make up £600, and of that we are nearly certain. Greenough, for example, has not yet answered. This amount I hope will be useful to Lonsdale, and will relieve us of the discredit of having for so many years employed such a man at a salary so very paltry. Darwin seems convinced that Buckland is right about glaciers in Wales. We have as yet no successor to Lonsdale.

538 Fitton, William Henry — To: H.T. De la Beche
Athenaeum.
17 March [1846]
I hope to see you tomorrow about 2 o'clock at Craigs Court. But if I should not be able to appear at that time, let me ask whether you can tell me anything as to the probability of "Reports" from Phillips & Mr. Blackwell? The latter has already sent one in on the late explosion in a colliery at Merthyr which, at least, I presume will be published soon.

539 Fitton, William Henry — To: H.T. De la Beche
Geological Society.
21 December 1847
Coming here to settle my account, I have enquired of Mr. Nicol respecting Mr. Saxby's Nautilus, of which the description was read by Mr. Morris at the last meeting. Nicol says he is not going to insert the figure in the next number which I should regret for several reasons. I should hope the Council will now order Mr. Saxby's paper to be printed forthwith.

540 Fitzwilliam, Charles William Wentworth,
3rd Earl — To: [H.T. De la Beche]
Grosvenor Place.
30 May 1848
You suggested a trial of some of my coals. Send for them, where they may be wanted.

541 Fleming, E.B.   To: [H.T. De la Beche]
at General Irving's,
Balmae,
Kirkcudbright.
8 January 1852

Lord Selkirk has handed me your note to him regarding my application to join the Geological Survey. I now make formal application for a position. I refer to Prof. Sedgwick, Robert Harkness of Dumfries and Dr Landsborough of Saltcoats as referees.
[See also Sedgwick, A. to De la Beche, H.T., 14 March 1852, **1896** and Selkirk, Earl of, to De la Beche, H.T., 31 December 1851, **1901**].

542 Forbes, Edward   To: H.T. De la Beche
5 November 1844

I commenced my duties on Friday and conferred with Baily. Where are the Blackdown fossils? Tomorrow I will lay my resignation before the Council Meeting of the Geological Society. Capt. Bethune left for Borneo on Sunday. A geologist will follow. You can probably furnish a man, for good coal in Borneo would make a new move in Eastern affairs.

543 Forbes, Edward   To: H.T. De la Beche
11 November 1844

Baily and I will proceed to work on the Warminster fossils. The absence of organic remains in the Evenlode shale is most important as giving a clue to the physical conditions of deposition. The President of the Geological Society renewed his proposition to me as consulting officer. I objected and stated my reasons. There is a strong feeling in the Society to give up their collections. They would come to the Survey. I have not heard from Phillips.

544 Forbes, Edward   To: H.T. De la Beche
17 December [1844 FJN]

Phillips' letter is very provoking. He should have consulted me. The organic remains have not been treated as they ought. His descriptions are notoriously defective. You should write the preface. I like your title of organic remains etc", and the omission of word "Ordnance". I wish to go to Edinburgh to collect my baggage. Playfair is anxious to join the staff.

545 Forbes, Edward   To: H.T. De la Beche
Geological Society.
[1844 FJN]

Received yours of 27th. Re Royal Institution. I shall give a conditional answer to the formal invitation. I plan a short, popular course there in preparation for a stricter course on the Survey. I have been with Baily. I have heard nothing from Phillips. I regret he is vexed. He should be more decided in his motions.

546 Forbes, Edward   To: H.T. De la Beche
Geological Society.
Tuesday [1844 FJN]

Phillips informs me that he refused to meet the Colonel on official grounds. The Colonel called on me today. Five minutes conversation showed you had defeated him. He asked me what to do. I replied he should leave it all to you and Phillips. You mistake me re Phillips. I wish to cooperate with him and value his advice greatly. Please give me directions on my official entry on duties.

547 Forbes, Edward   To: H.T. De la Beche
Geological Society.
Wednesday [1844 FJN]

I arrived here on Sunday. I met the Colonel on Tuesday morning. He requested a document from me on what assistance, draughtsmen etc. would be required. I refused. He asked if Phillips could be reached at York, so I put the Professor on his guard. I hope the matter is settled soon, or I'll fall back on my old fancy of Scotch University Chair. Is there no order for junction of Ordnance collections with Museum of Economic Geology?

548 Forbes, Edward
[1844 FJN]

The limestone from Egypt contains fossils in a very imperfect state. They appear identical with fossils in the Tertiary limestones of Malta and Greece. I should say they came from Tertiary beds of the Miocene epoch.

549 Forbes, Edward   To: H.T. De la Beche
Shetland.
1 August [1845 FJN]

We have entered Lerwick harbour after a fortnight's cruising during which we examined the great Ling Bank. I have dredged the tideways to find the effects of currents. 2 species thought extinct have been found living. We nearly capsized in a squall. We leave in 2 days.

550 Forbes, Edward   To: H.T. De la Beche
Wexford.
22 September [1845 FJN]

The best word for young zoophytes is "fry". I will attend to the despatch for Bowerbank about the microscope. I worked over the Hook fossils carefully. James showed me his Pleistocene fossils. What is the probable time of meeting you? I had a note from Phillips.

551 Forbes, Edward   To: H.T. De la Beche
Isle of Man.
26 December [1845 FJN]

I am glad to hear of the Trilobite purchase. As to Purdue, he does his best to make what he can of his collections. The 310 for £65 are worth the money. I should advise to offer £50 for all. We would save money by it in the end.

552 Forbes, Edward   To: H.T. De la Beche
[July 1846 FJN]

Your opinion of Oldham is as I expected. We are examining James' Silurian fossils. The volume is out. There are difficulties over providing for those requesting copies. I am going on with plans for plates of types of British fossil genera. Here are my plans for August. Have you seen Dilke's article on the Museum. A naturalist, young MacGillivray, has been found for Stanley.
[Cited: Herries Davies 1983, p.140; Herries Davies 1995, p.32].

553 Forbes, Edward   To: Sir H. De la Beche
Craig's Court.
28 August [1846?]

I have returned with capital results of service for our next publication. I will leave town next week. I must delay while my tailor makes new working breaches.

554 Forbes, Edward   To: H.T. De la Beche
6 Craig's Court.
2 October 1846

The shells you sent are interesting as identifying the drift. I have been working over my specimens and collections, selecting subjects for the Museum illustrations. I hear they are at work at the foundations of the Museum. I propose to join Ramsay at Bala. Your accepting the Presidency of the Geological Society will give great pleasure to all.

555 [Forbes, Edward]   To: H.T. De la Beche
Bala.
1 November 1846

I am still here and much interested in the country and fossils. Salter has returned to town, with a commission to inspect Purdue's collection. How he and Sedgwick contrived to draw the conclusions such as they are, set forth in the Geological Society's paper, is unaccountable to me. The Rhiwlas limestone is decidedly upper Silurian.
[Letter incomplete].

556 Forbes, Edward   To: H.T. De la Beche
Dorchester.
[1846 FJN]

I send back the proof correct. The printers have mislaid the Austen diagram. I have explored the Tertiaries here with Ramsay. If you undertake the Isle of Wight, a supplement can be concocted on the Lower Tertiary and Cretaceous paleontology. De Koninck has written asking when the Cystidean paper will be out.

557 Forbes, Edward   To: H.T. De la Beche
51 St. Stephen's Green,
Dublin.
9 March [1847 FJN]

I am going over our Irish collections. The Courtown limestone fossil specimens are problematic. Flanagan has gone to collect what he can. I am sorry you let Kane carry off the invertebrate collection. A protest from our department would have been offered. Kane's object will be a great source of trouble to us.

| 558 | **Forbes**, Edward | To: H.T. De la Beche |

51 St. Stephen's Green.
31 March 1847

Herewith two travelling bills. I am now in Dublin working on the collections. McCoy has sadly bungled the geology of the district. Silurian fossils are scarce in Ireland. My visit to Lord Enniskillen gave me material for my Hook report. I saw some fine sections through the Carboniferous. I observed Pentremites in situ. Also encrinites at Cleenish. I wrote to Salter and received a very satisfactory reply. The country here is very sadly off.

| 559 | **Forbes**, Edward | To: H.T. De la Beche |

6 Craig's Court.
26 July [1847 FJN]

I intend to marry Miss Ashworth and teach her geology in the Snowdon district. I will give you further details at Swansea.

| 560 | **Forbes**, Edward | To: H.T. De la Beche |

Kington.
5 September [1847 FJN]

Thanks for what you have done about specimens from MacGillivray. I offered to take charge of them believing it would secure the cooperation of the Admiralty. I hope you approve. Ramsay is still disabled. Salter is on his way to London. I visited the Dillwyns at Swansea.

| 561 | **Forbes**, Edward | To: H.T. De la Beche |

Llangollen.   [Ireland].
17 September [1847 FJN]

Oldham will have told you about my wedding. We went to Bournemouth, then to South Kington, staying with Austen on the way. I have written Hunt directions about the books. In your name copies should be sent to Lord Northampton and Lady Hastings both of whom sent specimens for figuring.

| 562 | **Forbes**, Edward | To: H.T. De la Beche |

Llangollen.   [Ireland].
22 [September 1847]

Thanks for your congratulations. Ramsay is here to look to the fossil points with me. Marriage is no impediment to geological doings. There is no need for Salter in the field this autumn. He can do registering of fossils for the new Museum.

| 563 | **Forbes**, Edward | To: H.T. De la Beche |

Church Stretton.
24 September [1847 FJN]

Jukes and Gibbs have found a fossil explaining the nature of Ischadites and relationship of Spheronites with the pseudocrinites. I have it with me and send you a sketch. Show it to Phillips.
[sketch enclosed].

| 564 | **Forbes**, Edward | To: H.T. De la Beche |

51 Stephen's Green,
Dublin.
6 November [1847 FJN]

A fine collection of fossils has been made here this Summer. I must go to the Isle of Man on special family business. My wife remains behind. On Wednesday week I intend to impress on the Dublin Geological Society the importance of the Survey fossil work. How are things in the Royal and the Club?

| 565 | **Forbes**, Edward | To: H.T. De la Beche |

Dublin.
27 December 1847

I suggest that the Botanical and Zoological plates in the memoir be numbered consecutively, and new ones substituted for the Palaeontological Series. I am more convinced that the Kildare fossils are equivalents of the Bala.

| 566 | **Forbes**, Edward | To: H.T. De la Beche |

[1847 FJN]

I have been sifting through the Cystideans here. We are fortunate that this material has come to the Survey. I have received a letter from Davidson in Paris about Cystideans. Another from De Verneuil asks for the 1st vol. and maps for the Geological Society of France which you promised.

| 567 | **Forbes**, Edward | To: H.T. De la Beche |

51 St Stephen's Green.
22nd [1847 FJN]

Collections made during the summer are considerable and important. There are some new forms among them. Those Chair of Kildare things only appear to be upper Silurian. I will send the results later. I am preparing my Cystidean paper for the press.

| 568 | **Forbes**, Edward | To: H.T. De la Beche |

2 June 1848

The willow you sent is Salix fusca, and the common barnacle is Balanus vulgaris. The articles in the Athenaeum on the Royal Society are by people ignorant of the real state of the society. I hope Brown gives way on Grove's appointment. Lord Morpeth gave a dinner last Saturday. We have been comparing the Bala and Kildare fossils. At the Geological Society meeting Owen told Mantell he had a better Iguanodon jaw fossil.

| 569 | **Forbes**, Edward | To: H.T. De la Beche |

Efford.
[1848 FJN]

The proofs came this morning. Yesterday and today I have worked on Lady Hastings' tertiaries. The collection is wonderful. When the Isle of Wight is to be done, it will be invaluable.

| 570 | **Forbes**, Edward | To: H.T. De la Beche |

26th [1848 FJN]

At the meeting of the Royal Society yesterday Lyell put your motion about Grove. The matter was postponed till next Council. A year or two of Grove would purge it of impurities. Gray has burnt both original and revises of his Chiton paper. At the Club awkward circumstances occurred about new members. Sykes withdrew his name in offensive terms. There is a conspiracy against the whole Council list organised at the British Museum.

| 571 | **Forbes**, Edward | To: H.T. De la Beche |

1 January [1849 FJN]

Concerning the coral decade of Milne Edwards, it is as it was, since Lowry is working on the trilobite part before the corals. Sir Philip has stated that he is ready to work in the autumn on his fish decade. Tomorrow I go to Swanage. I hope Jukes is successful in his canvas.

| 572 | **Forbes**, Edward | To: H.T. De la Beche |

West Lulworth,
Wareham.
28 September 1849

Bristow is still at Upway. I will say nothing to him about the possibility of his leaving for good. I had a letter from Jeffreys about selling his collection for £500. I asked him to prefer us to the British Museum. The decrease of cholera is a good sign.
[Cited: Secord 1986b, p.250]

| 573 | **Forbes**, Edward | To: H.T. De la Beche |

West Lulworth,
Wareham.
12 October 1849

I will get particulars of Jeffreys collection. Cuming has sent a series of fossils of the Isle of Man. Dixon's collection must be watched. Yesterday I examined Damon's stores. It is very fine in specimens. His septaria slabs are ideal for the walls [of the Museum].

| 574 | **Forbes**, Edward | To: H.T. De la Beche |

West Lulworth,
Wareham,
Dorset.
22 October 1849

I have received a document from Jeffreys about his collection. I asked him to prepare a statement of the number of species etc. Bristow is completing Dorsetshire before he leaves. He is exceedingly pleased at the arrangements you proposed for him. The Purbeck fossils is a curious subject. I will make it the subject of our next volume.
[See Jeffreys, J.G. to [Forbes, E.], 19 October 1849, **800**].

| 575 | **Forbes**, Edward | To: H.T. De la Beche |

West Lulworth,
Wareham.
8 November 1849

Very satisfactory results from our Purbeck enquiries. All the palaeontological work published on them will have to be done again as the result of our work. The so called Lower Greensands are Hastings sands. Bristow concludes the same on purely geological grounds.
[Cited: Secord 1986b, p.238]

576 Forbes, Edward To: H.T. De la Beche
West Lulworth.
2 December [1849 FJN]
Reeks will have told you of my wife's illness. She is now recovering fast. The village tailor has died of fever. I have laid down my work in the form of a section. Some capital finds, among others Gyrogonites in the Purbecks, completing the analogy between them and the tertiaries.

577 Forbes, Edward To: H.T. De la Beche
Swanage.
26 December [1849 FJN]
Christmas greetings. I am over my illness. I will be in town by the New Year. Gapper is here for the next three days. I shall have material for two decades of Purbecks.

578 F[orbes], E[dward]
[1849]
Decade 1. Persons who have contributed.
Professor Sedgwick, Mr Sharpe, Lady Hastings, Mr Stokes, Mr Bowerbank, Mr Dixon, Mr Bunbury, Mr Tennant, Mr Morris, Mr J. de C. Sowerby, Mr McCoy, Mr Woodward, Mr Waterhouse.

579 Forbes, Edward To: H.T. De la Beche
Dublin.
19 September 1850
I am glad to hear of your safe arrival. Oldham and I have been over the Malahide section. I hope Jukes' matters will go straight. The two incomes are essential to him, from the College and Survey.

580 Forbes, Edward To: H.T. De la Beche
Sandgate,
Kent.
28 December [1850 FJN]
Good news for the Museum. Clark, here, has donated the whole of his collection of Gault fossils, containing numbers of unique and splendid things. The owner has a great belief in the excellence of our new museum.

581 Forbes, Edward To: H.T. De la Beche
Swanage.
[No date, watermarked 1850]
I requested Baily to send to Prestwich for the fossils I collected in the Vale of Wardour. I was most fortunate both in facts and fishes. A complete series of sands and clays from Alum Bay might be made up for the Museum. I'm sure you'll be pleased with what we have done here. The material for the Purbeck Memoir is nearly ready. Our movements depend on instructions from you.

582 Forbes, Edward To: H.T. De la Beche
Saluber House,
Sandown,
Isle of Wight.
10 November [1852 FJN]
I can report very satisfactory progress here. We set off on a preliminary inspection of the Whitecliff bay tertiaries. Respecting the fluvio-marine portion of them, none of the freshwater limestones there are equivalent to those on the shore between Cliffend and Alum Bay. The whole will go to confirm our new classification. I had a long letter from Jukes who seems profoundly unconscious of the trouble he gives.

583 Forbes, Edward To: H.T. De la Beche
Sandown.
13 November 1852
Yours of 10th, with Jukes' enclosed arrived today. I am glad to hear that Jukes has written more judiciously. We are bewildered at the number of unrecorded strata here. I am sorry to hear of Mantell's death. He was a man of great talent in spite of his absurdities.

584 Forbes, Edward To: H.T. De la Beche
Sandown.
14 November 1852
Regarding the Presidency of the Geological Society, Mantell's death removes the ostensible objection to Owen. Please do not propose me if Prestwich, Sharpe, Austen or Hamilton object. They have prior claims. Your proposal for a microscopic examination of coal should be offered to Hooker first. The freshwater limestones of Bembridge Point & St. Helens are higher in the series than those of Whitecliff.

585 Forbes, Edward To: H.T. De la Beche
Sandown.
19 November [1852]
A storm has exposed strata here. Yesterday we connected the Bembridge and Whitecliff limestones, and proved them the same as those at St. Helens and Brading. Collections are thriving famously. Respecting Hooker and coal, can I speak of renumeration? Hopkins wrote concerning the Society; I repeated my views. Dr Stanger wants to set up a Geological Survey of Port Natal.

586 Forbes, Edward To: H.T. De la Beche
Sandown.
24 November 1852
We measured up the St. Helens section yesterday. Everything continues to come out according to our anticipations. Beneath the Bembridge series there is remarkable set of sands and sandstones. We shall show that more than half of the fluvio-marine is above the horizon of the Headon Hill white sand, contrary to Lyell etc.

587 Forbes, Edward To: H.T. De la Beche
27 November 1852
On Thursday we went to Seaview. The walls, houses and groins to the east of Ryde are built of a local stone which we found in situ. Their place in the series was ascertained from their fossil content. Herewith sketches illustrating series near Brading. The Cowes limestone may prove to be part of the Headon Hill series.

588 Forbes, Edward To: H.T. De la Beche
Sandown.
2 December 1852
We have been at work every day. The ground is not yet firm enough for measuring. A great mistake has been made by all who have described the Fluvio-marine of this end of the Island. Bristow and Aveline are getting on fast with the mapping. Next week we go to Gurnet bay etc. Please lend me the current number of the French Society's Bulletin, for Herbert's paper on Whitecliff.

589 Forbes, Edward To: H.T. De la Beche
Sandown.
4 December [1852 FJN]
I am glad you talked with Hooker. Respecting the Geological Society Presidency, Owen's claims should be recognised before others. I cannot think of taking office in preference to him. The Kensington movement folks and the scientific societies will have a battle, and my position as P.G.S. would be anomalous. Gibbs and I found the Hamstead cliff beds with characteristic fossils in terraces of mud in the Whitecliff freshwaters.

590 Forbes, Edward To: H.T. De la Beche
Sandown.
12 December 1852
On Monday we went to Cowes. The limestone is Bembridge, with St. Helens sands below. No Headon beds visible. Tuesday, Gurnet bay to Newtown. Wednesday, remainder of the coast to Yarmouth. The true relations of all strata are proved by superposition throughout. Thursday, Collwall bay section visited.

591 Forbes, Edward To: H.T. De la Beche
Sandown.
1 January 1853
Sorry to hear of your Influenza. Bristow and I gave a communication to Ryde Scientific Society on the freshwater tertiaries, to put our priority on record. I asked Lyell to come here and bring a set of Belgian fossils. All were exactly as surmised. He admits he did not understand these strata, and also agrees to my argument on their foreign equivalents.

592 Forbes, Edward To: H.T. De la Beche
Sandown.
14 January 1853
I shall prepare the report you requested. I am writing the fluvio-marine memoir, also Decade 5, and preparing my lectures. If Easter is decent, I wish to go to France to run over the corresponding beds. I shall assist with the working-men's lectures. Respecting the Irish Survey, Medlicott should also lay down fossils.

593 Forbes, Edward To: H.T. De la Beche
Dublin.
10 November 1853
I am getting the fossil collections in order here. I will be finished by

Monday, then to London. I will bring my notes on the Treasury document. The officers of the Survey find themselves in a false position under the new arrangements. Haughten at Trinity gave a paper on the Carboniferous Limestones of the Menai Straits in which he did justice to the Survey.

594 Forbes, James David　　　　　　　　　　To: H.T. De la Beche
British Coffeehouse,　　　　　　　　　　　　　　　　　Alpha Road,
Cockspur Street.　　　　　　　　　　　　　　　　　　Regents Park.
6 March 1832

There is something in the paper of Parrot on the figure of a gaseous body. I showed it to Avory. Can you lend me the original?

595 Forbes, James David　　　　　　　　　　To: H.T. De la Beche
Carlisle.
26 May 1846

I cannot doubt that the Royal Society of Edinburgh will consider the presentation of the Transactions in return for Geological Survey Maps.

596 Forbes, James David　　　　　　　　　　To: H.T. De la Beche
Edinburgh,
(Royal Society).
28 November 1851

Thanks for your letter answering enquiries about Scotland. I am glad no new authority is needed for Scotland, just money. Can the Survey collection be united with unseen collections here and a Museum opened at once?

597 Forchhammer, Johann Georg　　　　　　To: H.T. De la Beche
Copenhagen.　　　　　　　　　　　　　Museum of Economic Geology,
28 December 1849　　　　　　　　　　　　　　　　　　Craig's Court,
　　　　　　　　　　　　　　　　　　　　　　　　　Charing Cross,
　　　　　　　　　　　　　　　　　　　　　　　　　　　London.

You may have received the order through our minister. The delay was caused by the illness of our King and the absence of the Minister of Foreign Affairs. My congratulations on becoming a member of the order. Calding is going to England to study water works, public lighting and drainage. Please assist him.

598 Forchhammer, Johann Georg　　　　　　To: H.T. De la Beche
Copenhagen.　　　　　　　　　　　　　Museum of Economic Geology,
16 March 1850　　　　　　　　　　　　　　　　　　　Craig's Court,
　　　　　　　　　　　　　　　　　　　　　　　　　Charing Cross,
　　　　　　　　　　　　　　　　　　　　　　　　　　　London.

Your letter of 23 January arrived long ago. I have sent the large cross to wear on the coat and some ribbons for the button hole. I cannot come to Edinburgh this year.

599 Fox, Robert Were　　　　　　　　　　　To: H.T. De la Beche
Falmouth.
7 November 1837

My remittance of £50 received and given to Jordan. I have just sent off specimens of my laminated clay to the Electrical Society in London. My explanation of metallic veins is proved by experiment. I am also putting together the results of experiments on temperature in mines. This should be taken at deeper stations where possible.

600 Fox, Robert Were　　　　　　　　　　　To: H.T. De la Beche
Falmouth.　　　　　　　　　　　　　　　　　　　　　Tavistock.
23 November 1837

Thank you for your interesting letter. Herewith the results of measurements of temperature in killas at the bottom of Consols. Corresponding results were obtained in Levant Mine. I have sent a statement to the Philosophical Magazine showing that ratio of progressive increase in temperature diminishes as we descend into the earth. Herewith my experiments on clay, using electricity. The latter may be considered as the laminating force in strata.

601 Fox, Robert Were　　　　　　　　　　　To: H.T. De la Beche
12 June 1838

I have borrowed the shell converted with sulphuret of copper, and Lemon will bring it to you. Henwood is industriously attacking my views. Specimens of clay laminated by electricity were considered unequivocal at the Electrical Society. My results on mine temperatures have been inserted in the Philosophical Magazine.

602 France, Institut de　　　　　　　　　　　To: H.T. De la Beche
Secretary of the Academy of Sciences.
29 April 1853

The Academy notifies you that you have been elected corresponding member in the section of geology and mineralogy.

603 France, Société Météorologique de　　　　To: H.T. De la Beche
Paris.
28 December 1852

[Certificate of membership of H.T. De la Beche].

604 Garby, [?]John　　　　　　　　　　　　To: H.T. De la Beche
Treruffe,　　　　　　　　　　　　　　Museum of Economic Geology,
Near Redruth.　　　　　　　　　　　　　　　　　　Craig's Court,
25 March 1848　　　　　　　　　　　　　　　　　　Charring Cross,
　　　　　　　　　　　　　　　　　　　　　　　　　　　London.

Fragments of clay-slate, quartz, iron pyrites, copper pyrites, in the form of pebbles have been found in Tincroft Mine. Specimens will be sent for the Museum. I would buy specimens for your museum from miners returning from Lake Superior if I felt safe in so doing.

605 Gardiner, J.R.　　　　　　　　　　　　To: H.T. De la Beche
15 November 1851

Prince Albert requests that you recommend someone to make a special Survey of mines belonging to the Duchy of Cornwall. A Report is needed within a month, if possible. The Prince thinks the person recommended should not be acquainted with Mining Adventurers.

606 Geneva, Société de Physique et d'Histoire Naturelle
25 November 1819

[Certificate of election of H.T. De la Beche as an Honorary Member; signed by Stefano Moricand (President) and Colladon (Secretary)].

607 Gibbs, Richard　　　　　　　　　　　　To: [H.T. De la Beche]
Bishops Castle.
27 July 1845

The enclosed is the remaining account for the pony. The geology of this neighbourhood is very interesting. Caradoc sandstone found, and I have seen a quarry of columnar Trap. I need a Map of the Welshpool country.

608 Gibbs, Richard　　　　　　　　　　　　To: [H.T. De la Beche]
Brassington,
Near Wirksworth,
Derbyshire.
16 October 1849

Two more boxes of fossils have been sent. We shall be finished in 8 or 9 days. Derbyshire has yielded some fine fossils. Please inform me of my next move.

609 Gibbs, Richard　　　　　　　　　　　　To: [H.T. De la Beche]
Bala.
28 November 1849

Herewith a few lines on my progress at Bala, a list of some things found in the Rhiwlas Limestone and two sketches. Forbes says my collection was the best he has seen from any one place, with the exception of Dudley. Thanks for the copy of the book.

610 Godwin, George　　　　　　　　　　　　To: H.T. De la Beche
24 Alexander Square,
Brompton.
30 October [1846 FJN]

C.H. Smith and myself have obtained specimens of Caen stone for the purposes of experiments similar to those of the Government Commission. The analyses might be made in the Museum of Economic Geology, if you are disposed to aid.

　　　　　　Godwin-Austen, R.A.C. see Austen, R.A.C., **24** to **41**

611 Gossett, W. Driscoll　　　　　　　　　　To: H.T. De la Beche
Ordnance Map Office,
Southampton.
1 February 1854

The line of levelling passes 50 links west of Osborne, and 173.67 ft. above sea level. I am glad the sections pleased you.

612 Gower, J.R. To: H.T. De la Beche
187 Piccadilly.
27 October 1847
All correspondence including Fitton's letter, on the need for a geologist on Stokes' survey of New Zealand has been forwarded to the Admiralty. Were you to use your influence one would be permitted to go. There is excellent coal in New Zealand and it would be a successful enterprise.

613 Gradby, Henry To: Richard Turner
4 Bayham Terrace,
Camden Town.
15 August 1846
In reply to your note desiring to have my opinion of the propriety of using the Patent Marine glue, instead of putty, in the glazing of the Great Palm House at Kew Gardens, I am of the opinion that the Marine glue is by far the best for such a purpose of any cement that has yet been devised.

614 Graham, George To: H.T. De la Beche
General Register Office.
31 July 1846
The Andalusian Mining Company is about to break up. Can you recommend any Mining Agent to whom I can apply as medium of communication? A great deal of money wasted on the property in Spain.

615 Graham, James Robert George To: H.T. De la Beche
Whitehall.
17 April 1843
A Commission for preserving Public Health by improved Sewerage, more abundant water supply, and better construction of dwellings for the poor, has been set up. The Duke of Buccleuch will head it. I am anxious to secure your services also.

616 Granville, Lord To: H.T. De la Beche
London.
7 July 1849
I have now made arrangements for the management of my Staffordshire Mines. Woodhouse approves my choice of Agent.

617 Granville, Lord To: H.T. De la Beche
Board of Trade.
[1850 FJN]
We have a difficulty with regard to the Great Exhibition. Peel suggests Playfair as a travelling commissioner to supply the deficiency of equivalents of the French Prefects who guide the local juries. I have seen Playfair privately, but a formal proposal will be made only after your sanction.

618 Granville, Lord To: H.T. De la Beche
Foreign Office.
14 February 1852
Thank you for the lectures you sent me. I congratulate you on the successful progress of your labours.

619 [Greenough, George Bellas]
[November 1828]
Abstract of a paper on the Geology of Nice by H.T. De la Beche Esq. [In Greeenough's hand, this abstract appears to be the substance of that printed in *Proceedings of the Geological Society of London*, Vol 1, No 9 1828-29, pp.87-89].

620 Greenough, George Bellas To: H.T. De la Beche
Regents Park. Lyme Regis,
28 July 1829 Dorset.
Regarding your paper on Nice, I had many misgivings that you would not approve of my alterations. Buckland will withdraw his alterations as soon as he finds them disapproved by the author. Thanks for the letter from Florence, discussing the tertiary of Italy and France. With respect to your paper on Torquay, I have written to Broderip. You differ from my doctrine on whether the limestones are transition or mountain. Our views on the Serpentines of Italy are the same.
[Cited: McCartney 1977, p.27; Rudwick 1985, p.89]

621 Greenough, George Bellas To: H.T. De la Beche
[4 December 1834]
Your Bideford paper gave rise to a very animated discussion. Murchison led the attack. I affirmed you were not mistaken, adding that he had never seen the country. Lyell, equally ignorant, supported Murchison. Turner replied. Murchison repeated his conviction of your mistake about the Greywacke. I answered that experienced observers such as you & Weaver would not fall into error. Murchison then disagreed toto coelo declaring he would never use the term Greywacke again. Prepare your defence well. Austen has attacked you also. The Royal Medal was eventually given to Lyell. Everyone praises your book on theoretical Geology.
[Cited: Rudwick 1985, pp.99, 108, 109; Secord 1986a, p.94]

622 Greenough, George Bellas To: H.T. De la Beche
Regents Park. Ordnance Map Office,
11 November 1836 Tower.
I was glad to hear of your progress. The hare which Sedgwick and Murchison have started must now be hunted down, let her double as she will. Do you still adhere to your former opinion? We had a paper at the Geological Society from Strickland who accompanied Hamilton to Asia Minor. Darwin is an intelligent, agreeable fellow. Belcher left yesterday for Panama. I hope you have been consulted on Lyell's successor in the chair.
[Cited: Rudwick 1985, p.181]

623 Greenough, George Bellas To: H.T. De la Beche
Regents Park. Post Office,
14 April 1837 Falmouth,
 Cornwall.
Your letter and enclosure arrived. I have had a conference with Kerr. I disapprove of the conduct of your assailants. Buckland and I will not be absent from our posts to defend you in your absence. Colby has acted wisely. I see no reason why your original opinion should be abandoned. Continental writers are all on your side. You should not give up your appointment.
[Cited: McCartney 1977, p.31; Rudwick 1985, p.204; Secord 1986b, p.226]

624 Greenough, George Bellas To: H.T. De la Beche
Regents Park.
21 April 1837
Sedgwick has had a riding accident. His paper on Devon has been deferred. Colby's letter has made us more apprehensive of mischief than there is any occasion for. His course of action is the best for you and science. The question at issue is of very great importance to geology. Warburton is prepared to defend you in the House of Commons. The Silurian and Cambrian systems are local and not applicable to the structure of Europe. My map advances slowly.
[Cited: North 1939b, p.1053; Rudwick 1985, p.205]

625 Greenough, George Bellas To: H.T. De la Beche
Regents Park.
24 July 1839
I am grateful to you for Logan's introduction to me. He has done the most beautiful work for me. I shall be thankful to you for the boundaries of the Pennant sandstone. In regard to the "old story", it goes much against the grain to make all the metalliferous rocks of Cornwall either red sandstone or coal-measures. "Chaos is come again". There is a meeting of the Geological Society of Yorkshire on 5 September. You will confer with McLauchlan on the parts of the map you colour for me.
[Cited: North 1936a, pp.61-62; Rudwick 1985, p.314]

626 Greere, Francis To: [H.T. De la Beche]
Court Henry,
Llandeilo,
Carmarthenshire.
7 October 1844
I enclose a specimen of copper ore from Pantyglien Slate Quarry, hoping it may be acceptable for your forthcoming geological work. It assays out at 30¾% However I believe it would not warrant the outlay required. I shall be glad to answer any other questions.

627 Greig, Charles To: [W. Kay
Bristol Infirmary. 9 Caledonia Place,
27 June 1843 Clifton].
Herewith the number of fever cases admitted last year, their mode of admission, average expense per patient, rate of mortality, localities whence and malignancy, and other details.

628 Greig, Charles To: W. Kay
Bristol Infirmary. 9 Caledonia Place,
28 June 1843 Clifton.
We had 11 deaths of fever patients last year, not 10. Figures given of age and sex of victims, with distribution by parishes, and indicating areas of greater severity, etc. Any other information I will gladly give.

629 Grey, Charles To: James Hudson
Buckingham Palace. [Royal Agricultural
12 April 1850 Society].
Prince Albert wishes to draw your attention to his scheme for turning sewage (now the cause of disease and pestilence) into a source of national wealth by its application to the purposes of agriculture. The Prince has tried the operation on a small scale with apparent success. Would the Society carry out the necessary experiments?

630 Grey, Charles To: H.T. De la Beche
Windsor Castle.
25 October 1852
I have submitted your letter to Prince Albert, who thanks you for the Prospectus of your School for the ensuing years. He agrees that you are not rightly placed, but hopes that when Public attention is turned to the great exertions of other Countries in the promotion of Science as applied to Art and Industry, that we may be shamed out of our native apathy. He is pleased to hear that Playfair is to deliver a lecture on the subject; it may be a useful appendix to the Report of the Commissioners on the subject of the appropriation of the Surplus.
[Cited: McCartney 1977, p.38]

631 Grey, Charles To: H.T. De la Beche
Osborne.
4 December 1852
His Royal Highness thanks you for the copy of Playfair's introductory lecture. The foreign information is very interesting. I hope it shames us into efforts in the cause more worthy of the country. The various public bodies and Institutions will not be blind to their true interests. The opposition of the Daily News so bad and narrow.
[Cited: McCartney 1977, p.38]

632 Grey, Charles To: H.T. De la Beche
Buckingham Palace.
17 March 1853
The Prince thanks you for the second edition of your valuable work.

633 Grey, Charles To: H.T. De la Beche
Windsor Castle.
14 November 1853
I have submitted the copy of Forbes' lecture to the Prince, and he thanks you for proof that you do not forget the interest he has in your Institution.

634 Grey, George To: H.T. De la Beche
Home Office.
21 March 1849
Can I have a few minutes conversation with you?

635 Grey, George To: H.T. De la Beche
Home Office.
13 February 1852
Thank you for the Lectures, and for the encouraging report of the success attending them and the school.

636 Grey, Henry George, 3rd Earl To: H.T. De la Beche
Downing Street.
8 February 1847
Thanks for the information through Hawes as to Admiralty Coals investigations. I enclose one of the papers filled up. A duplicate will be sent with the coals. Has the form been properly filled up?

637 Grey, Henry George, 3rd Earl To: H.T. De la Beche
Colonial Office.
18 December 1849
Thanks for the maps and sections, specimens of the splendid work in progress under your charge.

638 Grove, William Robert To: H.T. De la Beche
28 January [1848 FJN]
We had some further conversation last night touching the President of the Royal Society. It is highly expedient that if your application to Faraday meets no success, you and Sabine should go and see Herschel and see if he will change his mind.

639 Haidinger, Wilhelm Karl von To: H.T. De la Beche
Vienna.
20 August 1847
Herewith I forward to you the first volume of a collection of Memoirs I have lately begun to edit by subscription as also the first annual volume of Berichte - Proceedings of a Society now forming in Vienna for promoting natural science. I would consider it as a particular favour if you would send us in exchange such of your contemporaneous publications as you would find to answer the purpose, particularly your Memoirs. Your great work the Maps of course are not called for, as being too far beyond the value of what we are likely to offer for the future.

640 Haidinger, Wilhelm Karl von To: H.T. De la Beche
Vienna. Government Geological
25 January 1848 Survey Office,
Craig's Court,
Charing Cross,
London.
I have not failed to point out the propriety of the Austrian Empire beginning to set to work at the production of a series of geological maps, to which your beautiful maps should serve as models. It is not yet however sanctioned by the entire Institute. The first step to take would be to send over to England, in particular to the Government Geological Survey office Mr. V. Hauer and Dr. Hornes this next summer to see and learn how you undertook this heavy task, as also to see some of the geological features of the country.

641 Haidinger, Wilhelm Karl von To: H.T. De la Beche
Vienna. London.
28 May 1849
I should long ago have written to thank you for the great matter of kindness conferred on my young friends Messrs. Hauer and Hornes. I always intended to give an account of their further progress, which owing to our unfortunate politicals was much retarded. They are off on a tour of Austria, before setting at the work itself. I hope to enlarge on the subject in a few days.

642 Haidinger, Wilhelm Karl von To: H.T. De la Beche
Vienna. Geological Survey Office,
8 July 1849 Craig's Court,
Charing Cross,
London.
Thanks for the many marks of kindness and the most material scientific assistance you have given Hauer and Hornes during their last summer trip in England. They returned on 7 October to political turmoil. At present thanks to our gallant army we again take breath and hope to see the disasters drawing to a close. I hope you have received the volumes we sent to the Museum. Among all other things here in Austria, your own Geological Survey and Museum of Practical Geology is the very thing that is wanted also in our own country.

643 Haidinger, Wilhelm Karl von To: H.T. De la Beche
Vienna. Government Geological
1 September 1849 Survey Office,
London.
I have lately been called upon by our present Government to lay before them the plan of an Institute for the Geological Survey of our country, similar to that existing in England. In all respects you will not be astonished to find yourself the very point of comparison for our present doings. I beg leave to lay before you some questions. How many are employed? What length of time before completion? How many maps given to the public? Do you plan lectures in the winter season?

644 Haidinger, Wilhelm Karl von To: H.T. De la Beche
Vienna.
10 May 1852
My thanks for The Geological Observer are long overdue. Today I ask you the particular favour of an introduction for Ceylon for a young friend of mine, or rather for a scientific expedition, which he is now preparing to study the geography of plants and the zoology of the whole island. Pray let me know your desiderata also. I have lately sent you some volumes published by the Institute here. On our Geological Survey, we nearly

finished Lower Austria last summer, and are proceeding to Upper Austria this summer.

645   Haidinger, Wilhelm Karl von To: H.T. De la Beche
 Vienna.
 25 July 1853

Thank you for the great cargo of articles and the second edition of The Geological Observer. You have the true spirit of science which caused you to work on steadily till you got your Government in the ranks of scientific enterprize in geology and the kindred departments. I may lay claim to have been endeavouring to follow your example, yet I must allow of being greatly distanced. What admirable works you have already executed and what immense wave of knowledge is set agoing, and increases daily. Our maps are not executed on the scale of one inch to the mile, but one inch to 2¼ miles, whereby particularly in the mountain districts many things become too crowded.

646   Hall, [?]Benjamin To: [H.T. De la Beche]
 Llanover.
 17 December [1851?]

Many thanks for your kindness in sending me a copy of the lectures, and I am very glad to hear that the school goes on so well.

647   Hall, Henry To: H.T. De la Beche
 21 Great Bridgwater Street,
 Deansgate,
 Manchester.
 28 June 1852

The Times today talks of "a course of lectures on gold" in the Metropolis. I am about to emigrate to Australia, and living in the provinces, I am anxious to know the contents of these lectures. Cheap books and the libraries of Mechanics' Institutions do not provide the necessary information.

648   Hall, Jos. To: H.T. De la Beche
 Marble Works,
 Derby.
 5 October 1849

The marble you speak of is available. Two good faces can be produced. The pilaster we made is of Ricklow Dale marble.

649   Hall, Lewis Alexander To: H.T. De la Beche
 Ordnance Map Office,
 Southampton.
 16 June 1848

The cost of engraving the whole of the Block plan of Chelsea and Westminster will be £252, or £14 each for the 18 sheets.

650   Hall, Lewis Alexander To: General [?]
 Ordnance Map Office,
 Southampton.
 9 December 1850

The correspondence concerning the reduction of County maps of Ireland to one inch scale came to me on 21 August. I reported costs of 15/- per square mile, not 20/- as quoted by Sir Henry De la Beche. If he wishes to get the point of the wedge in, he must not expect me to give it first blow.

651   Hamel, Joseph von To: H.T. De la Beche
 55 Davies Street,
 Berkeley Square.
 5 October 1844

I saw the Geological Map at the meeting of the British Association at York. If application were made to the British Government by the Emperor of Russia, would a Copy be presented to the Imperial Academy of Sciences at St Petersburg? It is not otherwise available.
[See De la Beche, H.T. to Hamel, J., 16 October 1844, **401**]

652   Hamilton, Charles William To: [H.T. De la Beche]
 37 Dominick Street,
 Dublin.
 18 June 1838

I have several queries on the Geology of this country. How can we account for the position of the rocks described in my sketch? Are not Geologists led astray by assuming often that rocks of different character are not coeval? Will the arrangement of strata in the sketch indicate the direction of force? Have you ever found porphyry overlying stratified arenaceous rocks?

653   Hamilton, William John To: H.T. De la Beche
 14 Chesham Place,
 Belgrave Square.
 13 February 1849

A letter to remind you of my request for a suite of shells from Jamaica, as many different species as possible. Don't worry about localities or packing.

654   Hammond, E. To: [?]
 Foreign Office.
 22 November 1849

Here are our regulations about Foreign orders. If adhered to they are fatal to Sir H. De la Beche. Murchison carried his point, but he ought not to have succeeded.

655   Harcourt, William Venables Vernon To: H.T. De la Beche
 40 Grosvenor Square,
 London.
 29 June 1843

Liebig has taken up the important subject of coprolite beds and phosphate of lime. It is of paramount importance to the fertility of soils. He wishes for Government aid, and Sir R. Peel is disposed to help. Can you help by sending further specimens to Giessen so as to provide Liebig with more correct information?

656   Hardwick, P. To: H.T. De la Beche
 21 Cavendish Square.
 19 January 1852

Capt. Denham, who is about to go to the Pacific, dines with us on 3 February. Will you also come? I am anxious to preserve the only gratification I derived from the Sewers Commission, your friendship.

657   Harris, William Snow To: H.T. De la Beche
 Plymouth.
 21 December 1834

Thank you for your present of books. My work on electricity is progressing. Come and spend a few days at Christmas and we'll talk over your affair on the force of currents. I have been ill-treated by the Navy respecting my lightning conductors. A false report on their operation has come out, and I am refuting it.

658   Harris, William Snow To: H.T. De la Beche
 Plymouth.
 24 May 1836

I understand you have had disappointment over the fracturing of thermometers, and suggest a method for taking the temperatures of rocks. Shall we see you at Bristol? Saxton has invented a machine for measuring the force of running water. I have received my Copley Medal from the Royal Society.

659   Harris, William Snow To: H.T. De la Beche
 Plymouth.
 17 November 1838

Herewith my record of observations on temperature, rainfall, winds etc., as requested. I believe this is all you require. I advise some other sources. Look at the Nautical Magazine for September last. The Admiralty are fitting my lightning conductor without consulting me. If anything goes wrong, the blame is mine. If right, the wisdom is at the Admiralty.

660   Harris, William Snow To: H.T. De la Beche
 Plymouth.
 30 November 1838

Herewith more results of my calculations on the temperature here etc. Also some corrections to my previous figures. The mean annual temperature is 52.5. If you need any other particulars, I will find them for you. The reason for my two statements differing was my applying an unnecessary correction. The gales have been heavy here.

661   Harris, William Snow To: H.T. De la Beche
 Plymouth.
 29 March 1840

Thank you for your note about the education of my boy at Woolwich. My present view is to send him to Cambridge. The report on my lightning conductors out. I explain how the matter stands. The efficacy of the device is proved beyond doubt. The Admiralty are refusing to pay me expenses. My friends will bring the question before Parliament.

662 **Harris**, William Snow   To: H.T. De la Beche
     Plymouth.
     14 February 1841

I have heard nothing in reply from you to my letter of over 12 months ago. Lord Eliot is about to bring my affair before Parliament. The Admiralty refuse to reimburse me for my lightning conductor. Lord Minto's conduct is perfectly dishonest. A little exertion on the part of my friends would induce the Government to meet the question.

663 **Harris**, William Snow   To: H.T. De la Beche
     5 March 1841

The debate which took place in the House has had a good impression on the Admiralty. They are manoeuvering to avoid paying me. If you can do me any good, I know you will. You see by the enclosed papers I have a strong case. If it again comes before the House of Commons and the House of Peers, Lord Minto will be terribly shown up.

664 **Harris**, William Snow   To: H.T. De la Beche
     3 December 1847

My domestic and private affairs have been pressing. I am anxious to propose Tomlinson as F.R.S. He would be a most valuable acquisition. Some important facts have turned up lately relative to my Conductors and the Navy. Four frigates have been saved from destruction by lightning in eight months.

665 **Harris**, William Snow   To: H.T. De la Beche
     Plymouth.
     2 May 1849

Thank you for your immediate attention to my letter. I accord with you over the difficulty of a Money Grant. The question as it stands is now not an Admiralty but a Treasury one. Do you think I might get the Government to give me an increased Pension? There are many things a Government can do besides Money Grants. Please advise me.

666 **Harris**, William Snow   To: H.T. De la Beche
     Plymouth.
     27 June 1850

Philipps has not replied to my letters claiming expenses for the Buckingham Palace commission. I think Lord Seymour's conduct paltry in the extreme. I have consulted my Legal Advisor and have been led to make a formal demand up on the Board. I have been unwell since I saw you. An old faithful family servant has died.

667 **Harris**, William Snow   To: H.T. De la Beche
     Plymouth.
     28 June 1850

Burn my last letter. Philipps has written to me quite a satisfactory letter. Perhaps I have been hasty. I thought Lord Minto might have been at the bottom of it all.

668 **Harris**, William Snow   To: H.T. De la Beche
     Plymouth.
     9 July 1850

I am sorry to hear of your daughter's illness. The sea air and boating would do her good. With respect to the Woods & Forests affair, I should be obliged by their settling the account as soon as possible. I have been at great expense lately in various ways. I have made a demand for expenses and £100 for responsibility etc. Still, having heard from Philipps I do not wish to press the point.

669 **Harvey**, George   To: H.T. De la Beche
     Plymouth.
     29 November 1830

There is something about currents of the Sargasso Weeds in the Monthly Notices of the Astronomical Society. You can obtain a copy from any member of the Society.

670 **Harvey**, George   To: [H.T. De la Beche]
     Tavistock Place.
     9 November [?]

I send you Horsburgh. I have not been able to obtain Beaufort's book. Also enclosed is a volume of Jameson's Journal, and one of Young's Natural Philosophy. There are errors in the former. Your name is entered at the Public Library in Cornwall Street.

671 **Harvey**, W.   To: H.T. De la Beche
     [?c.February 1840 FJN]   Cardiff.

Mr. Williams makes the height of the Garth 895.08 feet above the church at Cardiff, the distance from the church being 34868 feet. He cannot answer your question respecting the low water at Cardiff dock which cannot well be ascertained here.
[Cited: North 1936a, p.88]

Haskings, John
[See Anning, C.C. to De la Beche, H.T., July 1847, ll]

672 **Haskings**, R.   To: J.S. Enys
     Gerran Foundry.   Enys.
     2 May 1837

Enclosed is a sketch of a very simple pump for lifting water, and I believe it quite effective. It is a double action lifting force pump or more commonly called a Jack-head. You will see that clocks are dispensed with.

673 **Hastings**, R.G   To: H.T. De la Beche
     Elford.
     Wednesday [No date]

I hope to be with you about midday next Monday, and on Tuesday and Wednesday.

674 **Hawes**, Benjamin   To: H.T. De la Beche
     Colonial Office.   Economic Museum.
     22 July 1847

We understand you are experimenting on comparative qualities of coal as fuel. Lord Grey wishes to have two specimens tested. How should we proceed? Your enquiries are important, as steam ships will ply to Australia. Is there any hope of finding coal on the West Coast of Australia?

675 **Hayward**, A.   To: H.T. De la Beche
     Exeter.
     14 March [1841 FJN]

The case is fixed for Tuesday at nine but it will probably last three days. Both sides are very confident and we shall have a very hard fight. I will write the moment it is decided. At all events I shall never forget your zeal.

676 **Henslow**, John Stephens   To: [H.T. De la Beche]
     Hitcham,
     Hadleigh,
     Suffolk.
     18 June 1853

In Cambridge I applied to Deck about the red-frit. He gave me the enclosed for you, and said he would take a mass to show in London. He has another mass from the Roman Villa at Comberton. All are much the same thing and have been prepared in crucibles of which they retain the impression.

677 **Henslow**, John Stephens   To: H.T. De la Beche
     Hitcham,
     Hadleigh,
     Suffolk.
     24 February 1854

I enclose a bit of the clay from the bore at Coombs. They are now 800 feet through the chalk. Should they persevere? Also enclosed is a pottery fragment. What process is used to produce the ornamentation? I wish to place a specimen in the Ipswich Museum of Economic Geology.

678 **Henwood**, William Jory   To: H.T. De la Beche
     The Orchard,
     near Penzance.
     27 June 1850

A German gentleman from Bonn has offered me specimens of ores he is engaged in working. I advised him to send them to you, and take the liberty of forwarding them herewith.

679 **Herschel**, John Frederick William   To: H.T. De la Beche
     Collingwood,
     Hawkhurst,
     Kent.
     11 June 1846

Smyth informed me that you wished me to make enquiries about the terms on which Mme. Witte would part with her Model of the Moon. She replied

£600. However, now she would take £300. Are you desirous of making an offer?

680 Herschel, John Frederick William To: H.T. De la Beche
Collingwood. Museum of Economic Geology,
23 December 1847 5 Craig's Court,
Charing Cross,
London.

Lord Auckland has asked me to officiate as Editor of a Manual of Scientific Information for Officers employed in surveying etc., and stated it was his intention to ask you to do a set of instructions for Geology and Mineralogy in conjunction with Professor Sedgwick. Have you anything suitable in that department?

681 Herschel, John Frederick William To: H.T. De la Beche
Collingwood. Museum of Economic Geology,
5 January 1848 Craig's Court,
Charing Cross,
London.

Although it is impossible to separate Mineralogy from Geology, yet the phenomena of mineral veins could be pointed out as of special interest, together with blowpipe instructions, simple chemical tests, indications of coal, silver etc. I dismiss the fear of poaching to winds, or would recommend consulting Sedgwick. (Lord Auckland's "Man of Geology"). The whole book is not to exceed 200 pages.

682 Herschel, John Frederick William To: E. Sabine
Collingwood.
25 January 1848
Copy
From your letter I learn of Lord Northampton's intention to retire from Presidency of the Royal Society. Regarding your proposal as to my accepting the office for a period of 3 years, it is perfectly impossible for me to undertake it. The proof, though, of the good opinion of my friends will ever be regarded by me as one of the most gratifying circumstances of my life.
[See Sabine, E. to Herschel, J.F.W., 24 January 1848, **1842**
and Sabine, E. to De la Beche, H.T., 26 January 1848, **1843**]

683 Herschel, John Frederick William To: H.T. De la Beche
Collingwood.
13 March 1848
Your manuscript has arrived safely. Lord Auckland contemplates no elaborate and extensive volume, so perhaps it will have to be abridged, if we cannot get greater scope. There will also be delays in publication as some contributions have not come in, for various reasons.

684 Herschel, John Frederick William To: H.T. De la Beche
Collingwood.
5 December 1848
I have received your proofs. I presume you would wish to see the prints, which will be sent. Would you return them to Murray, or direct to the Printers.

685 Herschel, John Frederick William To: H.T. De la Beche
32 Harley Street.
5 November 1851
I regret that an adjourned meeting of the Cambridge University Commission will prevent my attending your inaugural lecture tomorrow. Many thanks for your invitation. My wife and daughter would have come were they not out of town.

686 Hibbert, George To: H.T. De la Beche
London. Lyme Regis.
20 May 1825
You may have heard that the legislature of Jamaica have taken into account the Botanic Garden and gave me the commission of finding a Curator at £400 or £500 per annum. I have agreed that James McFadyen should take the office. Can you speak with him before his departure. I therefore furnish him with this introduction to you.

687 Hibbert, George To: H.T. De la Beche
Billeter Court,
London.
4 May 1830
I am desired by the Commissioner of Public Accounts, Jamaica, to ascertain the expense of importing an engine capable of impressing the Island Arms on gold and silver coin for circulation in the Island. As I know you have recently had something of the kind to do, could you give me an introduction to the person who will give the requisite information?

688 Hibbert, W.J. To: H.T. De la Beche
London.
29 June 1841[?]
The cost of the medal to be made by Wyon can be charged to the Estate. I hope it answers your expectations, and helps McKinnon in getting job work taken on more reasonable terms than his predecessor could (or did) obtain for you. I am afraid property in Jamaica must for some time to come remain rather a name than a substance, especially if Lord John Russell should again attain a House of Commons majority. We should like to have your opinion on the purchase of stock for the Estate.

689 Hibbert & Co. To: H.T. De la Beche
London.
12 October 1849
We have 4 kegs of Tamarinds for you sent by Mr. McKinnon. To where should we forward them? We have received 133 measures of sugar and 44 Puns Rum this year from the estate but such is the wretched state of the market that we have not sold a cask of this or of any other of our importations. Last years are yet unsold. An average crop is expected at Halse Hall this season, but hitherto the weather has been untoward.

690 Hitchcock, Edward To: H.T. De la Beche
Amherst, Geological Society,
Massachusetts, London.
U.S.A.
28 December 1833
Though a stranger to you, I am not a stranger to your invaluable Geological Manual. Allow me therefore to beg your acceptance of the accompanying Geological Report to the Government of this State. I am sending one copy for the Library of the Geological Society, and also one each for Buckland and Conybeare.

691 Hitchcock, Edward To: H.T. De la Beche
Amherst, Ordnance Map Office,
Massachusetts, Tower,
U.S.A. London.
1 March 1837
I have received letters of thanks from you and the Geological Society for the casts and specimens of fossil footmarks. My interest in the subject has been recently reawakened by fresh discoveries, and I am thinking of preparing a paper for the Geological Society if pressure of work allows. None of your latest works which you have offered to send have arrived. I obtained a copy of your work on Theoretic Geology from Silliman, and have put it in the hands of publishers to prepare an American edition.

692 Hitchcock, Edward To: H.T. De la Beche
Amherst College, Ordnance Map Office,
Massachusetts, Tower,
U.S.A. London.
10 May 1837
I beg your acceptance of a copy of the American edition of your work on Theoretic Geology. The box containing it has been sent to the Hunterian Museum and also contains plaster casts and moulds of my best specimens of fossil footmarks as well as some specimens in the rock more distinct than any which I sent to the Geological Society. The Government of Massachusetts have directed that still further researches be made into its geology and natural history and I expect ere long to be engaged in this work. Not less than 10 States of the Union are now engaged in surveys of this sort, and probably their example will soon be followed by all the rest. What of European Governments?

693 Hodgkinson, Eaton To: [?]
14 Crescent,
Salford,
Manchester.
13 December 1844
Herewith a copy of the Manchester Guardian, 20 November. On the occasion of the fall of the Mill at Oldham, my paper on the strength of beams was used by the editor. The model, which you did me the honour to accept, would have been more carefully made if I had anticipated its high destination. Its defects might be easily corrected by comparing it with the

written descriptions and drawings in my paper. I am preparing answers to your questions.

694 **Holdsworth**, A.H.   To: H.T. De la Beche
18 Duke Street,
St James.
8 July 1851

The quarries referred to in this printed paper are mine. These opinions have been taken by me before. You will observe that the slate is on its edge. Good slate alternates with rough slate. There must have been some strange disturbance when it was formed. Do you conceive that these men are correct in their view that the slate would be found to a greater depth? I will call tomorrow about eleven for an answer if you can give me one.

695 **Holl**, Harvey Buchanan   To: H.T. De la Beche
Philadelphia.   Ordnance Map Office,
29 May 1842   Southampton.

American geologists are now uniting the results of the various surveys. I will forward to you any reports you wish, also maps etc. Logan has used the results of the Corps without sufficient acknowledgement. Likewise his claim concerning stigmaria beds under coal is not original.

696 **Holland**, George   To: H.T. De la Beche
Lyme,   Ordnance Geological Survey,
Dorset.   Swansea,
27 January 1838   South Wales.

I am glad to give you the required data. I have a set of uninterrupted observations for 20 years on temperature, rainfall etc. Your daughter will be able to arrange these as to periods. Winds also recorded. How may I communicate the results to some Society?

697 **Holland** & Sons   To: H.T. De la Beche
19 [?],
St. James.
7 September 1846

We heard at the Office of Woods that our account would be paid at your Office. To whom are we to apply?

698 **Hollins**, Peter   To: H.T. De la Beche
Great Hampton Street,
Birmingham.
5 November 1850

Thank you for your opinion on the material for the Peel Testimonial. As an artist, I also have noted the unsuitability of marble in the atmosphere. I will recommend bronze.

699 **Home**, Everard   To: H.T. De la Beche
Sackville Street.   Lyme,
23 September 1818   Dorsetshire.

The skeleton you sent is highly valuable and unique. It should have a place in the British or Hunterian Museum. Will you kindly add any further particulars as to its discovery how it came into the hands of Col. Birch of Thorpe Hall, and acquaint me with its final destination. It is one of the most extraordinary animals which inhabited the Antediluvial world.
[Cited: Torrens 1980, p.561]

700 **Hooker**, Joseph Dalton   To: H.T. De la Beche
Kew.
19 July 1846

I shall be ready to go down to Bristol very soon, and would be exceedingly obliged for any recommendations that could aid me in fossil hunting as I know of no one whatever either at Radstock or Forest of Dean. I have been making out a report of my doings in S. Wales and a few notes on the S. Welsh flora, particularly the geographical distribution of recent plants. For continuing the recent botany I need an aid, as coal pits are exclusive.

701 **Hooker**, Joseph Dalton   To: H.T. De la Beche
Warrington.
20 May 1847

I send you a formal recommendation of Bunbury's paper. I will be back by Saturday week. The Manchester and Birmingham Museums have been visited. My father is unwell at Hastings. I have also recommended that Bunbury's plates be published.

702 **Hooker**, Joseph Dalton   To: H.T. De la Beche
Kew.
22 July 1847

I am extremely obliged by your kind letter and the hopes held out therein. I saw Phillips the other day and mentioned the subject to him. He wished much that I should go as connected with the Museum of Economic Geology and Kew Gardens and is ready to back any recommendation. He did not see any difficulty in my retaining my position as Botanist to the Survey. This latter is an essential point with me, both because I like the Survey and my duties and because I have no other sure income of any kind whatever.

703 **Hooker**, Joseph Dalton   To: H.T. De la Beche
Kew.
17 August 1847

I have been urged very much to go down to Scotland and get some fossils from the coalfields near Glasgow where there are duplicates said to be at my service. I do not think I should go on the Survey account wholly. However as I would like a fortnight or three weeks holiday I should like your permission to be away from 25th September. If I get good fossils, so much the better. I will charge the Survey if the results are good. I have had very distant hints of being sent for 4 or 8 months into Tibet if I should go to India where an Embassy has gone and another will next year. This is very vague and quite private. You can rely on my telling you of anything affecting the Survey.

704 **Hooker**, Joseph Dalton   To: H.T. De la Beche
Kew.
12 September 1847

I am all ready to go into print, and will put the Mss. into the printer's hand forthwith. There are three papers. They might also be rolled into one, but that will depend on when Mr. Lowry will get the plates. Mr. Philipps informs me that my Survey pay cannot be forthcoming during my absence, but I thank you for your efforts. I find that I must go down to Scotland for a week, but shall get the Mss. into the printer's hand first. I have no definite answer yet from Lord Morpeth about the Garden salary.

705 **Hooker**, Joseph Dalton   To: H.T. De la Beche
Kew.
14 September 1847

I have put the Mss. into the printer's hands. The last plate is just turned out of hand by Mr. Bone and I think the most beautiful of them all. These are the first really good structural plates on Fossil Botany that have been published in London, and as such I am exceedingly proud of them. I concur with you that Bone should have his salary increased. Lord Morpeth has not decided yet upon the Indian affair.

706 **Hooker**, Joseph Dalton   To: H.T. De la Beche
Museum of Economic Geology.
15 October 1847

From what I remember of the East Indian fossils they were decidedly oolitic. They are further related to the Australian, Scarborough and other beds. I will go over them again and report more particularly. Can you tell me in what order the plates are to be numbered, as the printer is waiting for my references? I have not yet got the Treasury money. Lord Morpeth has been a most unintentional impediment, it having been very hard to make him understand the business. Philipps is most kind and Lord Morpeth too. With Lord Auckland I have arranged to spend some months in Borneo, proceeding there from India next winter.

707 **Hooker**, Joseph Dalton   To: H.T. De la Beche
Cairo.
8 November 1847

We have had thus far a very prosperous and most agreeable passage. In Malta I bargained for and purchased a vase cut from the limestone. I also acquired fossil fish teeth which I have packed away until I get to India. In Cairo I have looked at the petrified woods. Striking east we followed a very broad but more irregular valley, the cliffs of which were much inclined and disturbed. I found no trace of a fossil animal remain in the sand. We are much glorified here. Will Reeks tell my father I am well, as I may not have time to write?

708 **Hooker**, Joseph Dalton   To: H.T. De la Beche
H.M.S. Sidon.
5 December 1847

The bearer of this letter, J.H. Chalmers, will bring back to you the fossil teeth purchased at Malta. Will you kindly defray any small expenses he may have been put to in purchasing them from the natives? I have advised him not to exceed £1 and take tolerably perfect specimens. Should he be fortunate enough to get more perfect remains then I'm sure he will use his own judgement about purchasing them.

709 **Hooker**, Joseph Dalton To: H.T. De la Beche
Darjeeling, Craig's Court,
Sikkim Himalayah. Charing Cross,
26 July 1848 London.

My last letter to the Survey was addressed to Forbes and written during my passage down the Ganges. Since then I have been residing in the Himalaya and north of Calcutta, my time occupied almost wholly with botany and studying geography etc. I have myself collected little in geology beyond rock specimens. In Calcutta, gneiss is the prevailing rock. Tibet is a dry tableland with probably as much elevation as 15000 feet, north of the snow. Congratulations on your C.B. Long may you live to enjoy the honour.

710 **Hooker**, Joseph Dalton To: H.T. De la Beche
Govt. House,
Calcutta.
5 April 1852

I present to you the Prime Minister of Nepal, his two brothers and other gentlemen of the Nepal Mission. He is particularly interested in science and its useful applications. Thus your Museum should be studied by him. I am personally indebted to him for permission for my trip in Nepal.

711 **Hooker**, Joseph Dalton To: H.T. De la Beche
Kew.
12 November 1852

I cannot answer your question positively. That I found underclays loaded with Stigmaria I can confidently affirm, for I turned them over by thousands hunting for structural specimens; but though I have no recollection of seeing underclays without Stigmaria, they may very well have occurred. The fact of underclays being void of Stigmaria does not appear to me to be of any moment, nor to militate in the least against the general rule, that that organism is characteristic of underclay.

712 **Hooker**, Joseph Dalton H.T. De la Beche
Museum of Economic Geology.
[No date, watermarked 1847]

On arrival at Kew last night I found a letter announcing another arrival of fossil plants from Van Diemen's Land. They are presented to me by a Lieut. Breton, whom I knew in that colony. He offers any quantity and size of specimens and does not even ask the expenses of packing to be defrayed, which is no joke.

713 **Hooker**, William Jackson To: H.T. De la Beche
Royal Gardens,
Kew.
2 June 1849

Dr. Hooker has by Lord Dalhousie's requirement obtained the Raja of Sikkim's permission to visit the great passes into Tibet. The Mission to Borneo is now undesirable. The grant from the Treasury for India ceased in November. I have now written to Lord Carlisle to ask that the Borneo funds be transferred to the Himalaya Mission.

714 **Hooker**, William Jackson To: H.T. De la Beche
Royal Gardens,
Kew.
30 June 1854

Thank you for the pine seeds. They are very acceptable. Commend us to any of your friends in that region. The Jamaica wood will also be acceptable to our Museum.

715 **Hopkins**, Evan To: H.T. De la Beche
Barrington Road,
Brixton.
3 October 1849

I have been lately inspecting and valuing a mineral property in Wales, between Llanharry and Llanharan, with the view of putting it into immediate operation, and in order to expedite my investigations I obtained your beautiful map No.36 with all the sections. I found two faults not mentioned on your plan which have a very considerable practical bearing as to the value and mode of working to be adopted. I enclose a sketch to draw your attention to the above, and have advised the interested party not to spend any money. I am willing to correct the above sheet by sections.

716 **Hopkins**, William To: H.T. De la Beche
Cambridge. Ordnance Survey.
8 July 1837

I think you are aware that I have been working on the limestone and coalfield of Derbyshire for the last three or four years. I wish to do in Derbyshire what you have done in Cornwall and suggest that a geological survey be made at the same time as the Trigonometrical Survey. Would it accord with your plans? Could it be carried out under my direction? I do not wish to interfere with your appointment. The principal miners are anxious for a geological survey to be carried out.
[See Sedgwick, A. to De la Beche, H.T., 10 July 1837, **1878**]
[Cited: Secord 1986b, p.237]

717 **Hopkins**, William To: H.T. De la Beche
Cambridge. Ordnance Map Office
17 July 1837 Tower,
London.

Thank you for your letter from Falmouth. Should you delegate to me the geological survey of Derbyshire I would explain to Colby the circumstances of the case. My work on Derbyshire has been most laborious and detailed. When you shall have received an answer from Colby, I may of course expect a further communication from you. I would still have University obligations, especially in the winter half of the year.

718 **Hopkins**, William To: H.T. De la Beche
Cambridge.
10 October 1846

The Government has established a Railway Board for the general superintendence of railway matters. Lyell has been very recently applied to recommend someone as the scientific member of this Board, the qualifications needed being mathematics and geology. He was given to understand there was no chance of the appointment going to Babbage. I have the required qualifications, and request your assistance in obtaining the post.

719 **Hopkins**, William To: H.T. De la Beche
Cambridge.
19 October 1846

Buckland, who wrote to Lord John Russell on my behalf touching the Railway Board, has just forwarded his Lordship's answer, from which it appears that it is the intention of Government to appoint a practical Engineer Officer. Everybody knows how many Engineers there are very incompetent to grapple with general questions which may arise, and that all of any character would turn up their noses at an appointment which would require them to give up their profession. I consider the question entirely settled as far as I am concerned, but would be grateful should you remember me for any future post of public utility.

720 **Hopkins**, William To: H.T. De la Beche
Cambridge.
19 April 1848

I have looked carefully over Lubbock's paper. The earth being a spheroid, if it were set spinning round an axis not coinciding with the axis of the spheroid, the axis about which it would subsequently revolve would be constantly changing its position. This supposes no resistance to the rotating motion from any external cause such as a resisting medium like that supposed in the paper.

721 **Hopkins**, William To: H.T. De la Beche
Cambridge.
25 January 1850

On my return to Cambridge I found your note giving me the information that the Geological Society had done me the honour of awarding me the Wollaston Medal. How should I reply to your note? I consider this award as one of the most gratifying compliments I could have received. The medal is presented, I presume, on the anniversary of the Society. Should the recipient say a few words, or bow his thanks and retire?

722 **Horner**, Leonard To: H.T. De la Beche
2 Bedford Place.
7 February 1846

Thanks for your presentation copy to the Society. I wish to notice in my Anniversary Address advance made in Geology by your institution, more particularly since the late extension of it. Can you supply me with particulars?

723 **Horner**, Leonard To: H.T. De la Beche
Embley Park.
20 September 1846

I cannot give up hope you will be my successor at the Geological Society.

Murchison and Lyell agree. I mentioned your objections in conversation with Phillips and Forbes. They think your standing with the Government Department will be enhanced, not the reverse. Lyell would take over after you.

724 Horner, Leonard   To: H.T. De la Beche
30 Dickinson Street,
Manchester.
29 September 1846

I am glad you've yielded to my solicitation. I shall cooperate with you as far as I can be of use. I will advise you as to Vice-Presidents. Please sketch out your plan of what the Foreign Secretary should do.

725 Horner, Leonard   To: H.T. De la Beche
30 Dickinson Street,
Manchester.
5 October 1846

After receiving your letter, I wrote to Murchison, Lyell, Forbes, Phillips and Bunbury, telling them of your acceptance. You are a lucky man to have your official duties so high and so delightful a calling.

726 Horner, Leonard   To: H.T. De la Beche
30 Dickinson Street,
Manchester.
10 October 1846

Thank you for the Memoirs of the Survey. There is matter of the highest importance throughout. I shall do my best to give an account of the volume in my Presidential address, though overwhelmed by the quantity of matter.

727 Horner, Leonard   To: H.T. De la Beche
30 Dickinson Street,
Manchester.
8 March 1847

Keilhau's article on the Gneiss of Norway is not suitable for our Journal. Various other notices in Periodicals seem to be of importance. Can you send them to me?

728 Hoskins, J. Elliott   To: [?Sir James Clark]
[No date]

[Part of letter only] Lines of white are found by scraping away the gravel as soon as laid on so as to give the building the appearance of being built of hewn yellow stone with blue copings.

729 Hosmer, John   To: H.T. De la Beche
14 Lombard Street Chambers,
City.
5 November 1849

As one of the Commissioners of Sewers, I draw your attention to the enclosed pamphlet describing a system of self-flushing, and suggest that it be tested in one street in the Metropolis. The Commissioners have a model at their office.

730 Howell, Henry Hyatt   To: [H.T. De la Beche]
3 Euston Place,
Leamington.
1 October 1850

I acknowledge with thanks your letter of 29th, appointing me an Assistant Geologist on the Geological Survey of Great Britain and I shall endeavour so to discharge my duties, as to give you no cause to regret having made the appointment.

731 Howell, Henry Hyatt   To: [H.T. De la Beche]
Tewkesbury.
13 November 1852

On Monday last I went to see Hull at Winchcomb to compare notes with him and to see how our lines agreed. I had two days with him in the field, where he showed me good sections of the Inferior Sand, Fullers Earth and Great Oolite which I had not seen before. I hope to be able to leave here some day next week. The weather has been much against me or I should have finished before this time.

732 Howell, Thomas   To: H. Brand
6 Eaton Place West.
11 June 1850

Can you obtain for me Sir Henry De la Beche's permission for my son Henry to be attached as an amateur to one of the parties now particularly engaged on the Geological Survey under Sir Henry's directions? He has been a student under Professor Ansted at Kings College but wishes to get a knowledge of the mode of conducting the Geological Survey preparatory to proceeding to Australia. Can you assist him?
[See Brand, H. to De la Beche, H.T., 12 June 1850, **99**]

733 Hume, Joseph   To: H.T. De la Beche
Bryanston Square.
2 December 1847

I am anxious to know the results of your experiments on coals, now two years in progress. Has a report been made? What of progress with your new buildings in Piccadilly?

734 Hume, Joseph   To: H.T. De la Beche
Bryanston Square.
10 March 1848

Can you let me have 5 or 6 copies of the first Report on Coals, to send to interested parties?

735 Hume, Joseph   To: H.T. De la Beche
Burnley Hall,   Jermyn Street.
[East Somerton],
Norfolk.
19 December 1851

I have requested Mr. Dakin to close the box and send it to you free that you may see what it contains. A letter from Meerut states that it contains earths to be analysed and it refers to another letter which has not come to hand. Perhaps it may be in the box. But you will see.

736 Hume, Joseph   To: H.T. De la Beche
18 May 1853

Two years ago I sent you box of specimens from India for analysis. A letter from Meeruth reminds me to ask you if you have any results. If not, please put their examination in hand. Please ask Lyell whether specimens have been sent for the geological lectures at Madras.

737 Hunt, Robert   To: H.T. De la Beche
6 Craig's Court,
London.
21 July 1846

Our volume of Memoirs is now before the Public. No copy has been received at the Museum. The Daily News contains a notice on the Museum. Forbes wishes copies for Murchison and Horner. What about societies?

738 Hunt, Robert   To: H.T. De la Beche
6 Craig's Court.
31 July 1846

I have just received the enclosed from Gorden. I have given Bone the sheet of paper sent. I will inform you of the result, and you will perhaps communicate with Southampton. Kane is in town, but he has not come near the Museum.

739 Hunt, Robert   To: H.T. De la Beche
Liverpool.
29 November 1849

I was also startled by the charges for the serpentine. I have written strongly on the subject to Organ. I hope I've done some good by seeing the smelters. I hope to get the lead produce of the Kingdom from Mather of Glyn Abbot. I am improved in health, and will be back on duty on Monday.

740 Hunt, Robert   To: H.T. De la Beche
27 September 1850   Museum of Practical Geology,
Jermyn Street,
St James's.

Our cheques are signed and payable on Monday. Apsley Pellak and his brother have placed the vase at our service. I have promised Weale I will give him pages describing our museum for his guide book. Have you anything to highlight? Have you seen the letter in the Times about this place as a Mining School.

741 Hutton, Robert   To: H.T. De la Beche
Putney Park.
18 June 1847

We wish some distinguished man of science to present the prizes in the Faculty of Arts at University College. Will you oblige?

742 **Ibbetson**, Levett Landon Boscawen   To: H.T. De la Beche
     George & Railway Hotel,   Museum of Economic Geology,
     Bristol.   Craig's Court,
     8 October 1845   London.

I send you a section of a Pennant quarry at Fishponds near Bristol on the Gloucester Railway. It shows the commencement of a coal seam and a stratum of fossil plants. I think Forbes or Morris should come and examine them on the spot. I have finished the Gloucester line some days.

743 **Ibbetson**, Levett Landon Boscawen   To: H.T. De la Beche
     Dr Percy,
     56 New Hale Street,
     Birmingham.
     3 May 1846

I enclose you Mr. Lycett's letter. It is really worth having. Let me know what you think about it. Here is my account of the geology of the railway from London to Birmingham. Will you be good enough to give a helping hand with your friends for Armstrong's election as F.R.S. I hope Forbes is not ill again.

744 **Ibbetson**, Levett Landon Boscawen   To: H.T. De la Beche
     Slough,
     Windsor.
     20 March [1847]

I send you a resume of my last quarter's work. I hope in the course of this summer to get all the sections of the new railways into such a state of progress that I shall be able to examine each section of railway throughout the Kingdom at the critical moment of its being available which is no easy task. Incidental expenses will not exceed £10, but personal expenses have been very heavy, from 15/- to 18/- daily.

745 **Ibbetson**, Levett Landon Boscawen   To: H.T. De la Beche
     Clifton House,
     Old Brompton.
     24 March 1848

I shall be happy, under the present state of the finances of the Government, to continue the appointment as surveyor of the railway sections, at the clear sum of £100 per annum, so as to enable me to proceed with the same.

746 **Inglis**, Robin H.   To: H.T. De la Beche
     7 Bedford Square.
     8 May 1852

I have been commissioned by the Board of the Charity of the Sons of the Clergy to look out for some person or persons fit to be employed in investigating the prospects of their estate in Northamptonshire as relates to the ironstone recently discovered there. Can you aid me in this matter?

[[?] **James**, Henry see also **2266**]

747 **James**, Henry   To: H.T. De la Beche
     Wotton-under-Edge.
     7 December 1842

I have had two most agreeable days with Phillips. We left Gloucester and drove to Cam. then across country to Stone. The Silurian rocks are full of fossils. We saw Mr. Greswell's collection which I will try to get. The railroad runs nearby, and I will examine it.

748 **James**, Henry   To: H.T. De la Beche
     Wotton-under-Edge.
     8 December 1842

I called on Mr. Greswell yesterday but was unsuccessful in my attempt to get the fossils, as they belong to the Rev. George Cooke. I will write to him if you wish. I examined the railroad cutting yesterday. The trap rocks are overlying Silurian strata. The trap is vesicular. I saw one piece of agate 6 inches in diameter. This will all be covered up in very short time.

749 **James**, Henry   To: H.T. De la Beche
     Wotton-under-Edge.
     9 December 1842

I received your two notes this morning and a note from Larcom. He says they are printing Cap. XIII of Portlock's Report. Examined tunnel and cuttings at Wickwar yesterday and had a treat both as engineer and geologist. Lonsdale's work is quite right.

750 **James**, Henry   To: H.T. De la Beche
     Davenham Hall.
     27 December 1842

I am exceedingly obliged for the too flattering manner in which you have spoken of me. I am satisfied with the arrangement, whether I go to Ireland or not. I will be in town next week.
[Cited: Herries Davies 1983, p.107].

751 **James**, Henry   To: H.T. De la Beche
     Ordnance Survey Office,
     Dublin.
     15 April 1843

Portlock sails for London on Monday. The fossils will be sent on Wednesday. There is a good deal yet to do in the Cabinet here, so I will remain longer. Portlock sails for Corfu on 1 May. Do not let him put his hand on the fossils. The spirit of confusion dwells in him, and there abideth. Flanagan is a very useful fellow and a capital collector.
[Cited: Andrews 1975, p.169; Herries Davies 1983, p.129].

752 **James**, Henry   To: H.T. De la Beche
     Ordnance Survey Office,
     Dublin.
     19 April 1843

Seven boxes of fossils are on their way to you. The remainder will follow. They were collected by officers on the General Survey from all parts of the Kingdom, not all by Portlock's geological party.

753 **James**, Henry   To: H.T. De la Beche
     Ordnance Survey Office,
     Dublin.
     22 April 1843

I have forwarded 8 more boxes of fossils today, all mentioned in the Fossil Table of Portlock's Report. Also a few electrotypes to see how we manage matters. Larcom goes on leave soon, and I shall have to do his duty here.

754 **James**, Henry   To: H.T. De la Beche
     Phoenix Park,
     Dublin.
     27 April 1843

Amongst the Irish fossils sent over you will find Portlock's new genus Koleoceras. K. Ballii is nothing more than a Pleurotomaria enveloped in a zoophyte. The geology of Portlock's book is zero, but I thought he was stronger in fossils.

755 **James**, Henry   To: H.T. De la Beche
     Wexford.
     9 November 1845

Wet weather hinders my work. Mrs. James is delivered of a boy. I will complete my work here before going to her. Murphy has made an excellent collection of Tertiary shells. Flanagan is making good collections. I have directed him to New Ross.
[Cited: Herries Davies 1983, p.134].

756 **James**, Henry   To: H.T. De la Beche
     Dublin.
     4 February 1846

The Council of the Geological Society have elected you vice president for the year. Mallet is the president. I met Griffith who complains of Hamilton's attacks on him. Penny has reduced the greater part of the Wexford map to the one inch scale. The mild weather of January has set the potato rot going again.
[Cited: Herries Davies 1983, p.154].

757 **James**, Henry   To: H.T. De la Beche
     Dublin.
     20 February 1846

We have reduced the six inch map of Wexford to the one inch scale, a most laborious operation. Please write to Colby to find out when we may expect maps of Waterford, Carlow, Kilkenny and Wicklow on the inch scale. Adare is at work again on the Memoir for Ireland.
[Cited: Herries Davies 1983, p.136].

758 **James**, Henry   To: H.T. De la Beche
     Dublin.
     20 February 1846

I have found the place for the Museum: the Exchange. About £300 is all that is needed. The vestibule has statues and must be open to the public. There is plenty of room for storage and laboratory work. I have said nothing to Kane. This will beat your new museum by chalk.

759 James, Henry   To: H.T. De la Beche
Dublin.
23 February 1846
I sent you a tracing of part of Wexford showing the faults in the Carboniferous Limestone at Taghmon. I send today a tracing of Knockmahon to show the manner in which the Old Red is there inlaid by faults. Please ask Lowry to make engravings of these.

760 James, Henry   To: H.T. De la Beche
Dublin.
23 February 1846
The Exchange affair is advancing. We may get it for a nominal rent, one tenth of that calculated for Stephen's Green. I have seen Kane, Burke and Papworth to urge them on.

761 James, Henry   To: H.T. De la Beche
Dublin.
10 March 1846
I intend to go to Gorey on Thursday to see Wyley's work, and examine the coast near Tara. Willson leaves with me. McCoy is ill. I hope Forbes comes soon to examine our fossils. I have tried to find Kane, but he is absent. Burke enquired about the Museum. McCoy will work at the fossils for public exhibition until they are finished.

762 James, Henry   To: H.T. De la Beche
Gorey.
18 March 1846
I intend to locate Mrs. James here, that I may freely move on Survey duties. I have no desire to have anything to do with the Museum in Dublin. I intend to produce collections of fossils, minerals etc. I have made a section which I suggest you substitute for that part of your long section from Croghan to Arklow.

763 James, Henry   To: H.T. De la Beche
Dublin.
19 March 1846
I send you my note to Griffith and his reply concerning the fossils drawn for him by McCoy. We have shown him consideration he does not deserve. But you must put a stop to his seeking to employ McCoy or Salter on a new matter. It is a great impertinence for him to ask. To consent would be suicidal.

764 James, Henry   To: H.T. De la Beche
Dublin.
31 March 1846
Enclosed I send an abstract of the charges for yourself, Forbes and Smyth against the grant for Ireland. We will be within £1500 for the year. I have asked the assistants for their accounts. If people are in distress here, jobbers can make a harvest for themselves. I have not seen Kane for some time.

765 James, Henry   To: H.T. De la Beche
Dublin.
13 April 1846
I forward accounts and an abstract of expenditure for 1845-46. The grant of £1500 will fall short of our demands. Our party is too small, and no provision has been made for soil collection. An extra £500 per annum is needed. We must get more money or reduce our establishment forthwith.

766 James, Henry   To: H.T. De la Beche
Dublin.
15 April 1846
I received a note from Col. H. Jones, Chairman of the Board of Works, asking my assistance on an inspection of sites for piers. I have no personal objection, and mentioned that Canning's approval is needed. If you object you can stop it in London. I have a section ready, also one from McCoy. I will send them tomorrow. We find difficulty in making the sections from the quantity of drift and Tertiary marls and sands.

767 James, Henry   To: H.T. De la Beche
Dublin.
17 April 1846
I forward Kane's letter about soil collections. Don't move until the letters for collections for the Colleges are received, then do so effectively for a larger grant. I have told Kane our funds are inadequate, but we will cooperate as much as possible. Herewith McCoy's sections, still confoundedly stiff.

768 James, Henry   To: H.T. De la Beche
Dublin.
26 April 1846
I was not consulted in any way about the collections for the Colleges. Supposing Canning does want me to inspect pier sites, my local directorship is not questioned, any more than your position when you are frequently employed in like manner. I wrote to Canning for an answer to give to Col. Jones. I hope you get more money for the Survey. You've not answered Kane's letter about the soils; the whole thing is a farce.

769 James, Henry   To: H.T. De la Beche
Dublin.
27 April 1846
I have received the catalogue of the Geological Survey publications. You must be successful in your application for funds, as the Survey will not proceed at a proper rate. I am glad Smyth is coming over. I have said nothing about reduction to Wyley, until I hear about your application for funds. Let Salter do English collections of fossils, McCoy the Irish; they will differ horribly, but let Forbes give decisions over all.

770 James, Henry   To: H.T. De la Beche
Dublin.
30 April 1846
Lincoln has sent me Canning's letter about my employment on the inspection of pier sites. I presume Jones will now find someone else. You have sent me no directions for answering Kane's letter about soil collection. We must get authority for the extra expense. Wyley ought to be transferred to England.

771 James, Henry   To: H.T. De la Beche
Dublin.
30 April 1846
We are selecting fossils and preparing them for England. We look for a reciprocal advantage. You already have a perfect set from the Survey under Portlock. I hope you will give directions for similar arrangements in England. Please give this note to Forbes.

772 James, Henry   To: H.T. De la Beche
Dublin.
4 May 1846
I have given written directions to the party, so that if I am employed on the piers problem for a few weeks, there is no inconvenience to the Survey. Smyth goes to Waterford. McCoy to Newton Barry. Willson to New Ross. Wyley is in Wicklow. Penny goes on making one inch maps. Flanagan goes to Rathdrum. My remarks on Survey matters have been made to Kane and Hamilton only.

773 James, Henry   To: H.T. De la Beche
Dublin.
7 May 1846
I enclose a copy of the correspondence on piers. You should have no fears about conduct or progress of the Survey. We can keep up one third of the progress in England, in proportion to our means. Col. Jones seems vexed at the delay.
[See also Jones, H.D. to James, H., 2 May 1846, **808**] and Lincoln to Canning, 4 May 1846, **859**].

774 James, Henry   To: H.T. De la Beche
Dublin.
15 May 1846
Concerning building stones from the Engineer Department, herewith a correct account of the correspondence. On 20 June I asked for the Engineers' help in making collections. Authority was given on 25 June. Kane was informed afterwards. Kane must have stated inadvertently that the assistance of the Engineers was given "at his request".

775 James, Henry   To: H.T. De la Beche
Dublin.
16 May 1846
I understand that the question of scale for the Ireland map is before the Board of Ordnance. Colby is not recommending the expected one inch scale. It is important we receive early information. Please write to Canning and have him ascertain the Ordnance's intentions.

776 **James**, Henry To: H.T. De la Beche
Dublin.
17 May 1846
Col. Jones has given me instructions to proceed to Cork. Board of Works crowded, clerks and draughtsmen working in all the passages, notwithstanding Griffiths and Larcom doing all their road and drainage work at the Ordnance Survey Offices. You have not answered Kane about the soils.

777 **James**, Henry To: H.T. De la Beche
Dublin.
19 May 1846
I am to leave for Waterford, then Youghal and Cork. Penny will redirect letters from this office.

778 **James**, Henry To: H.T. De la Beche
Dublin.
19 May 1846
I had a consultation with the stone polisher. We can get one pair of each of our cubes polished for 1/6. Would you like the London set polished here at that price?

779 **James**, Henry To: H.T. De la Beche
Cork.
25 May 1846
I will keep your note on the collection of soils for the Colleges and show it to Kane. Dr ApJohn is analysing marls in Wexford. Kane does not know this. My report on piers is completed. Burke informs me that the Stephens Green house is available. I presume I shall arrange our transfer there when this work is finished.

780 **James**, Henry To: H.T. De la Beche
Kenmare.
17 June 1846
My pier inspection is finished. I met a Cornish mine captain who will send me some ores. I have received no letter from you concerning my successor. He should take charge as soon as I have closed the quarter's accounts. I hope I am not posted to a hot climate.

781 **James**, Henry To: H.T. De la Beche
Dublin.
21 June 1846
Col. Matson tells me you have spoken to him about my leaving the Survey. Herewith my official letter of resignation. Please inform me of my successor as soon as possible. Hamilton is making laughable mistakes as an observer.

782 **James**, Henry To: H.T. De la Beche
Dublin.
23 June 1846
I have received yours of 16th. I see I have done the right thing in sending the official [resignation]. I saw Kane this morning. I need money to close the accounts.

783 **James**, Henry To: H.T. De la Beche
Dublin.
27 June 1846
I have directed the assistants to send up their accounts as early as possible. When they are closed I'll hand over to Oldham. I will be in London next week, if you are not here before.

784 **James**, Henry To: H.T. De la Beche
Dublin.
2 July 1846
A feud has for some time past existed between Oldham and McCoy. At the Geological Society meeting some angry letters between them were read. Oldham's appointment as local Director makes McCoy's position awkward. I suggest he be removed to England.
[Cited: Herries Davies 1983, p.141; Herries Davies 1995, p.33].

785 **James**, Henry To: H.T. De la Beche
Dublin.
2 July 1846
Penny asks again about his expenses from London to Liverpool. Can this be introduced into your English account?

786 **James**, Henry To: H.T. De la Beche
London.
11 July 1846
Herewith the receipt for the balance due to me on 30 June. I cannot understand why you think I am not due for time in this quarter. I will call on Morpeth. I will return Portlock's documents if they are considered public.
[See De la Beche, H.T. to James, H., 18 July 1846, **407**]

787 **James**, Henry To: H.T. De la Beche
London.
20 July 1846
I will send the book belonging to Survey. McCoy has one or two of mine and I should be glad to have them sent over.

788 **James**, Henry To: J.C. Moore
Portsmouth. Secretary to the
25 March 1847 Geological Society.
I return the section exhibited on 24 February, and a reduction for publication. Please preserve the original. The excavation is now 8ft below low water. Please alter the figure in my description.

789 **James**, Henry To: H.T. De la Beche
Portsmouth.
10 September 1847
At the meeting of our Iron Commission I pointed out the importance of correct analyses of ores. I suggested your assistance should be obtained. A letter has been sent to Woods and Forests. It bears directly on economic geology. Our Commission is unpaid, so there will be no pecuniary advantage to my old colleagues.

790 **James**, Henry To: H.T. De la Beche
Ordnance Map Office,
Southampton.
10 August 1854
The hill drawing of the Isle of Wight has been redrawn on the 6 inch scale. In my report on the one inch map for Scotland I have insisted on contour lines. I therefore agree about same for Isle of Wight, but we have no authority. Contours are immensely important for engineering, mining, geological and statistical purposes.

791 **James**, Henry To: H.T. De la Beche
Liverpool.
25 September 1854
As regards the progress of the one inch map of Ireland, Wexford is engraved, Wicklow all but finished, Kilkenny and Waterford in progress. Please represent on an Index map the progress of the Geological Survey so I may direct the engraving of the required sheets. I am going to Ireland soon, to push on those parts you require. Sedgwick feels he has lost much concerning Silurian and Cambrian boundaries by not publishing earlier.

792 **James**, Henry To: H.T. De la Beche
Ordnance Map Office,
Southampton.
16 October 1854
I have received the diagram showing the progress of the Geological Society in Ireland and given directions to push forward the sheets required. With regard to the Berwickshire coast, a party has been ordered to survey the portion required. The 6 inch map of Lancashire and Yorkshire engraved. I send a diagram of our progress and changes.

793 **James**, Trevor Evans To: [H.T. De la Beche]
St. Clears. Haverfordwest.
31 August 1841
I have now carried the boundary of the Old Red and Silurian to less than three miles of Carmarthen. Fossils collected include Encrinites, Orthoceratites, Orthis, Graptolites. I leave for Carmarthen on Thursday.

794 **James**, Trevor Evans To: [H.T. De la Beche]
28 April 1844
How or where our section is wrong I cannot conceive as both the commencement and the end appear to be correct. As soon as I can get some paper from Bristow I will repeat it though Aveline and myself as we thought worked it very carefully both assisting one another in laying it down on paper. Enclosed I send you an extract from my note book of the different angles we took between Dundry church and the Hotwell road. Bristow proposes to be here on Thursday.

795 **James**, Trevor Evans To: [H.T. De la Beche]
Derby.
28 June 1846
On the subject of my leaving the Survey, it is a case of bread and cheese. The reason for the small amount of work completed is that I have been engaged on Old Red Sandstone districts. As for the rest I am not far short of the amount done by others.

796 **Jameson**, Robert To: H.T. De la Beche
21 Royal Circus, Care of Treattel &
Edinburgh. Wurth, Booksellers,
17 April 1831 Soho Square,
London.
When do you expect to publish your "Manual of Geology?" I have recommended it to my class at the University, and am daily pressed for information about it. The discovery of bones in caves and breccia in New Holland is noticed in the Edinburgh Philosophical Journal. You ought to have an agent here for the sale of your works.
[Cited: McCartney 1977, p.28]

797 **Jameson**, Robert To: H.T. De la Beche
College, Athenaeum Club House,
Edinburgh. London.
7 May 1831
I am glad your work is so far in advance. With regard to the encrinal limestone, it occurs in beds alternating with lower members of coal deposits in this quarter. I hear a sharp attack has been made on Elie de Beaumont. Buckland tells me the Society has received another case of New Holland fossil bones.

798 **Jameson**, Robert To: H.T. De la Beche
College Museum,
Edinburgh.
December 1850
I understand I can bring to your notice the request from the Museum of our University for specimens from the Museum of Economic Geology. The Museum here is miserably poor in English and Welsh rocks. Your absence from the British Association meeting last August was universally regretted.

799 **Jameson**, Robert To: H.T. De la Beche
21 Royal Circus,
Edinburgh.
24 March 1851
Thank you for the copy of the "The Geological Observer". Such a work was much wanted and no living Geologist could have executed it in so accurate and interesting a manner as yourself.

800 **Jeffreys**, John Gwyn To: [E. Forbes]
Norton,
Near Swansea.
19 October 1849
Herewith a description of my collection of 600 species. There are at least 50 specimens of each species. It includes the collections of Clark, Humphreys, King, Barber etc. The only reason for my disposing of it is that I cannot complete the arrangement of it. As payment of the £525 purchase price is to be deferred, it must be secured and have interest.
[See Forbes, E. to De la Beche, H.T., 22 October 1849, **574**].

801 **Jenkyns**, John To: H.T. De la Beche
Bolley Hill. Geological Survey Office,
24 October 1847 51 St. Stephens Green,
Dublin.
Yours of the 20th inst., has reached me here rejoicing in leisure and throwing away money in farming and what I call improvements. The £250 was invested in Consols 3 weeks ago. I am sorry to say that the Irishman dishonoured the bill we gave for the remainder. We have not let go the security, and I have no doubt we shall get it in time.

802 **Jenkyns**, John To: H.T. De la Beche
14 Red Lion Square. Bangor.
3 June 1848
You may now consider yourself as possessed of £1190 3% Consols, standing in the names of Mr. Dillwyn and myself. Shall I obtain a power of Attorney to enable you to receive the dividends from time to time? Keep this in your mind and we will talk of the matter when you come to Town.

803 **Jenkyns**, John To: H.T. De la Beche
14 Red Lion Square. Jermyn Street.
25 January 1850
We purchased a further sum of £50 Consols on the 17th inst., as agreed on, as soon as the stocks opened. But I have not to call on you for any cheque for I found that the half year's dividend in July was not received. We had enough money therefore to make the purchase without your aid and have still in hand some money. Enclosed is an account showing £1300 shares.

804 **Johns**, Charles Alexander To: H.T. De la Beche
35 Park Street, R. Fox's,
Plymouth. Falmouth.
7 January 1837
I have been exploring the strata of the district, and have found fossils at the junction of the sandstone and greywacke slates. The greywacke adjoining is in some places entirely composed of organic remains. I do not intend leaving Plymouth for Dublin for 10 days, and if I can do anything for you at either place I shall be most happy.

805 **Johns**, Charles Alexander To: H.T. De la Beche
Grammar School,
Helston.
22 March 1847
I am about to become a candidate for the Principalship of one of the Government Normal Schools, and request your recommendation as to my scientific knowledge. Please address a line to Shuttleworth at the Privy Council Office directly.

806 **Johns**, Charles Alexander To: H.T. De la Beche
Merther,
Truro.
3 January 1848
I am obliged by your mentioning the Chaplaincy of Cirencester Training College. It would not be suitable now that I have a wife and family. I have resigned my school at Helston and will move near to London, where I will take private pupils.

807 **Jones**, Calvert Richard To: H.T. De la Beche
Heathfield [House],
[Swansea].
17 June 1850
Have you further considered the proposal you made me with respect to taking photographic views of the North Wales mountains in return for expenses? The Lewis Dillwyns are prolonging their stay in Radnorshire, and might be persuaded to join us.
[Cited: Jones 1990, p.170]

808 **Jones**, Harry D. To: H. James
Office of Public Works,
[Ireland].
2 May 1846
The difficulty in procuring professional men induces me to ask for your services on Reports on Pier Applications, notwithstanding what De la Beche has said concerning inconvenience to the Geological Survey.
[See also James, H. to De la Beche, H.T., 7 May 1846, **773**].

809 **Jones**, Rev R. To: Dr.[?W.H.B.]
E.I.C.,
Hertford.
8 February 1847
I send you the belt dug up as a curious pebble. Edward Childs, the donor, was employed in digging foundations of the railway station at Hertford. Also included, a bone, Echinus, and fossil walnut. Is is not time you geologists produced a new theory of the earth? The Glacier one is at least 6 years old.

810 **Jones-Bateman**, M. To: H.T. De la Beche
Folkestone.
23 May [1850 FJN]
We regret you cannot come on 3rd, and hope you can manage Friday June 4th at 7 o'clock.

811 **Jukes**, Joseph Beete To: H.T. De la Beche
Ysbyty Ifan,
Near Pentrefoelas,
Denbighshire.
3 September 1847
Do you recollect a paper of mine on Australian geology? I would like to

have it back, in order to add new information, condense the previous work, and if not acceptable to the Geological Society I will try Jameson's Journal. Nicol will write to you about it.

812 **Jukes**, Joseph Beete To: H.T. De la Beche
Halesowen,
Near Birmingham.
2 July 1849

I am to marry Augusta Meredith the week after the meeting of the British Association. Besides my salary, I shall have only £120 per annum.

813 **Jukes**, Joseph Beete To: H.T. De la Beche
Llangollen.
20 September 1850

What you say of the cheapness of living in Ireland greatly reassures me. I am careless about money matters, and this was the sole cause of my delay in giving you an answer. I accept your offer with many thanks, and will do my best to give satisfaction as Director for Ireland.
[Cited: Herries Davies 1983, p.158]

814 [**Jukes**, Joseph Beete] To: H.T. De la Beche
Dublin.
10 March 1851

I have looked out some sheets that have a good lot of work on them and send them for exhibition. Medlicott tells me you also have in London copies of the counties Wicklow and Kildare on 6 inch scale. If any remarks are made about slowness of work, the following are the reasons: 1st. Difference of maps. 2nd. Collection of soils, taking a large percentage of time and money. 3rd. Frequent change of Directors, involving going back over old work. 4th. Previous Directors had each some other employment and did not give up their whole time. Had we not better try to get rid of the soil affair?
[Part of letter only].
[Cited: Andrews 1975, p.177]

815 **Jukes**, Joseph Beete To: H.T. De la Beche
Waterford.
25 May 1851

Getting even the commencement of the one inch map is good news indeed. The contouring of Kilkenny is finished. With contoured lines it is impossible to engrave the drift. I see no other way than publishing separate drift maps. Shall I write Medlicott to have him purchase maps of Tipperary?
[Cited: Andrews 1975, p.141]

816 **Jukes**, Joseph Beete To: H.T. De la Beche
Monkstown,
Cork.
30 October 1851

I fully concur in your decision regarding private surveys. I recall Ramsay having done something of the sort formerly, but that was a public matter. The principle laid down is exactly the one I should have laid down myself in the matter.

817 **Jukes**, Joseph Beete To: H.T. De la Beche
51 Stephens Green.
4 December 1851

I have come up for the winter, but do not propose to come to London, unless you want me. There is work to be done in the office, which is in a state of chaos. The men will do sections in January. We now have the spare room over mine as a work room, but need furniture.

818 **Jukes**, Joseph Beete To: H.T. De la Beche
Dublin.
18 December 1851

Capt. Cameron of the Ordnance Office is still awaiting an order to commence the one inch map. He also wants copies of our coloured maps and sections. Are any Wexford sections prepared for engraving? I found Portlock's old work rotting and spoiling in the cellar. I will put the office in order, but will miss the Royal Hammerers dinner this year.

819 **Jukes**, Joseph Beete To: H.T. De la Beche
Dublin.
17 January 1852

I hope the accompanying official contains all you want. It is all I have time to send by return of post. I will hold hard the Waterford map. It can wait till next year if there is any chance of the one inch map catching up.

820 **Jukes**, Joseph Beete To: H.T. De la Beche
[Undated]

I made a mistake in the number of soils etc., sent up last year. The figure ought to have been 408, not 370.

821 **Kane**, Robert To: H.T. De la Beche
51 Stephens Green, London.
Dublin.
11 April 1841 [sic, but ?1847]

I have only just received your note and extract from Trueman's Journal. It seems to be the usual town gossip, and I contributed no information. These mistakes are common ones. I will be happy to set them to rights with your cooperation. No hint from you about maps, no sections, no intimation that Wicklow was published.

822 **Kane**, Robert To: H.T. De la Beche
Laboratory, Cecilia Street,
School of Apoth. Hall.
16 October 1845

Have only just received your note on arriving in town this morning. I shall be here tomorrow between 11 and 1, and absolutely at your disposal.

823 **Kane**, Robert To: H.T. De la Beche
Royal Dublin Society. London.
16 March 1846

I forward you, through Longman, a copy of a work on the Turbine which I have edited, as a testimony of my respect for your Scientific eminence.

824 **Kane**, Robert To: H.T. De la Beche
London.
8 September 1846

I hoped to catch you, but missed you by a few minutes. Oldham will have told you about the arrangement I've made for his entering on his apartments and removings. I hope to be back in 5 or 6 weeks.

825 **Kane**, Robert To: H.T. De la Beche
51 Stephens Green.
14 November 1846

Please send me set of forms for Bills, Receipts etc., which you use in the London Museum as I want to see them before I get any printed for myself.

826 **Kane**, Robert To: H.T. De la Beche
Museum Irish Industry,
51 Stephens Green.
26 August 1847

The Mining Co. of Ireland have not made out returns for me. Their Secretary has promised to send you their report for 1845-46, which he says contains what you want.

827 **Kane**, Robert To: H.T. De la Beche
51 Stephens Green. Geological Survey,
21 March 1848 Craig's Court,
  Charing Cross,
  London.

The bearer Mr. Hugh Maguire is a surveyor and engineer who is anxious to get employment under you on the Survey of London. He has been well trained on the Ordnance Survey of England and has since been under the Board of Works here, where he gave full satisfaction. My own impressions of him are very favourable and I should feel very much obliged to you if you could do anything for him.

828 **Kane**, Robert To: H.T. De la Beche
51 Stephens Green,
Dublin.
25 July 1848

Penny's son has been employed under the Board of Works for the last 18 months as Surveyor and Draughtsman. Through cessation of works, he is now out of employment. Can you put him in the way of employment on sewerage or Survey duties in London?

829 **Kane**, Robert To: H.T. De la Beche
Queens College, Geological Survey Office,
Cork. Craig's Court,
13 November 1849 Charing Cross,
  London.

I enclose the letter of thanks. All are pleased that we are to get Falconer's

830 **Kane**, Robert To: H.T. De la Beche
Museum of [Irish] Industry,
Dublin.
12 November 1852

Thank you for your note with proposals of cooperation. Copper and tin collections from Swansea and Cornwall will be most useful. Iron industry specimens would be welcome immediately as I am anxious to advance that at once.

831 **Kane**, Robert To: H.T. De la Beche
Museum of Irish Industry, Jermyn Street.
Stephens Green,
Dublin.
23 June 1853

Thank you for the specimens. I now hope you will continue your inestimable assistance. Iron stones, iron products, collection of Austrian steels etc. are all welcome.

832 **Kendle**, John To: H.T. De la Beche
Tavistock. Swansea.
12 December 1838

The absence of our agent in Ireland has delayed my reply to your letter. Here is his reply, which I hope will suit your purpose.
[Enclosure:] We are still working the Culm Mines near Bideford but for the last year little has been done. Accounts of workings in the recent past give some indication of successes, but no authentic accounts of the ancient workings are available.

833 **Kennedy**, John Studdert To: [H.T. De la Beche]
51 Stephens Green.
24 August 1854

I have always and do now prefer England to India. As I thought England was out of the question, I felt my choice was between Ireland and India; settling on the latter and borrowing money I feel I must go on with it. I regret any inconvenience.

834 **Ker**, Charles Henry Bellenden To: H.T. De la Beche
5 January 1835 Tiverton,
 Devon.

I have proposed to Knight to publish How to observe in parts. 1. Geological, including sketching and country surveying. 2. Fine Arts, painting, sculpture etc., 3. Botany, agriculture, weather, 4. Men and morals, and general statistics. As soon as I get his answer I will let you know and go quam cito to press. Greenough rejoices in your doing the geology.

835 **Ker**, Charles Henry Bellenden To: H.T. De la Beche
Park Road, Tiverton.
London.
3 April [1835]

Greenough made 3 or 4 verbal amendments in the first pages not really worth sending to you. He sent back the proofs saying that the ideas as he could judge were all sound. To business. I am the Editor, and mean to write a Preface, but I do not mean to put my name to it, because it would be a humbug, as my name would of course be no guarantee. But I should like you to add 2 or 3 pages as to how empirically to lay down a country on the plan you explained to me.

836 **Ker**, Charles Henry Bellenden To: H.T. De la Beche
Park Road.
[?]2 August [?1835]

The time draws near when we must begin to think of How to observe. I have spoken with Knight on the plan of the work, and he has agreed to publish it. You may have from 40 to 60 pages of small 8vo., and you must put your best part foremost and make the matter very popular and attractive, valuable to a traveller both in England and abroad. There must be sections etc. I should also like to have sketches.

837 **Ker**, Charles Henry Bellenden To: H.T. De la Beche
Lincoln's Inn. Falmouth.
12 April 1837

I lose no time in replying to your letter. I have sent it to Greenough for his advice. I heard from various quarters some time ago that they were persecuting you and suspected that Lyell and Murchison were at the bottom of it. Greenough told me you were safe, that Spring-Rice had been informed of the truth so that you would be well looked after, and so I thought it but necessary to write to you. My opinion is that Colby is overdoing it. I know he is not in very good odour and I suspect he is wishing to appear busy. You should write to the Ordnance forthwith. Some theory must be added to a survey, that others agree with you, that you would defend what you had done when your report was published. Write a strong note to Colby saying that this is what you proposed to do.
[Cited: North 1939b, p.1053; Rudwick 1985, p.204]

838 **Ker**, Charles Henry Bellenden To: H.T. De la Beche
Lincoln's Inn. Falmouth.
14 April 1837

I have seen Greenough. He says that you have been very ill used, that he and others are doing all they can to defend you. He agrees with me that you should take the matter up on a high ground and write immediately to the Master General, and require enquiry - the best enquiry being a select number of scientific men. Stress the necessity of doing parts theoretically unless they authorise an actual field survey. On geological questions there must be doubts and different opinions. Send a copy to Colby, but I suspect he wishes to show his activity. Col. Fox is unfortunately at St.Annes, but I shall see him on Sunday, and will take care that he knows the whole case. Greenough tells me the matter is to be debated on Wednesday. I am out of favour with him ever since the map business, but he has been very civil in this matter for your sake.
[Cited: Rudwick 1985, p.204]

839 **Ker**, Charles Henry Bellenden To: H.T. De la Beche
17 May 1837 Map Office,
 Ordnance Office,
 Tower.

I sent you Col. Fox's letter and his account of what Sir H. Vivian had said. I talked over your matter with Colby and told him that I and Fox had advised you to go the fountain head at once, and that Sir. H. Vivian had stated he had heard no complaint etc., and that when he did, every possible attention should be paid to you. I thought it as well that Colby should know this. He was very civil, but I don't think he liked the notion of the fountain head. He said he had proposed your getting a portion of Cornwall finished and before the public before you went into Wales, and in this I acquiesced, as far as I could do so, not knowing whether there were particular circumstances that made this impossible or inconvenient. I hope to hear that you have got into smooth waters again, and are contentedly going on with your work.

840 **Ker**, Charles Henry Bellenden To: H.T. De la Beche
Lincoln's Inn.
27 March [?1848]

You were so good as to say you would send to Jamaica for some plants and ferns. If you can do this I shall feel much obliged. If they come by the mail steamers they might be packed up in dry moss and would come alive as they would not be more than 15 or 20 days en route. Pray do not forget what you can do for me.

841 **Ker**, Charles Henry Bellenden To: H.T. De la Beche
Lincoln's Inn.
10 November [?1850]

I have been in high exchange of correspondence with Mr. Dillwyn Llewellyn. I went to see his beautiful place but he was not there. I could not learn where your daughter was, my old friend, so I did not see her. Now do not be such a humbug but get me the orchids from Jamaica, at least as you promised.

842 **Ker**, Charles Henry Bellenden To: H.T. De la Beche
[No date]

I have great pleasure in sending the enclosed. I dare say General Fox will do what you wish. He is rather capricious at times. When your labours are over will you fix what day after next week you will come to us.

843 **Killaly**, Richard G. To: H.T. De la Beche
8 Talbot Street, Ivy Cottage,
Dublin. Lodge Road,
11 December 1832 Regent's Park,
 London.

Enclosed Weinbohla as I saw it at the beginning of this year. Also a sketch of a Quarry near Grillenburg. I gave Greenough some sections which you could see. The Weinböhla quarry limestone is by no means pure. The Grillenburg quarry is of sandstone.

844 **Kington** V.F. To: H.T. De la Beche
Ilsington, Post Office,
[Devon]. Ivybridge.
21 May [No year] Beechwood,
Plymouth.
[Readdressed to:] Plympton.
[and again to:] Royal Hotel, Plymouth

I am glad you are in Cornwall carrying on your important work. I shall be glad to be of assistance, and receive you when you establish headquarters in this neighbourhood. I will meet you if you inform me of your arrangements.

845 **Kleville**, C.L. To: [H.T. De la Beche]
Boulogne.
28 May 1849

I cannot leave Boulogne without expressing to you my fond memories of my voyage in your country and particularly the feelings of recognition for all the kindness shown me. I have few claims to such kindness, or rather only one - the friendship and esteem of Elie de Beaumont. I will be seeing him in Paris.

846 **Knight**, Charles To: H.T. De la Beche
[London]
24 February 1836

Account current with Chas. Knight for Geological Manual and Geological Researches.

847 **Krigstatscher**, Jos. To: [?H.T. De la Beche]
32 Old Compton Street,
Soho,
London.
10 May 1842

I have discovered the true nature of light and cause of motion. My discoveries are greatly at variance with established theories. Please inform the Society of the contents of this letter. Astronomical and geological phenomena are connected, so that the changes in the planets explain the changes on the earth.

Labouchere, Henry To: H.T. De la Beche
23 December 1850
[see **2179**]

Labouchere, Henry To: H.T. De la Beche
11 July 1852
[see **2180**]

848 **Landale**, D. To: Capt. William Ramsay R.N.
Edinburgh.
28 June 1850

The field of coal on this property has never been explored. The coal is of anthracite type, and trap distorts the strata considerably. It may be worth working if it is of the quality Sir Henry has given from the small specimen. I will be glad to have Sir Henry's report and hope it will warrant an order for 1000 tons, and lead to the opening out of this coalfield.

849 **Leach**, William Elford To: H.T. De la Beche
[?]Hoxton. Lyme,
28 September 1826 Dorset.

Thank you for your letter received today. The box of shells also arrived. Please send the ones which are to be drawn by Sowerby and I will arrange it. I am in the country with Abernethy. If you have not published your remarks on the temperature of the Lake of Geneva, allow me to arrange it for the Annals of Philosophy, as the Edinburgh Philosophical Journal is a mere jest with our continental brethren.

850 **Lemon**, Charles To: H.T. De la Beche
46 Charles Street. Athenaeum.
25 February [?1838]

I return your prospectus and give you my views on your proposed school of mines. The objects of your plan are excellent. Difficulties come in England from the fact that mining is a private speculation; thus Government interference not called for. Science is clearly connected with mining but not its economical pursuits. Your school would resemble the Academy at Lyons, a private association under the protection of Government.

851 **Lemon**, Charles To: H.T. De la Beche
Carslow,
Penryn.
14 October [1848 FJN]

I had little time before the Meeting. I would be obliged by your looking over what I have written and tell me what to alter before it is printed in the Report. Some curious specimens were produced at the meeting. Peach brought a trilobite which has been pressed laterally till it is flat.

Lemon, Charles To: H.T. De la Beche
Carslow.
4 January [1852 FJN]
[see **2267**]

Lemon, Charles To: H.T. De la Beche
Carslow.
10 October [no year]
[see **2268**]

852 **Lemon**, Charles To: H.T. De la Beche
Carslow,
Penryn.
15 August [No year]

We shall be short of matter at the Geological meeting at Penzance on 16 October. Can you give us something? What about extracts from the notes from the Books of the Bishop of Bath & Wells, relating to Mendip Mines? Murchison is coming at the time, and I wish that you could meet him.

853 **Leopold**, King of the Belgians
2 September 1852

Nomination of H.T. De la Beche Chevalier de l'Ordre de Leopold.

854 **Le Play**, Pierre Guillaume Frederic To: H.T. De la Beche
Ministere des Travaux Publics,
Commission de Statistique de l'industrie Minerale,
Paris.
3 January 1848

The Commission on the Statistics of Mineral Industry present to you a series of 14 volumes published up until the present day by the Commission. You will find much of interest in it, and perhaps a model of what could be stated about Great Britain. Ordnance maps and your own publications would be acceptable here. We are completely at your disposal for anything you may desire.

855 **Levrault** & Co To: H.T. De la Beche
Paris. Ordnance Map Office,
15 April 1840 Tower,
London.

Since writing to you last year concerning the reduced sale of books on geology, we now find that only a few copies of your Manual remain, and we would like to prepare the new edition so that it would appear at the end of October, or sooner if possible. Are there any changes or additions you would like to make? In case of difficulties we could have recourse to Elie de Beaumont who would certainly find it a pleasure to help this work by his advice.

856 **Lewis**, George Cornewall To: H.T. De la Beche
Treasury.
10 December [?1851]

The Chancellor of the Exchequer thinks that the expense of making the Ordnance Map of Ireland ought to be included in the Ordnance estimates. Please communicate with them on the subject. Thank you for the inaugural lecture.

857 **Lincoln**, Henry Pelham Fiennes Pelham Clinton,
Earl of To: H.T. De la Beche
Office of Woods etc.
9 June 1842

I am anxious to learn your opinion as to future superintendence and management of all mining affairs of the Duchy of Cornwall, persons to be employed, their employment limitation, salaries etc. No one better than yourself can judge.

858 **Lincoln**, Henry Pelham Fiennes Pelham Clinton,
Earl of
24 September 1844
[Copy of Minute]

De la Beche's proposal is very liberal and disinterested, but I cannot let the matter be so arranged. True, I did not think the whole expense was to exceed £150, but time is another question. A short report should go to the Treasury.

859 **Lincoln**, Henry Pelham Fiennes Pelham Clinton,
Earl of   To: Viscount Canning
Irish Office.
4 May 1846
Your letter and De la Beche's have been sent to Capt. James to show them to Col. Jones, and ascertain if some other competent person could be found for the piers inquiry. Please accede to my original application for James to be appointed.
[See also James, H. to De la Beche, H.T., 7 May 1846, **773** and Jones, H.D. to James, H. 2 May 1846, **808**].

860 **Lincoln**, Henry Pelham Fiennes Pelham Clinton,
Earl of   To: H.T. De la Beche
Whitehall Place.
17 August 1846
Thank you for the First Volume of the Memoirs of the Geological Society and Museum of Economic Geology. I feel as if I stood somewhat in the relationship of a stepfather to both institutions.

861 **Lincoln**, Henry Pelham Fiennes Pelham Clinton,
Earl of   To: H.T. De la Beche
Warrens Hotel.
2 January 1849
Thanks for the Second Volume of the Memoirs of the Geological Survey. Excuse the delay in acknowledgement. My desire to assist your valuable labours was not forgotten.

[See also **Newcastle-under-Lyme**, Henry Pelham Fiennes Pelham Clinton, 5th Duke of, **1040**].

862 **Lindley**, John
1832-33
[Lecture note-book, containing a Syllabus of Five Lectures upon Fossil Botany, six fossil lectures. Letters of E. Turner to J. Lindley, 11 March 1836; C. Lyell to J. Lindley, 1 October 1833 and E.H. Bunbury, to [?]J. Lindley, 28 January 1833, preserved on last pages].

863 **Loftus**, William Kennett   To: [H.T. De la Beche]
Southampton.
29 January 1849
On leaving England, my thanks for favours shewn. You shall not regret the interest you have taken in me.

864 **Loftus**, William Kennett   To: [H.T. De la Beche]
Diyarbakir.
30 March 1849
I am on my road to Mosul down the Tigris. I have formed some idea of the geology of the Taurus. In my hurry to leave England, I did not obtain some books. Could you send them to me? Concerning specimens, have you a copy of the official Foreign Office letter?

865 **Loftus**, William Kennett   To: [H.T. De la Beche]
Baghdad.
5 November 1849
I am delayed in Baghdad. Fever is rife. Since the great plague of 1830 such an unhealthy season has been unknown. I visited Babylon, the shrines of Sheah, the Mosque at Meshed Ali, etc. I plan sending a large box of specimens and sword-blades in about month from Bombay.

866 **Loftus**, William Kennett   To: [H.T. De la Beche]
Baghdad.
26 December 1849
I have despatched a box of specimens to the British Museum and the sword-blades previously mentioned, enamel on kiln-brick, and pebbles collected in Mesopotamia. We march on tomorrow from Baghdad. Our stay has been a waste of 8 precious months.

867 **Loftus**, William Kennett   To: H.T. De la Beche
S. Base of Mt. Ararat,
Persia.
9 September 1852
[continued:] 2 November 1852
Baghdad.
The term of the Turko-Persian Frontier Commission is almost run. In five days' time we shall reach Bayazid, and our labours there terminate. My last letter was dated from Kirrind in the summer of last year. You may have heard through Dr. Brown that I was engaged in the geology of the district. I took the liberty of sending two boxes of rock specimens to your care from Ararat through Col. Williams. As soon as I get home I shall have to prepare my geological report to be sent in with the others. There is plenty of work cut out for me.

868 **Logan**, William Edmond   To: H.T. De la Beche
Cambrian Place,
Swansea.
14 April 1838
My visit to the Pyrenees was made in July 1834. I saw sulphuret of copper associated with anthracite on Mount Marconton. Anthracite constituted the chief body of the bed, which was conformable all round the mountain. The best specimens are at the Institution.

869 **Logan**, William Edmond   To: H.T. De la Beche
Swansea.   Athenaeum,
19 June 1838   London.
I have tried a test or two on the specimens as collected around our copper smelting furnaces. One is undoubtedly arsenious acid. A second is probably carbonate of alumina. A third is pure copper. If I should leave Swansea before you return I shall bequeath the land of Gower to you leaving Benson as my executor. I am at a loss what thickness to give to the Gower limestone.

870 **Logan**, William Edmond   To: H.T. De la Beche
4 New Broad Street,   Rhyddings,
London.   Swansea.
21 June 1839
Herewith an extract from your work by a friend of mine who wants information on Jamaican minerals near the Swift River. I saw Phillips at St. Thomas. He is much pleased with his nearly completed laboratory at Craig's Court.

871 **Logan**, William Edmond   To: H.T. De la Beche
4 New Broad Street,   Swansea.
London.
11 July 1839
In order to prevent the appointment of Keeper of Mining Records slipping out of your hands I think Mr. Edmond had better accept the situation for the present, and think later. I could not call on you at Swansea. Edmond will not be required for 6 months to come. So much the better. It will give me time to replace him.

872 **Logan**, William Edmond   To: H.T. De la Beche
4 New Broad Street,   Bridgend,
London.   Glamorgan,
27 November 1839   South Wales.
Thank you for the consideration towards my cousin in the matter of Keeper of Records. W.P. Struvé, is your man. He will make his fortune at it. I am annoyed at my frequent trips to London. Things will not be as perfect as they should be. The protractor is a capital sort of implement both in and out of the field.

873 **Logan**, William Edmond   To: H.T. De la Beche
4 New Broad Street,   Collenna,
London.   Cardiff.
1 August 1840
I have yours of 24th July. I must go to Canada for 6 weeks. On my return I will put the dry proof to rights. I shall put you in my will for all my sections. I have built an accurate column of measures. My brother in law, Mr. Gower, offers you his home in Pembrokeshire when you examine that part of the country.

874 **Logan**, William Edmond   To: H.T. De la Beche
Montreal.   Museum of Economic Geology,
5 October 1840   Craig's Court,
                                                   London.
I intend to return to Britain in November, when I will put my lines on the dry proof of the Ordnance map. I have been in the field in three states bordering on Canada, and gathered specimens for the Museum and Swansea Institution. I have discovered vertebrate fossils in previously supposed transition rocks. It is difficult to account for anthracite in cracks in limestone. I will send or bring specimens.

875 **Logan**, William Edmond   To: H.T. De la Beche
Liverpool.   c/o L.L. Dillwyn,
19 October 1841   Swansea.
I paid a visit to the coalfields of Pennsylvania and Nova Scotia. Both

regions confirm observations in South Wales. I now consider it truth universally applicable that immediately below coal will be found a bed of fireclay containing Stigmaria ficoides. I mentioned it to Lyell in New York, who wishes to quote me in his next edition. I must visit Scotland. I will peep into collieries there in search of underclay.
[Cited: North 1936a, p.52]

876   **Logan**, William Edmond   To: H.T. De la Beche
       4 New Broad Street,   Cardigan,
       London.   South Wales.
       3* December 1841

What Lyell has said of the Stigmaria beds of Pennsylvania is what I wrote him from Halifax. I am pleased he quoted my name in a laudatory way. But it is your commendation I need, as the Canada Geological Survey will be entrusted to me on your reference. Buckland will give me a character reference. Before I go next summer I'll read papers on Canada at the Geological Society.
[*According to Logan, W.E. to De la Beche, H.T., 11 December 1841, 877, this date should be 8 December].
[Cited: Winder 1965, p.112; Dott 1996, pp.130, 139]

877   **Logan**, William Edmond   To: H.T. De la Beche
       4 New Broad Street,
       London.
       11 December 1841

My specimens will be sent to Swansea and displayed. Those for the Museum can be sent afterwards. Concerning the Canada survey, I thought that it would not be a bad thing that its director was approved of by the head of the Survey of this Country. Could you educate an assistant for me? Did you see my flint gravel at Cardigan? Buckland said he would read my paper on it at end of one of his own at the Geological Society.

878   **Logan**, William Edmond   To: H.T. De la Beche
       Swansea.   Museum of Economic Geology,
       3 February 1842   Craig's Court,
                                                                  Charing Cross,
                                                                                  London.

Thanks to you and Buckland in respect to the letter from the Bishop of Oxford to Sir Charles Bagot. I should be particularly obliged by your testimony in my favour. I have been working hard on my sections and maps, and have employed Struvé.

879   **Logan**, William Edmond   To: H.T. De la Beche
       Swansea.
       26 July 1842

I leave for Canada from Liverpool on 4th August. Where shall I leave the Ordnance dry impressions on which I have put my lines and sections? I'll perfect them when I return towards the end of the year. I should like to see you for a few hours before I leave. Murchison sent me a letter of introduction to Bagot. You promised me one to Rawson. Murray is with me here, and when I start out, shall send him to you for practice.

880   **Logan**, William Edmond   To: H.T. De la Beche
       Liverpool.
       4 August 1842

I received your letter with its enclosure addressed to Rawson. He has already left for Canada. I give you my address in Canada and the frequency of mails. Concerning my Penclawdd section, I think it ought to be well examined before a positive opinion is pronounced. My pocket books are with Edmond. Buckland's word to Bagot's brother has worked wonders in Canada.

881   **Logan**, William Edmond   To: [H.T. De la Beche]
       London.
       24 April 1843

I am anxious to carry out the Geological Survey of Canada as well as possible and wish to have assistance of the Director of the Geological Survey of Britain, especially in the palaeontological department. Canada might become the measure of comparison between the continents of Europe and America. The main object of the Canada Survey is the determination of mineral riches. A system of Annual Reports appears to me objectionable. What is your opinion?
[Cited: Harrington 1883, pp.136-138]

882   **Logan**, William Edmond   To: H.T. De la Beche
       Halifax.   6 Craig's Court,
       31 May 1843   Charing Cross,
                                                                           London.

I landed in Canada yesterday, and start northward walking to Gaspé to look at the Triassic, Permian, and Carboniferous gypsiferous rocks. I expect to be on my ground in about a fortnight. There is said to be coal on southern boundary of the Gaspé district. If there are no gypsiferous rocks to the north of it, I fear it will not be the productive part of deposit. I wrote to Struvé about the Caswell Bay section.

883   **Logan**, William Edmond   To: H.T. De la Beche
       Montreal.   6 Craig's Court,
       20 April 1844   Charing Cross,
                                                                                London.

How come you have never written to me? I have reduced all my measurements from fieldwork to a Carboniferous column 14,590 feet thick. It may be useful in your coal report. Stigmariae occur under coal beds. I have worked all summer on the Gulf of St. Lawrence. I collected fossils from Devonian and Silurian rocks. I have studied the New York geological collection at Albany. What is the consumption of lithographic stone from Kelheim? I would like to show the Canadians they have something of value. I enclose a description of Brontes Logani, and a tracing.
[Cited: Harrington 1883, pp.181-184]

884   **Logan**, William Edmond   To: H.T. De la Beche
       Montreal.
       11 November 1844

I have spent three and a half months in the Forest, in primitive conditions. Did you get my Joggins section? Will it be useful to you in the Welsh coalfield? The eastern section of this Province will not be a replica of the Western. The New York geologists cannot have made any serious mistakes. I will refer to nobody's system in arriving at my conclusions. My only conclusions will be a description of strata and their corresponding fossils. They can do what they like with them.

885   **Logan**, William Edmond   To: H.T. De la Beche
       Montreal.
       12 May 1845

I inform you of the position of my campaign in Canada. There was not a dissenting voice in the House of Assembly when a Bill I drew up was passed. No changes were made in it. We are forced to map topographically. The micrometer is a useful instrument. Canada is an enormous territory. I hope one day to survey it in the employ of the British Government under you. I have been informed of coal in Saskatchewan and Oregon. During the coming season, Murray will return to Gaspé, and I to Ottawa.
[Cited: Harrington 1883, pp.229-236; Zaslow 1975, pp.46, 50]

886   **Logan**, William Edmond   To: H.T. De la Beche
       Montreal.
       27 December 1845

I send you by this mail a report on the progress of my Survey, ordered to be printed by the House of Assembly. In these reports I intend to give fact not theory. I have introduced my Joggins section as a fact. I have sent a copy to the President of the Geological Society. I have just returned from exploration and mapping near Lake Timiskaming. I have found pre-Silurian rocks.

887   **Logan**, William Edmond   To: H.T. De la Beche
       Montreal.   Museum of Economic Geology,
       10 December 1846   Craig's Court,
                                                                 Charing Cross,
                                                                           London.

I have returned to winter quarters after a hard summer and autumn on Lake Superior. My investigations have been of the economic kind. Copper finds are the main event. Herewith a detailed description of ores and locations. Silver is associated with the copper. I have given a letter of introduction to you for Forest Shepherd a mineral surveyor. I have sent you a copy of my 1844 Report of progress. Verneuil examined our fossils in my absence. There are new genera and new species among my lower Silurians.

888   **Logan**, William Edmond   To: H.T. De la Beche
       Montreal.   Museum of Economic Geology,
       7 January 1848   Craig's Court,
                                                                 Charing Cross,
                                                                           London.

On examining fossils from localities in the Green Mountains, I shall have to raise their strata from Lower to Upper Silurian. I expect the formation will run all the way down to Gaspé. This will make a sweeping change in the geology of the Eastern States of the American Union. I will visit the south side of the St. Lawrence next season, and endeavour to follow the limestone as far as Gaspé (300 miles).

889 Longman & Co. To: E. Forbes
Paternoster Row. 6 Craig's Court,
19 May 1847 Charing Cross.

Thanks for yours of 30th ult., and one from De la Beche. There are difficulties from the present incomplete state of the Geological Survey so the work should be brought out up to present state of information. Conybeare's objections can be met or superceded, and we can have another more perfect edition later. What course do you propose?

890 Lonsdale, William To: H.T. De la Beche
Geological Society,
London.
22 November 1834

Herewith a copy of Lindley's letter containing information on the Devon fossils. Have you heard that Sedgwick has a preferment to a stall at Norwich? Thank you for a copy of Theoretical Researches. Babbage spoke highly of it.
[Enclosure: Copy J. Lindley to W. Lonsdale 21 November 1834] The following species, as far as I can make them out, are contained in fossils sent by De la Beche. They are all coal measure species.
[Enclosure cited: Rudwick 1985, p.97]

891 Lonsdale, William To: H.T. De la Beche
Geological Society,
Somerset House.
30 November 1835

Your Weymouth paper is at last ready. How many copies do you wish to have? Buckland has taken his 40 copies. Agassiz has gone but hopes to be back in two years. Beck of Copenhagen has been here. He does not think Deshayes is a first rate conchologist. I have been working on Chalk fossils. Have you found fossils in the old rocks?

892 Lonsdale, William To: H.T. De la Beche
Geological Society. Truro,
22 December 1835 Cornwall.

I saw the map at the Tower and Robe allowed me to take it home. I have examined it with great pleasure and it contains much matter for serious thought, especially the problem of the trap rocks in the granite. The box of organic remains arrived safely yesterday. I send a copy of your Weymouth paper and a copy of Hopkins' paper on cracks and fissures.
[Cited: Rudwick 1985, p.229]

893 Lubbock, John William To: H.T. De la Beche
Mansion House Street.
3 May 1848

I am sorry that I cannot make out your writing but I think that in the case you describe the periods would probably be too short to account for the observed phenomena. The main object of my communication is to point out the limitations under which the mathematical theorem of the invariable position of the earth's axis of rotation holds true.

894 Lutke, F. To: H.T. De la Beche
St. Petersburg.
17 May 1847

His Imperial Highness the Grand Duke Constantine in recognition of the attention which you showed him during his visit to the Museum of Practical Geology, desires to offer to the said Museum a collection of miners' tools used in Russian mines. Please acknowledge receipt when they have arrived.

895 Lyell, Charles To: H.T. De la Beche
23 June 1831 Athenaeum Club,
  Pall Mall.

Curious Extinct Species, Anoplotherium palaeotorium - Weaponless Old Tory [Waggish account of extinct Tories] I have put in the citation from King correctly now, or at least as nearly so as will suit the joke. Hoping you will strike while the reforming irons are hot.

896 Lyell, Charles To: H.T. De la Beche
2 Raymond Buildings, Lyme Regis,
Grays Inn, Dorsetshire.
London.
19 August 1831

Thanks for your volume. I have just returned from an examination of extinct volcanoes of Eifel. I have quoted you in my work, on tropical rains. Have you any more definite information on the Port Royal earthquake? Have the submerged houses disappeared? Similar events in the 1814 Cutch earthquake. What in the world can a wave rushing inland have to do in invalidating a subsidence?

Lyell, Charles To: J. Lindley
16 Hart Street,
Bloomsbury Square.
1 October 1833

You stated the other day that there were some remains in our English coal fields which satisfied you that the plants had not been drifted from great distances. Are the Baffin Bay coal plants really of tropical form? Can you ascertain the furthest limit to which true coal plants have been found in the North?
[See Lindley, J., 1832-33, **862**]

897 Lyell, Charles To: H.T. De la Beche
2 January 1835

I met a friend of yours, Mr. Donne, in the coach from Hastings, who expressed regrets at not seeing you so often. I read Boase on way down, a book after Greenough's own heart. I read yours on way back, infinitely the best thing you have done. Defer further combat by letter on the Devon coal plants. There was nothing for you to complain of in the debate, but Greenough got nettled. I have a letter from de Beaumont who has been to Somma and Etna.
[Cited: McCartney 1977, p.30; Rudwick 1985, p.112]

898 Lyell, Charles To: H.T. De la Beche
13 March 1835 Tiverton,
  Devon.

Your proposition was laid before the Council. A debate on the duties of the Foreign Secretary ensued. The general conclusion was that the reading of notices on foreign works would interfere seriously with the business of meetings. Have you ever thought of publishing a journal to embody such? Would Knight fancy it?

899 Lyell, Charles To: H.T. De la Beche
Hart Street,
Bloomsbury.
3 June [1835]

I must request you to write short notice on the map such as may be put in the Proceedings. You may speak about the map but let there be a text or there is no keeping any order in our business. I must even exclude one of Sedgwick's because of the press of papers.

900 Lyell, Charles To: H.T. De la Beche
London. Truro,
5 April 1836 Cornwall.

Concerning the £25 for Deshayes, a letter of thanks was received. He has accumulated an immense body of facts on the subject of time and variations in species. I wish he were richer as he is an enthusiast who would rather explain the philosophy of his science than slave for booksellers. The Geological Society is the first to ever give him aid.

901 Lyell, Charles To: H.T. De la Beche
16 Hart Street. Ordnance Office,
1 December 1836 Tower.

Herewith some information for you as Foreign Secretary of our doings at the Geological Society. Two new secretaries should be appointed. Our new President is a difficult point. I asked Whewell to allow me to nominate him. Greenough, Murchison and others are equally favourable. I was glad to hear from Sedgwick that you are working well and successfully in Cornwall.
[Cited: Rudwick 1985, p.201]

902 Lyell, Charles To: H.T. De la Beche
16 Hart Street, Ordnance Map Office,
Bloomsbury. Tower.
2 February 1837

I require a clarification of your position on the Devon culm-measures for my anniversary speech. Is the following a fair account of the controversy? I believe Weaver once thought he had discovered coal-plants as low down in Ireland but he has since most openly and studiously recanted. If you are prepared also to say you have changed your mind, give me the opportunity

of saying it on your behalf in my speech.
[Cited: North 1939b, p.1054; Rudwick 1985, p.195-196]

903 Lyell, Charles   To: H.T. De la Beche
2 August 1843

Phillips has told me you are in Bristol. I am coming myself to get up the geology of the Magnesian Limestone and dolomitic conglomerate. Your survey is questioning accepted interpretations. Can you show me the way with your preserves? A letter from Logan confirms my reading of the Fundy Bay gypsiferous formation.

904 Lyell, Charles   To: H.T. De la Beche
11 Harley Street.
2 November 1846

I was relieved to hear from Horner that you have acceded to our wishes. Bunbury said modestly he could not do justice to the office of Foreign Secretary, but his scruple might be overcome. Neither the President nor I can with propriety stir in the matter, but must leave it to you.

905 Lyell, Charles   To: H.T. De la Beche
11 Harley Street.
15 June 1847

Herewith Horner's address in Germany. If I am not at Council please let a bill of Dinkel's for lithographs be paid, as he is afraid of our parting for vacation without getting his cash, now that he has decided not to follow Agassiz to the U.S.

906 Lyell, Charles   To: H.T. De la Beche
1 Harley Street.
20 February [1848 FJN]

I had a discussion with Brown having learnt from Faraday that he approved. Brown is favourable to limited Presidentships of 2 or 3 years. He wishes to try Herschel again. He suggested Lord Burlington. Sir J. Clark will vote for Brown if we do.

907 Lyell, Charles   To: H.T. De la Beche
11 Harley Street.
13 March [1848 FJN]

Heschel deprecates the necessity of the pain of declining. It would be unfair to him as well as useless to go as a deputation. Please tell Grove and others.

908 Lyell, Charles   To: H.T. De la Beche
11 Harley Street.
15 March 1848

I agree with Smyth that Brown is our man. More acknowledged all over Europe than any one F.R.S. His Knighthood from the King of Prussia shows how differently foreigners estimate him.. Roget is bitter against him. The objection against Brown is not sound.

909 Lyell, Charles   To: H.T. De la Beche
11 Harley Street.
20 March 1848

I will attend Council and Evening Meeting, but I cannot preside at the Club. Ask Greenough or someone else to do so. Lubbock assents to my proposing him as F.G.S.

910 Lyell, Charles   To: H.T. De la Beche
11 Harley Street.
2 May 1848

I will do duty at Council, Club, and Evening regularly in your absence. I think the affair is an important one and will take pains while you are away.

911 Lyell, Charles   To: H.T. De la Beche
11 Harley Street.
25 May 1848

Concerning my election at the Royal Society, Brown begged me to put off my proposal. I presevered and was seconded by Sabine. Brown moved an amendment and I postponed my own. Owen thinks we should persevere and get Grove. All agree that Grove and Owen would be best.

912 Lyell, Charles   To: H.T. De la Beche
10 June 1848

I presume you will be in the chair on Wednesday. The last meeting saw our Museum report referred to us to cut it down for publication. At the Royal Society Brown started Owen again, but I have got it right at last. Brown recommends a 4 year term for all officers.

913 Lyell, Charles   To: H.T. De la Beche
11 Harley Street.
19 June 1848

Brown proposes Bell as Secretary of the Royal Society, vice Roget. I regret this turn of the affair, as Bell was against us in everything, wanting a certain Duke as P.R.S.

914 Lyell, Charles   To: H.T. De la Beche
11 Harley Street.
31 May 1849

You were right when on retiring from Council you warned us about the Foreign Secretary. Show my letter to Phillips. Sabine perseveres with unscientific nominees, and yet he says the British Association must have a scientific President.

915 Lyell, Charles   To: H.T. De la Beche
11 Harley Street.
31 October 1849

My father's illness calls me away to Scotland. Attend the Council if you can. None of the candidates for Assistant Secretary were geologists, so I have proposed J. Morris. Nicol can serve until Xmas. Anything would be better than an inferior candidate who would be a fixture for life.

916 Lyell, Charles   To: H.T. De la Beche
11 Harley Street.
6 March 1850

I received news of my mother's death this morning. I have asked Murchison to chair the Council of the Geological Society meeting, or you if he cannot. Rupert Jones is the best man for Assistant Secretary. If the Commission require more work, Playfair should be invited to do it professionally.

917 Lyell, Charles   To: H.T. De la Beche
11 Harley Street.
4 December 1851

The last sheet of the new edition of my manual is to be printed next week. Logan's Canada tortoise is mentioned. You would do me and Logan a good turn if you could let me put in a sentence or two on his latest discoveries, Lower Silurian bone and footprints.

918 Lyell, Charles   To: H.T. De la Beche
11 Harley Street.
19 June 1854

Have been looking in your "Geological Observer" to find your views on the age of beds containing Thecodont Saurians. They used to be regarded as Permian. Have we to change that? Is the question settled? I hope to send you copy of my American Report in a day or two.

919 Lyon, J.W.   To: H.T. De la Beche
Spring Garden.   Swansea.
10 August 1848

Jones, my agent at Swansea has sent the enclosed results of a trial of Lyon's patent fuel against Warlick's. Lyons fuel is by far the best to use.
[Enclosure: statement by Jones].

920 Macfadyen, James   To: H.T. De la Beche
Hope,   Post Office,
St. Andrew's,   Plymouth.
Jamaica.
16 August 1830

I will gladly make the collection of Jamaican ferns which you wish to procure. I can promise you a parcel by the first vessel next season. Botany takes up my time so that I neglect Geology. Our Society is in a flourishing condition. Thank you for procuring a reading for my paper at the Linnaean Society.

Macfadyen, James
[no date, watermark 1821]
[see **2269**]

921 Mackenzie, Sir George Stuart   To: H.T. De la Beche
Dingwall.   V.P. Geological Society
11 April 1836   Care of Charles Knight Esq,
                                              22 Ludgate Street,
                                                        London.

I have derived much pleasure from your Theoretical Researches and welcome the acceptance of Hutton's views. The old Huttonians used the term 'bed' for igneous rock and 'stratum' for sedimentary. I believe that the vast masses of sediment have been formed by one operation. All the trap rocks of Iceland, Greenland and the Hebrides etc., constitute a system. I suggest that a steamer should be hired to visit these.

922 Maguire, Hugh J.   To: H.T. De la Beche
120 Lucas Street,
Commercial Road,
London.
4 April 1848

Agreeably to your wishes I beg to state that I was appointed to the Ordnance Survey in July 1839 and left it in September 1846, during which time I was employed spirit levelling. I refer you to Yolland and Cameron as to my qualifications as a leveller. As you asked me to name a sum as salary I shall say 30/- per week, but would much rather leave that to yourself.

923 Maguire, Hugh J.   To: [H.T. De la Beche]
98 Great Russell Street,
Bedford Square,
London.
2 June 1851

I have to apologise for not sending this to you earlier. I have taken it up several times intending to make it more complete but without making studies from nature, may I beg your acceptance of it in its present state and hoping you will yourself give it a few masterly touches to make it more worthy of possession.

924 Mallet, Robert   To: H.T. De la Beche
London.
28 July 1850

Have you heard of Kaulbach's new method of Fresco painting? It is one very illustrative of the intimate relations of science and fine art. The colours have great force and purity. There is an example in the new Museum of Art in Berlin. The subject is the dispersal of mankind at Babel.

925 Mantell, Gideon Algernon   To: H.T. De la Beche
[1844 FJN]

I do not think it too late to obtain a slab of calcspar from Brooker. Would you also like a slab of Kentish rag with molluskite for your Museum?

926 Mantell, Gideon Algernon   To: H.T. De la Beche
19 Chester Square,
Pimlico.
27 December [1847 FJN]

Should you be coming, do give me a call and see the splendid collection of fossil birds' bones brought from New Zealand by my eldest son. I have put the novelties into Owen's hands for the Zoological Society, but I can draw up notes for our Society and put on an exhibition.

927 Mantell, Gideon Algernon   To: H.T. De la Beche
19 Chester Square,
Pimlico.
24 January 1848

Please accept the accompanying volumes as mark of my respect. Together with my Geology of the Isle of Wight, they have been published since you were installed in the President's chair. But they will be my last publications on this Science.

928 Mantell Gideon Algernon   To: H.T. De la Beche
21 February 1848

Some corrections of localities will have to be made in your notice of my son's researches in New Zealand. Letters from him since my former paper was read make an important supplement to the former announcement.

929 Mantell, Gideon Algernon   To: H.T. De la Beche
19 Chester Square,
Pimlico.
Tuesday [13 June 1848]

I propose bringing the jaw of the Iguanodon for exhibition tomorrow night. I will also send a brief notice on some recently discovered Weald fossils.

930 Mantell, Gideon Algernon   To: H.T. De la Beche
Chester Square.
Monday [1851 FJN]

Thank you for your noble volume. It is a most splendid and important contribution to the Science. My eldest son says that your "How to Observe" taught him more than all other works on the subject.

931 Mantell, Gideon Algernon   To: H.T. De la Beche
19 Chester Square,
Pimlico.
[No date]

Although you will have copies of my papers on Belemnites and the Iguanodon in the Phil.Trans., I offer further copies for your acceptance as you may have correspondents to whom they would be of some interest.

932 Marcus, Otto   To: [?]
37 Soho Square,
London.
24 August 1848

Enclosed I take the liberty to send you a criticism on a new publication on chemical and physical geology by Gustavus Bischof, Professor at the University, Bonn, and would be happy if you would allow it a place in your esteemed journal.
[Enclosure: review of above work].

933 Martin, John   To: H.T. De la Beche
30 Allsop Terrace,
New Road.
31 May 1847

More recent modifications of my plan for the use of refuse of Towns seem to have escaped your examination. I have recently patented a mode of constructing pipes, traps, etc., which may be of interest to you. Please come and see my models.

934 Martin, John   To: H.T. De la Beche
Lindsey House,
Chelsea.
24 August 1850

I notice that the Plan of drainage for Kent and Surrey is close to my own. It falls short, however, in two particulars. Sewer water is not utilized, and the river is still polluted. I have suggestions as to remedying these defects.

935 Martin, L.C.   To: H.T. De la Beche
H.M.S.O.
14 May 1847

The cut for p.200 was sent to Clowes yesterday. Hooker's engravings have been ready for some time. Am I to send him proofs?

936 Martin, L.C.   To: H.T. De la Beche
H.M.S.O.
15 January 1850

The Wedgwood vases are available for the Museum. I have purchased one for you. I will keep it myself if you do not wish to have it.

937 M'Coy, Frederick   To: H.T. De la Beche
17 Ash[?] Terrace,
Cambridge.
9 October 1848

I have just received the volumes of your Survey reports which you were so kind to give me. If we had plenty more of such memoirs geology might begin to hold its head up. The memoirs of Forbes and Hooker seem beyond all praise. I sat up nearly all night to read them. I have never seen anything so exquisite as the illustrations in these volumes.

938 McKinson, George   To: H.T. De la Beche
26 Eldon Square,
Newcastle-on-Tyne.
16 December 1851

Thank you for your encouraging letter together with copies of addresses delivered at the Museum of Practical Geology, which shall be duly delivered. They are admirably adapted to popularize the study of science for practical objects. A principal motive for soliciting the favour of your attention to our local institution was the hope that your influence might succeed in inducing the Government to prescribe certain courses of study for persons desirous of becoming colliery managers or mining engineers. I will bring the subject before the attention of the Newcastle College of Practical Science. Such a step has long been demanded by public opinion.

939 McLauchlan, Henry   To: H.T. De la Beche
Tregony.   Ordnance Geological Survey,
6 November 1836   Helston.

I have received both your letters, and have written to Captain Mudge stating what I did to you. I have industriously sought for fossils, but have not found them. I saw weathered portions of the limestone at Polcreek, but cannot say there are fossils. I saw Mr. Watson as I came through Penrhyn. He took me to his study and I saw your bust, and one he is making of Sir C. Lemon.

940 **McLauchlan**, Henry
St. Austell.
13 November 1836
To: H.T. De la Beche
Ordnance Geological Survey,
Helston.

I arrived here on the 10th, and have finished my coast observations as far as Charlestown near to this place. I conceive it better now that I should fix a few points among the mines, perhaps the steam chimneys, so as to form myself, and also enable you to form, an opinion as to the neccessity of a minute geographical survey of this district, before you commence your geological researches.

941 **McLauchlan**, Henry
St. Austell.
28 November 1836
To: H.T. De la Beche
Ordnance Geological Survey,
Helston.

I have found the mines extending over a less distance than I expected. I have no doubt we shall complete the affair by February even should it be bad weather. I find there are not many cross courses, or elvans that you will have to notice in this country, but the great question of the geography comes, and on this head I am not able to say yet, for I have been occupied in making points to adopt any method that may be necessary.

942 **McLauchlan**, Henry
St. Austell.
14 December 1836
To: H.T. De la Beche
Ordnance Geological Survey,
Post Office,
Falmouth.

I am gratified that my observations are not thrown away, and that the geological ones are noticed. Not that I am desirous they should be chronicled, but that some of the questions that arise from them should be answered, as the principle pleasure one takes in these pursuits is the novelty of the phenomena. The Whigs have abolished the office of Treasurer of the Ordnance. Mr. Henwood mentions the lodes at St. Just as being opposed to Fox's theory.

943 **McLauchlan**, Henry
St. Austell.
19 December 1836
To: H.T. De la Beche
Ordnance Geological Survey,
Post Office,
Falmouth.

I forgot in my last letter to thank you for the invitation to dine on Xmas day. I have only seen one north and south elvan here, which is near the town, where it joins the granite; on leaving which it is heaved twice in a short space, as if the Killas had been harder or of some different nature to that further off, where it continues a course nearly at right angles to the other elvans more south. I can find lodgings for you in the house of Mr. Geach here.

944 **McLauchlan**, Henry
St. Austell.
6 January 1837
To: H.T. De la Beche
Ordnance Geological Survey,
Post Office,
Falmouth.

I came from Falmouth without the pieces of impression you promised me, and shall be glad if you will send me a few spare fragments relating to this country that you may have. Since my return I have been occupied in making some little corrections and additions to the detail about the mines, so as to facilitate your examination. I have distributed Mr. Fox's questions you gave me and some say they will answer them.

945 **McLauchlan**, Henry
St. Austell.
22 May 1837
To: H.T. De la Beche

Tomorrrow you will receive a parcel containing the book of the line of road and other papers which Mr. Austen gave you relative to steam engines. Also a piece of map with the result, in red figures, of my expedition to St. Just to ascertain heights, I hope sufficient. The blue parcel contains some Athenaeums for Mr. Still, which perhaps you will send with the Magazine of Natural History.

946 **McLauchlan**, Henry
East Looe.
4 June 1837
To: H.T. De la Beche
Ordnance Geological Survey,
Falmouth.

I arrived here on the 24th, as planned. Since this I have been occupied in triangulating and have accomplished the outdoor observations. I have now data to examine the relative position of the two sheets 31 and 24, which I shall more immediately do at their junction. I have found fossils at Pencarrow, fragments of shells larger than any I have seen in the county before. They are imperfect specimens, but I think one is like a Trigonia.

947 **McLauchlan**, Henry
Fishguard.
24 July 1838
To: H.T. De la Beche
Ordnance Geological Survey,
Swansea.

I received your letter and forwarded the receipt. The rocks in this vicinity are Lyell's hypogene. The slate beds appear the same system from Cardigan to St. Davids. I find very few fossils. I have not yet ascertained the angle of intersection of the Hypogene and Slate. I hear Raspe examined this district 40 years ago. There is much evidence of denuding causes on the coast. Our one inch map will be looked on in future, as we look on our predecessors.
[Cited: North 1936a, pp.60, 61, 94]

948 **McLauchlan**, Henry
St. Davids.
17 October 1838
To: H.T. De la Beche
Ordnance Geological Survey,
Swansea.

I see by the papers that you have resumed Geological Survey work at the termination of the Commission. Congratulations on your daughter's marriage. I believe Still is finishing the Pembroke sheet. I looked at Culm ground at St. Brides Bay. They are rather deficient in good specimens of plants. I have confined my work to the coast.

949 **McLauchlan**, Henry
Fishguard.
25 December 1838
To: H.T. De la Beche
Ordnance Geological Survey,
Lilliput,
Swansea.

My accounts for the Quarter are made up. I have completed my coast line Survey. I have been tracing slates in conjunction with lofty sandstone. The admiralty are charting the coast. I find some good junctions of trap and slate along the shore. Have you seen report of the Cornish Geological Society, in which your surface geology is looked forward to?
[Cited: North 1936a, p.60]

950 **McLauchlan**, Henry
Fishguard.
19 August 1839
To: H.T. De la Beche

I am sending a box containing the jasper specimen, a piece containing concretionary forms, crystals, and a black substance. I have been lent Murchison's Silurian System. The structure of the Precelly Range agrees with the alternate eruptions and depositions theory, as given in Murchison's work.

951 **McLauchlan**, Henry
Fishguard.
26 September 1839
To: H.T. De la Beche

I enclose a receipt for horse allowance. I await your account of the specimens I sent. I continue to read Murchison's work. I have found trilobites and goniatites at Abereiddy Bay. Speculation on slates. I consider my present position more a geographical than geological one.

952 **McLauchlan**, Henry
Fishguard.
13 December 1839
To: H.T. De la Beche
Ordnance Geological Survey,
Swansea.
[Readdressed to:] Bridgend.

I have requested leave to spend Xmas in London. The season is very much against progress here. The geological character of Preseley is as described by Murchison. There are veins of quartz north of Preseley. Trials for coal have been made in a black shale, but there is no chance of coal or culm though the people would not be easily convinced of it. More black substance has been found. Mortimer was pleased with your letter of thanks.

953 **McLauchlan**, Henry
Accrington,
Lancashire.
5 September 1841
To: H.T. De la Beche

I received your letter yesterday with much pleasure, for I cannot boast of that apathy which could wait with indifference for a reply, from the 20th July to the 2nd September, particularly from one who has permitted such friendly relations to exist so long, and from whom I have received much information and great civility. Your answer is quite satisfactory. I

| 954 | **McLauchlan**, Henry | To: H.T. De la Beche |

Accrington,
Lancashire.
17 September 1841

I am pleased you give my papers so good a reception, but do not expect too much from them. The formation is extremely difficult to my conception, and though I took great trouble about the Precelley range I have not done it to my satisfaction. I shall look forward with great interest to your map and description. I am glad you are about to take up my old quarters at Fishguard. Herewith some advice on where to stay. I will send you shortly a few words more about Pencaer.
[Cited: North 1936a, p.81]

| 955 | **McLauchlan**, Henry | To: H.T. De la Beche |

Accrington,
Lancashire.
24 September 1841

I enclose you two fossils I find here in the coal shale, named as I find them in the Manchester collection. If you have not the Pecten in Wales you will be glad to see it. Have you seen the article on Whewell's Inductive Sciences in the last Quarterly. I have made several extracts which I thought would interest you, and will send them to you if you have not seen the article.

| 956 | **McLauchlan**, Henry | To: H.T. De la Beche |

Accrington,
Lancashire.
September 1841

I did not reply to your note in detail because I wished at the time to give you every information about Precelley, otherwise I should have told you that we are about to commence on the contour line system similar to your map of Glamorganshire, and in this coal district I can readily imagine how beautiful the thing must be. I am glad you have found other shells in the Encrinite bed. I saw fragments and hinges but could make out no shell. I have noted some passages from the article in the last Quarterly on Whewell's Inductive Sciences.

| 957 | **McLauchlan**, Henry | To: H.T. De la Beche |

Accrington,
Lancashire.
8 October 1841

Carn Llanllawer is the Nipple Rock, the prominent rock seen from Penslade. Perhaps I owe you some apology for not perpetuating the name on the map. The heights down the coast to St.Davids were generally deduced from Plumstone I think, occasionally proved by a shot at some island along shore, but they were never intended to stand your levelling process, and the coincidence at Llanllawer must be accidental. The observations in blue were from the Navy. They had a good instrument.
[Cited: North 1936a, pp.88-89]

| 958 | **McLauchlan**, Henry | To: H.T. De la Beche |

Accrington,
Lancashire.
23 October 1841

The author of the Silurian System in remarking on the two great lines of elevation through Wales says the influence of the southern line is impressed on the strata considerably to the north of the general line, and mentions Felindre, or College, as a place where the southern dip may be observed. I took some pains to find out this place but without success. It is suprising how much an idea of the geological formation of a country adds to the pleasure of our labours.

| 959 | **McMahon**, T. | To: H.T. De la Beche |

Halse Hall,
Jamaica.
11 February 1832

c/o G.W. Hibbert & Co
Billiter Court,
London.

In the eventful times here your people have behaved in a most exemplary manner. There may be some difference in quantity and quality of sugar. I fear the drafts from breeding cattle will be inconsiderable for some time. Martial law was proclaimed here from 31 December to 9 February. I expect to send home a good crop, nonetheless.

| 960 | **McMurray**, Amelia | To: H.T. De la Beche |

Brome Hall,
Scole,
Norfolk.
[1849 FJN]

Thank you for acknowledging my specimens. The hills of Melibocus and Auerbach are acknowledged to be granite. The Old Red Sandstone would not come between granite and other primitive rocks. What do you think?

| 961 | **Meade**, Thomas | To: [H.T. De la Beche] |

Chatley,
near Beckington,
Somerset.
22 April 1844

The coal impressions will be forwarded to London by tomorrow's train. I hope Phillips will be able to look over my collection and find specimens worthy of engraving.

| 962 | **Medlicott**, Henry Benedict | To: H.T. De la Beche |

Monghyr.
8 June 1854

Museum of Practical Geology,
Jermyn Street,
London.

I have not yet heard officially of the Roorkee appointment, nor has Oldham. No doubt the loss of a man would cripple my chief, but the independent position is priceless. The Roorkee settlement is just cut out for me. It would be to their interest if the Government in India would let me have the £200 outfit. I wish my brother's name had been given in instead of mine.

| 963 | **Medlicott**, Henry Benedict | To: H.T. De la Beche |

Monghyr.
20 June 1854

Museum of Practical Geology,
Jermyn Street,
London.

The offer of Roorkee has at last reached me. I send you copies of the secretary's letter and my reply. The conditions of Mr. Allen's letter are unacceptable to me, and are actually inferior to those I hold on the Survey, and much inferior to the prospects Oldham holds out to me. If you could insert my brother's name in a second advice, you would be conferring a service.

| 964 | **Medlicott**, Henry Benedict | To: H.T. De la Beche |

Monghyr.
25 July 1854

It has turned out as I expected. Allen has made a second and third offer to me, with £100 for outfit as well, and I have accepted. I have exchanged letters with the people at Roorkee, and am beginning to have nightmares about the introductory lectures etc. I must not try to do too much and lose sight of my speciality. My sincere thanks for the appointment.

| 965 | **Melville**, J.T. | To: H.T. De la Beche |

Upper Harley Street,
London.
19 December 1839

Penllergare,
Swansea.

Thank you for your replies concerning coal supplies, and site for our works. Buckland's opinion has been found satisfactory. Your objections are entirely from a mineralogical point of view.

| 966 | **Meyendorff**, [?]Alexandre von | To: H.T. De la Beche |

Paris.
4 April 1840

Geological Society,
London.

I have asked Murchison to give you the enclosed specimens. You may write to me at St. Petersburg.

| 967 | **Meyendorff**, [?]Alexandre von | To: H.T. De la Beche |

St. Petersburg.
29 April 1840

Geological Society,
Somerset House,
London.

You would do me a kindness in giving to Murchison for me the drawings and sections relating to the workings of coal mines, and the work of Phillips on the Carboniferous formation of England. I will pay Murchison when he arrives in St. Petersburg early in June. I will obtain for you any specimens from Russia that you may want.

| 968 | **Meyer**, V. | To: H.T. De la Beche |

Buckingham Palace.
19 June 1847[?49]

According to the Prince's commands, of which I found an opportunity already a few days ago of informing you orally, I have the honour of herewith forwarding to you the portrait of Humboldt of which H.R.H. begs your acceptance.

969 Meyer, V. To: H.T. De la Beche
Buckingham Palace.
4 July 1848

I have the honour to transmit to you the best thanks of H.R.H. Prince Albert for the Copy of your Anniversary address to the Geological Society. H.R.H. hopes soon to find some hours of leisure to peruse your work from which he is sure to derive a very essential assistance in his study of a science in which he has always taken a great interest and with which he is impatient to become more profoundly acquainted.

970 Meyer, V. To: H.T. De la Beche
Osborne.
29 August 1848

I enclose the official acknowledgement of your valuable present, which I received this morning and laid before H.R.H. soon afterwards. I was in hopes for several days I might obtain leave for going to Ireland next month and meeting you at Dublin, but it now seems more probable that I shall have to go to Germany.

971 Meyer, V. To: H.T. De la Beche
Windsor Castle.
18 December 1848

Many thanks for your kind note. Also H.R.H. the Prince is most thankful to you for having directed his attention upon the interesting geological fact represented by Section No.1. I laid before him your note and the publication it accompanied last night. He will have great pleasure in receiving Mr. Robert Hunt's "Poetry of Science". The Court will be at Windsor Castle until next Wednesday when they will go to Osborne for 3-4 weeks.

972 Meyer, V. To: H.T. De la Beche
[Buckingham] Palace.
28 April 1849

I have been commanded by H.R.H. Prince Albert most thankfully to acknowledge the receipt of the copy of your Presidential address to the Geological Society. He admires it as a new specimen of the well known power you possess to convert an infinity of scientific details into the short contents of a highly interesting speech.

973 Meyer, V. To: H.T. De la Beche
[Buckingham] Palace.
31 May 1849

H.R.H. Prince Albert was highly pleased with the new publication Memoirs of the Geological Survey containing the First Decade of Figures etc., by its scientific import as by the artistical execution of the plates. He desires me to express his best thanks, and likewise to thank you for the explanations contained in your note.

974 Michelotti, Jean To: H.T. De la Beche
Turin. Somerset House,
22 May 1839 London.

I have just received your letter of the 3rd inst., which however was not posted on that date or perhaps was written on the 14th. The Society has badly interpreted my letter if it thinks I wish to be made a corresponding member out of turn and before others. In a few months I will send a small work to the Society on palaeontology, with illustrations. I will also add another memoir I have published. May I suggest you publish another edition of your Manual? But it would be necessary to adopt the divisions of Silurian and Cambrian.

975 Millbourn, William To: R. Murchison
241 Heneage Street, President, British Association,
Birmingham. Institution,
17 May 1847 Poultry,
London.

I send particulars of my Plan for ventilating mines by means of a fan driven by the Wimsey engine. I have not the means to bring it into operation. Should British Association use their influence to bring it into use I should feel much gratified.

976 Miller, J. To: [H.T. De la Beche]
4 November 1831

Are there any remains of buildings beneath the sea at Port Royal supposed to have sunk down during the great earthquake of 1692? If there are any such remains, what do they consist of? How deep are they beneath the surface of the sea? Are any parts of the walls upright? How far are they distant from the shore? If there are could not a plan of their position relatively to the present town be made? Was not the old town of Port Royal situated on a bank of sand?

977 Miller, Johann Samuel To: H.T. De la Beche
Bristol Institution. Geological Society,
12 November 1829 Somerset House,
London.

Thanks for your letter of 23 October. I have presented in your name von Buch's map of the Lake of Lugano, and your lithographic plate of the Havre crocodile. Regarding your share in the Bristol Institution, I have to say that many are for sale. Still if you make out the form I will fill in the details if I can find a buyer, and tell you the purchaser's name.
[Cited: North 1956, p.141; Taylor & Torrens 1987, p.139]

978 Milne, A. To: H.T. De la Beche
Office of Woods.
4 September 1840

In case we should want a good practical Mineral Surveyor to survey and report on Mines of Duchy of Cornwall, can you suggest a good name. I had thought of Taylor or Sopwith. Write your opinions in confidence to me.

979 Milne, A. To: H.T. De la Beche
Office of Woods.
17 April 1843

You are instructed to make an appointment with Barry and Smith to inspect the stone being used in the building of the New Houses of Parliament to ascertain whether or not it is of the quality and description of the stone you described in your 1839 report as being most fit and proper for the purpose.

980 Milne, David To: H.T. De la Beche
10 York Place,
Edinburgh.
16 April 1847

The experiment of establishing an Agricultural Chemistry Association in Scotland has been successful, but we wish to obtain permanent Government aid. Arrangements for the Geological Survey of Scotland might harmonise with ours. Before I present this memorial to Government, I should be glad to hear your views on mutual aid.

981 Milne, John To: H.T. De la Beche
R[oyal] A[gricultural] C[ollege].
22 May [1847 FJN]

There are two points which I have omitted in the Report but of both of which I think some notice should be taken. They relate to the assistance and liberal conduct if Mr. G. Kenner is giving the use of his boiler etc., and to the great advantages rendered to the investigation by being permitted to carry on the trials at the College of Civil Engineers at Putney. What are your opinions on these subjects?

982 Milnes, William To: H.T. De la Beche
Marble Works,
Ashford,
Near Bakewell.
15 March 1848

I take the liberty of calling your attention to the many extremely beautiful marbles this country produces, especially the splendid specimens I have sent to the Museum of Economic Geology. This neighbourhood abounds with an endless variety of fine marbles, some of which are unique and peculiar to this country and cannot be surpassed by any marbles from any part of the world.

983 Mitchell, James W. To: [H.T. De la Beche]
17 Upper Wimpole Street.
19 May [1847 FJN]

Is it probable that Artesian wells will be successful on the south side of the island of Jamaica? I propose various possible sites. Such water would render properties situated in dry weather parishes extremely valuable. Can I talk with you in London?

984 **Mitchell**, James W.   To: [H.T. De la Beche]
     17 Upper Wimpole Street.
     21 July [1847 FJN]

I am travelling to Jamaica and will have opportunities with regard to geological observations. Shall I break up the floor of the cave at Portland point and collect fossil shells? I will also go to Cuba to visit sugar growing districts. Any suggestions from you will be welcome. Also what is the effect of an electric current on a semifluid or viscous mass while cooling?

985 **Moggridge**, Matthew   To: [H.T. De la Beche]
     Burrows Lodge.
     31 May 1838

The specimen of lead ore is from Woodfield Pits, the highest worked in Monmouthshire. The immediate neighbourhood is unusually faulty.

986 **Monteagle**, Thomas Spring-Rice, 1st Baron
     Exchequer.   To: H.T. De la beche
     15 December 1851

Thanks for the copy of your inaugural and introductory lectures at the School of Mines in the Museum of Economic Geology. I am pleased with this new development to the powers of usefulness by an establishment in which I have been interested since its infancy.

987 **Moore**, John Carrick   To: H.T. De la Beche
     Geological Society,   Jermyn Street.
     Somerset House.
     30 May 1851

Capt. Nelson has offered to send home a large specimen of Coral Reef from the Bahamas. The Geological Society cannot afford the funds or space. Do you wish such a specimen for the Museum of Practical Geology?

988 **Moore**, William   To: H.T. De la Beche
     Lyme,
     Dorset.
     15 March 1848

I have a specimen of Ichthyosaurus for sale at £15.0.0. Mr. Roberts thinks it will suit you. We have also other fossils available.

989 **Morley**, Edmund Parker, 2nd Earl   To: H.T. De la Beche
     Kent House,
     Knightsbridge.
     20 September 1850

You will much oblige me if you will give me by letter, the same character of the Morley Clay as you gave me by work of mouth on Tuesday last. The fact is that I am anxious that some parties with whom I am in treaty should have some authority for the quality of the article which I know that you are enabled to supply, and a word from you will have its weight.

990 **Morley**, Edmund Parker, 2nd Earl   To: H.T. De la Beche
     Kent House.
     10 June 1852

The enclosed paper which was drawn up for the purpose of being put into the care of the Lee China Co., in the exhibition will give you the analysis of the China clay by Mr. Richard Phillips which you require. If you wish for any further information upon the subject I shall be too happy to give it to you.

991 **Morley**, Edmund Parker, 2nd Earl   To: H.T. De la Beche
     Kent House.
     19 June 1852

I enclose you a letter which I have received from a Mr. Williams, a great connoisseur in Old Plymouth China, and a gentleman on whose authority I can rely. You will see what he says and perhaps act accordingly.

992 **Morley**, J.   To: H.T. De la Beche
     Labuan.   Museum of Economic Geology,
     29 March 1850   Jermyn Street,
     London.

I fear I may seem a little ungrateful for your kind assistance, but the delay in writing is due to my wishing to include some description of the neighbouring coast of Borneo. What I want is liberty to examine the coal. My visit to Borneo, however, seems to be postponed, and I hasten to send you all the information I can collect from this island alone.
[Enclosure: Map of Labuan to accompany description].

993 **Morley**, J.   To: H.T. De la Beche
     Labuan.   Museum of Economic Geology,
     22 August 1853   Jermyn Street,
     London.

My letter of 18 months ago was returned through the dead letter office. I will send it together with a box of fossils. We have made great progress here and the yield of coal has been increased. We are now working the small seam, a specimen of which was brought home by Mr. Wise. The colonial surgeon here who is a little bit of a chemist has discovered that nearly all the pyrites here contain a proportion of cobalt. Is not this unusual? I have had a Director appointed over me and am thinking of writing and enquiring about other jobs.

994 **Morpeth**, George William Frederick Howard, Lord
     23 September [1846]
[Minute]
Our department would receive credit from Sir Henry's acceptance of Chair of the Geological Society in addition to that we already derive from his acknowledged qualifications for it.
[See also Philipps, T.W. to De la Beche, H.T., 23 September 1846, **1320**]

995 **Morpeth**, George William Frederick Howard, Lord
     30 April 1847   To: H.T. De la Beche

I have ventured to tell Mr. William Ramsden that you will give him a short audience.

996 **Morpeth**, George William Frederick Howard, Lord
     1 December 1847   To: H.T. De la Beche

Will meet you at the Jermyn Street side of the Museum at 2.30 if that suits you.

997 **Morpeth**, George William Frederick Howard, Lord
     17 January 1848   To: H.T. De la Beche

Following your recommendation, the salaries of Selwyn and Jukes can be increased from £150 to £200 per annum.

998 **Morpeth**, George William Frederick Howard, Lord
     21 May 1848   To: H.T. De la Beche

I want to modify the text of the prohibition of smoke clause. Do you see any objection to changing the list of exceptions as indicated?

999 **Morrell**, George F.   To: H.T. De la Beche
     149 Fleet Street.   President of the
     9 December 1851   Geological Society.

In the lecture you delivered at the last meeting of the Society of Arts you alluded to the manufacture of plumbago into pencils as a new and valuable discovery. Such is not the fact. I have been using a process for upwards of thirty years. This class of manufacture should have received more attention at the Exhibition, and the above mistake would have been prevented.

1000 **Morton**, Owen   To: H.T. De la Beche
     27 Southampton Row,   Whitehall.
     Russell Square,
     London.
     9 November 1849

Not having received any answer to the prospectus of the Female Aid Society, I again solicit your attention and aid for this great work of love. Anonymous donations, or otherwise, are welcome. All should unite to subdue the great evil in our land, juvenile depravity.

1001 **Morton**, Samuel George   To: H.T. De la Beche
     Secretary,
     Academy of Natural Sciences,
     Philadelphia.
     18 September 1832

Herewith your diploma as a Corresponding Member of this institution. Thank you for the favourable notice of my "Synopsis" in your Geological Manual. A new edition of my Synopsis of Organic Remains is being prepared.

1002 **Moscow**,
     Societas Caesarea Naturae Curiosorum Mosquensis.
     13 November 1852

[Certificate of membership (ordinary member) of H.T. De la Beche, signed by Vladimir Nazimoff (President), G. Fischer de Waldheim (Vice-President) and F. Auerbach and Renard (Secretaries)].

1003 Moscow, Société Impériale des Naturalistes de Moscou
19 January 1853   To: H.T. De la Beche
London.

The Society wishes to add your name to those of its members and has decided, at the meeting of 25 November, to grant you the title of active member. The Society requests that you communicate to it from time to time the results of your interesting and important researches.
[Covering letter of certificate **1002**]

1004 Moseley [?]   To: [H.T. De la Beche]
Wandsworth.
12 November [No year]

I did not think you were in Jermyn Street or I should not have failed to seek an interview with you on Thursday and to thank you for having assigned a place to our apparatus cabinets in your Museum. I entirely agree in both your recommendations and have taken steps to have them acted upon.

1005 Mountenry, Barclay de   To: H.T. De la Beche
Pelham Road,
Thurloe Square,
Brompton.
18 May 1850

Excuse the liberty of my applying for an order to see the Museum at Piccadilly. I have visitors from Manchester to whom I would like to show it.

1006 Mudge, Robert   To: H.T. De la Beche
30 November 1835

I have long arrears to make up and a long story to tell, but first let me acknowledge the receipt of Sheet No.30. I am amazed at the quality of the work. What a job you must have had. My daughter had written to me of a discovery of bones in a cavern, where as a child I used to play. I visited the cavern, and wrote to Buckland. He has not answered yet.

1007 Mudie, R.   To: H.T. De la Beche
19 Jay Lane.
Tuesday [1840 FJN]

Thank you for the note containing your kind promise, of which I shall avail myself.

1008 Murchison, Kenneth   To: H.T. De la Beche
2 Marlbro Street,
Bath.
19 December 1847

Ramsay and Aveline have left the Survey for the Christmas holidays. I ask the same indulgence for my nephew, Roderick. Despite the unfavourable nature of Ramsay's report on him, I hope you will still keep him on.

1009 Murchison, Roderick Impey   To: W. Buckland
[13 February] 1830   Christ Church College,
Oxford.

I have arranged that on Friday next we will have the Basin of Baya instead of Weymouth. We now have plenty of stuff for the session. Boué has sent lots of maps and sections. I really thought him stronger, or I should not have proposed him for F.G.S. Alfred Thomas alters part of De la Beche's map of Pembrokeshire.

1010 Murchison, Roderick Impey   To: H.T. De la Beche
3 Bryanston Place.
19 March [1832]

The bearer of this letter, Mr. Kaltner, is really the torment of my life. It appears that about £15 would provide his return to Germany. As you are the one who received his recommendations, you are especially called to exert yourself. I have given him a guinea already. Put my name at the top of the list for another.

1011 Murchison, Roderick Impey   To: H.T. De la Beche
Up. Park,   Athenaeum Club,
Petersfield.   London.
12 December 1833

Herewith for the correction of my grauwacke views in your French edition of the Manual a correct view of the whole matter. I wish to illustrate the whole mass of strata from the mountain limestone downwards. I must help you rewrite all the appropriate article in your 4th edition. Please stress that the middle and lower members of the Old Red are fossiliferous. The word 'sub-groups' will not do. Such a mass is entitled to the word 'Formation'.
[Cited: North 1939b, p.1053; Rudwick 1985, pp.82-83; Bassett 1985, p.233].

1012 M[urchison], Roderick Impey   To: H.T. De la Beche
London.   Tiverton,
19 December 1834   Devon.

I believe I did speak out against you more than Lyell. Stokes made a few observations in his usual quiet and judicious manner. Had I his temperament, or worked less among these cursed old rocks, I should have avoided the reproof I got from the Chair, for the President, after reading out your rejoinder, interdicted all observations thereon. Fitton criticised him for stopping free discussion, and your opponents for being too vifs. I should not have slept without writing to you.
[See also Murchison, R.I. to De la Beche, H.T., [1834] **1013** for enclosure].
[Cited: Rudwick 1985, pp.107, 109; Secord 1986a, p.99].

1013 Murchison, Roderick Impey   To: H.T. De la Beche
[1834]

I was sorry to learn that my comments on your Bideford memoir prompted a written reply from you. I could not let your solecism pass by unheeded. I do not contest the facts; but they may be interpreted in a different manner. You will convince me only if you produce the orthoceratites and other fossils characteristic of the true transition rocks. Weaver has quite abandoned his case. The age of rocks can not be determined by mineral character alone.
[Cited: Rudwick 1985, pp.109-110, 116; Secord 1986a, pp.93,99]

1014 Murchison, Roderick Impey   To: H.T. De la Beche
2 Eccleston Street,   Map Office,
Belgrave Square.   Tower.
6 April 1838

I have resolved to use the word 'Protozoic' to apply to the whole fossiliferous series below the Old Red Sandstone. In adopting this word I wish to eschew theory. Wealden and Purbeck beds have been discovered near Elgin. I don't know if Sedgwick will use my term 'Protozoic' as he seems to prefer 'Primary', a word I cannot endure as productive of confusion.
[Cited: Rudwick 1985, p.242]

1015 Murchison, Roderick Impey   To: H.T. De la Beche
9 March 1841

At the last meeting of the British Association in Glasgow, it was proposed that Conybeare be invited to preside over Section C at the next meeting in Plymouth. A unanimous council decision in favour was taken last week, and Conybeare was invited. He has written, proposing yourself instead. I sincerely concur, and would like your reply by return of post.
[Cited: Morrell & Thackray 1981, p.455; Rudwick 1985, p.373].

1016 Murchison, Roderick Impey   To: H.T. De la Beche
16 Belgrave Square.
20 March 1841

The Council will name you President of Section C, and I reiterate my pleasure in seeing you in the chair. Ramsay came two nights ago, and will be at your disposal on Thursday. At the meeting on Wednesday I think I will swamp the glacial heresy in its extended form. Lastly I have recommended your knowledge of Jamaica to Lord John, regarding a question of copper ores.
[Cited: North 1936a, p.79; McCartney 1977, pp.25, 60; Stafford 1984, pp.7, 14; Rudwick 1985, p.373; Stafford 1989, p.83]

1017 Murchison, Roderick Impey   To: H.T. De la Beche
Belgrave Square.
27 November 1846

Samarsky is the name of the Chef 'ad interim'. He is a stick and not worth a good stone, and speaks only Russian. It is however art to be civil to this 'homme de bureau'.

1018 Murchison, Roderick Impey   To: H.T. De la Beche
Belgrave Square.
23 December 1846

May I allude in a general way to the palaeontological facts stated by Forbes the other evening, when I make a statement opposing Sedgwick's dismemberment of my Silurian System. If he were to prevail my poor Silurian would be driven from Scandinavia and Russia, and cooped up in the Baltic, as in this country only a band would remain if Sedgwick cut away the larger half.
[Cited: Secord 1986a, pp.190, 197]

1019 Murchison, Roderick Impey To: H.T. De la Beche
Belgrave Square.
7 January 1847

I was so thoroughly vexed to see my old friend Sedgwick bring forward a proposition this year so at variance with his published memoirs of last year. It is to you I look for the fair and impartial finale of all the North Welsh chaos. Sedgwick is in love, and has lost his head. So says rumour. May you long unite Whigs and Tories in supporting the good cause of geology.
[Cited: Secord 1986a, pp.184, 191].

1020 Murchison, Roderick Impey. To: H.T. De la Beche
16 Belgrave Square.
9 February 1847

In 1825 I explored the Isle of Wight and the coast from Swanage to Lyme. Fitton gave me my first field lesson in geology, and I attended evening lectures of Brand. With such preparation I had little difficulty in colouring the Ordnance map & prepare my memoir of Western Hants. It was the happiest year of my life. My wife was instrumental in bringing me to geology from a military career and foxhunting. Buckland is opposed to a Swansea meeting until the rail is finished.
[Cited: Secord 1986a, p.62]

1021 Murchison, Roderick Impey To: H.T. De la Beche
15 Pall Mall.
29 April 1847

Enclosed a letter from Capt. Vicary. You will read when and how to expect the fossils. Please look at them and have Forbes work on them before Falconer leaves for the East. Your promptitude in having the paper read will please the General and the Captain.

1022 Murchison, Roderick Impey To: H.T. De la Beche
Innsbruck. Craig's Court,
21 August 1847 London.

Keyserling has sent copies of his work and map to England for distribution among Societies, Universities, etc. Will you please look after that when they arrive at the Geological Society? I have been studying the Prague Basin with De Verneuil. We went also to Vienna. There is an excellent suite of models available (of the salt limestone at Salzburg), which you ought to buy for the Museum. Write to me in Venice and I will arrange the purchase.

1023 Murchison, Roderick Impey To: H.T. De la Beche
Up. Park.
Tuesday [1847]

I shall be at the Club and Society tomorrow to do my duty & have invited Mr. Napier to be present. Please look at the note I have drawn up and allow it to be read as a prelude to Vicary's paper.

1024 Murchison, Roderick Impey To: H.T. De la Beche
[1847 FJN]

I have written to Gyde to have the 1835 woodcut printed as it is, and with no changes to make it look better, as this seems what you wish. I should have asked you do dine yesterday.

1025 Murchison, Roderick Impey To: H.T. De la Beche
Rome. Craig's Court,
21 February 1848 Charing Cross,
London.

I set out for Naples tomorrow, but am not prepared yet to send you anything on the Alps or Italy. Sedgwick has written. The political situation is unsettled here, and I shall be forced to abandon my visit to Sicily. The Geological Journals arrived. They are unknown here, as the Italians see through French spectacles. Please remember my servant Haley, and arrange employment if possible.

1026 Murchison, Roderick Impey To: H.T. De la Beche
Up. Park,
Petersfield.
11 October 1848

My memoir on the Alps is so long, I will only be able to read part of it on Wednesday. My main object is to explain the true relations of the Cretaceous and supra-Cretaceous deposits. I wish to discuss the Eocene in the presence of Rogers before he returns to the U.S. I should like to produce the second part on January 17.

1027 Murchison, Roderick Impey To: H.T. De la Beche
Bedford Hotel,
Brighton.
1 November 1848

May I have the night of 13 December, for my "Notes on the Alps and Apennines"? I expect De Verneuil over and would like him to be present. I am anxious to get it into print. Strzelecki showed me a note from Lord Grey to say he has been made a C.B. for his services in the Irish famine.

1028 Murchison, Roderick Impey To: H.T. De la Beche
1 February 1849

Herewith the missing parts of my memoir on the Alps. Please look at both of them. In the part on the Miocene and Pliocene of Italy I specially call your attention to the passage upwards from an upper Numulitic zone. Also my warning against using words like 'Molasse'. My object is the establishment of Older Tertiary in S. Europe and the East.

1029 Murchison, Roderick Impey To: H.T. De la Beche
16 Belgrave Square.
29 June 1849

I wish to contribute some items to your new Museum in Piccadilly. One specimen is from the Devonian rocks of Russia, another a model of a great lump of malachite, and thirdly a set of ores from Australia. Long may you have health to conduct your gigantic and most important concern.

1030 Murchison, Roderick Impey To: H.T. De la Beche
16 Belgrave Square.
1850

Phillips tells me that neither yourself nor Forbes will be able to come to the meeting. Please send a note to the Chairman saying you wish to see a President of Section C whom geologists wish to have and who was understood to be the person at Birmingham.

1031 Murchison, Roderick Impey To: H.T. De la Beche
16 Belgrave Square.
3 April 1851

I have been reading your Geological Observer, and am delighted with its perspicuity and value. It will be the favourite vade mecum of geologists. You seem to favour Lyell's piddling school, more than of old when you drew the cartoon of Frank Buckland denuding the valley.
[Cited: McCartney 1977, pp.52-53]

1032 Murchison, Roderick Impey To: H.T. De la Beche
16 Belgrave Square.
15 July 1851

The bearer has the small box of Swedish ores sent by Baron Rehausen. A letter of thanks should be sent to the Swedish Minister. Don't forget to send Alexander MacCallum a trifle. Shall I ask the Bavarian to write directly to yourself or Lord Seymour for what he wants?

1033 Murchison, Roderick Impey To: H.T. De la Beche
4 Circus,
Bath.
2 January 1854

I enclose the proposed dedication of my 'Siluria'. In the preface I show, whether it be the map which is simply an index to your detailed and real work, or by the fossils of your Museum, how it is a hand book to lead all enquirers to the true basis on which the whole thing rests. My only object is to convey to the world my sense of importance of your services and those of capital associates.

1034 Murchison, Roderick Impey To: H.T. De la Beche
Lichfield House,
Richmond.
2 August 1854

My brother died yesterday. Please try and get enclosed to Elie de Beaumont who is in Town only for a day or two. I sent him a copy of my book 2 months ago, but fear he never got it.

1035 Murchison, Roderick M.
[March] 1847

College for Civil Engineers and of General Practical and Scientific Education
[Report by M. Cowie, Principal for Lent term on Roderick M. Murchison, nephew of R.I. Murchison].

1036 Murray, Alexander — To: H.T. De la Beche
Liverpool.
19 April 1843
Our day of sailing has arrived, and I write to thank you for your kindness and assistance. I could not obtain a copy of your Manual. Would Logan have the goodness to bring me a copy? Ramsay is in great spirits.

1037 Murray, Peter — To: W. Buckland
Scarborough.
10 April 1849
On behalf of an acquaintance of mine, a wood engraver who has invented an improvement which may be transferred to copper by electrotyping, I am anxious to have it examined by an eminent Chemical philosopher. May I have your counsel as to whom I should go?

1038 Napier, Charles James — To: H.T. De la Beche
Karachi, — Geological Society,
Sindh. — London.
4 September 1847
Thank you for your letter conveying the Geological Society's thanks for assisting Captain Vicary in geological researches. I have lately helped Dr. Stock, an able botanist, to see the country here. As I have resigned the Government of Sindh I fear I shall no longer have it in my power to be of use to science.

1039 Nasmyth, James — To: H.T. De la Beche
Bridgewater Foundry, — London.
Patricroft,
near Manchester.
2 July 1842
I regret missing meeting you at the last British Association meeting. I am preparing a new set of specimens of iron for your Museum. May I suggest the following arrangement of ores, metals, coal, etc., so as to show by eye the actual quantities used in the process of reduction.
[Cited: McCartney 1977, p.36]

1040 Newcastle-under-Lyme, Henry Pelham Fienne Pelham Clinton,
5th Duke of — To: H.T. De la Beche
17 Portman Square.
28 February 1852
Thank you for the copies of the lectures. Your account of the attendance and attention of the working men is very gratifying.
[See also Lincoln, Earl of, **857-861**].

1041 Newton, Charles Thomas — To: H.T. De la Beche
British Museum.
11 July 1847
In Arrian's Periplus of the Erythrean Sea, Vincent's edition p.74 is the following passage. Among the imports into this country (the African shore of the Red Sea a little north of Yemen) are copper compounds which the natives use for ornaments. Is there any kind of reason for supposing that copper could be prepared with honey? Perhaps you will kindly give me your opinion about the meaning of the passage.
[See Smyth, W.W. to De la Beche, H.T., 12 October 1847, **1951**].

1042 Nichols, Charles — To: H.T. De la Beche
Geological Society, — President,
Somerset House. — Geological Society.
21 June 1848
I ask permission to take my vacation on Friday next, for a week, and then for a further week on Mr. Charlton's return, and the remainder on the return of Mr. Nicol.

1043 Nicol, James — To: H.T. De la Beche
Geological Society.
27 September 1848
We are not very rich in papers for the next Session, but still have enough to begin with. Sharpe's on Slaty Cleavage should be read in full. Murchison has a lot of material, including a letter from Hall in New York about the Orthoceras. Nesbit's on Phosphate of Lime will do for the second night. The Journal is now getting on very well.

1044 Nicol, James — To: H.T. De la Beche
Geological Society.
[?1848]
Darwin wishes to have his last paper to alter one sentence slightly and to insert two notes, with the date. I suppose you will have no objections to my letting him have it for a few days.

1045 Nicol, James — To: H.T. De la Beche
40 Princes Street,
Edinburgh.
15 August [1849 FJN]
I have observed in the newspapers my appointment to Queen's College, Cork, and have since received official intimation. My success is owing to your recommendation. I regret leaving the Geological Society, but the change is for the better. My resignation has been sent to Lyell and a Council is to appoint a successor.

1046 Nicol, James — To: H.T. De la Beche
Cork.
20 April 1851
Thank you for the present of the volume. I have only a small class this year, yet the number is double that of McCoy in Belfast. Kane has had several differences with his Professors. There is a great want of field sections for teaching geology here. I was disappointed at not seeing you in Edinburgh for the British Association meeting. Is there no hope of you taking Scotland under your charge?

1047 Noaks, E. — To: H.T. De la Beche
24 June 1846
I enclose a receipt for £23.18.0, the amount of Miss [Rosie] De la Beche's account to midsummer.

1048 Noaks, E. — To: H.T. De la Beche
[December] 1847
Tuition bill for Miss [Rosie] De la Beche £46.6.0. Studies will re-commence on 25 January 1848.

1049 Norrey, Stephen — To: H.T. De la Beche
House of Commons.
14 March 1851
Lord Seymour has taken up my motion relative to the Survey. Do not fancy that the object I had in view was disrespectful to you. On the contrary I believe you to be too useful. I think, if allowed, the Government could use you in every possible purpose but your Survey. Now in my view the Director General of the Survey has quite enough to do in his own department, if left to that, which you are not. I believe Lord Seymour will confirm my motion and explain it to you.

1050 Norris, J.P. — To: H.T. De la Beche
Privy Council Office.
15 June 1853
I am planning a prize scheme for my yearly examination. What would be Jermyn Street's terms for exhibitioners? Is there provision for board and lodging? Can aid be expected from the Government? I will bear in mind your caution that in these matters it is well to be rather under than over the mark.

1051 Northampton, Spencer Joshua Alwyne Compton,
2nd Marquis of — To: H. T. De la Beche
Torquay.
December 1844
I was sorry not to have found you in London when I was there last week. I called at the Museum to see if I could be of use to you in this out of the way place. There is a specimen of an inlaid table which I think worth buying for the Museum, showing economic use of coralline marble. Can I be of use in buying specimens of madripores?

1052 Northampton, Spencer Joshua Alwyne Compton,
2nd Marquis of — To: H.T. De la Beche
Castle Ashby.
7 December 1846
I infer you will be at the Council next Thursday. With respect to the question of the Charter Committee, I do not think we should bring it up on Thursday. Amendments should be considered at the third Council to come to a decision. I have written to Horner and others to make the same suggestions.

1053 Northampton, Spencer Joshua Alwyne Compton,
2nd Marquis of — To: H.T. De la Beche
Castle Ashby.
15 October 1847
The time is approaching for the affair of the Wintringham Bequest. The

choice should be made between astronomy, chemistry and geology. I intend to write to Airy and Faraday for their advice on the first two. On the last I ask you to recommend a subject.

1054 Northampton, Spencer Joshua Alwyne Compton, 2nd Marquis of   To: H.T. De la Beche
Piccadilly.
29 November 1847

I had certainly heard a rumour of the kind you mention, but I never heard that you had anything to do with the matter. I am only just arrived in Piccadilly having had business in town. I hope to see you at the Council tomorrow.

1055 Northampton, Spencer Joshua Alwyne Compton, 2nd Marquis of   To: H.T. De la Beche
Castle Ashby.
7 February 1848

I wrote to you when I was last in London, but have not heard from you. I want to say that I have mislaid your questions about the Wintringham Bequest and ask you to send them again to me. I shall not be able to be present next Thursday at the Royal Society, but I hope to preside the following one.

1056 Northampton, Spencer Joshua Alwyne Compton, 2nd Marquis of   To: H.T. De la Beche
Castle Ashby.
13 February 1848

So many Fellows have given me their opinions against the Wintringham medal, I think it not right to proceed in the matter without consulting my present Council, and it is too late for that now. I had intended to summon a Council for 2 March. Perhaps you will think that too far off for the object of your letter to me. I am going tomorrow to London and you can write to me there. What is the day of the Geological anniversary?

1057 Northampton, Spencer Joshua Alwyne Compton, 2nd Marquis of   To: H.T. De la Beche
Piccadilly.
16 February 1848

In consequence of our conversation the other day, I wish to tell you the reasons why I think that though it is not right that the duration of the Presidency of the Royal Society should be unlimited, yet still it should not be short like that of other Societies. Three or four years is too short and will make for laxity, apart from other reasons depending on his ex officio duties to other Societies and Institutions.

1058 Northampton, Spencer Joshua Alwyne Compton, 2nd Marquis of   To: C. Lyell
Piccadilly.
15 March 1848
[Copy]

I am more sorry than surprised at Herschel's answer. I doubt much Brown's being agreeable to the Presidency. It would place him in a delicate and awkward position at the British Museum. As an officer there, he would also become an official trustee, thus acting as judge in his own case, and causing great jealousy among his fellow officers. The President of the Royal Society is considered by other trustees, as the representative as it were of that part of the scientific world that is interested in the natural history collection, and it would be a great evil to lower his importance in that capacity.

1059 Northampton, Spencer Joshua Alwyne Compton, 2nd Marquis of   To: H.T. De la Beche
Piccadilly.
21 March 1848

I hope nothing will prevent my having the pleasure of seeing you at my Council dinner on 1st April, which is the night of one of my soirées.

1060 Northampton, Spencer Joshua Alwyne Compton, 2nd Marquis of   To: H.T. De la Beche
Royal Society,
Somerset House.
[1848 FJN]

There is no other room in our apartments except that occupied by the Society of Antiquarians. Perhaps if we had applied first for it we might have had it, but it is now too late. The Society of Antiquarians might be prepared to lend us their rooms when they did not require them. There is great doubt as to the advisableness of this change. I think that tea and coffee on ordinary nights should be paid for by the Society.

1061 Northumberland, Algernon Percy, 4th Duke of   To: H.T. De la Beche
Alnwick Castle.
31 July 1850

Will you pay us a visit here on your way to the Edinburgh meeting? There are some scratches on Ratcheugh Craig rocks (3 miles from here) which indicate passage over them of what?

1062 Norwich, [Edward Stanley] Bishop of   To: H.T. De la Beche
Palace,   Metropolitan Buildings,
Norwich.   6 Adelphi Terrace.
23 October 1848

I send you a box of specimens of shells from Australia dredged up by my son Capt. Stanley, HMS Rattlesnake. Perhaps Forbes should see them. I have others which I could send for inspection.

1063 Oldham, N.A.   To: H.T. De la Beche
18 Pembroke Road.
27 November 1849

Thank you for signatures of self, Buckland and Peel. Dr. Buckland's seems written for some other purpose than that of being read.   Enclosed, the promised pair of mittens.

1064 Oldham, Thomas   To: H.T. De la Beche
9 Suffolk Street,
Dublin.
15 July 1845

Allow me to introduce a friend of mine, Du Noyer, who is visiting London. He may settle there. He possesses powers as draughtsman, and would make a most valuable aid.
[Enclosure: card of G. Du Noyer]

1065 Oldham, Thomas   To: H.T. De la Beche
Kildare.
27 June 1846

I have only a few moments to catch the post, so excuse my hurried acceptance of the office. I am sure there will be no difficulty at Trinity. Thanks for your kindness and flattering recommendation.

1066 Oldham, Thomas   To: H.T. De la Beche
Geological Survey Office,
Dublin.
4 August 1846

Concerning the house in Stephens Green, Kane says he does not wish to press the lawyers, as if things do not go fair, we will be blamed afterwards. I think he welcomes the delay himself. I will go to the country next week. Wilson has shifted to his home district.

1067 Oldham, Thomas   To: H.T. De la Beche
Geological Survey Office,
Dublin.
5 August 1846

Penny was anxious to speak to you about £1.13.0 travel expenses from London to Liverpool. I told him I was writing a private note to you about it. Please give directions. Kane has not shewn yet.

1068 Oldham, Thomas   To: H.T. De la Beche
Geological Survey Office,
Dublin.
6 August 1846

Herewith a long report. What is this talk of a Committee to investigate the position of the Ordnance Survey? Larcom has been appointed fourth Commissioner of Public Works here. Something is brewing. Kane is at home. I will write again if I succeed in seeing him.

1069 Oldham, Thomas   To: H.T. De la Beche
Custom House.
6 August 1846

I have seen Kane. Hamilton & Co. say the delay on the house is on the other side. Phillips says it is with you. The only difficulty is that official consent is needed from the First Commissioner. Can nothing be done?

1070 Oldham, Thomas   To: H.T. De la Beche
Enniskillen.
11 August 1846

I came here for a friend's wedding. Griffith's map is curious. I don't think it was all done by his own men, and quite agree in suspecting treason. The new movement has been got up secretly. Some of the Committee are upright men, and will not lend themselves to any job.

1071 Oldham, Thomas   To: H.T. De la Beche
    Ashford,
    Co. Wicklow.
    17 August 1846

News of an extra grant for next year and transfer of a balance now is most welcome. Could we press for one inch maps as well now? I fell from the mail last Thursday and injured my face. Owen and Agassiz are in Dublin. I am certain someone has peached giving Griffiths material for his map.

1072 Oldham, Thomas   To: H.T. De la Beche
    Tinahely,
    Co. Wicklow.
    26 August 1846

I have recovered from my accident, working the last three days. The section of Croghan does not give a fair view of the structure, being in an unfavourable line. The financial news is good. We may go ahead. Penny's account is settled. What about Southampton?

1073 Oldham, Thomas   To: H.T. De la Beche
    Geological Survey Office,
    51 St. Stephens Green,
    Dublin.
    23 November 1846

Orders received, and I meet Kane tomorrow to effect the transfer. I will settle Flanagan's matter. Ramsay tells me you have received notice to quit Craig's Court. Thank you for the money. The weather is stormy and wet. Ask Reeks for stationery.

1074 Oldham, Thomas   To: H.T. De la Beche
    Geological Survey Office,
    51 St. Stephens Green,
    Dublin.
    24 November 1846

Kane tried dodge after dodge, but all were defeated. What about things not under the head of "collections". We need the stores, presses, drawers, etc. Let me know by return. For our fossils we require 3 times the drawer room.

1075 Oldham, Thomas   To: H.T. De la Beche
    51 Stephens Green.
    2 January 1847

I enclosed the quarterly returns. I will be in London on 9th or 11th. I have been looking over Lord Enniskillen's collection. He has a glorious collection of fish. I got a lot of things from him for Trinity College, Dublin. Everything goes smoothly with Kane. We are checking everything.

1076 Oldham, Thomas   To: H.T. De la Beche
    Geological Survey Office,
    Dublin.
    24 February 1847

I have been anxiously awaiting the sections we agreed should be sent over. The accounts you hear of destitution in Ireland are not exaggerated. They say here, and act on the saying, that Government will provide. Thus the land is left untilled and lying waste.

1077 Oldham, Thomas   To: H.T. De la Beche
    Geological Survey Office,
    Dublin.
    4 March 1847

The Ovoca map and coal districts section arrived this morning. The weather is improving and I am preparing for the country. I wish we had the inch map. I cannot find a good assistant. All the competent people are engaged in poor relief. Conditions are more wretched every day. Phillips asks me to be Secretary of the Geological Section of the British Association at Oxford. Will you be President?

1078 Oldham, Thomas   To: H.T. De la Beche
    Geological Survey Office,
    Dublin.
    22 March 1847

It would be politic to present sheets of sections to the Lord Lieutenant. Please send copies over if you agree. It will shew that we, distinct from the Museum, are at work.

1079 Oldham, Thomas   To: H.T. De la Beche
    51 Stephens Green.
    22 March 1847

The Board of Works have made a 20% manpower cut, so we might be able to get a good man. Portlock's wife has died, so he may come home sooner than intended. I meet Forbes in Armagh tomorrow. Flanagan is at Courtown. Mallet has offered the facilities of his works and laboratory.

1080 Oldham, Thomas   To: H.T. De la Beche
    Geological Survey Office,
    Dublin.
    27 March 1847

The man I was treating with at the Board of Works to come to us has had his salary doubled, so this is at an end. My "Secretary" is anxious to do field work. Forbes and I returned from Dungannon last night. One should be able to claim travel allowance for such trips, even if outside the district.

1081 Oldham, Thomas   To: H.T. De la Beche
    51 Stephens Green.
    3 April 1847

Herewith an Official letter asking for an extra 6d a day for Flanagan. The accounts are ready. Forbes is doing good service, examining the Chair of Kildare yesterday. I am anxious not to lose the opportunity of doing the railway cuttings before they are covered up. Can we get money for it?

1082 Oldham, Thomas   To: H.T. De la Beche
    Geological Survey Office,
    Dublin.
    5 April 1847

The difficulties about the markings for the boundaries on the one inch maps. I propose these markings, as the simplest way to deal with the problem. Alterations can be made in the first proofs. I will send them over with the necessary corrections, if alteration can be made.

1083 Oldham, Thomas   To: H.T. De la Beche
    Geological Survey Office,
    Dublin.
    7 April 1847

No documents from Reeks, so the accounts will have to come after me. I shall put the railway works in train. Please send a pattern to work by. There are errors in McCoy's work on the Courtown limestones. It will have to be gone over carefully.
[Cited: Herries Davies 1983, p.142; Herries Davies 1995, p.34].

1084 Oldham, Thomas   To: H.T. De la Beche
    Courtown Harbour,
    Gorey.
    7 May 1847

Wicklow is the first County to map. We will have to give an index of colours on the sheet. The district here was so badly done it will have to be examined again. James' work is faulty. I intend to push ahead with Wicklow so as to be ready by end of summer, then re-do Wexford.
[Cited: Herries Davies 1983, p.142; Herries Davies 1995, p.34].

1085 [Oldham, Thomas]   To: H.T. De la Beche
    Courtown Harbour,
    Gorey.
    10 May 1847

[Part letter only]
I have had three applications for appointments as assistants, two from persons quite ineligible, the third from Mr. Du Noyer, whom you know. I think he would make a good and useful assistant. I have engaged him to do some of the railway sections, and he has applied for a permanent place on the staff. I find a little difficulty in getting proper aid from the engineers of the lines.

1086 Oldham, Thomas   To: H.T. De la Beche
    Baltinglass,
    Co. Wicklow.
    6 June 1847

I am glad you take the same view about the porter's wage as I do. I told Kane I thought we were bound to 5/- per week, not to half of what he chose to engage someone for. I have been working hard around here. This is a very complex district. Willson is working well but slowly. What are your arrangements for a visit to Ireland?

1087 Oldham, Thomas To: H.T. De la Beche
Bray.
23 July 1847

I asked Medlicott to sort out the ores for Kane. I will check them tomorrow and get rid of them. I am glad to hear your Welsh rocks bear closely to our Irish ones. I have found a considerable patch of trap ash, throwing light on the origin of slates.

1088 Oldham, Thomas To: H.T. De la Beche
Geological Survey Office,
Stephen's Green,
Dublin.
1 November 1847

I enclosed a receipt. I also paid 10/- for the glass ordered for Flanagan. I have a bronze sword which I hope to secure for you. I have discovered two heads of Reindeer, the first found in Ireland.

1089 Oldham, Thomas To: H.T. De la Beche
Geological Survey Office,
Stephen's Green,
Dublin.
22 November 1847

The bronze sword is still available, but the collection must be bought entire. Various people to take different pieces have still to be organised. Only one new murder in the papers today! I hope to send sections in a few days.

1090 Oldham, Thomas To: H.T. De la Beche
Geological Survey Office,
51 Stephen's Green,
Dublin.
16 December 1847

I am preparing the accounts for this quarter. We shall want about £200. Forbes is working on the Kildare Silurians and Cystidean research. I am checking the map to send to the Park for engraving. The weather is bad and sickness is prevalent.

1091 Oldham, Thomas To: H.T. De la Beche
Geological Survey Office,
Dublin.
14 February 1848

Our move to College is an important step, and will affect the University here. It is a triumphant proof of the liberality of our Colleges as distinct from English Universities.

1092 Oldham, Thomas To: H.T. De la Beche
Dublin.
20 March 1848

The 17th most quiet. The party was cowed by preparations to suppress any attempt at even a riot. There is a meeting today to express sympathy with France. A dangerous phase of public agitation. All here are well.

1093 Oldham, Thomas To: H.T. De la Beche
51 Stephen's Green.
30 March 1848

I enclose receipts for travelling. Everything here going on steadily. I shall be off to the field soon. Our map progresses and will look well. Does the Committee on the Woods and Forests affect us?

1094 Oldham, Thomas To: H.T. De la Beche
Geological Survey Office,
51 Stephen's Green,
Dublin.
1 April 1848

The cash will be sufficient for the present. A copy of the address for the Geological Society has arrived and will be presented. I am pushing on the map, but they are slow at Mountjoy. Things look bad here. The feelings of the people are deeply excited. The results must be a collision. It will be much more a religious war. I am fearful of very serious consequences.

1095 Oldham, Thomas To: H.T. De la Beche
51 Stephen's Green.
5 April 1848

I enclose the usual quarterly return. The accounts are all ready, except for Wyley's. The country here is in a fearful state. Science is at a standstill, as the College is made a barracks, and the Royal Dublin Society also occupied as troop accomodation etc. All are arming. I am not anxious for myself, but for Wyley, who goes about freely giving out his opinion.

1096 Oldham, Thomas To: H.T. De la Beche
Geological Survey Office,
Dublin.
17 May 1848

Herewith 2 copies of the plate of Wicklow. The colouring is uneven, but Bone will smooth it all out. Concerning details of the map etc., three particular spots need attention. I wish you could come for chat about them. Let me know if you could arrange a visit.

1097 Oldham, Thomas To: H.T. De la Beche
7 June 1848

The coloured map arrived, and I have done all the corrections. I have sent off to the Ordnance Office the final proof sent for examination. There are some difficulties on the map regarding drawing the line near Ashford, and the Rathdrum river. I was sorry you could not visit us here. I go to Lambay Island next week, where Flanagan is finding good things.

1098 Oldham, Thomas To: H.T. De la Beche
5 Trinity College.
21 June 1848

The map will be ready in a day or two, when we will arrange printing. Would you kindly look at the section on which Lowry is working. I have a dry proof of Kildare, and am putting on the lines for the engraver. We shall want £200. I had an announcement of the F.R.S. election, for which my thanks to you and others.

1099 Oldham, Thomas To: H.T. De la Beche
Geological Survey Office,
51 Stephen's Green.
26 June 1848

I am posting a coloured copy of our Wicklow map, and a clean and a coloured copy of Sheet 4 of the sections. In future engraving might be done here in winter. I think we ought to produce our map and sections at Swansea. Things are stagnant here, and money scarce.

1100 Oldham, Thomas To: H.T. De la Beche
Naas,
Co. Kildare.
2 July 1848

I have received your notice of £150 from London. It will keep us afloat for this quarter. The vote must pass soon now. I paid the assistants by drawing on my own funds. I enclose travelling receipts. Reports are rife that the Queen is coming over.

1101 Oldham, Thomas To: H.T. De la Beche
Geological Survey Office.
21 July 1848

I have carefully gone over the map inserting all omissions, and have sent it back to Hunt. Touching the section, the evidence is very good for granite. I have begged Hunt to make the line of junction less distinct. I have had no reply from Cameron on the electrotyping of Kildare and Carlow. I believe we are doomed to eternal uncertainty in this country. I wish I was well out of it.

1102 Oldham, Thomas To: H.T. De la Beche
Carlow.
7 November 1848

Wyley and Du Noyer get on well now, without annoyance. I am at a loss to know what to do with Wyley. Everywhere I go after him, he is looked upon as demented. He seriously impedes fair progress. Forbes has commenced the fossils in earnest. Willson's work is going ahead.

1103 Oldham, Thomas To: H.T. De la Beche
Geological Survey Office,
51 Stephen's Green.
14 November 1848

Concerning your remarks about reading papers to the Geological Society here, I do not understand them. Such notices (not papers) increase the value of our final publications. We must carry the public with us. The only person with a right to offer an opinion is the First Commissioner. The practice of hoarding results has prevented Ramsay from attaining his due

standing in the Geological world. Murchison's Silurian System was increased in value by the previous publication of its parts. I am satisfied that this injunction does not originate from you.
[Cited: Secord 1986b, p.239]

1104 Oldham, Thomas To: H.T. De la Beche
Geological Survey Office,
51 Stephen's Green.
30 November 1848

Official letter received. I am glad the matter is so settled. I regret you thought me wanting in proper respect and courtesy. The objections against the permanent adoption of such measures should be freely stated. I suppose you will have a regular row at the Royal Society. I hope the matter will be settled by having both Grove and Bell as Secretaries.

1105 Oldham, Thomas To: H.T. De la Beche
Geological Survey Office,
51 Stephen's Green.
2 December 1848

Do you want copies of the six inch geological maps? I have just sent to the Park the Carlow map for engraving. A proof of Kildare will be ready shortly. Du Noyer moves to Dublin. If you have no objection, will submit the Dundalk railway cutting to the Geological Society next week. It will fix our discovery of the Old Red.

1106 Oldham, Thomas To: H.T. De la Beche
Geological Survey Office.
4 June 1849

A coloured copy of the Kildare map is being sent by post today. Under Bone's hands it will come out nicely. Kane had the Lord Lieutenant here on Saturday, but didn't bring him near our part of the house. We must do something to prevent this. Reeks says the first volume of decades is ready. Please send one for us and one for the Lord Lieutenant.
[Cited: Herries Davies 1983, p.152].

1107 Oldham, Thomas To: H.T. De la Beche
Balbriggan.
11 June 1849

The "needful" touching the Royal Dublin Society will be done. Medlicott will forward the maps. I have been working over some very puzzling ground, slates, volcanic matter, and intruded greenstones. Your arrival next month is prepared for. My brother Charles got a scholarship to College. Wyley is unwell. Willson is working in Wexford.

1108 Oldham, Thomas To: H.T. De la Beche
Enniscorthy.
29 June 1849

I enclose a letter from the Ordnance regarding payment for the Wicklow Map. What is to be done? It would be simpler to have all these things charged to one account.

1109 Oldham, Thomas To: H.T. De la Beche
Wexford.
3 October 1849

Limestone bands crossing this corner form a link between Hook and the more northerly beds of Carlow. The weather is bad. Medlicott's second brother, my student at Trinity, is doing better work than Wyley, whom we cannot afford to keep. I enclose the usual receipts.
[Cited: Herries Davies 1983, p.142].

1110 Oldham, Thomas To: H.T. De la Beche
Wexford.
12 October 1849

This County is turning out a trump card. Closer work is essential, so it will not be completed until Spring. The preliminary field work of Dublin county is done. The sections are so bad, we must not venture to publish until they are revised. I will adjust my plans to suit your visit.

1111 Oldham, Thomas To: H.T. De la Beche
Wexford.
12 October 1849

If the official authority has been procured, it would be desirable to send maps, decades etc., to the Queen's Colleges of Cork, Belfast and Galway, as they will shortly open. Did you get authority to send Griffith the rest of the maps?
[Cited: Harper & Parkes 1996, p.235]

1112 Oldham, Thomas To: H.T. De la Beche
Enniscorthy.
8 November 1849

I was glad to hear of your safe arrival. We have had wet weather since you left. Thanks for handing Reeks the £5. Nothing new here.

1113 Oldham, Thomas To: [?]
5 Trinity College.
18 February 1850

In reply to your enquiries regarding your cousin, he studied well at the School of Engineering here, answered me well in examinations, and deserves a trial on the Geological Survey.

1114 Oldham, Thomas To: H.T. De la Beche
Geological Survey Office,
Dublin.
6 March 1850

I have been trying to ascertain opinion here on the one inch map. All think it desirable. Now that I am not President of the Geological Society here, a memorial from them would promote the map. Can the same be obtained in London? We will be so delayed if we cannot obtain the one inch map.

1115 Oldham, Thomas To: H.T. De la Beche
18 Pembroke Road,
Dublin.
7 March 1850

Regarding the opening in India, I am grateful for the offer. The appointment is not one I would like to decline without consideration, but I am disposed to say no. I would much like to go, but I have many dear ties here. I derive no great emolument from my Survey appointment. I will give you a definite answer at once, when any further information of the arrangement is available.

1116 Oldham, Thomas To: H.T. De la Beche
Dublin.
12 March 1850

After a good deal of thought, I have decided to refuse the offer of the India appointment under the terms stated. But if they give me £1000, and two assistants of my own choice, and travelling costs, etc., I will accept.
[Cited: Stafford 1984, pp.8,9]

1117 Oldham, Thomas To: H.T. De la Beche
Dublin.
13 March 1850

Please send us cash to meet demands for pay etc. What shall I pay H. Medlicott who is standing in for his brother? All progresses here. I intend to complete Wexford and get it out of the way.

1118 Oldham, Thomas To: Sir H. De la Beche
Dublin.
20 March 1850

Manderson called today expecting payment for the tazza of £57.8.6. I have been experimenting on maps using engraved shading to shew the extent of alterations around intrusions. It interferes with other engraved tints and I have not adopted it.

1119 Oldham, Thomas To: H.T. De la Beche
Geological Survey Office,
Dublin.
27 March 1850

I have despatched a copy of the Carlow and Kildare maps to Reeks; also 50 clean copies of Kildare to be coloured as our stocks are low. Severe weather here. Any more on the Indian matter? I hope you like the new Commissioner, Lord Seymour, and that he appreciates the value of the Survey.

1120 Oldham, Thomas To: H.T. De la Beche
Dublin.
8 April 1850

[Part of letter only]
I sent the accounts to Reeks on Saturday, and hope you will find them all right. We are greatly in want of additional drawers for our rapidly accumulating fossils. We have had four men in the field the past year. If I can get two or three fine days here it will complete all my Dublin work in the field, and then polishing off for Wexford. I am heartily tired of the

1121 Oldham, Thomas　　　　　　　　To: H.T. De la Beche
Foulkes Mill,
New Ross.
15 May 1850

Has anything been done about the Indian matter? There is an unpleasantness about suspense. Besides, I wish to invest in a horse if I do not go, as it may save money, given the absurd regulation of the Survey, of travel within 15 miles of station. I have been working on some intricate trap hills here. Wexford will be ready for the Association meeting.

1122 Oldham, Thomas　　　　　　　　To: H.T. De la Beche
Wexford.
24 May 1850

I am satisfied that the previous estimate of 10/- per square mile for the one inch map will be sufficient. Herewith a new estimate calculated on 5/-, not 7/- as the selling price. The cost to the Survey would be the same for publication as at present. Regarding India, I asked the Board to secure an assistant's pay, not to provide passage and outfit. A £1200 salary would allow me £200 to implement my views.
[Insert, previous estimate]
　　10/- per square mile for one inch map.
　　371 copies need to be sold to cover expenses.
[Cited: Stafford 1984, p.8]

1123 Oldham, Thomas　　　　　　　　To: H.T. De la Beche
Knockmahon,
Kilmacthomas,
Co. Waterford.
11 July 1850

I have been working on the coast here for some days. Willson is at work between here and Tramore, and Du Noyer in the hills. I must return to town, being a judge at exhibition of Manufactures at the Royal Dublin Society. I will be glad to hear the arrangements for your visit here. Anything further on India?

1124 Oldham, Thomas　　　　　　　　To: H.T. De la Beche
Geological Survey Office,
Dublin.
16 July 1850

I neglected to add a memo on the accounts to explain that the differences in charges are due to the fact that return charges are cheaper. Nothing not absolutely necessary will be spent on furnishings. You do not say when you will be here.

1125 Oldham, Thomas　　　　　　　　To: H.T. De la Beche
Clonmel.
25 September 1850

The East India Company have consented to my not going until December. I will lecture at College during the coming term. The yellow sandstone here is difficult to differentiate from the Old Red.

1126 Oldham, Thomas　　　　　　　　To: H.T. De la Beche
Geological Survey Office,
51 Stephen's Green.
12 October 1850

Herewith the travelling bill and vouchers. The quarter's accounts are going to Reeks today. I hope to see you on Thursday.

1127 Oldham, Thomas　　　　　　　　To: H.T. De la Beche
16 October 1850

The specimens selected are packed to be sent today. I have put in others from the Newry district. Wake, the Comptroller of Stationery here, has said that special authority is needed for copies of the Tipperary map. Serious delay will result from such procedure each time we ask. The matter is likely to come to you. It is desirable that it be settled at once.

1128 Oldham, Thomas　　　　　　　　To: H.T. De la Beche
Wooden Bridge,
Arklow.
20 October 1850

Herewith a copy of an official letter from Wake at the Stationery Office regarding maps. It will be ruinous if we have to pay for 6 inch maps. Probably a good lever could be made out of this expense to aid the one-inch project.

1129 Oldham, Thomas　　　　　　　　To: H.T. De la Beche
18 Pembroke Road,
Dublin.
17 November 1850

A section will be ready on Tuesday and will be sent over. College have made me offers which may induce me to remain in this country. Yet it would not be worthwhile for Jukes if both College and Survey work were not available to him. Let me know if the Survey would still be available to me.

1130 Oldham, Thomas　　　　　　　　To: H.T. De la Beche
Geological Survey Office.
18 November 1850

The accounts shall be prepared as desired. I have sent over the pieces of Mexican pottery you said you wanted to have. I am preparing to go eastwards, and shall be soon beyond the fogs and damps of this climate.

1131 Oldham, Thomas　　　　　　　　To: H.T. De la Beche
Geological Survey Office.
20 November 1850

I am sorry you thought there was personal hostility in my letter of resignation. I have expressed the same opinions frequently to you, and you have not taken them so. I could not leave the Survey without putting the facts on record, in the hope that Lord Seymour will make the arrangements workable by modifying them.

1132 Oldham, Thomas　　　　　　　　To: [H.T. De la Beche]
[No Date]
[Part letter only]

I have sent the map to the engravers, leaving only a very few additions to be made which can be put in afterwards. These new postal regulations about Sunday are a horrid nuisance. I think if ever I was tempted to become an agitator it was now. Let not 24 hours in the week be snatched from everyone else to satisfy their puritanical absurdities. It riles one as the Yankees say.

1133 Oldham, Thomas　　　　　　　　To: [H.T. De la Beche]
[No date]
[Part letter only]

I think a little arrangement of Sir Robert's plan would render it quite as easy for him to allot us the three new rooms as any others, thus placing us much in the same state as now, where all our rooms are right over each other. I hope to be able to send the map of Wexford to the engravers on Monday or Tuesday. The last two days have been broken, but today was very fine. Best thanks for all your trouble about India. I shall write again concerning it.

1134 Ord, Major　　　　　　　　To: H.T. De la Beche
Edgbaston,
Birmingham.
10 November 1842

A specimen of stone from Frampton Cotterell has been sent. Is it suitable for paving? Also specimens from Ceylon, which are at Sir Henry's service, if worth his care.

1135 Ormerod, George Waring　　　　　　　　To: Sir H. De la Beche
75 Mosley Street,　　　　　　　　Geological Survey Office,
Manchester.　　　　　　　　51 Stephens Green,
7 September 1847　　　　　　　　Dublin.

I shall be happy to give you a paper in December or January. I have a long one ready on the saltfield of Cheshire with the intention of printing it as a small volume. If it will suit, I will cut out everything but the geology and get it in order. I am busy getting ready for the next session of Parliament.

1136 Owen, Richard　　　　　　　　To: Chairman of Committee
6 Park Cottages,　　　　　　　　of Illustration of
Park Village East.　　　　　　　　British Fossil Reptiles
19 February 1842

Regarding the application of the grant of £250 voted by British Association, a great outlay attends one plan, without corresponding advantage. I propose to combine with Plates of British Fossil Reptiles fuller descriptions with wider applications, in a work on British Fossil Reptiles. Estimated cost, £1000.

1137 Owen, Richard   To: H.T. De la Beche
College of Surgeons.
23 May 1842

You might report the following progress on the Publication of Drawings of British Fossil Reptiles. 30 plates have been lithographed or zincographed and 250 impressions of each printed off. A few more will be finished before the meeting. I shall be glad to draw the remainder of the grant.
[Cited: Gardiner 1993, p.1]

1138 Owen, Richard   To: H.T. De la Beche
Royal College of Surgeons.
1 June 1842

You have not told me how I am to send you, or with whom else I may deposit receipts for expenditure on the Illustration of the British Fossil Reptiles. I must sell out to continue the work if the rest of the grant does not soon come to my relief.

1139 Owen, Richard   To: H.T. De la Beche
Royal College of Surgeons.
15 June 1842

£110 has arrived safely. It enables me to keep my artists in employ. I propose that the £250 granted me for Illustrations of Reptiles should go towards a 4 volume work I plan on British Fossil Reptiles. My leave of absence will not allow me to stay the whole time at Manchester.

1140 Owen, Richard   To: H.T. De la Beche
Royal College of Surgeons.
3 August 1842

Thank you for your note, and evidence of your sympathy with my workings. I will begin the preparation of part 1 for the Press, and will not hesitate to ask your advice when needed.

1141 Owen, Richard   To: H.T. De la Beche
College of Surgeons.
26 April 1843

Thank you for remembering me while so busy. I also am busy with lectures, and am anxious to work in earnest on Fossil Reptilia. Illustrations for the first two numbers are now completed.

1142 Owen, Richard   To: H.T. De la Beche
College of Surgeons.
1 November 1843

The second box of bones from New Zealand has arrived. It contains the great tibia of Dinornis gigantens, 2ft 11in long. I have enough specimens to establish 5 species of different sizes. I suspect a family of fern eaters.
[Cited: Rupke 1994, p.127]

1143 Owen, Richard   To: H.T. De la Beche
College of Surgeons.
23 May 1847

I have recovered and sent the packet to Nicol. I propose a paper on Mammals, including a new genus, for the meeting of 9th June, if time is available.

1144 Owen, Richard   To: H.T. De la Beche
College of Surgeons.
30 September 1847

Lady Hastings has taken back the crocodile fossils. It ought to have been accurately lithographed for our publications before it was returned. Funds should be allocated for such work.

1145 Owen, Richard   To: H.T. De la Beche
College of Surgeons.
7 April 1848

Thank you for your Address, and kindness in noticing my share of the animal labours.

1146 Owen, Richard   To: H.T. De la Beche
Royal College of Surgeons.
9 December 1848

I have completed comparison and description of Saurian remains submitted to me by Prof. Rogers, and could give results to the Geological on Wednesday next. Can time be made available?

1147 Owen, Richard   To: H.T. De la Beche
Royal College of Surgeons.
30 December 1848

Thank you for your donation to the Library of the College, and to my own. The volumes will do good service in the reading room. The official acknowledgement will come from our President.

1148 Owen, Richard   To: H.T. De la Beche
Royal College of Surgeons.
25 April 1849

Thank you for your Address. Touching Belemnites, the essence of them is the phragmocone. The spathose appendix is an accessory. I came to my conclusions by going carefully over the ground with Pearce and Cunnington. Mantell has merely reproduced fallacies. Time will prove me right.

1149 Owen, Richard   To: H.T. De la Beche
Royal College of Surgeons.
17 October 1849

The library of the Royal College of Surgeons would be enriched by copy of Fauna Antiqua Sivalensis, sent from the Treasury.

1150 Owen, Richard   To: H.T. De la Beche
Royal College of Surgeons.
29 November 1849

I wish to have your opinion on the administration of a grant for the Illustration of a Description of British Fossil Reptiles. I thought it was similar to Agassiz's grant for Fishes. I have integrated the services of the Palaeontographical Society. Charlesworth has complained. You will know what testimony to arm me with.

1151 Owen, Richard   To: H.T. De la Beche
Efford,
Lymington.
20 June 1851

Can you spare a morning to come down here to look at Lady Hastings' collection of Eocene mammals. You could have the pick of this fine and useful collection.

1152 Owen, Richard   To: H.T. De la Beche
College of Surgeons.
13 January 1852

Having spent Xmas holidays with the Bucklands, we found your card on our return. I will call on you at Piccadilly. Have you heard any more from Logan on Potsdam footprints? Buckland is improved in health.

1153 Owen, Richard   To: H.T. De la Beche
Royal College of Surgeons.
20 December 1853

Thanks for the Decades of British Organic Remains. The text is of high scientific value, The plates of extreme beauty and accuracy. A lasting and honorable example of British Science and Art.

1154 Papworth, Edgar   To: H.T. De la Beche
17 Newman Street,
Oxford Street.
6 January 1847

A friend of mine, Oswald Laurence, is seeking employment, after his family's losses in Jamaica as result of the Emancipation Act and Duty on Jamaica produce. The alteration in your establishment suggests to me that there would be an opening for an Assistant.

1155 Parish, Woodbine   To: H.T. De la Beche
26 March 1850

I have ordered the cast of my antique vase to be taken to the Museum. 2½ guineas is not an unreasonable price for it. With respect to the glass, you can have duplicates for the Museum, but the Public must pay for them.

1156 Peach, Charles William   To: H.T. De la Beche
Goran Haven.   Conductor of
24 December 1838   Ordnance Geological Survey,
     Tower,
     London.

Perceiving that your work on Cornwall is not yet published, I have taken the liberty of sending this to inform you that I have met with organic remains at the undermentioned places. Madrepores, Encrinites and broken shells abound at Fowey and Polruan but are not so plentiful at the other places. My visits to most of these places have been of such a hasty nature that I have not examined them so closely as I could have wished. I will gladly make a collection of specimens for the Ordnance Museum.

Summaries *Peach - Peel*

1157 Peach, Charles William To: H.T. De la Beche
Goran Haven. To the care of Col. Colby,
10 February 1840 Tower, London.

I have now extended my discoveries to Gerrans Bay, where in the quartzose rock I have found trilobites, and many other interesting fossils near Polruan. I have traced a continuous line from the Van to beyond Polruan. Henwood tells me that the Government are to employ someone to describe Cornish fossils. If so what should you think of my being employed to find and collect the specimens?

1158 Peach, Charles William To: H.T. De la Beche
Goran Haven.
14 April 1840

Here are the details of my present pay and conditions in the Coast Guard Service. I have no wish to remain in it. If I could get a position in the Custom House I should be better off. I should like to get a situation where I could follow my interest in Natural History. I leave myself entirely in your hands. If necessary I will send references. I wish you or John Phillips could see my fossil collection. I am proud of it. Thank you for the copy of the Report on Cornwall.

1159 Peach, Charles William To: H.T. De la Beche
Goran Haven.
29 December 1840

The situation of Principal Coast Officer in the Customs at Spalding in Lincolnshire will soon be vacant. Could you befriend me on this occasion, as I am very desirous of obtaining the post. It is only thirty miles from my native home, and a far more respectable position than my present one.

1160 Peach, Charles William To: H.T. De la Beche
Customs,
Fowey.
4 August 1846

The Collector of this port died last night after only two days illness. His death will be the cause of many changes. I am desirous of being promoted to the situation of Comptroller. Your intervention would help. I shall study to prove that I was not unworthy of your consideration.

1161 Peach, Charles William To: H.T. De la Beche
Fowey.
22 April 1848

Thank you for your kind notice of me in your address at the Geological Society meeting. I am anxious to examine the rocks between Falmouth and St. Michael's Mount. My duties leave no spare time. I must look out the collection for Prince Albert this summer. There is no kindred soul near me here.

1162 Peach, Charles William To: H.T. De la Beche
Fowey.
2 October 1849

Thank you for the volumes of Organic Remains, a really valuable present to me. I am engaged with a timber vessel from Norway and other duty. I am preparing a paper for the Geological meeting at Penzance. I will inform you of any new thing.

1163 Peach, Charles William To: H.T. De la Beche
Custom House, London.
Peterhead.
20 September 1850

I have obtained a Graptolite from a quarry near the Black head in Cornwall. Professor Forbes of Edinburgh confirms my identification. This is a new leaf in the pages of Cornish geology, and may help to explain Murchison's Lower Silurian. I also met Hugh Miller and Adam Sedgwick at Edinburgh. The latter did not think any of the specimens agreed with the fossil fishes of the Silurian System. I have sent you three boxes, but Hunt says they have not arrived. I hope by now you have seen them.

1164 Peach, Charles William To: H.T. De la Beche
Peterhead.
23 September 1850

I have learned from Ramsay and Forbes that there is an opening for a fossil collector on the Geological Survey. I have asked Ramsay, to use his influence on my behalf. My son William is most anxious to obtain that position. I can recommend him as a steady persevering fellow, and has many other qualifications. I hope this application will meet with your favourable consideration.

1165 Peach, Charles William To: H.T. De la Beche
[No date]
[Part letter only]

I am hoping to meet you in August. I have found some china clay intermingled with chalk flints in a quarry, but not sufficient from an economic point of view. I have done a little among the Zoophytes here and my list is getting a long one. I hope as the summer advances to add greatly to this list. Herrings have come in shore very early this year.

1166 Pearce, James To: H.T. De la Beche
2 Percy Place,
Bath.
11 February 1848

Herewith details concerning my late son (1811-47). Educated Guys Hospital. Took up medical practice with me. His collection of Saurians was the best in England. If it is to be disposed of it will be offered to British Museum, but I would prefer it to go to your Public Institution. Please send a copy of the Proceedings of the Geological Society when printed.

1167 Peel, Robert To: W. Buckland
26 January [1844 FJN]

I have had great pleasure in sanctioning a vote for £250 for the present year, for the object mentioned to you by De la Beche. The Cut from the Bourne is not more than 50 years old.

1168 Peel, Robert To: [H.T. De la Beche]
Whitehall.
9 May [1846 FJN]

Thank you for the box etc. I have just received a lump of marl from Drayton Manor which I will now be able to test with the means provided by you.

1169 Peel, Robert To: [H.T. De la Beche]
Drayton Manor.
10 September [1846 FJN]

Thank you for the volume sent a short time ago.

1170 Peel, Robert To: H.T. De la Beche
Whitehall.
10 September [1847 FJN]

You are invited to Drayton Manor for the weekend, from 23 September till the following Monday.
[On this invitation, De la Beche records his impressions of the stay and an anecdote heard on the subject of Catholic Emancipation].

1171 Peel, Robert To: [H.T. De la Beche]
Drayton Manor.
[9 November 1847]

You are invited to dinner next Monday. The 2 o'clock train will get you here in time. Buckland is also invited.

1172 Peel, Robert To: [H.T. De la Beche]
Whitehall.
26 November [1847 FJN]

I am obliged by the Returns you have sent me.

1173 Peel, Robert To: H.T. De la Beche
Whitehall.
4 December [1847 FJN]

You are invited to dinner next Wednesday.

1174 Peel, Robert To: [H.T. De la Beche]
Whitehall.
13 December [1848 FJN]

Excuse my delay in thanking you for your note and the volume. I will take it to Drayton to read. Can you come to dinner on Tuesday?

1175 Peel, Robert To: [H.T. De la Beche]
Drayton Manor.
25 December [1848 FJN]

Thanks for the warning in your note. We have acted on your suggestion, though we have no previous indications of lead in the water. Probably wooden pipes are the safest. I shall be curious to know if your analysis confirms the purity of water from the Spring.

1176 Peel, Robert
Whitehall.
14 May [1849 FJN]
To: [H.T. De la Beche]

This is a specimen of marle from the last opened pit at Drayton Manor.

1177 Peel, Robert
Whitehall Gardens.
To: H.T. De la Beche

Invitations for Saturday 15 June, 9.30 p.m., "to have the honour of meeting the King of Saxony".

1178 Pellatt, Apsley
Staines.
21 June 1851
To: H.T. De la Beche

With my brother's consent, I shall be happy to let you have any specimens you like from the Great Exhibition when it closes, for your National Institution.

1179 Pennethorne, James
Elms Court,
Highgate.
22 October 1849
To: H.T. De la Beche

I am anxious to get one of my boys into Carshalton, the preparatory establishment for Woolwich. Can you use your good offices with the Master General of the Ordnance to have his name placed on the list?

1180 Pentland, Joseph Barclay
Claverton Manor,
Bath.
21 April 1848
To: H.T. De la Beche

Thanks for the copy of your anniversary address. I have written to ask my friend at the head of the French Hydrographic Department to exchange publications with the Museum of Economic Geology. Their publications consist of some twelve hundred charts and plans. Would you authorise me to send the new volume of Museum memoirs to Elie de Beaumont, Dufrenoy, de Verneuil, Arago, etc., as I know they are anxious to have them as soon as possible?

1181 Pentland, Joseph Barclay
Trafalgar Square.
[1848 FJN]
To: H.T. De la Beche

I have just returned from Paris. Before I left Dufrenoy requested me to apply to you about maps of the Geological Survey which have not arrived. If they have not been sent could they be forwarded by the ordinary conveyance. My friend de Verneuil also requests copies of the Geological Memoirs published by your department, as he is now engaged on the Palaeozoic deposits of the west of England in connection with those of Belgium and Spain.

1182 Pentland, Joseph Barclay
Hotel Brighton,
Rue de Rivoli,
Paris.
31 January 1850
To: H.T. De la Beche

I have obtained the necessary permission of the Minister to have copies made of the documents in the Ecole des Mines relative to the organisation and administration of that establishment, as well as the courses studied. Please authorise £15 for the cost of copying. At my request the Minister of Public Works has given me for the Museum of Practical Geology a collection of every printed document connected with mines and geology of which there were copies at the Ministry. Many other publications require your authorisation of the expense to send to London a large case.

1183 Pentland, Joseph Barclay
Lilford Hall,
Oundle.
12 January 1851
To: H.T. De la Beche

The discovery of a considerable mass of iron ore in the Oolitic series in Northamptonshire is exciting much attention because Staffordshire smelters have made tempting offers to some of the proprietors who, in the present state of the corn markets, are glad to catch at anything. If you can give my Northamptonshire friends any encouragement or information you would greatly oblige me.

1184 Pentland, Joseph Barclay
Verona.
10 November 1852
To: H.T. De la Beche
Director of the
Museum of Practical Geology,
Jermyn Street,
London.

Here is a list of maps and publications for which you gave me a general commission to obtain in Italy. If you will let me know your wishes I will attend to them. As Murchison will tell you, you should include Passini of Venice in your distribution of Memoirs of the Geological Survey and Decades of fossils.

1185 Percy, John
Birmingham.
22 July 1845
To: H.T. De la Beche
Museum of Economic Geology,
London.

I send you herewith samples of English bronzes, made by Mr. Lane of this town. Lane is probably the largest and most successful manufacturer in the world. You will find the English superior to the German bronzes. They are made from alloys of copper. You will be pleased to consider these specimens as presented to the Museum by Mr. Lane and not by me. I have obtained also a series of specimens of the glass-button manufacture which I will send shortly.

1186 Percy, John
Birmingham.
1 December 1848
To: H.T. De la Beche

Messrs. Chances' address is Glass Works, Spon Lane, near Birmingham. So Bell has won by a large majority! As soon as I receive your next note I will confer with my friend Mr. Evans.

1187 Percy, John
Birmingham.
18 March 1846
To: H.T. De la Beche
Museum of Economic Geology,
London.

Under my marriage contract I am obliged by my father-in-law to reside near Birmingham. On my return from London I had a conversation with him on the subject of our interview. He is prepared to consent to my living in London, an event which I have long desired. The position would give me great pleasure, and in the event of my being connected with you, I would devote all my energies to my duties. I see a magnificent field for investigation, especially with the aid of students.

1188 Percy, John
Birmingham.
7 April 1846
To: H.T. De la Beche

We shall now be very happy to see you at any time when you can conveniently run away for a few days. I promise you that you shall not return empty in respect to the Museum. I am at present at work at the crystallized slags of the iron furnaces. I have made one quantitative analysis which I am now repeating. I have now fairly abandoned physics for my first love chemistry.

1189 Percy, John
Birmingham.
28 June 1846
To: H.T. De la Beche

Forbes and I are progressing well with the analyses of the crystallised slags, and we are anxious to present to the Association as complete a series as possible of the crystallised iron slags. If you have any such specimens in the Museum, which have not been analysed, I should be much obliged to you to provide us with a small quantity of each for the purpose of analysis, if they are not being analysed in the Museum.

1190 Percy, John
Birmingham.
4 December 1846
To: H.T. De la Beche

I have already enquired about what you desire. I hope to be able to send you the plans in a day or two. Plans have already been made by order of one of the coroners, who, I am sure, will immediately let you have them. I am sorry to say I do not know anything of the Lancashire matter. I hope to write again on Monday next.

1191 Percy, John
Birmingham.
13 December 1846
To: H.T. De la Beche

I have not received the information which I anticipated. If I had had the time, I would have gone over myself and collected the materials for you. The excitement is considerable and there is a general desire that something should be done by Government. I think you will find it necessary to examine the Staffordshire mines especially with reference to prospective legislation. More men perish by the falling of coal than by fire-damp. We want skilful engineers to superintend the working of the mines with a view to the protection of human life; and there will be no opposition to such superintendence on the part of the mine proprietors.

1192 Percy, John   To: H.T. De la Beche
      Birmingham.
      17 March 1848
Accept my best thanks for your valuable and interesting Report, which, I am sure, will be found to be of great practical importance. I was delighted to see by the Morning Chronicle a few days ago so favourable a notice of the Museum, and especially of the scheme which you contemplate. I have since I saw you put every specimen which I possess into perfect order, and I hope that at a future day they will be placed in your Museum. I hope you will one day cause an investigation to be made in connection with your Institution of all our British iron ores, and varieties of iron.

1193 Percy, John   To: H.T. De la Beche
      Birmingham.
      2 April 1848
Thank you for your interesting and valuable Geological Report. I am not quite satisfied about Ebelmen's results. I think they require confirmation. I am afraid I shall not be able for some time to come to work at the slags, for I am engaged on a work on medical subjects.

1194 Percy, John   To: H.T. De la Beche
      Birmingham.
      29 October 1848
To obtain the information you desire I waited for the return of my friend Mr. Evans, who constantly hears about the Staffordshire potters. I saw him last evening and enquired if he knew anyone capable of taking the direction of a manufactory of earthenware. He replied that at present he is acquainted with several persons every way competent for such a position. But he should know the precise nature of the situation to which you refer. Have you seen a savage and in some respects very stupid attack upon your Coal Report in the "Chemical Times".

1195 Percy, John   To: H.T. De la Beche
      Birmingham.
      2 March 1849
This letter will introduce to you my friend Mr. Arnoux of Toulouse. He was a manufacturer of hard porcelain, and has distinguished himself by several inventions of importance especially relating to colours. He is about to be connected with Mr. Minton of Stoke. I am sure he will be able to assist you in the way of specimens, and that you will be glad to give him information upon one or two geological points which he desires. Did you see the notice in the Literary Gazette of last week concerning your Museum and projected Mining school?

1196 Percy, John   To: H.T. De la Beche
      Birmingham.   Matlock Bath,
      28 September 1850   Derbyshire.
Henry and I propose to be with you and Smyth early on Tuesday morning, and to spend a week with you. I shall take my camera, and, if the sun be favourable, show you how important an instrument it may become in the hands of the Geological Surveyor. We should like to be in lodgings near you, and request you find them for us. We will drive direct to your lodgings. We shall have a multitude of things to talk over.

1197 Percy, John   To: H.T. De la Beche
      Birmingham.   Matlock Bath,
      15 October 1850   Derbyshire.
I have been requested to give my opinion on the most suitable material for the statue of Sir Robert Peel to be erected in this town. The choice is between bronze and certain varieties of marble. What is your opinion? I requested Smyth to tell you that I had determined to make some experiments with particular reference to geology and mineralogy on the solvent action of water at high temperatures. I will not proceed if you are doing something similar, and will hand over the apparatus to you.
[See also Hollins, P. to De la Beche, H.T., 5 November 1850, **698**].

1198 Pernolet   To: [H.T. De la Beche]
      Paris.
      26 November 1851
It was indeed a misfortune for me not to have met you on my recent voyage in England. Dufrenoy had given me a letter of introduction which turned out to be of no use as a result of your absence. Your well-known zeal for progress in geological science and extreme liberality to strangers led me to hope I might find you interested in my researches into veins and the mode of distribution of minerals in seams.

1199 Petherick, John   To: H.T. De la Beche
      Knockmahon Lodge.
      31 October 1847
In reply to your letter of 28th inst., I hasten to state that at the request of Sir Robert Kane I am preparing a collection of ores from these mines for the Dublin Museum of Economic Geology. This I presume would suit your purpose, and I would be happy to prepare a similar series of the ores for you. Thank you for the copy of the lead returns of the United Kingdom.

1200 Pethoud, H.   To: H.T. De la Beche
      Geneva.   c/o G.W. & S.Hibbert,
      10 February 1830   London.
I have been ill and in financial difficulties. Can I count on your protection? My good old mother presents her respects to you. I trust I have not taken too great a liberty in writing to you.

1201 Philipps, Trenham W.   To: H.T. De la Beche
      Office of Woods.
      22 September [1835]
I sent Baring's letter to you some time ago. The proper date to be quoted is 11 August. I have communicated the substance of your last cheering letter to Duncannon.

1202 Philipps, Trenham W.   To: [H.T. De la Beche]
      Office of Woods.
      21 January 1836
I would be fearful of your sending the parcel you mention by Post. I recommend you send it by Coach. Duncannon suffers from rheumatism, and has not moved out. He approves of the suggestions made in your former letter. With regard to naming the Institution I will write again in a few days.

1203 Philipps, Trenham W.   To: H.T. De la Beche
      Office of Woods.
      8 February [1836]
I have opened the case of specimens received on Friday. Duncannon tells me he will look at them. I have considered the question of a name for new the Museum and suggest "Economic" instead of your "Economical". The latter won't make objects of the collection understood. Will you look over the enclosed, and when you've changed it as you judge best, I will have it engraved on Copper.

1204 Philipps, Trenham W.   To: [H.T. De la Beche]
      Office of Woods.
      31 March 1836
Watson delivered his letter today. The messenger told me that he tried some of the Porphyries and found them very hard. I have applied to Sylvester from day to day respecting the letters. I have not received the specimens stated to have been despatched on 26th. Mudge called here a week ago.
[See also Watson, M.L. to De la Beche, H.T., **2080**]

1205 Philipps, Trenham W.   To: [H.T. De la Beche]
      Office of Woods.
      28 May [1836]
The case of specimens arrived today and are in good preservation. I have settled the matter of Civil Contingencies with Baring. Your money will be paid at the Treasury, and all we'll have to do is examine and certify your accounts. Do with us as you do with the Ordnance.

1206 Philipps, Trenham W.   To: H.T. De la Beche
      Office of Woods.
      2 June [1836]
Adopt the course you have suggested and apply for a sum of money on account. Consider this as settled and act accordingly. Also include sufficient to cover all my small expenses here, about £25 for the last 12 months.

1207 Philipps, Trenham W.   To: H.T. De la Beche
      Office of Woods.
      June [1836]
I forwarded your last letter to Baring. He may issue money without

returning it for a report. Your letter of the 10th refers to some brilliant specimens. The lock and key will be necessary. Petherick's letter will be placed before the Board. The suggestion has been made to us already.

1208 Philipps, Trenham W. To: H.T. De la Beche
Office of Woods.
16 July [1836]
Duncannon has sent the Report to the Treasury with a line to Baring asking for an issue of £100. Milne, though not an advocate, willingly does anything requisite to keep us alive. Perhaps you would like to cast an eye over the Report. I trust you will not criticise it, as I am alone here, with no one to resort to for assistance. I am to see Baring on Monday.

1209 Philipps, Trenham W .To: H.T. De la Beche
Office of Woods.
20 July [1836]
You will be gratified to see by the enclosed how little disposition there is at the Treasury to overlook you and your good cause. Baring is a well-wisher. I asked Duncannon if there was any objection to nominating you Director. He deferred the matter, but I am going to take the first opportunity of getting it settled.

1210 Philipps, Trenham W. To: H.T. De la Beche
Office of Woods.
23 July 1836
I enclose an abstract of expenditure. There are some claims for taxes which I am endeavouring to fight off. Are the certificates satisfactory?

1211 Philipps, Trenham W. To: H.T. De la Beche
Office of Woods.
27 July 1836
Duncannon wishes to use your discretion over the vase. I should retain some of the £100 imprest to pay demands in London. I have a book to use as an entry-book for correspondence. Respecting the Ordnance maps, it will be necessary to apply to the Treasury direct. Lord Duncannon will sign a letter as soon as I have your reply.

1212 Philipps, Trenham W. To: H.T. De la Beche
Office of Woods.
28 July 1836
I have been to the Treasury respecting the money. You will have to draw a Bill payable to me. The dates you mention are the quarter dates of the Department. Accounts should be made up according to dates of payment etc.

1213 Philipps, Trenham W. To: H.T. De la Beche
Office of Woods.
9 September [1836]
Enclosed are the accounts I have settled. I propose trying to see Konig tomorrow morning. Sylvester has woefully delayed the completion of receipt forms. Our Solicitor has advised that we must yield to the Poor rate Collector. I am now disputing the Land tax.

1214 Philipps, Trenham W. To: H.T. De la Beche
Office of Woods.
10 September [1836]
I enclose a letter from Fox. I am glad we are to get models. I intended to see Konig today, but have been showing Dr D.B. Reid our collection. He holds you and your endeavours in high honour. Drummond has written to say that Colby will send a copy of the Irish Survey as it proceeds. Duncannon told me to write to him about the Directorship.

1215 Philipps, Trenham W. To: H.T. De la Beche
Office of Woods.
16 September 1836
Enclosed are letters from Kenwood and Lavier. I have written to Duncannon respecting the Directorship, having reason to know you stand well with Baring. For the rest I am satisfied of Duncannon's good feeling and honourable principle. We have received a complete set of the yet published Ordnance Maps. How do you wish to have them bound?

1216 Philipps, Trenham W. To: [H.T. De la Beche]
Office of Woods.
7 October 1836
Sylvester has at last furnished me with receipts. What are we to do when stamps are required? There are five books. Would it be worthwhile to have one stamped? I have been over the vouchers and sent your letter to Duncannon. Please ask Colby to send me the maps intended for the Museum. If they go to Milne they are delivered into store immediately.

1217 Philipps, Trenham W. To: [H.T. De la Beche]
Office of Woods.
10 November 1836
Your accounts are correct, and satisfactory to Duncannon. If you are to have an Appointment I recommend a formal course. Duncannon, Baring and Rice will be in town on the 20th, to decide. I have seen Konig and he asserted the right of British Museum to the Collection. Cases are needed for displaying specimens. We must pay Land-tax.

1218 Philipps, Trenham W. To: H.T. De la Beche
Office of Woods.
12 November 1836
McClellan has been with me today and says he made three cases for Oxford for £115. This appears to me heavy. He recommends flattened glass instead of plate, thus making a saving. Ours would cost about 30 guineas each. What do you think of this?

1219 Philipps, Trenham W. To: H.T. De la Beche
Office of Woods.
19 November [1836]
It is no use discussing the subject of cases with Duncannon as a separate charge. I would advise the construction of one common case, which will be of future use for housing specimens. Duncannon will be here on Wednesday to speak for himself.

1220 Philipps, Trenham W. To: H.T. De la Beche
Office of Woods.
3 January 1837
Things have been busy here. The cost of stamps was small. I have made no other payments for the Museum during the last quarter. Land tax demand is above £5. The two cases arrived safely yesterday. Duncannon wants to know what size the Mining dial will be? Baring has been out of town.

1221 Philipps, Trenham W. To: H.T. De la Beche
Office of Woods.
16 January 1837
I have been unable to get anyone to sketch the vases, so I have done it myself. I am fearful they are not what you require, as they should be on the scale of working drawings. Duncannon has heard the unexpected news of the death of his brother. He has given authority for you to purchase the Mining dial.

1222 Philipps, Trenham W. To: H.T. De la Beche
Office of Woods.
10 March 1837
Your letters are sent by this night's post. Don't lose time in sending up the proper authority for money. Before I make any payments I will let you know the whole account due. The packet from Lyme Regis has not yet arrived. White has sent a case of good things.

1223 Philipps, Trenham W. To: H.T. De la Beche
Office of Woods.
15 March [1837]
I have unpacked the case from Devonshire, and also opened the one from Lyme Regis. I forwarded Taylor's letter the day it reached me. Sylvester has brought me a finished proof of the Letter. I have got your seal and will send it as you direct.

1224 Philipps, Trenham W. To: H.T. De la Beche
Office of Woods.
22 March [1837]
I approve of the proposed Postscript for the second edition of "How to Observe" although you should change the wording to make it clear that as the result of your suggestions, the Government undertook the Museum of Economic Geology. I am delighted at your account of the models. Your seal is being sent through the Ordnance Office. Refer Watson to me.

1225 Philipps, Trenham W. To: H.T. De la Beche
Office of Woods.
25 March 1837
Our money has been received today, and I remit you the first halves of 10

£10 notes. The accounts for the present quarter are ready, total £33. The box from Lyme arrived some days ago. White has sent some fine chalcedony from Hemel Hempstead.

1226 Philipps, Trenham W.　　　　　　To: H.T. De la Beche
Office of Woods.
19 April [1837]

With this you will receive a Memorandum of money paid by me on account of the Museum. With reference to the attack on your Devon work, I am convinced the proper course is silence. They'll never dispossess you of your appointment. Sedgwick has refused the Bishopric of Norwich, preferring Geology to Theology.
[Cited: McCartney 1977, p.31; Rudwick 1985, p.204]

1227 Philipps, Trenham W.　　　　　　To: H.T. De la Beche
Office of Woods.
16 May 1837

I send you the amended account. I am unwilling to open any cases of models, unless you particularly desire it. Visitors to Craigs Court are very few. Tell me how to describe the models and I will mention their arrival when sending in the accounts. Can you suggest a way to bind the maps?

1228 Philipps, Trenham W.　　　　　　To: H.T. De la Beche
Office of Woods.
19 May [1837]

I found Case A in No.5 Craig's Court and unpacked the specimens into the back room. Baring is to see me on the subject of small contingencies. I hope we can keep clear of Parliament this session, as we would wish to start with a larger sum, one year's being a check for the ensuing year. The specimens have travelled well.

1229 Philipps, Trenham W.　　　　　　To: H.T. De la Beche
Office of Woods.
27 May 1837

The accounts for March will go to the Treasury tomorrow. Formal sanction will be no trouble after they have read your statement. Lord Duncannon is quite satisfied with your management, but in ticklish times like these he is unwilling to anticipate the Treasury. Colby called some days ago with a friend. I am looking to you for directions respecting the maps.

1230 Philipps, Trenham W.　　　　　　To: H.T. De la Beche
Office of Woods.
26 June 1837

I sent you a few days ago the last letter of the Gentleman who wants to place his son under you. I sent you today the accounts for the quarter. In future, I should like to be neither in advance, nor to have balance in hand. I explained to Baring the cause of your leaving London without seeing him.

1231 Philipps, Trenham W.　　　　　　To: H.T. De la Beche
Office of Woods.
30 June [1837 FJN]

I enclose Tindall's receipt for the amount of his account. A box of specimens from Mr Harding arrived yesterday. Another from Mr Fleetwood. Also two specimens of sulphate of lime from Mr Rowbotham, a friend of mine.

1232 Philipps, Trenham W.　　　　　　To: H.T. De la Beche
Office of Woods.
8 July 1837

Thank you for the other half of the £20 note. Your view on a man for chemical analysis coincides with mine. I have put your letter in Duncannon's box, and the holiday weekend will allow him to give his deliberate opinion on the plan. A case of stone from Gloucestershire has arrived, sent by Hopkinson.

1233 Philipps, Trenham W.　　　　　　To: H.T. De la Beche
Office of Woods.
19 July [1837 FJN]

Herewith John Church's address. Our plan for a Chemist will not do, as Duncannon is not disposed to add staff for some time. I think we'd best consider the Craig's Court house as place of deposit and continue to show it to persons disposed to befriend us.

1234 Philipps, Trenham W.　　　　　　To: H.T. De la Beche
Office of Woods.
3 August [1837]

I received your letter enclosing your bill for £100, and took it to the Treasury. It will be Monday before I remit you the money. Your suggestion about the first account accords with my last.

1235 Philipps, Trenham W.　　　　　　To: H.T. De la Beche
Office of Woods.
3 October [1837]

The Treasury have ordered the £100 you applied for. Please send the usual paper and it will be forwarded.

1236 Philipps, Trenham W.　　　　　　To: H.T. De la Beche
Office of Woods.
7 October 1837

I enclose two small demands which I have paid for the Museum. Herewith a list of stores in Craig's Court. Another completed show case has arrived. What was cost of vases you had made for the Museum? I would like the same for my own mantlepiece.

1237 Philipps, Trenham W.　　　　　　To: H.T. De la Beche
Office of Woods.
21 October 1837

Having received no instructions in reply to my enquiry, I send the halves of 20 £5 notes. The remainder will follow on Monday. Our Irish Survey Maps are all bound. The English ones bother me.

1238 Philipps, Trenham W.　　　　　　To: H.T. De la Beche
Office of Woods.
12 December 1837

Despite having a fire waiting for me at Craig's Court, I was unable to go in. I trust to be more at liberty when Parliament is up. 2 cases arrived last evening.

1239 Philipps, Trenham W.　　　　　　To: H.T. De la Beche
Office of Woods.
27 December 1837

Herewith a memorandum of monies owing from the Museum. Woolcott's case has been delivered for some time, but you had models to pay for. Dalgleish's bills have been accumulating. Demand is in for Land tax. Cases have been opened, for airing, but the models cannot be taken out, as we have no room for them.

Philipps, Trenham W.
1837
[see **2270**]

1240 Philipps, Trenham W.　　　　　　To: H.T. De la Beche
Office of Woods.
2 January 1838

The money arrived in this morning's post. I will pay some of the bills. Application shall be made for another £100. How much more have you to pay for models? Drummond spoke of your concern for the Museum only as an "on dit". Robe repudiated the notion. You are deemed a utilitarian in geology, but persevere!

1241 Philipps, Trenham W.　　　　　　To: H.T. De la Beche
Office of Woods.
30 March [1838 FJN]

I enclose a letter from Mr Wortley and some designs from Mr Papworth. I forgot to give you the dates for the Treasury letters before you left.

1242 Philipps, Trenham W.　　　　　　To: H.T. De la Beche
Office of Woods.
23 July 1838

I have received your letter addressed to Barry, but am holding it until you reply to my present report of a conversation I had with him. He has made no engagement with Smith, and wishes you to know that he proposes to take a working mason with him. He wishes to discuss and arrange everything with you. You must come to town. There are incalculable advantages for us in it.

1243 Philipps, Trenham W.　　　　　　To: H.T. De la Beche
Office of Woods.
23 September [1838]

Dr. Smith received the money for the first journey from Barry, but as to the second I cannot tell you. I took Simpson the engineer into the shop this morning. I am quite sure that your presence in town for three months would advance the concern in any instance twelve months. Herbert's final account shows an easily justifiable exceeding.

1244 Philipps, Trenham W.   To: H.T. De la Beche
Office of Woods.
28 November [1838]

You will have heard from Barry that Daniel and Wheatstone had entered upon the duties of their commission. He set them to work without the consultation with Baring you recommended. But Barry assures me that there are sufficient funds to meet the expense. Baring will be pleased if whole enquiry is paid for within the original sum. Barry is working systematically and well.

1245 Philipps, Trenham W.   To: H.T. De la Beche
Office of Woods.
3 December 1838

I enclose the proceeds of the sale: a wretched affair. I would have given twice the money for them myself. Swinfield should have asked me before he parted with them.

1246 Philipps, Trenham W.   To: H.T. De la Beche
Office of Woods.
24 April [1839]

Barry has brought fresh specimens from Bolsover. Staunton has got up a regular row concerning Limerick Stone and is printing documents. It is scarcely fair for the Chancellor of the Exchequer to let his son put his name to any document of the kind, as Ireland formed no part of your enquiry. Phillips has taken a fancy to the upper or skylight floor of the Museum. Cannot this be done?

1247 Philipps, Trenham W.   To: H.T. De la Beche
Office of Woods.
7 May 1839

Duncannon would be pleased if you would reply to these as soon as you can. He wishes to be of service as long as he can, and the Government is out. Hence the formal tone of these references. These affairs will rest with his successor unless attended to now.

1248 Philipps, Trenham W.   To: H.T. De la Beche
Office of Woods.
8 May [1839]

Do you know of anyone suitable as Messenger or Porter to the Museum. If I do not hear from you by return of post, Duncannon will have had some proper person selected. The new Government will be formed in a very short time.

1249 Philipps, Trenham W.   To: H.T. De la Beche
Office of Woods.
13 May 1839

Duncannon and I read your letter as saying that Barr would cost us £1560 [sic] a year as Keeper. I dare say we are wrong, but set us right as soon as you can. You would gain rather than lose power by detaching yourself from the Committee. I foresee trouble with that concern. Perhaps your meaning about Barr was "not more than £180", but Duncannon will not have him unless he is wholly part and parcel of this concern.

1250 Philipps, Trenham W.   To: H.T. De la Beche
Office of Woods.
20 May 1839

Duncannon thinks £180 not too much for a Keeper of Mining Records. His objection to Barr is his Committee connections. The Treasury will probably concede only £150 for the first year. I cannot tell you what pleasure it gives me to see the bantling of 1835 already grown into so promising a child. Would Phillips admit a youth into the Museum Laboratory? My godson wishes to continue his education there.

1251 Philipps, Trenham W.   To: H.T. De la Beche
Office of Woods.
21 May 1839

Duncannon tells me that he has found a very good man for porter at the Museum. How this may be we shall see. I should have been glad of a man from the Polytechnic. Our Clerk of Works at Hampton Court speaks highly of him.

1252 Philipps, Trenham W.   To: H.T. De la Beche
Office of Woods.
22 May [1839]

Duncannon quite approves of your answer to the Royal Institution of South Wales, and you may officially announce his approval of their offer.

1253 Philipps, Trenham W.   To: H.T. De la Beche
Office of Woods.
28 May 1839

Respecting the fitting up of the laboratory, I find that we can do all that is wanted. Phillips will find matters in progress on his return. I have drawn up a memo from your letter for Lord Spencer, by Lord Duncannon's desire. Let me have the Estimate, as I have no copy.

1254 Philipps, Trenham W.   To: H.T. De la Beche
Office of Woods.
3 June 1839

I am scarcely sorry that the negotiation with Barr has failed. I should have doubted his ability to steer clear of difficulties arising from his divided interests. Duncannon thinks it best you look elsewhere in the Country to find a party free from connections in any particular quarter. I have always viewed the Committee with suspicion. Phillips will see Faraday tomorrow about the laboratory fittings. Have you further advice about my godson, as Phillips seems disposed to take him on.

1255 Philipps, Trenham W.   To: H.T. De la Beche
Office of Woods.
12 June 1839

I think I can explain why Barry has done nothing in the matter to which you refer. He and Smith visited Bolsover. A fine specimen of stone has come from Lord Bathurst's quarry, and from his alone. Barry should have answered your letter. It seems to me he has no intention of altering the terms of the original Report. He and Smith are to make a short supplementary report.

1256 Philipps, Trenham W.   To: H.T. De la Beche
Office of Woods.
14 June 1839

Phillips was with me respecting his salary which he thinks ought to have commenced on 7th March. I told him you thought about 5th April. You will be glad to see the enclosed from Lord Spencer. Sir Benjamin Stephenson is dead, so that one of our opponents is out of the field. Gore from the Home Office succeeds him. I think we shall find him an ally.

1257 Philipps, Trenham W.   To: H.T. De la Beche
Office of Woods.
20 June 1839

I'm afraid I was not clear in my letter about Phillips' salary. The papers you have with you, and they will explain his meaning. My recollection is clearly that it should commence in April, but if your understanding differs from mine, let me know and I will do what I can.

1258 Philipps, Trenham W.   To: H.T. De la Beche
Office of Woods.
27 June 1839

You have not returned Lord Spencer's letter to Duncannon respecting the collection of soils. Duncannon wishes to show it to Rice. I have suggested to Phillips that he writes to you about his salary.

1259 Philipps, Trenham W.   To: H.T. De la Beche
Office of Woods.
28 June 1839

I have seen Baring today by Lord Duncannon's desire, and he will not consent to £180 a year for a Keeper of Mining Records. He even objects to £150, and wishes the original proposition of someone for one day a week to stand. Baring wishes you to see if you can find someone in Government service, or completely independent of private parties, to undertake the duty.

1260 Philipps, Trenham W.   To: H.T. De la Beche
Office of Woods.
1 July 1839

I saw Baring on Saturday about repairing No.5. The estimate is £500. This will never do. Baring is averse to doing anything more. I have ordered cases from Herbert. The estimate I sketched out with Chawner was £1800. Duncannon's hair stood on end when I suggested it to him. I will write further when Baring has been over it. Our difficulties will be surmounted.

1261 Philipps, Trenham W.   To: H.T. De la Beche
Office of Woods.
17 July 1839

I enclose the account for copying in our Department. Your £100 will not go far at this rate, and we must hold our hands about cases. The books you wrote about have been printed some time ago. I hope we will not be asked to open until the beginning of 1840. Phillips has seen our porter, and a rare good man he seems to be. Do not worry yourself respecting Edmonds.

1262 Philipps, Trenham W.   To: H.T. De la Beche
Office of Woods.
24 July 1839

I send you two more sheets of Proof, and hope you do not grudge the trouble it is giving you. We shall have difficulty in getting our two sets of cubed specimens, or indeed either of them, from Barry. He is so obstinate. I am unwilling to put Baring in motion until the last thing, and will try Duncannon first. Phillips is not expending attention on the public needs of the Museum, because of pressing private business.

1263 Philipps, Trenham W.   To: H.T. De la Beche
Office of Woods.
30 July 1839

Baring has been here to ask me to forward to you the enclosed. There is some explanation at the back. I received the proof from Swansea this morning. Baring protested against so many capitals, but changed his mind on finding that such is the practice in all printing at the Queen's Printers.

1264 Philipps, Trenham W.   To: H.T. De la Beche
Office of Woods.
2 August 1839

The printers were pressing for the Report yesterday. I delivered the proof received from you. The Queen's printers will not depart from their system of capitals. Barry has tried very hard in the duplicate stone affair. Duplicates still a question. I augur not very favourably from Baring's visit in the matter of adapting No.5. I hope to produce fresh circulars respecting quarries. Barry has objected to the case for the stone on the ground that is not airtight.

1265 Philipps, Trenham W.   To: H.T. De la Beche
Office of Woods.
19 August [1839]

Baring has sanctioned the employment of Edmonds at £150 a year. Two copies at least of our stone document might be of use to you for your friends. Baring has recommended Wood to apply to Phillips respecting the soap question. I am to see Baring tomorrow regarding No.5, and he promises to decide soon. Rice is going to the Lords.

1266 Philipps, Trenham W.   To: H.T. De la Beche
Office of Woods.
26 September 1839

Chawner has given us a small box of cubes of building stones, for which a letter might be sent him immediately. Smith delivered the remainder of the Houses of Parliament Stone cubes on Tuesday. I have made application for £100. The Ordnance Survey [map] of Kildare arrived today.

1267 Philipps, Trenham W.   To: H.T. De la Beche
Office of Woods.
21 October 1839

Herewith a selection of Parliamentary papers which might be useful to you. Regulations fix the winter season as commencing 1st Nov. and summer 1st March. Perhaps you will settle the matter with Phillips. Herbert is making a grand push about the House. Please do something about the Bramah Press.

1268 Philipps, Trenham W.   To: H.T. De la Beche
Office of Woods.
24 October 1839

Two more boxes have arrived. I saw Baring, and instructions were given for money for you, and Herbert, and Chawner. While Phillips was on the Soap enquiry, Gordon, the new Secretary of the Treasury, called and examined the place.

1269 Philipps, Trenham W.   To: H.T. De la Beche
Office of Woods.
4 June 1842

We have been calling in tenders for Guernsey granite, and the demand has mounted. The Board would be glad to know if you can give them a hint of any material cheaper and equally good, or nearly so, as a substitute.

1270 Philipps, Trenham W.   To: H.T. De la Beche
Office of Woods.
16 June 1842

Your accounts will be put in course of examination. We are in a mess about the soils. Lincoln's correspondence has got into print. Phillips complains you have not acknowledged his letter. I wrote you some time ago about Colenso and Cornish matters. This was to prepare you for Lincoln's referring to you. I will try the money affair again tomorrow, but I am not listened to, being regarded as a partner in crime with you.

1271 Philipps, Trenham W.   To: H.T. De la Beche
Office of Woods.
18 June 1842

Your application for money was again mentioned at the Board yesterday, but again deferred. Lincoln afterwards said that our Estimates must come on in about a fortnight, and that he should prefer your waiting till they were through.

1272 Philipps, Trenham W.   To: H.T. De la Beche
Office of Woods.
15 July 1842

I delivered your letter to Lincoln, but have heard nothing since. Hume visited Craig's Court, but only to abuse it. He complained of the beggarly character of everything. He will never assist us, depend on it. None of our friends were in the place. Egerton was not there. Lemon was at Bristol. Lincoln has just told me he will give your letter to Gladstone.

1273 Philipps, Trenham W.   To: H.T. De la Beche
[July 1842]

The Vote for the Museum came on last night. Hume was the only one to speak about it. His objection was to its smallness etc. Pusey was in Bristol. Hume's objections are pertinacious. Lincoln went to Peel yesterday morning. I have constantly recommended Playfair to Lincoln. A report for the Treasury has been prepared requesting £1000 on account. You have been extremely patient.

1274 Philipps, Trenham W.   To: H.T. De la Beche
Office of Woods.
1 August 1842

No money is forthcoming. Your application has been deferred till next Board-day. If none comes then, I will say that Jordan will have to dismiss his men. I showed Lincoln the letter of Buckland and your own. He said Peel would not hear of two chemists. Hudson is publishing instructions on soil analysis. Is he competent?

1275 Philipps, Trenham W.   To: H.T. De la Beche
Office of Woods.
11 August 1842

I will try to get you a plan of the Forest tomorrow. There are several points on which you might be of great service to us. I think it would help Reek's position if he were paid by draft and included in our general salaries list.

1276 Philipps, Trenham W.   To: H.T. De la Beche
Office of Woods.
22 August 1842

Will you do Lincoln the favour of looking at the enclosed papers. He is at present in the Country, but asked me to see what you could do on this subject, which is of great importance to the interests of the Forest. How long are you likely to remain in the neighbourhood of Dean?

1277 Philipps, Trenham W.   To: H.T. De la Beche
Office of Woods.
27 August 1842

I enclosed to Lincoln your letter of the 24th inst. His Memo on it asked me to assure you he appreciates the courtesy and liberality of your offer to inspect the Railways etc. of Dean Forest. I was aware of Reeks's application for leave. I tried to impress on him the need to make himself essential, and that only by sticking to his post can a man make himself so.

1278 Philipps, Trenham W.   To: H.T. De la Beche
Office of Woods.
29 August 1842

Poor Knipe is in want of money apparently. Will you answer me about him as soon as you can. I gave your letter, with Lawley's enclosed, to Lincoln who asks to keep them for a time. I am satisfied he means well to the Museum, in reference to Playfair especially. It is fair enough when he says Peel is prepared to double the Curator's salary, yet has he not the right to dispose of Phillips?

1279 Philipps, Trenham W.   To: H.T. De la Beche
Office of Woods.
30 August 1842

I will try and manage your money in a day or two, but the Commissioners are obdurate. You will be glad to hear that the British Association have at length got the Observatory at Kew. I wish Lincoln would write you a letter as the prime mover in the affair. Doubts have arisen here as to your opinion on the relative merits of Colenso and Taylor.

1280 Philipps, Trenham W.   To: H.T. De la Beche
Office of Woods.
1 September 1842

Since you asked me for a Map of Dean, I have had to write to Sopwith for a print of his Index map. Boundaries are being added and coloured. It comprises now only part of our property. I forwarded your letter respecting analyses for the Fine Arts Committee to Lincoln. I wish he would give a few days to matters of this nature.

1281 Philipps, Trenham W.   To: H.T. De la Beche
Office of Woods.
6 September 1842

I shall send you Captain Morrison's survey of the proposed railway from Dean Forest to Gloucester. Lincoln wants your opinion on this plan. The company want Government money for making the line. The Government are chary about it, and are interested in the underground wealth of the Forest. Write to me for any information you wish.

1282 Philipps, Trenham W.   To: H.T. De la Beche
Office of Woods.
20 September 1842

Herewith some information on the traffic of the railways in Dean Forest. Sopwith contributes willingly. Apply direct to him. Langham, the Assistant Deputy Surveyor, will give you local information. You may find him very useful.

1283 Philipps, Trenham W.   To: H.T. De la Beche
Office of Woods.
27 September 1842

Herewith Rendal's Reports and Plans. He was considered to have exceeded his instructions, and his Bill was not paid. A sum was offered, which he accepted, and he left the documents with us. Use them with caution. Lincoln would like your Report by 12th October.

1284 Philipps, Trenham W.   To: H.T. De la Beche
Office of Woods.
6 October 1842

The Great Western people are tied up for want of funds, but there is no doubt that their line to Gloucester will soon be completed. Protheroe has insinuated you are favouring the Stroud line. Lord Lincoln is alarmed, as he engaged to consider any proposal made by Protheroe. Frank, the head man in Jordan's workshop is ill. Will you allow him pay while he takes two weeks holiday? I don't think Reeks does as much work in the Laboratory as he ought. I consider his chemistry everything.

1285 Philipps, Trenham W.   To: Sir H. De la Beche
Office of Woods.
12 October 1842

I received your Report and all the papers this morning. I am sure Lincoln will feel sensible enough, and that you've bestowed pains on this knotty question. I do not mean old Milne to get hold of my copy of your Memo. A case has arrived at the West India docks, supposed to be soils.

1286 Philipps, Trenham W.   To: H.T. De la Beche
Office of Woods.
13 October 1842

I acknowledge your two letters of this morning. Lincon has returned, but has just heard of the death of his brother in China. I will try to get your letter respecting Playfair under his notice tomorrow. He is too distressed this evening.

1287 Philipps, Trenham W.   To: H.T. De la Beche
Office of Woods.
3 November 1842

I have received two letters from you since I left the Forest, but have been unwell as a result of the soaking I got at Ross the day I left you. I return Taylor's letter. If anyone could do something with Phillips it is he, but Richard is an obstinate little fellow, blind to his own interests. Thanks for your desire to benefit Reeks. Fawcett will get his copy of the Report.

1288 Philipps, Trenham W.   To: H.T. De la Beche
Office of Woods.
14 November 1842

The Chancellor of Exchequer is asking for an estimate for 1842/3. What do you say to £2000? If you go beyond, please state your reasons. Do not press Phillips too hard for an answer, as Lincoln wishes him to act entirely from his own impulse. Do not hurry him.

1289 Philipps, Trenham W.   To: H.T. De la Beche
[November 1842]

We are out of money, but application will be made. I have put £2000 into the Estimates. Lincoln told me that Playfair's business was all off. I've made a suggestion to Phillips which may be of service. Have you any objection to his being absent part of the day for the next fortnight? I go for a short visit to Lord Lincoln on Friday. He promises to write strongly to Sir Stratford Canning respecting Warington Smyth.

1290 Philipps, Trenham W.   To: H.T. De la Beche
Office of Woods.
5 December 1842

Captain Morrison had not given you his address when he wrote the first time, and I was obliged to leave town suddenly. I will write in a day or two, but this is just to set your mind at rest concerning the Captain's papers.

1291 Philipps, Trenham W.   To: H.T. De la Beche
Office of Woods.
8 December 1842

The order for a fresh issue of money has arrived, and your wants will be attended to without delay. I know Merewether, the Dean of Hereford well. His letter shall be sent to you. The first time I have a favourable opportunity of speaking to Lincoln, I will mention your fears respecting the stone used at Westminster. Other matters press at the moment.

1292 Philipps, Trenham W.   To: H.T. De la Beche
Office of Woods.
17 December 1842

Mr Sergeant of the Treasury has deducted income tax there, whereas this is done here quarterly. The matter I hope will be adjusted. I've mislaid your application of 19 November. Please write another. Lincoln thinks Phillips should receive ten guineas for his labours in Payne's affair. Merewether has written acknowledging receipt of the Report.

1293 Philipps, Trenham W.   To: H.T. De la Beche
Office of Woods.
27 June 1843

The enclosed will answer you. Lemon, Smirke and Hope discussed your Memo with me. I showed them your rejoinder. They all state that the Thames is not affected by anything you state in your paper, which applies exclusively to estuaries. All I want to do is enforce caution. I send you Jordan's letter. I asked Page to note the apparent inconsistency respecting velocity and other parts of his evidence.

1294 Philipps, Trenham W.   To: H.T. De la Beche
Office of Woods.
31 July 1843

I can ascertain nothing respecting the Queen's movements. I was a little surprised at the difficulty made in supplying Tucker with the plan of the garden at Frogmore. Phipps is providing for the completion of the plan. You ought to come soon, if you desire to have the Queen away. I gave Lincoln your address some days ago for him to write to you by Sir Robert's desire in Harcourt's affair.

1295 Philipps, Trenham W.   To: H.T. De la Beche
Office of Woods.
25 July 1844

Within an hour of reading the debate in the House, I sent Lincoln a Memo recommending Page as Secretary. Lincoln has promised to name him to Sir Robert. In the Times of Tuesday, they have abused the Sanitary Report, all in good set terms. They say the Commissioners tell how cabbages decay in gutters in Drury Lane. The article is worth reading.

1296 Philipps, Trenham W. To: H.T. De la Beche
Office of Woods.
27 July 1844

I have been disturbed by Dick Phillips telling me he has had to employ his son, because Reeks gives him no assistance in the laboratory. He seems now to set up a claim for assistance. I maintain he has no such claim. We must have a return of what he gets from other Departments. He has gone to Derby to work for the Excise.

1297 Philipps, Trenham W. To: H.T. De la Beche
Office of Woods.
29 July 1844

I saw an advertisement for the sale of the Egyptian Hall in Piccadilly, and I got Pennethorne to value it. I then informed Lincoln and got him to ask Peel about it. Since the sale (it was bought in for £6,250) the Auctioneer has been about it to Pennethorne, and I think the purchase could be effected. Peel gave his consent, the Chancellor was against, and Lincoln is indifferent. What are your views?

1298 Philipps, Trenham W. To: H.T. De la Beche
Office of Woods.
26 August 1844

I have found another spot near Regent Street. It backs on to Jermyn Street so two entrances could be made. The Egyptian Hall, for which they demand £6500 for the remaining lease, is thus now No. 2 on the list. I enclose a copy of my letter to R. Phillips, the result of a conversation with Lincoln.

1299 Philipps, Trenham W. To: H.T. De la Beche
Whitehall Yard.
1 September 1844

I have sent your letter respecting the transfer of the Survey to Lord Lincoln this morning. The direction of the whole should be placed with the Chief Commissioner only. I have urged on Lincoln that if these things are to be effected you are pre-eminently the instrument for bringing them about. The particulars of Piccadilly have been sent to the Isle of Wight, but nothing has been heard since.

1300 Philipps, Trenham W. To: H.T. De la Beche
Office of Woods.
8 September 1844

I send these off to get your opinion on the arrangements. There are no rooms for the office keeper, no workshop, no water closets, no private room for yourself, and if the Survey comes to us, no working room for your men. What is your opinion as to our wants? I have in my possession all the correspondence respecting the Geological Survey for perusal. You must obtain the sole responsibility of the Chief Commissioner.

1301 Philipps, Trenham W. To: H.T. De la Beche
Office of Woods.
15 September 1844

Please look over these tracings, so that I can get everything in shape for an estimate. I think the affair of the Geological Survey is going on fairly, and I should advise your not coming here or appearing in the affair. If you see Buckland at York, say nothing. Reeks has been ill.

1302 Philipps, Trenham W. To: [H.T. De la Beche]
[Office of Woods.
September 1844?]
[Part of letter only]

If Phillips does not intend lecturing all the year round, and the benches remain empty, we'll have applications of all sorts for permission to use the facilities, and Phillips will never be able to call the place his own. All furniture should be moveable, and chairs hired for the occasion of lectures. Talk it over with Phillips.

1303 Philipps, Trenham W. To: H.T. De la Beche
Office of Woods.
20 October 1844

I have heard nothing from Lincoln about the transfer. Peel is in Brighton, and all seems dormant. All is in haste and excitement here. You have no reason to fear about the subject of rooms in Whitehall Yard. Howard has a son, whom he wishes to place in the Museum for a foundation in agricultural chemistry. I can't encourage him by lying about the place. Barry is ill.

1304 Philipps, Trenham W. To: Lord Lincoln.
Office of Woods.
26 October 1844
[Memorandum]

The present letter arises out of Howard's application to place his son in the laboratory of the Museum for a complete course in agricultural chemistry. I could not recommend him to do so. De la Beche was consulted and his reply is the first practical abandonment of the Institution by its founder and only active and zealous supporter. The next will probably be his resignation.
[See also Philipps, T.W. to De la Beche, H.T., 27 October 1844, **1305**].

1305 Philipps, Trenham W. To: H.T. De la Beche
Office of Woods.
27 October 1844

Thank you for remembering Howard. I fear the course you suggest is the only one. I sent all the correspondence to Lincoln, together with a Memo from me, written in a temper. Peel is the only one who can help us. Lincoln retained the letter intended for Howard, and this affords a glimmering of hope. I send you a copy of my Memo.
[See also Philipps, T.W. to Lord Lincoln, 26 October 1844, **1304**].

1306 Philipps, Trenham W. To: H.T. De la Beche
Whitehall Yard.
October 1844

I received your letter respecting Colby and his proceedings. Lincoln made no use of it when at Peel's, but will be seeing Sir Robert again on Monday. The estimate for the building is reduced to £15,000. If you expect to get into the heaven of our department, you must expect to undergo purgatory with the Ordnance. They are very accomplished tyrants and you make a very indifferent slave.

1307 Philipps, Trenham W. To: H.T. De la Beche
Office of Woods.
21 November 1844

You ought to come to Town as soon as you can. I've told Lincoln you may be here on Saturday, but Monday will do at the latest. You are now expected. The Museum affair is done at £28,000. I have not bothered with details.

1308 Philipps, Trenham W. To: H.T. De la Beche
Office of Woods.
9 August 1845

I must have my say about Playfair and his travelling for enquiries on Mines' explosions. I do not consider such a Museum affair, and if allowed, then I say that all that you and Lincoln have urged against R. Phillips has been gross injustice. Graham has written suggesting an enquiry into Explosions in Mines. Is it a Museum affair, or not? I am heartily sick of the concern, and not a little with yourself for the bother you occasion. I will watch out for its interests, but am not disposed to get mixed up in sad departures from the spirit of its beginnings.

1309 Philipps, Trenham W. To: H.T. De la Beche
25 August 1845

Thank you for your letters despite my lack of encouragement. I have little heart for work under Lincoln. He has made both Smoke and Gas enquires Museum affairs. I have sent him today letters of instructions for you on each of them. Money for the Survey is still delayed. You may apply again if you think well.

1310 Philipps, Trenham W. To: H.T. De la Beche
Office of Woods.
16 September 1845

I send you copy of a letter from which you will see your wishes have been attended to. Kane will want some of this money. Cheffling's account of the Great Western Railway is to be published, and he wants to put in the Bristol Coalfield, anticipating your report. Thanks for the letters from Playfair.

1311 Philipps, Trenham W. To: H.T. De la Beche
Office of Woods.
23 September 1845

Please send me separate drafts for the moneys advanced to you and Playfair. I suggest you offer Lincoln the first payments out of the £2000. I see something like a termination to the money difficulty by Treasury adjudication. What should be amount of imprest for the Smoke enquiry?

1312 Philipps, Trenham W. To: H.T. De la Beche
Office of Woods.
24 September 1845

You are now in charge of a large concern, and must work at home as well as abroad. You may be sure that the Board are in no humour to repeat the error of July, so you had better have drafts here by the 10th. The Commissioners are to answer the ultimatum of the Treasury, after which we are to open an account in the name of Lincoln. You must not be frightened at paying away your present money.

1313 Philipps, Trenham W. To: H.T. De la Beche
Office of Woods.
8 October 1845

Thank you for the Tamarinds which arrived yesterday. I told Lincoln before he left that you would be short of money. The Treasury care not what inconvenience we are suffering, you may feel assured. I agree with you as to the state of Playfair. He is evidently a very anxious person. Still you do better to consolidate the position in Ireland so that James cannot injure by indiscretion or blundering.

1314 Philipps, Trenham W. To: H.T. De la Beche
Office of Woods.
22 October 1845

Enclosed is your authority to receive the Collection. Lincoln expects them to go to Dublin Museum. I shall shortly move to town and will be glad to renew our strolls. James speaks cordially of you. Work cordially with Kane, and you will have got over your difficulties. Dick Phillips is about to have an operation for hydrocele.

1315 Philipps, Trenham W. To: H.T. De la Beche
Office of Woods.
5 December 1845

Lord Lincoln refers to you the Reports and specimens of coal from Borneo, in order that with Playfair you may report whether the coal is of sufficiently good quality to warrant the outlay required.

1316 Philipps, Trenham W. To: H.T. De la Beche
Office of Woods.
11 July 1846

Please tell Kane I will send him a reply about his accounts. If there is still difficulty about Kane getting possession of the house in Stephen's Green, assure him the fault is not here. I got Gardiner to write to Hamiltons. I wish you would see Burke and get him to bully Hamilton if he is hanging fire.

1317 Philipps, Trenham W. To: H.T.De la Beche
Office of Woods.
25 July 1846

Pennethorne's Report for the building in Piccadilly is now before the Board, with fair prospect of moving on. Canning's leaving at this juncture is a loss, as indeed it would have been at any time.

1318 Philipps, Trenham W. To: H.T. De la Beche
Office of Woods.
31 July 1846

Herewith James' explanation. It is a poor document, and grievously uncandid. I send it to you for your observations in the first instance. Kane complains of the manner of dealing with the soils question. I am disposed to think him right. Let me tell Morpeth what the cost on the limited scale suggested by Kane would be, as the Treasury are pressing.

1319 Philipps, Trenham W. To: H.T. De la Beche
Office of Woods.
31 July 1846

Two points in Playfair's letter cannot pass unnoticed. First, that you have power to grant leave of absence for a fortnight, and second that he is required to provide for public business sent to laboratory only. The second is the more important to refute. There is a want of candour in your letter. We must therefore get as much recorded in writing as we can.

1320 Philipps, Trenham W. To: H.T. De la Beche
Office of Woods.
23 September 1846

You should accept this office, which you should have filled long ago. I sent Morpeth a Memorandum upon which I hoped he would record his opinion, so that you might see what I had said on the subject. He has kept mine, and sent one of his own, a copy of which I enclose.
[See Morpeth, 23 September [1846], **994**].

1321 Philipps, Trenham W. To: H.T. De la Beche
Office of Woods.
16 August 1847

Your draft for £350 arrived this morning. I arrived in London from Frankfurt on Saturday afternoon. I am too sick of London and its concerns to write more. You will hear from me soon again.

1322 Philipps, Trenham W. To: H.T. De la Beche
Office of Woods.
21 October 1847

It is impossible that we should work for Commissions with money. Money must pass through our public account at the bank. I agree with Playfair in principle. The proper course will be to find assistance for the Iron Commission. The same goes for Chadwick's Health of Towns Commission. Morpeth must be kept straight against a possible public enquiry.

1323 Philipps, Trenham W. To: H.T. De la Beche
Office of Woods.
31 July 1848

The Treasury have issued £4000 in addition to the £2000 already issued. I have calculated the proportions, so that there should not be any further mistake between yourself and Kane. Kane will be entitled to £1830, and you to £4170. Our Committee is altogether up, to be called together next Session.

1324 Philipps, Trenham W. To: H.T. De la Beche
Office of Woods.
3 April 1849

Carlisle wishes to know whether you can undertake the enclosed without prejudice to other matters. I discussed your estimate today, and Carlisle is adverse to making the last year of your Craig's Court occupancy the subject of an increase upon the last two years. Jermyn Street will be another matter.

1325 Philipps, Trenham W. To: H.T. De la Beche
Office of Woods. Craig's Court.
10 November 1849

I can now communicate with Pennethorne respecting the removal of cases in Craig's Court, but I shall not do so until he answers my letter of yesterday. The delay is his, not mine.

1326 Philipps, Trenham W. To: H.T. De la Beche
Office of Woods.
2 June 1851

Playfair states that he is about to arrange a scheme for the division of duties between Percy and himself. Seymour will not officially communicate with you until that scheme is settled. If you get the line of division settled, there will be no delay here. I am anxious about your health.

1327 Philipps, Trenham W. To: H.T. De la Beche
Office of Woods.
17 January 1853

I hear that Smyth's acceptance of an appointment under the Commissioners of Woods has been the subject of communication with the Treasury. Smyth is in the employ of the Commissioner of Works and Buildings, and is not free to make his services available to other Government Departments without the sanction of the Commissioners. I wish to discuss all of this with you. While the Department does not wish to interfere with Science, it has it other duties in the Public Interest.

1328 Phillips, John To: H.T. De la Beche
York. 58 Jermyn Street,
30 January 1830 London.

I have received your letter requesting me to publish my lists of organic remains which I found in the Green Sand and Kimmeridge Clay. My paper will be short. I anticipate much pleasure from your paper on the oolitic fossils.
[Cited: Morrell 1988, p.12]

1329 Phillips, John To: H.T. De la Beche
Yorkshire Museum. 58 Jermyn Street,
13 March 1830 London.

I am glad my paper is likely to be of service to you. Where can I find a

description of Gryphaea virgula? Here are the synonyms of my fossil plants in Brongniart's Prodrome, in order of the plates found in my Geology of Yorkshire. I shall be happy to furnish any other notices you may think useful or necessary.
[Cited: Morrell 1988, p.12]

1330 Phillips, John   To: H.T. De la Beche
1 April 1835
I have been arrested by the unexpected coincidence of our views in Geology. Your views have been found in the impressive communion with nature which many of our book men have not enjoyed. The desire of my life is to produce a book on organic remains. Your discovery of Carboniferous plants in greywacke is interesting and I am not surprised for many reasons.
[Cited: Rudwick 1985, p.121; Secord 1986b, p.250; Morrell, pp.11, 12]

1331 Phillips, John   To: H.T. De la Beche
York.
4 November 1837
I will set to work with zeal on your Petherwin fossils as soon as Robe sends them. You and I think on the same parallels as to the divisions of the Silurians. The flora of the coal formation found in rocks of older date is fact to be introduced into the induction, not an anomaly to be frightened at. The Newton Bushel things are strangely analogous to the Carboniferous Limestone.
[Cited: Rudwick 1985, p.223; Morrell 1988, p.16]

1332 Phillips, John   To: H.T. De la Beche
3 April [1838]
W. Smith's map was published in 1815 and dedicated to Sir Joseph Banks. I am glad you have revisited the country near Ashburton and found further facts to reason upon. If enough fossils are found, I will wager them against all sections, but two or three silent witnesses are not to be trusted. Sedgwick does not appear to wish to state what difference of opinion still exists between you and himself. From all I've seen of fossils of the upper Barnstaple series, the culmiferous beds appear to belong to Carboniferous system.
[Cited: Rudwick 1985, p.241]

1333 Phillips, John   To: H.T. De la Beche
25 April 1838
There is a movement here in King's College somehow or other connected with your and Sir C. Lemon's plan for a Mining School. It would be better to have a Government scheme if practicable. If not, the mineral proprietors should have much influence in the establishment. Sedgwick has sent me a box of Devonian fossils.

    Phillips, John
    1838
    [see **2271**]

1334 Phillips, John   To: H.T. De la Beche
York.
26 January 1839
Herewith a statement of what I have done in the examination of Devon and Cornish fossils, in order to help you make your report. You will see how far it is honest to recommend the Treasury to continue the grant. How do you find the divisional planes in the South Wales coal field?
[Cited: Morrell & Thackray 1981, p.439; Rudwick 1985, p.272, Morrell 1988, p.14]

1335 Phillips, John   To: H.T. De la Beche
York.
22 February 1840
I intend to leave for Devon and Cornwall on 10th March, passing through London. Before leaving I will pack up all the fossils for return to the owners. Are there any special points you wish me to attend to? I can devote 2 months to my walks, and am a fair walker over rougher roads than those of N. Devon.
[Cited: Rudwick 1985, p.362]

1336 Phillips, John   To: H.T. De la Beche
Barnstaple.
28 March 1840
Herewith an account to date of our North Devon survey. Why did you not give a real coast section? The problem of the specimens found in a quarry near Combe Martin has been solved. We got good fossils at Lynton. We shall stay here until 1st April. The contrast between North and South Devon fossils is very great. Is there no analogy with North Devon in South Wales?
[Cited: Rudwick 1985, p.362]

1337 Phillips, John   To: H.T. De la Beche
Launceston.   Craig's Court,
6 April 1840   Charing Cross,
  London.
We have made a successful search for fossils at Baggy Point and Croyde. A seam at Croyde yielded trilobites. Pilton quarry was carefully examined. I think we are now ready to go to South Devon. It would suit me very well to be regularly employed for part of the year on your Survey, as I am giving up my official position in York and at King's College.
[Cited: Morrell & Thackray 1981, p.442; Rudwick 1985, p.362; Morrell 1988, p.23]

1338 Phillips, John   To: H.T. De la Beche
Chudleigh.
20 April 1840
Hennah's collections are mostly in the hands of Murchison and Sedgwick. I am glad we found things of our own. I have so many new things, I wonder what your young men were about in not supplying you with more organic fossils. Sanders sends remembrances.
[Cited: Rudwick 1985, p.362; Morrell 1988, p.23]

1339 Phillips, John   To: H.T. De la Beche
Lyme Regis.
22 April 1840
We have come to see the landslip here. The upturned beds divide the low water like a long little mountain crest. The sketch will tell its tale. Buckland and Conybeare were also on the ground, and I mentioned the great novelty. Does it not seem a puzzling affair?

1340 Phillips, John   To: H.T. De la Beche
Taunton.
27 April 1840
We have found at Pilton and Baggy great novelties of the Crinoidal group. Also plenty of plants. New forms found in both carbonaceous limestone and lower beds. I will prepare the plates in York, but would not figure the plants.
[Cited: Rudwick 1985, p.362; Morrell 1988, p.23]

1341 Phillips, John   To: H.T. De la Beche
Taunton.
27 April 1840
Here is my notion of the expense which would be entailed in the description of organic remains collected by the Geological Survey. I have resigned my responsibilities at Kings College and Yorkshire Museum, so can devote my full time to the task. £250 yearly would be needed, without my undertaking the responsibility of Curatorship. Surveyors would be helped in the field.

1342 Phillips, John   To: H.T. De la Beche
16 June 1840
I send you a few notes on some polyparia to give you a notion of what I am about. You might be tempted to accelerate the preparation of the book. It would delight me to collect in Pembroke, but I am so desperately poor, having to settle my Uncle's affairs and maintain his half mad widow.
[See also Phillips, J. to De la Beche, H.T., June 1840, **1346**].
[Cited: Rudwick 1985, p.362; Morrell 1988, p.23]

1343 Phillips, John   To: H.T. De la Beche
Tenby.
23 June 1840*
Having reduced all the measures at Marloes Bay to true thickness, I send the sections in that form. The trap fault may be as much as 1000 ft. horizontal. Nothing appears in this section lower than Haverfordwest beds. Ramsay can give you sections of Stackpole and Skrinkle. Be sure to criticize the preface I purposely send.
[*Although clearly dated 1840 by Phillips, the contents of this letter suggest that it should be dated 1841.]

1344 Phillips, John   To: H.T. De la Beche
Tenby.
25 June [1840*]
From my sketch you will be able to see the exact area I assign to the

Lydstep black shales. Also I am now confident of the shape of basin. The breadth from Old Red to Shale series is about 2,300 ft. There is an odd bit of limestone NE of Lydstep. Mind it in colouring!
[*The contents of this letter suggest that it dates from 1841.]

1345 Phillips, John   To: H. T. de la Beche
25 June 1840

I send you a lithograph of 3 species of Lonsdale's Turbinolopsis. Is there no engraver for such things in your department? Regarding Prof. Shepard's specimens, I find it hard to believe they are all from the same quarry as they include both Silurian and Mountain Limestone forms.

1346 Phillips, John   To: H. T. De la Beche
June 1840

Notes on the distribution of some Polypiaria of Devon & Cornwall. 15 or 16 genera. I am prepared to speak about 3 generic groups only, Lonsdale's, Common, and Goldfuss's.
[See also Phillips, J. to De la Beche, H.T.,16 June 1840, **1342**].

1347 Phillips, John   To: H.T. De la Beche
York.
30 September 1840

Having completed the figures and descriptions of organic remains connected with the Geological Survey of Cornwall, Devon, and West Somerset, I request that £250 due to me be paid when convenient.
[Cited: Rudwick 1985, p.362]

1348 Phillips, John   To: H.T. De la Beche
St. Mary's Lodge,   Ordnance Geological Survey,
York.   Cardiff.
12 October 1840

I am paying the last of my Uncle's debts, and will then revise all the descriptions for printing. I must not omit introducing into the certificate of the completion of the work the fact that copying is still needed to fit it for the press. I am anxious about Murchison's Old Red fossils in Russia. Do they mean that all below them is Silurian?

1349 Phillips, John   To: H.T. De la Beche
Royal Hotel,
Manchester.
20 October 1840

I hope to give you Ms. and drawings for 'Fossils of the Palaeozoic strata of Devon and Cornwall' on 1 December. I would undertake publication by subscription. It may, however, go through the Stationery Office. I have resolved to accept your proposal to describe organic remains of South Wales, and shall resign from the York Philosophical Society on 31 December. What sort of Crinoidal stems are there in Magnesian Limestone?
[Cited: North 1936a, p.75; Rudwick 1985, pp.371, 372; Morrell 1988, p.23]

1350 Phillips, John   To: H.T. De la Beche
20 November 1840

If you make a better heading or distribution of columns for the Catalogue of Devon & Cornish fossils, I shall be glad of a copy. I think we should use local names for groups of strata. I estimate we have 225 distinct forms to describe.
[Cited: Rudwick 1985, p.371]

1351 Phillips, John   To: H.T. De la Beche
4 December 1840

I have written to Goldfuss regarding the Eifel fossils & perhaps he may send me a package. I have asked him to address it to the Museum of Economic Geology. Is it all right?
[Cited: Rudwick 1985, p.372]

1352 Phillips, John   To: H.T. De la Beche
St. Mary's Lodge,
York.
8 December 1840

With regard to the plates for the palaeontology of Wales, I think 10 reams of paper will be needed, but I am a mere novice in the stationery department. R. Phillips tells me that money is still due to my uncle from the Civil Service Pay Office. I will start field work for 1841 in Pembroke.

1353 Phillips, John   To: H.T. De la Beche
St. Mary's Lodge,
York.
30 January 1841

In reply to yours of 29th, regarding my doing the Palaeontological survey to accompany the other geological work for the Ordnance, I have put myself in a position to act efficiently in the affairs you propose to confide in me. It is perfectly my resolution to abide by your proposition and not to desire any change therein.
[Cited: Morrell & Thackray 1981, p.442; Morrell 1988, p.23]

1354 Phillips, John   To: H.T. De la Beche
31 January 1841

Proofs of other sets of plates have been sent to me. In 16 plates there are above 100 figures. In all we shall have 600 or more figures. I am quite determined to work according to the plan proposed by you. If the authorities approve your plan, you have a ready and willing agent.

1355 Phillips, John   To: H.T. De la Beche
19 February 1841

I received your announcement of the successful termination of your effort to put me in a position to be useful to the geological department of the Ordnance Survey, and will make arrangements accordingly. I will be in town for the Association Council meeting on 27th.
[Cited: Morrell 1988, p.22]

1356 Phillips, John   To: H.T. De la Beche
3 Salmons Inn,
Usk.
1 April 1841

I have arrived safely after a wet traverse of the Malverns. James will send the fossils to me tomorrow, and I will examine them. I have visited Preston to see Gilbertson's collection, and also examined the Museum at Birmingham.

1357 Phillips, John   To: H.T. De la Beche
3 Salmons,
Usk.
2 April 1841

James has guided me today along the junction of Old Red and Lower Palaeozoics. The micaceous character of Old Red descends some distance. Fossils diminish towards the Old Red. The fossils should go to the Museum of Economic Geology. The red peroxide has destroyed the mollusks.
[Cited: Secord 1986b, p.250]

1358 Phillips, John   To: H.T. De la Beche
Usk.
8 April 1841

I have only to decide about the huge box of fossils collected by James, and then can leave for Bristol by the steamer with Sanders. Can the box be examined in Cardiff, Bristol or Tenby? I am much interested in the limestone here.

1359 Phillips, John   To: H.T. De la Beche
Usk.
9 April 1841

I think your Caldy section capital. We found some interesting things on the left bank of the Usk, in particular a fault near the southern end of the limestone. I have written to Sanders to ask if there is a steamer to Tenby later than Tuesday, as I need three more days here.

1360 Phillips, John   To: H.T. De la Beche
Usk.   Tenby,
15 April 1841   South Wales.

I have seen Conway's collection which is very rich. Sanders has joined me here and we have looked at all the Silurian grammar, except the Llandeilo beds. James will take the boxes of fossils to Bristol. We leave for Bristol, Llandeilo and Tenby, to be there as soon as we can.

1361 Phillips, John   To: H.T. De la Beche
[Tenby].
27 April 1841

Yesterday we found a layer of limestone nodules such as in Yorkshire lie beneath Millstone grit. From one we extracted Goniatites. Ivy Tower yielded several good fossils.

1362 Phillips, John   To: H.T. De la Beche
Tenby.
21 May 1841
Sanders has asked me to tell you he has received the maps and sent off a receipt. I propose to go with Ramsay to Skrinkle tomorrow. Soon I shall come to Haverfordwest.

1363 Phillips, John   To: H.T. De la Beche
Tenby.
22 May 1841
We have finished the Skrinkle section. Ramsay has done it very well. It is like the Caldy section. Could you give me an extract of the thickness of beds at West Angle Bay? I have looked at Austen's fossils, but I can't introduce more species in the book without spoiling the regularity of arrangement.

1364 Phillips, John   To: H.T. De la Beche
Tenby.
24 May [1841]
I am just going to Stackpole with Ramsay, and shall stay tonight at Pembroke. He will then go to Haverfordwest, and I to Tenby. I have received your section of West Angle.

1365 Phillips, John   To: H.T. De la Beche
Tenby.
28 May 1841
Your plan of numbers and colours will do exceedingly well. I hope Ramsay has told you of our Stackpole work, including his swim. Murchison is in a state of fluctuation about Old Red fossils in Mountain Limestone Shales. Here is my calculation of the height of Castle Hill. I am glad Sedgwick has written to you re Henwood.

1366 Phillips, John   To: H.T. De la Beche
Tenby.
31 May 1841
I shall carefully measure the heights you mention. All species collected will number 274. Can you add anything to this sketch of De Saussure? I shall be in Haverfordwest on Wednesday evening at the latest.

1367 Phillips, John   To: H.T. De la Beche
Dale.
17 June 1841
The notes were received and distributed. Ramsay has finished the work you gave him. James' series is very good. The Procellaria had hurt its feathers, and I made a new arrangement. I am melancholy at the dispersion of our party.
[Cited: North 1936a, p.76]

1368 Phillips, John   To: H.T. De la Beche
Dale.
18 June 1841
Thank you for the packages. I will be in Tenby as soon as possible. Fossils are packed in two barrels, one to go to Bristol, the other to Tenby. The printing of the Devonian fossils is advanced. A new notion at Dale has us as chartists. They fancy all chartists are miners, and all miners chartists.
[Cited: Rudwick 1985, p.372]

1369 Phillips, John   To: H.T. De la Beche
Tenby.
21 June 1841
Captain James is a good aid. I have been examining the remaining sheets of the volume, and am sending you the preface. I propose to go to Devonshire on 30th. Here is my scheme for the Devon fossils.
[Cited: Rudwick 1985, p.372]

Phillips, John   To: H.T. De la Beche
23 June [1841]
see **1343**

1370 Phillips, John   To: H.T. De la Beche
2 Marine Terrace,
Tenby.
24 June 1841
The money should be sent here for me to transmit to York. I suppose Rapley can publish in 14 days or so, and Monkhouse will send me an account of his charge. We must discuss the principle for the selection of new plates. I am about to draw in the Lydstep shales.

Phillips, John   To: H.T. De la Beche
25 June [1841]
see **1344**

1371 Phillips, John   To: H.T. De la Beche
Tenby.
26 June 1841
The money order has arrived, and I will cash it in Bristol. The Plymouth meeting will occupy 42 days, after which I will return to the Survey in Bristol. The plan for the future arrangement of illustrations of organic remains must include fossils of the lower rocks of Wales, and Limestone shales.
[Cited: Rudwick 1985, p.372]

1372 Phillips, John   To: H.T. De la Beche
16 Richmond Terrace,
Clifton.
18 August [1841]
I enclose Cypridiform shells as found in Caldy. I found them in the Lower Limestone shales of the Avon bank. The limestone seems in general not rich.
[Cited: North 1936a, pp.76-77]

1373 Phillips, John   To: H.T. De la Beche
16 Richmond Terrace,
Clifton.
22 August 1841
As you'll be here soon, I won't go into too much detail about the proposed change in plan. I suggest 6 months in the field, and the remainder of the year in York processing fossils etc. I do not find the present arrangement very satisfactory.
[Cited: North 1936a, pp.77-78]

1374 Phillips, John   To: H.T. De la Beche
16 Richmond Terrace,
Clifton.
[August 1841]
My sister is obliged to return to York to take possession of our house, quitted by the temporary tenant. I wish to go also, and plan to leave Friday or Monday. From York I plan to join you in Wales.

1375 Phillips, John   To: H.T. De la Beche
St. Mary's Lodge,
York.
3 September 1841
I enclose Buckland's letter and enclosures re Railway Cuttings. Add my assent to any modified view you now think it right to adopt. I wish to look at Chester, Bala, and Cardigan on my way to the Towey.

1376 Phillips, John   To: H.T. De la Beche
York.
10 September 1841
I hope you received my last with Buckland's letter enclosing matter on Mr. Craig and his Railway Sections. I will be in Haverfordwest on 15 September, and will not bring my library this time. I have been examining Harding's collection from North Devon.
[Cited: North 1936a, pp.75-76]

1377 Phillips, John   To: H.T. De la Beche
Blue Boar Inn,
St. Clares.
29 September 1841
The road was flooded on today's voyage. I found black shale on road near Commercial Inn, as also at Ffynnon. This seems a good centre. Tell Ramsay I am sorry to hear of his accident.

1378 Phillips, John   To: H.T. De la Beche
Blue Boar,
St. Clares,
Caermarthen.
4 October 1841
No. 1
I have spent two successful days in Llanddowror and Laugharne district. Murchison's Map is utterly at fault. Today we go to Pendine, tomorrow to Tavernspite. What delays Aveline? There is here a neater section than Robeston Wathen.

1379 **Phillips**, John   To: H.T. De la Beche
4 October 1841
No.2
Your note and fossils just received. The fossils are Silurian. James has toothache. Mr. James of Robeston Wathen is in some difficulty and his things are to be sold today. I thought of buying his pony.

1380 **Phillips**, John   To: H.T. De la Beche
St. Clares.
7 October 1841
James joined me at Pendine and we examined the limestone boundaries. The rain was incessant. I shall require a day at Laugharne. Your Dinas section is very curious. Murchison has not given well that patch of limestone west of Laugharne.

1381 **Phillips**, John   To: H.T. De la Beche
St. Clears.
11 October 1841
Your patience and fortitude in working these conglomerates will be rewarded by facts of importance. We have got a trilobite out of the black shale. Thanks for your care of the parcel. They cook capitally here.

1382 **Phillips**, John   To: H.T. De la Beche
Ivy Bush Inn,
Caermarthen.
16 October 1841
The trap hills have been done. James has been ill as result of cold. I have completed several sections. The Panteg section is full of organic remains. We have reason to think that since Murchison's 1837 map there has been a volcanic eruption near Caermarthen.
[Cited: North 1936a, p.76]

1383 **Phillips**, John   To: H.T. De la Beche
Caermarthen.   Fishguard,
21 October 1841   Haverfordwest.
Thank you for your sections of Prescelly. I am perfectly incomplete in my view of the real connexion between Pembrokeshire and Llandeilo. The question of the black shales between St. Clears and Caermarthen is a plague.

1384 **Phillips**, John   To: H.T. De la Beche
Llandeilo,   Fishguard,
Caermarthenshire.   Haverfordwest.
26 October 1841
I have made a copy of my letter to Harcourt accompanying yours to me. If this thing ends in 1842 I shall be still grateful to you, and whether in or out of office shall be too happy to count myself your friend and fellow labourer.
[Cited: Morrell 1988, p.27]

1385 **Phillips**, John   To: H.T. De la Beche
Llandeilo.
27 October 1841
I spent yesterday evening with Lord Cawdor and family. The sections hereabout wear the aspect of true Silurianism. Can you give me careful measures of strike and dip in cases where cleavage and bedding coincide. We shall have to work here for ten more days.

1386 **Phillips**, John   To: H.T. De la Beche
Llandeilo.
2 November 1841
Sorry to hear you are ill. I propose to come to Fishguard at the end of the week, leaving James to fossilize for a further week here. How will you colour Llandeilo? If I were you I would colour a greater number of beds than did Murchison. I stayed a day or two at Golden Grove.

1387 **Phillips**, John   To: H.T. De la Beche
5 November 1841
I am glad to hear you are well again, and have cleared up the mystery of the limestone boundary. I am happy to confirm Logan's accuracy, in one or two places remarkably. Should I come to Fishguard?

1388 **Phillips**, John   To: H.T. De la Beche
Llandeilo.
5 November 1841
James and I crossed a ridge five miles N of Llandeilo, by two lines. Herewith the results. Your letter has now determined my moves. At the end of December I wish to go to York. Today we got Llanarthney.

1389 **Phillips**, John   To: H.T. De la Beche
8 November 1841
Buckland's letter is very satisfactory and creditable to Peel's habit of business, consideration for science and promptitude. I suppose you get the wedge rightly placed before you try to impel it. I plan to finish field work at the end of December.
[Cited: Morrell 1988, p.27]

1390 **Phillips**, John   To: H.T. De la Beche
Llandeilo.
13 November 1841
Ramsay and I will carry out your excellent plan of operations. £300 is sufficient for me. I will quit Llandeilo, tomorrow by Caermarthen and join you on Monday at latest. The thickness of the Old Red is 3,500ft, due to a fault.

1391 **Phillips**, John   To: H.T. De la Beche
Haverfordwest.
23 November 1841
I have paid Potter, and here is the receipt. Arrangements have been made with Mrs. Gibbs to send a tub of specimens to the Museum of Economic Geology, provided you write to instruct her so.

1392 **Phillips**, John   To: H.T. De la Beche
Narberth.
28 November 1841
I am moving to St. Clears, on foot. I have cleared up the mystery of the black shales. There are three beds, two of which are graptolithic. I find fossils in the Narberth section. Could you visit the limestone spots mentioned on your map by Still and McLauchlan?

1393 **Phillips**, John   To: H.T. De la Beche
Caermarthen.
29 November 1841
Today from St. Clears by Meidrim has given me a regular section of Silurians overlying graptolithic shales, & containing Asaphus Buchii etc. I thought you would like to know this new thing. It will be to us what an imaginary centre of stability would have been to Archimedes.

1394 **Phillips**, John   To: H.T. De la Beche
Llandeilo.
4 December 1841
I will write to Rees and have him send you the piece of map. In a few days you can also have piece still more south. Herewith a sketch of Silurian formations near Meidrim. I plan to go to Lampeter when the rain eases off.

1395 **Phillips**, John   To: H.T. De la Beche
Llandeilo.
9 December 1841
I think as regards the museum you must receive the letters again, and enclose my note with the others to Buckland. Llanarthney field work is in a satisfactory state. Please look at the section south of Caermarthen which is not satisfactorily surveyed.

1396 **Phillips**, John   To: H.T. De la Beche
Llandeilo.
13 December 1841
I mean to go to Lampeter to see what help can be given by inspecting the ground. Could you spare one of the pieces of the Caermarthen sheet? I have settled James at Llandovery. Grits and shales remain an awful problem. I have seen Sedgwick's attack on the Palaeozoics.

1397 **Phillips**, John   To: H.T. De la Beche
Llandeilo.
14 December 1841
Regarding Murchison's critique of my work, I have no doubt the onset will be rather vehement. I will not allow it to affect my friendly feeling to him. I agree with you about the Meidrim Limestone. The sheet will be complete by December.
[Cited: Rudwick 1985, p.382; Morrell 1988, p.26]

1398 **Phillips**, John   To: H.T. De la Beche
Llandeilo.
15 December 1841
James has gone to Llandovery, and the weather prevents field work. I will

gladly look at the fossils Lyell has brought back from Pennsylvania. If he is not pressed for time, better send them to York. I expect Rees will be here by the end of the week.

1399 **[Phillips**, John]   To: H.T. De la Beche
Castle Inn,
Llandovery.
18 December 1841
[Part letter only]
Yesterday we walked over nearly all that remained of the Old Red boundary S of Llandovery to the edge of the sheet, because the snow had begun to fall heavily. Today there is more snow, but I am determined to try to run what little is left of that line into the Brecon sheet. Now as Rees and Aveline will be at Llandilo next Monday and stay, say 4 or 5 days, and then unless you countermand, come over here to help our finishing blows.

1400 **Phillips**, John   To: H.T. De la Beche
Castle Inn,
Llandovery.
19 December 1841
The Old Red boundary is completed to the Brecon sheet. In reply to your query about how far north my work extends, herewith a description of the line north of which I have not surveyed. Fossil gathering has been very limited. James is writing to you.

1401 **Phillips**, John   To: H.T. De la Beche
Llandovery.
21 December 1841
For us to do the Sawdde section together on 28th, the early coach from Ivy Bush will bring you to Llangadock by ten. Perhaps your view of the series of beds in the line from Carmarthen to Newcastle will apply to the country farther east, but I doubt it. My sister thanks you for the gift of the medal.

1402 **Phillips**, John   To: H.T. De la Beche
Llandovery.
24 December 1841
The package from the Geological Society contains a portion of Murchison and De Verneuil's paper. It is a long list of organic remains. I shall try to get to Pumpsaint tonight, and wish to go home to York as soon as possible after January 1.

1403 **Phillips**, John   To: H.T. De la Beche
Llandovery.
25 December 1841
Your merry letters dissipate many a cobweb from my dusty brain. The likenesses are all capital. I have completed a capital section from Lampeter to Llandovery. Not one fossil in all the country yet seen. The best section at Llandeilo, is in the Swansea Road, east of Golden Grove.

1404 **Phillips**, John   To: H.T. De la Beche
Llandovery.
29 December 1841
I enclose my receipts, amounting to £9. Jones of Cardigan overcharged me for car to Fishguard. I will leave here Saturday morning for York. There is a Council Meeting of the Yorkshire Philosophical Society on Monday. I have shown Rees part of the country here so that he can act efficiently.

1405 **Phillips**, John   To: H.T. De la Beche
Llandovery.
1 January [1842]
Yesterday I struck the last blow for 1841 on the hills halfway between here and Trecastle. Our young friends can complete the work, using horses. I recommend Jones, Lloyd & Co., as bankers here. I have no doubt that it will be useful to use orderlies in our parties, on the moderate plan you suggest. Lewis, the Cooper, is not to be had as he will not leave his wife and shop. I am now leaving for Gloucester.

1406 **Phillips**, John   To: H.T. De la Beche
St.Mary's Lodge,
York.
16 January 1842
Deep snow here. I will put the geology on Rees's section, which arrived safely. I have made progress on the examination of the Freshwater fossils. The East beds enclose Upper Silurian fossils. I will next do the Marloes Bay sections. A good collection of fossils is ready to send to town when you choose to have them.

1407 **Phillips**, John   To: H.T. De la Beche
St. Mary's Lodge,
York.
28 January 1842
No.1 Regarding Clog y fran section, I ought to have said that the colouring was put in to give you a notion of the parts. The Series below the Old Red is divisible into more arenaceous and more argillaceous groups. Ramsay says he is better. James and Aveline have nearly completed work to the Brecon sheet.
No.2 Sedgwick has given me a good account of Griffith's inquiry in the south of Ireland. His mind is working on a direct path for a larger contemplation of the 'Palaeozoics'. I augur much good, and have replied in the same spirit of frankness.

1408 **Phillips**, John   To: H.T. De la Beche
St.Mary's Lodge,
York.
4 February 1842
I have given James and Aveline work which will keep them until Ramsay joins them in Cardigan. I have resigned from the Philosophical Society here, and more time is available to you. I'll be charmed to see Hamilton's fossils. Joy to Elie de Beaumont, Dufrenoy and Von Buch, three very worthy men. My grateful feelings towards those who have procured such a mark of distinction for my publications.
[Cited: Rudwick 1985, p.389]

1409 **Phillips**, John   To: H.T. De la Beche
St.Mary's Lodge,
York.
15 February 1842
I am not sorry Murchison has broken ground in front of the Magnesian Limestone, which will certainly not be turned out of the Palaeozoic Union by any of the analogies he may bring from Russia. I have been poring over the Marloes Bay fossils. James has been told to write to you for instructions regarding Cardigan. What are your future plans, as I need to place my sister, having let the house again.
[Cited: Morrell 1988, p.27]

1410 **Phillips**, John   To: H.T. De la Beche
St.Mary's Lodge,
York.
22 February 1842
I have begun a treaty for letting my house from 6 April. Herewith my suggestions for a plan of Survey activities: work from Llandovery to the Severn, join Williams' work at the Forest of Dean, establishing my headquarters at Ledbury. Younger hands can traverse Brecon. I am busy with Marloes Bay fossils, but have no results that can be trusted.

1411 **Phillips**, John   To: H.T. De la Beche
St.Mary's Lodge,
York.
23 February 1842
On reading Sanders' letter, I think his disappointment is considerable. He does not seem to be aware of the contingent nature of your offer. I'm afraid it is a settled business. My Ledbury scheme seems to fit in with your plan of operations for 1842. I have let my house from 6 April, so that I am a free knight of the hammer.
[Cited: Morrell 1988, p.27]

1412 **Phillips**, John   To: H.T. De la Beche
[St.Mary's Lodge,
York.
24 February 1842]
I left Sanders' letter out of the envelope yesterday. James and Aveline are now up to the Old Red. Can Rees protract the Marloes Bay section to the scale of Ramsay's Skrinkle Bay section? If so, I would like references to the localities of fossils included. The Athenaeum has informed me of my election.

1413 **Phillips**, John   To: H.T. De la Beche
St.Mary's Lodge,
York.
5 March 1842
Rees' sections are all safe and the drawing of the sections measured in Marloes Bay also. There is a want of accordance in respect of fossils in North and South of Gateholm Island. Sanders' note of this morning shows

I was wrong in ascribing only disappointment to him over the Survey. He is hurt about it. I hope you won't forget the extension of the Brecon section into N. Wales.

1414 Phillips, John   To: H.T. De la Beche
St.Mary's Lodge,
York.
15 March 1842

I do not know what to do about examining Logan's fossils. In the field this year I must try to deposit and describe fossils in groups soon after obtaining them, or else my department will fall behind. What are your Surveyors to do under this new Income tax? Horner tells me you have engaged Morris as Botanist to the Survey. If it be true, I shall be very glad.

1415 Phillips, John   To: H.T. De la Beche
St.Mary's Lodge,
[York].
18 March 1842

James is right about the fossiliferous character of the beds he speaks of. It will be a clear spot in the sky of Bristol if you can set Sanders on his feet again. I am getting ready to go to Malvern. My finances are being recruited by my giving a course of lectures here and at Manchester.

1416 Phillips, John   To: H.T. De la Beche
St.Mary's Lodge,
York.
28 March 1842

I enclose the receipts for the quarter. My house has been let for 9 months, so that next winter I can resume operations with every prospect of completion of the Wales work. I have no Malvern sheet. Horner's paper on these hills appeared good to me, and I've seen a lot of him lately.

1417 Phillips, John   To: H.T. De la Beche
St.Mary's Lodge,
York.
31 March 1842

The half of £50 note arrived safely. Ramsay's merry letter also received. I'm glad you think the future course of Survey is clearer and safer. Murchison has not written to me since my uncomplimentary letter on his return from Russia. I think he might tolerate a plain spoken friend. I don't suppose I will receive merit for placing the Magnesian Limestone among the Palaeozoics.
[Cited: Secord 1986a, p.134]

1418 Phillips, John   To: H.T. De la Beche
Malvern,
Worcestershire.
8 April 1842

I find it a convenient situation here on the slope of the Sienite. Some of the Sienites are very ordinary. I forget whether you have hammered the country. The small spirit level I have brought with me will save time for short sections.

1419 Phillips, John   To: H.T. De la Beche
Malvern.
10 April 1842

Sorry about the botheration caused by my not putting a proper stamp on the receipt. The country here is more curious and complicated than I was led to expect. Ramsay has found fossils N of Caermarthen; so have I. We must discuss the meaning of all this fossiliferous stuff.

1420 Phillips, John   To: H.T. De la Beche
Malvern.
14 April 1842

The map is very small for the work here but seems pretty accurate. The metamorphism of the Silurians is very slight. Fossils abound. The coaches are convenient. Soon you will be able send me Sheet 43. Glad you had occasion to prove to De Verneuil and Count Keyserling the care with which your Survey proceeds.

1421 Phillips, John   To: H.T. De la Beche
Malvern.
17 April 1842

Bank notes safely received. I congratulate you on your still favourable view of the bearings of the Survey. I hear from Rees of his mapping of the anticlinal N of Brecon. Truth is our pole star and we will ever keep the bright deity in view.

1422 Phillips, John   To: H.T. De la Beche
Burford House,
Malvern,
Worcestershire.
18 April 1842

A letter from De Verneuil speaks of his pleasure at the news of your knighthood. If correct, then we of the Ordnance will be delighted. He compares the Ordnance geological maps favourably with that of France, as to exactitude and detail. Your work will confer a real benefit on the scientific and practical applications of geology.

1423 Phillips, John   To: H.T. De la Beche
Malvern.
19 April 1842

Congratulations on your knighthood. I trust everything will go to your heart's content, and that this branch of the public service will make sure and rapid progress. You, and none but you, have laid its foundations.

1424 Phillips, John   To: H.T. De la Beche
Burford House,
Malvern.
27 April 1842

I'm glad you've resumed field work. I have begged Ramsay to map the bit of country insufficiently seen by me. I write today to have James and Rees send to you their notes on the mines near Llandovery. I was glad to hear Sedgwick was on the right side and wish he would come and work with us for a week.

1425 Phillips, John   To: H.T. De la Beche
Burford House,
Malvern.
6 May 1842

I am glad to hear of your view of the affinity of the Caermarthen conglomerates with those of Dinas. I send you an outline of the boundary from the Towy to Pen y coed. James and I searched in vain for fossils W.N.W. of Caermarthen. I hope Ramsay was more fortunate.

1426 Phillips, John   To: H.T. De la Beche
Malvern.
7 May 1842

May own work between St. Clares and Llandovery is exact and complete only south of the line described. If you send the sheet named, I will colour it. The whole will be consistent, and Murchison desperately wrong. There is a series of trap hillules not noticed at all in Murchison's map, on the west flank of Malvern.

1427 Phillips, John   To: H.T. De la Beche
7 May 1842

I have been amusing myself with microscopizing coal and have taken a fancy to look at what is unattainable here, stone coal or culm not in natural state but in the light flocculent earthy stuff which it leaves. Will you send me a sample, the best you can, in a little box or something that won't be crushed?

1428 Phillips, John   To: H.T. De la Beche
10 May [1842]

I have received both your letters and will reply in full tomorrow. I'm delighted to find you hold firm to the idea that the Carmarthen and Cardigan grits are equivalent.

1429 Phillips, John   To: H.T. De la Beche
Malvern.
11 May 1842

I have been colouring the map all morning. I wish I had been at St. Clears with you. I have found the band of gritstone south of Llampeter Velfrey in other places along that line of disturbance. While there, please go and see the Meidrim section.

1430 Phillips, John   To: H.T. De la Beche
Malvern.
12 May 1842

I am sending today the map and a long string of remarks. I cannot colour the alluvium of the rivers. Rees, James, Aveline and Co., did the work. I suppose you will receive their maps.

1431 Phillips, John   To: H.T. De la Beche
Malvern.
12 May 1842

I now return the map coloured as my own knowledge enables me to do so prudently. The map shows only supposedly accurate results. Beds in downward order are Mountain Limestone, Old Red, Purple Band below Old Red, Silurian boundary to North, Great Shale. Are the Caermarthen conglomerates of the Black Shale system?

1432 Phillips, John   To: H.T. De la Beche
Malvern.
18 May 1842

Perhaps no geologist but you could laugh and be joyous over these horrible convolutions and conglomerates which disgrace South Wales. At present I do not know more than two conglomerates certainly, but there may be three. These names 'Silurian' etc. are a great plague. There is certainly a fault at Lampeter Velfrey.

1433 Phillips, John   To: H.T. De la Beche
Burford House,
Malvern.
21 May 1842

The whole of the region north of Murchison's assumed boundary is new to the geologcal world and well worth the labour. Either the lower Palaeozoic fossils descend far below their supposed limit, or, because mineral characters have changed and Mydrim formations are only a Silurian colony, my system has no limit. My mind is full of doubts.

1434 Phillips, John   To: H.T. De la Beche
Burford House,
Malvern.
23 May 1842

You are right about laws of possible variation in mineral deposits of a given geological age. The mental delusion of Werner was to make a Saxon mountain the world, and an English oolitic hill the measure of a cycle of time. I wish we could get up a talk at Manchester, with Sedgwick, on these points.
[Cited: Rudwick 1985, p.389; Secord 1986a, p.149]

1435 Phillips, John   To: H.T. De la Beche
24 May 1842

Enclosed are notes on 2 mines I saw which may be of use. It is said that about these mines the Romans settled. Would it be practicable to make a distinction between ash and fused rock?
[Enclosure]: Mines East of Llangadoc and Mines East of Carmarthen.

1436 Phillips, John   To: H.T. De la Beche
Malvern.
27 May [1842]

Logan has replied at last. I have not replied to him, but will wait till you send the letter back. I will be free after the Manchester meeting. This tract will be fully mapped before I leave it. A letter from Murchison slightly notices the address, but I have not seen it. I am glad Williams answers all your expectations.
[Morrell 1988, p.26]

1437 Phillips, John   To: H.T. De la Beche
Malvern.
30 May 1842

A map is the most positive-looking assertion that was ever invented. I share your doubts as to the age of the Llandilo limestones. Perhaps two colours might be justified. I'm afraid we cannot paint a theory. I have found a fish bed in these Silurians. The ashes you sent contain vegetable tissues.

1438 Phillips, John   To: H.T. De la Beche
Malvern.
4 June 1842

I have not yet moved to Manchester, and shan't get there until 7th or 8th. I was anxious you should see the Sawdde section measured as a 'steadier'. The country near Llandilo gives valuable facts in regard to gritstones similar to those of Garn goch. If you have time look at the points recommended. The ground movements here are curious, consisting of two interfering parallels of undulations.

1439 Phillips, John   To: H.T. De la Beche
Royal Hotel,
Manchester.
11 June 1842

Ramsay's statements as to dips are all true and right. The northern dips are real, and I could find no fallacy in them. If these lower things be Cambrian then there is a fault. I fear the whole region as far as Mydrim, Abernant, Taliaris etc. will hardly lend itself to the Murchisonian form at all.

1440 Phillips, John   To: H.T. De la Beche
Royal Hotel,
Manchester.
20 June 1842

I thought your sketch section exceedingly near the truth. I confess I could never see any real or essential characters to separate Cambria from Siluria. The colouring remains a problem. It seems that all our views are drawing towards a centre of conformity. I don't at present believe the Bala limestone to be a year older than some of that in the Vale of Towy.
[Cited: Secord 1986b, p.246]

1441 Phillips, John   To: H.T. De la Beche
Link Farm,
Malvern.
2 July 1842

Mr. & Mrs.L. Dillwyn called, but my sister was unwell. We will cross the hills in a day or two and pay our respects. As regards fossils, Ledbury would be the best place for James' and Aveline's station. I don't think of working the Woolhope tract minutely at present. I hope you got safely to Llangadoc.

1442 Phillips, John   To: H.T. De la Beche
Link Farm,
Malvern.
8 July 1842

Aveline has reported, with my barometer. Can you send a duplicate copy of the map? Ledbury is the best point for James and Aveline. This morning my sister and I visit Brand Lodge. Aveline tells me there are several boxes of fossils at Brecon. Did you see the eclipse this morning?

1443 Phillips, John   To: H.T. De la Beche
Link Farm,
Malvern.
10 July 1842

James reports his intended arrival at Ledbury with Williams on Monday. I propose to make a reconnaissance of the Woolhope country, as well as the country towards Newent. I expect a visit from Hugh Strickland, and another from Murchison and Keyserling. I hear doubts of the exactitude of Griffith's tables. Woodward is my informant.

1444 Phillips, John   To: H.T. De la Beche
Link Farm,
Malvern.
13 July 1842

I have had no letter from you since our Manchester Meeting, and need a map to give James and Aveline. I have been at work in drawing up lists of fossils collected in May. Tomorrow I hope to go towards Newent, and so finish the N.E. corner of the sheet.

1445 Phillips, John   To: H.T. De la Beche
Link Farm,
Malvern.
17 July 1842

Your welcome letter arrived this morning. You do not seem to find much hard solid sandstone at Noeth grug? Here all is the same as in Salop, and it leads to reflections concerning the ancient ranges of sediments and distribution of organic life. Sedgwick is at Cheltenham. The map has come from Cary.

1446 Phillips, John   To: H.T. De la Beche
Link Farm,
Malvern.
21 July 1842

My Woolhope trip is very satisfactory in regard to the lines of strata and occurrence of organic remains. I have not obtained S.E. & S.W. quarters of sheet 43. I have obtained fossils of Keuper sandstone in some perfection. Ramsay writes merrily.

1447 Phillips, John   To: H.T. De la Beche
Link Farm,
Malvern.
26 July 1842

I have been examining some astonishingly well preserved scratches and diggings on the surface of the Bone Bed of the Severn. Can you give me any facts bearing on the enquiry? The wonder you must see with me is a real movement and wave-like elevation of the moved mass. There are carbonaceous fragments and real plants in the Upper Ludlovians of Stoke Edith. I have seen Sedgwick and can hardly sleep for thinking of Bala.

1448 Phillips, John   To: H.T. De la Beche
Link Farm,
Malvern.
27 July 1842

The cheque for £35 has arrived safely. I begin to see daylight through this N.E. corner of the Sheet. The mapping of Woolhope will be very easy. The sections only will cause delay. The country is uncommonly woody, and running a whole line will be impossible. I should like to put hairs in the theodolite.

1449 Phillips, John   To: H.T. De la Beche
Link Farm,
Malvern.
29 July 1842

The ordnance sheets have arrived safely. I have killed and despatched in one day Murchison and Keyserling. They are now gone to Ledbury. I want you to see some things on the Wych road. Murchison had not seen them. Next year we may reach Builth and Bala. Ramsay is getting Caradocian fossils at Pumpsaint.

1450 Phillips, John   To: H.T. De la Beche
Swan Inn,
Staunton.
[5 August 1842]

I have been tracing the boundary of Lias of the Corse Hills. The Ordnance map is far from good. My sister has discovered Silurian fossils in abundance on the west side of the mountains in felspathose conglomerates. They are recomposed Malvern rocks. This should be published in the Philosophical Magazine.

1451 Phillips, John   To: H.T. De la Beche
7 August 1842

I am glad you have reached Ross. The Malvern section is the best you can see in the land. The Woolhope map is advanced but I have left undone the part nearest to Ross. I recommend the inn at Tarrington. James and Aveline were here yesterday with Williams.

1452 Phillips, John   To: H.T. De la Beche
Link Farm,
Malvern.
21 August 1842

I propose to use the pony and settle the lines with Rees tomorrow. I wish to take a good lot of depressions from the Herefordshire Beacon which may be well connected with the Worcestershire one. We can arrive at probable heights of our own stations. In Murchison's address, the data are not right, the object a phantom, and the conclusion an error.
[Cited: Rudwick 1985, p.389]

1453 Phillips, John   To: H.T. De la Beche
23 August 1842

Rees appeared and stayed during the afternoon. The theodolite has come and Rees is at work at Malvern Wells. I have told James about the best fossil haunts at Tarrington. I am anxious to finish the Keuper well, as its novelty will be worth something in our map.

1454 Phillips, John   To: H.T. De la Beche
Link Farm,
[Malvern].
28 August 1842

Rees has finished two sections already. Murray is very useful and understands the whole thing well. I wish to set up altitudes for all these hills, independent of the map. Thank you for Sheet 44. I hope to finish the Keuper by the end of this week. After that I think of riding over to Woolhope.

1455 Phillips, John   To: H.T. De la Beche
Newent.
1 September 1842

I have come to Newent because I thought it a good station to finish up that quarter sheet on which I have been so long engaged. The Spa here is a sulphur spring of ancient power and fame, but now neglected and deserted. I hope to surprise you with my Keuper map. Touching hill markings on Ordnance maps, I would say nothing about them.

1456 Phillips, John   To: H.T. De la Beche
Link Farm,
[Malvern].
2 September 1842

To the enclosed note from Murchison I have replied giving your address. I sent a short protest against the 'judgement' of the President of the Geological Society and I should make public any further of that sort necessary. Your note has arrived. I always put the white sandstone which lies over the Upper Silurians with them.
[Cited: Rudwick 1985, p.389]

1457 Phillips, John   To: H.T. De la Beche
Link Farm,
Malvern.
11 September 1842

The more I see of Capt. James the better I like him. If in any future day he be called to direction in Ireland, I think the united progress of our Survey will be accelerated. I am taking him today to Howlers Heath. Sanders' letter is a long one. As to the map, there remains nothing but to make a formal demand of it, or a friendly request for it.

1458 Phillips, John   To: H.T. De la Beche
Link Farm,
Malvern.
13 September 1842

Yesterday we packed up the fossils at Brand Lodge. Today I suppose Williams will arrive, and then Capt. James and I go north. Tomorrow I propose to arrive in Tarrington. Good gracious, what a letter from India! It made me stare prodigiously.
[See also Bright, B. H. to H.T. De la Beche 24 August 1842, 110]

1459 Phillips, John   To: H.T. De la Beche
20 September 1842

I do not esteem it worthwhile to copy my scrawl about that shelly conglomerate bed. Taylor will print it for better or worse in October Philosophical Magazine. My sister has just found some fossils in the New Red Marls. I was not able to get to Tarrington till Sunday morning.

1460 Phillips, John   To: H.T. De la Beche
Link Farm,
Malvern.
28 September 1842

I am sorry to hear about your sprain. Murray, with James and Aveline will help me across from Ledbury to Woolhope in the long section. James is most in want of Trigonometry. Williams' accounts are a problem. I have measured a beautiful section north of Malvern.

1461 Phillips, John   To: H.T. De la Beche
Link Farm,
Malvern.
29 September 1842

We measured a base the other day, of 3,000 ft, on Welland Common, and took angles to the highest points of the Malverns. The accuracy of my barometer is confirmed. A long letter from Sedgwick sketches a scheme for Palaeozoic 'Systems' very like what we concur in approving.

1462 Phillips, John   To: H.T. De la Beche
Malvern.
2 October 1842

Thank you for the £50. I am glad to have your second opinion on levelling to the Worcester Beacon. Gloucester offers certainties in preference. I propose we level only from the Gloucester Beacon. If you like the scheme in general, I will write on some special aspects of the process.

1463 Phillips, John   To: H.T. De la Beche
Link Farm,
[Malvern].
10 October 1842

The Woolhope section turns out well and exact, the points being fixed by triangles and the distances measured by staff. Rees will send the drawings he has in hand. James will forward fossil collections hither. Will you choose one point of departure at Gloucester for ground levelling? I hope your foot is no longer out of sorts.

1464 Phillips, John   To: H.T. De la Beche
Link Farm,
Malvern.
12 October 1842

James writes that he was to bring the 5 inch theodolite. I propose to send over mine to you tomorrow. As soon as you can spare it, it should go in for repair. I am glad Ramsay has come up, as your foot is not yet perfect.

1465 Phillips, John   To: H.T. De la Beche
17 October 1842
No.1
The weather has been too dull to use the theodolite on Malvern Beacon. If you think it worth the cost and time, send over a measuring staff. The coal structures grow daily more clear to my eye. Do you wish a paragraph from me for your Report on Dean Forest?

1466 Phillips, John   To: H.T. De la Beche
17 October 1842
No.2
All in vain. Nothing to be seen from this hill. The Old Red boundary is exact from the N.E. end of Sheet to Caermarthen. I will reply in detail about other parts of map. I have given Rees my only spare copy of the Malvern sheet. Can you spare one?

1467 Phillips, John   To: H.T. De la Beche
Tarrington.
22 October 1842

I am wet through. This is a strange piece of ground. In regard to the staff, your Trigonometry clears its equation except on two points. An optician can make the telescope with hairs in it, but we shall have to graduate the staff ourselves. Get well, that we may meet shortly.

1468 Phillips, John   To: H.T. De la Beche
28 October 1842
No.2
I have caught a cold from being soaked at Woolhope. I am likely to have to attend a Meeting of the Council of the British Association about 10 November. It was a capital hit to find Pentamerus Knightii; 'twill make old Murchy stare a bit. I will set Capt. James to look for Terebratula navicula which has no right to be far from it.

1469 Phillips, John   To: H.T. De la Beche
Link Farm,
Malvern.
13 November 1842

I completed my London business and came here last night. Harding's fossils have been found in Jordan's care. Wheatstone says the calotype process cannot be employed for outlines without consulting Fox Talbot, the Patentee. Murchison in fumes about Charlesworth and his claim to succeed Lonsdale. I am glad Curtis and Owen were awarded pensions.

1470 Phillips, John   To: H.T. De la Beche
Gloucester.
28 November 1842

Thank you for the funny sketch. The hill shading on the west bank of the Severn is most wickedly and deliberately falsified. I have finished two of the districts accessible from here. Murchison has made a sad blunder marking as Lias what is half Keuper and Red Marls. I return you Murchison's note. It will be dangerous to the Society if the Council are outvoted.

1471 Phillips, John   To: H.T. De la Beche
Gloucester.
29 November [1842]

Yesterday I walked as far as Bulley to see the sort of work offered. The Keuper will require a sharp and patient investigation. Huntley will be the best station. I am dead tired. I shall work by Coach from Gloucester for a few days. I have received long letters from Sedgwick and Murchison.
[Cited: Secord 1986a, p.163]

1472 Phillips, John   To: H.T. De la Beche
Post Office,
Gloucester.
30 November 1842

I should progress well on this quarter sheet in the next three or four days. Yesterday I found my own Yorkshire Lias section in the hills there. This is excellent. Can Ramsay tell me the height of the hill S.E. of Gloucester called Robin Hood's [Robin's Wood] Hill?

1473 [Phillips, John]   To: H.T. De la Beche
Camp Standish.
3 December 1842
No.1
Five square miles on Gloucester sheet surveyed. Formations traced to the Severn. East side of the river completed. The Keuper, Liassic and Oolitic lines made out.
[Letter in form of pseudo-military report].
[Cited: Herries Davies 1983, p.110]

1474 Phillips, J.   To: H.T. De la Beche
Gloucester.
3 December 1842
No.2
I have received your note and will transfer our work to the right bank of the Severn. Lonsdale's map is a good guide. I have found some nice things regarding Lias fossils, and am glad you can spare James and Aveline for Westbury to work the Lias.

1475 Phillips, John   To: H.T. De la Beche
Gloucester.
6 December 1842

I have shown Capt. James the Lias and Oolite sections near Wotton under edge, and was much in distress at leaving the Tortworth tract. The Caradoc is the grand formation of Tortworth. Lonsdale's lines are of good value. Tomorrow I get down to Westbury.

1476 Phillips, John   To: H.T. De la Beche
Victoria Hotel,
Newnham.
7 December 1842

I could not get lodging at Westbury, so came on to Newnham by coach and am now housed in the very room where William Smith and self stayed twenty years since for some weeks. The old red and white cliffs are the same, but the hotel has changed its name.

1477 Phillips, John   To: H.T. De la Beche
Gloucester.
9 December [1842]

The Keuper is a most felon-like business and the Lias abominable. I shall have to go to Malvern tomorrow evening, but will return to work on the lines on Monday. I hope then to go home on Saturday the 17th. This map has been for me a subject of laughter, tears and vexation.

1478 Phillips, John   To: H.T. De la Beche
Link Farm,
Malvern.
12 December 1842

I have received the dry and wet proofs and will lose no time in working at them. If you have a blank sheet of Woolhope, I will complete it for you. I will leave the Keuper lines near Bulley and Huntley to Ramsay and James. The Lias is done. I shall not be able to leave for home before Saturday or Tuesday.

1479 Phillips, John   To: H.T. De la Beche
Link Farm,
Malvern.
16 December 1842

I have now finished the colouring and lining of the Malvern sheets. The lines of Wenlock limestone are not complete, owing to the woods in which they lie. They can be finally drawn when I come this way next year, or I can draw them now as probable. My sister says you are going to send us a selection of your amusing sketches.

1480 Phillips, John   To: H.T. De la Beche
St. Mary's Lodge,
York.
17 December 1842

I am not going to write you a letter, but merely authenticate my arrival. I have opened a box of Devonian fossils, amongst which are new species, and (what is better) most beautiful specimens of old ones.

1481 Phillips, John   To: H.T. De la Beche
St. Mary's Lodge,
York.
21 December 1842

The map came safely and I will lose no time in putting in the dip arrows. If the Old Red turns out to be 6000 feet thick at Mitcheldean, I hope it will be less than 15000 at Brecon! Everything tends to show that still North Devon is an area of greater interest to philosophical geology than almost any other.
[Cited: Rudwick 1985, p.389]

1482 Phillips, John   To: H.T. De la Beche
St. Mary's Lodge,
York.
23 December 1842

I have now put in dips over the South Malvern sheet, but have not often used the sign for verticality. The horizontal mark is not used at all. If you send me the Hereford quarter sheet, I can mark the western Woolhope boundary for you.

1483 Phillips, John   To: H.T. De la Beche
St. Mary's Lodge,
York.
24 December 1842

I send by this post the South Malvern sheet. The dip arrows are clumsy because of shading and other lines, but correct as to direction and position. I shall next send you the North Malvern sheet. The perry and cider in my cellar are better than what I tasted in Herefordshire.

1484 Phillips, John   To: H.T. De la Beche
St. Mary's Lodge,
[York].
29 December 1842

I am glad you are resolved to make an immediate move to consolidate the two Surveys. A central directorate is necessary. As for my part, drawer space will be necessary. We can train our own artist. I would resign all other engagements if the Palaeontology were entrusted to me.
[Cited: Secord 1986b, p.257]

1485 Phillips, John   To: H.T. De la Beche
York.
10 January 1843

The Narberth map herewith sent requires no explanation to you. I never could be sure whether the black shales at Whitely were above or below the limestone. They are here drawn below. I am not clear as to the areas to be coloured as Caradocian and Cambrian.

1486 Phillips, John   To: H.T. De la Beche
York.
10 January 1843

I enclose the sheet on which the Lydstep synclinal shales and sandstones are marked. The line appears to me inexact. Shale with goniatites near Tenby is also marked. Tomorrow I will send the Lampeter sheets. £50 has arrived safely.

1487 Phillips, John   To: H.T. De la Beche
York.
18 January 1843[?]

I send you coloured Clog y fran section, with observed and calculated dips in order. I have sketched in pencil the probable reverse dip in Taf valley. Quetelet sends good news on Palaeozoics from Belgium.

1488 Phillips, John   To: H.T. De la Beche
York.
19 January 1843

I send by this post a package of drawings to show the process I use in arranging fossils for publication. I hope your application is successful, as we must give publicity to the Ordnance Geological Survey collections. Please send acknowledgement of safe receipt.

1489 Phillips, John   To: H.T. De la Beche
York.
21 January 1843

You can pay the sum through Barnett's Bank to my account with the York City Bank. I am glad you received the pictures and hope you succeed in using them. Is Buckland well again? Sedgwick was here for two days, and quite well for him.

1490 Phillips, John   To: H.T. De la Beche
St. Mary's Lodge,
York.
23 January 1843

Buckland has altogether misunderstood our position. We aim to collect more fossils than ever before done, and to produce drawings of carefully selected specimens, complete monographs of every species on a national scale. What do you think of this?
[Cited: Secord 1986b, p.250]

1491 Phillips, John   To: H.T. De la Beche
York.
27 January 1843

The bank has received your £30 remittance. The balance can be sent by P.O. order. Your account of McCulloch's quick conception of our plan is cheering. James writes from Usk and sends some fossils (of fish) found in red sandstone.

1492 Phillips, John   To: H.T. De la Beche
York.
1 February 1843

It always appeared to me that all Wales should be included in one system of Survey and Colouring. There ought to be a good general and real section run up to Menai before we can be sure of our grand lines. The same reasoning applies to fossils. It is gratifying to find Murchison taking a right view of our toilsome work.

1493 Phillips, John   To: H.T. De la Beche
St. Mary's Lodge,
York.
10 February 1843

As a result of the appropriation of the rooms and Ordnance willingness to ask for cases, I have been preparing to cooperate. It will take months to arrange the fossils - necessary before preparing any publication. Can you give me the dimensions and arrangement of the rooms, so that I can plan the drawers?

1494 Phillips, John   To: H.T. De la Beche
16 February 1843

Here is my notion of the drawer system, with ten drawers in a series, the top being strong as a valuable table. A glass case could eventually be placed on it. Trays are required in which to put fossils. I would get a moulding and plinth to suit the room.

1495 Phillips, John   To: H.T. De la Beche
York.
22 February 1843

Lowry's plates are very well engraved, yet in places the drawings want a sign of the naturalist's knowledge. I prefer competition at present among the known engravers, and not favouring one. I do not see clearly where you want me next year, nor what you want me to do about these organics.

1496 Phillips, John   To: H.T. De la Beche
York.
11 March 1843

I have sent all the Ordnance fossils to the Museum of Economic Geology in despair. I can now finish a few drawings, and will be ready to leave. I will not let my house until I know what the future holds.

1497 Phillips, John   To: H.T. De la Beche
21 March 1843

I send 5 drawings as a fair selection. You will see by A. Buchii what may be expected from our plan. I think it best to accompany each species with a full page of description, so as to have very little engraving on the plates. 4 more casks of fossils have been sent.

1498 Phillips, John   To: H.T. De la Beche
Gloucester.
9 April 1843

Having done small bits of work on my Malvern sheets, I will pass by Newent to Ledbury and then to Wotton. Have you a spare map of the

Wotton tract? If so Ramsay would bring it. Was the interruption I left in the Keuper west of Gloucester filled up?

1499 Phillips, John   To: H.T. De la Beche
Ledbury,
12 April 1843
Malvern 14th

I have found small trilobites in the lowest shales of Malvern below the fossiliferous beds commonly known, and below all of May Hill and Woolhope! I shall send notice of this to R. Taylor as it shows the difference of our work from that of others of less leisure.

1500 Phillips, John   To: H.T. De la Beche
Link Farm,
Malvern.
14 April 1843

I am glad of the opportunity of looking as a spectator at the country which cost me so much labour and thought, and verifying some perplexing points. As the Keuper is not traced about Huntley and Bulley, I will set about it. Ramsay has a good fortnight or more work.

1501 Phillips, John   To: H.T. De la Beche
Newnham,
Gloucester.
19 April 1843

I have at last found scales and bones of fish in the Newnham Keuper. The maps arrived safely. There is nothing to alter, only some bits to add. I have had a letter from H.D. Rogers. He is coming to the York Meeting in 1844. I have sent 16 more tubs of Malvern fossils to the Museum.

1502 Phillips, John   To: H.T. De la Beche
Newnham.
20 April 1843

I suppose you remarked on the singular junctions of Old and New Red south of Newnham? The Lias clays of Awre are very extensive. The Keuper lines are more interesting than I imagined. The map shading is laughably bad towards the Severn. Whoever did it was no surveyor.

1503 Phillips, John   To: H.T. De la Beche
Wotton under edge.
29 April 1843

I arrived on 27th and find Ramsay and Bristow working the Silurians. Williams' map seems good. Sanders' district is done as well as any man can. Lonsdale's work is an excellent approximation. Better fossils than we can collect are available in private collections. It will be Xmas before these two sheets are completed.

1504 Phillips, John   To: H.T. De la Beche
Wotton under edge.
4 May 1843

The 'Map of the Cotswolds' is an able sketch, but the lines are wrong in certain places, and we cannot adopt them. Bristow is getting on very well. I like him much. Ramsay is very happy, working on the oolites. We shall soon clear the north end of this sheet, leaving only the S.E. corner to be corrected.

1505 Phillips, John   To: H.T. De la Beche
Wotton.
8 May 1843

I always thought Lowry would do well. I have no drawings here. It may be best to give him two or three, and he could try these, though they are below average as to style and labour. We are getting on well. Bristow will be a first rate hand.

1506 Phillips, John   To: H.T. De la Beche
Wotton under edge.
11 May 1843

We will proceed with tracing the oolite boundaries. Lonsdale's map is a capital guide, and highly creditable to him. Could James bring the theodolite from Ross? Sanders is finishing his map, and I have told him your words.

1507 Phillips, John   To: H.T. De la Beche
Wotton under edge.
14 May 1843

All goes well here. I should have published a paper on the Coprolitic bed to secure credit for the Ordnance Geological Survey. Williams is now in possession. More specimens will be needed for Agassiz when he comes out of the ice. We must press on with Wales, else the Ordnance Geological Survey lose the only crown remaining which English geology has to offer.
[Cited: Secord 1986a, pp.164-165]

1508 Phillips, John   To: H.T. De la Beche
Wotton under edge.
15 May 1843

With this note, I send a short paper which may be sent to Colby if you think fit. Generally speaking, I think the error in Portlock's Report is entirely innocuous. It should be treated as a very excusable scientific error, or any criticism will rebound on us.

1509 Phillips, John   To: H.T. De la Beche
Wotton under edge.
18 May 1843

Sanders has lost his mother, so we are now without him. You would highly approve of Gibbs' skill in cleaning fossils. He should receive more than 2/6 per day. He begins to read and comprehend the wisdom of Lyell and De la Beche. If you have any spare money, please send some to me.

1510 Phillips, John   To: H.T. De la Beche
Wotton under edge.
19 May 1843

If I find your section of the Garden cliff in my papers, I will enclose it. Your management of the rooms is the best in the world, but the Ordnance are slow in providing drawers. We must press ahead with the classification. It will take all the summer to do even the Silurians. Gibbs and James are the best collectors of fossils.

1511 Phillips, John   To: H.T. De la Beche
Wotton under edge.
21 May 1843

The money was safely received and I now have whole of balance owing to me on the account of Ordnance Geological Survey for the last quarter. Playfair is here, a right merry and exceeding clever fellow. He gives answers to my questions which show he can solve some of our geological puzzles. Sanders is suffering.

1512 Phillips, John   To: H.T. De la Beche
Wotton.
26 May 1843

James has arrived. I will reserve two distant trips for when you arrive. Today I go down to Berkeley and Purton. What do you think of my journey to Town? We can finish here in a month, but not the Sections.

1513 Phillips, John   To: H.T. De la Beche
Wotton under edge.
27 May 1843

I saw the Silurian at Purton Passage yesterday. I found the Malvern lines of Upper Ludlows and bone beds. This part of the Silurians I will now take in hand. Ramsay is south of Berkeley. James has done the oolites, and Bristow has done well his piece of ground.

1514 Phillips, John   To: H.T. De la Beche
Wotton under edge.
9 June 1843

A letter from Buckland addressed to you here has arrived, and I have received one from Favre who expects to meet you here. So I suppose we will have to catch frogs by the dozen as well as ichthyosauri.

1515 Phillips, John   To: H.T. De la Beche
Wotton under edge.
20 June 1843

I have seen Badminton, and shown Ramsay and Bristow the easy boundary of the Oolite there. The map of Berkeley is completed, and the section line marked on Bristow's map. Tomorrow I leave for Stroud, and thence to York and London. Gibbs is going to Wickwar.

1516 Phillips, John   To: H.T. De la Beche
Link Farm,
Malvern.
23 June 1843

After two days' work round Ridborough and Stroudwater, I passed by Gloucester, and took the coach to Worcester. Bristow can finish the

Badmington Survey, and Ramsay and James should take up a station south of Sodbury.

1517 Phillips, John   To: H.T. De la Beche
St. Mary's Lodge,
York.
29 June 1843

I have been colouring the sheet on which I have been working and the Malvern quarters, but will remain here only a few more days. I have been comparing Silurian and New York Asaphi with interesting results. I am glad you think of coming to Town, as we might get into quicksands.

Phillips, John   To: H.T. De la Beche
Athenaeum.
23 July 1843
[see **2272**]

1518 [Phillips, John]   To: H.T. De la Beche
Old Whitehall Yard.
24 July 1843

As long as the Survey is in your hands I see a place for me in its progress. I wish to give my anchor a firm hold on Ordnance ground, since in that way I can render a fuller service to Science. The two or three years on the Survey so far have restored my health, and my programme should ensure success.

1519 Phillips, John   To: H.T. De la Beche
Whitehall Yard.
26 July 1843

I see as plain as can be that a noble work may be made upon our plan, and have the greatest desire to do my share in it, because somehow it suits my style of mind. I shall leave work here on Saturday and meet you at Bristol.

1520 Phillips, John   To: H.T. De la Beche
16 Grand Parade,
Cork.
9 August 1843

There is no difficulty whatever about this country, merely a set of East and West rolls. Lyell told me he should pass by Bristol on his way to the meeting in Cork, and I advised him to stop and see you about Magnesian Conglomerate, of which Murchison has made a fallacious use in his last but one address.

1521 Phillips, John   To: H.T. De la Beche
Cork.
10 August 1843

Logan corrects himself in his paper as to the position of the gypsiferous beds, and leaves Murchison rather oddly located. Why is coal all of one age, any more than gypsum? Answer me that, Lyell & Co. The Meeting here will be pleasant, but hardly large. More space is needed for our fossils.

1522 Phillips, John   To: H.T. De la Beche
Dingle,
Co. Kerry.
31 August 1843

I have made a successful sortie so far, as it appeared to me important for our views in Wales that one of us should see the peculiar strata called Cambrian, Silurian, and Old Red in these promontories of Cork and Kerry. In the interpretation of all this I find the South Wales sections very valuable and suggestive. I am moving to Holyhead, and thought to go over the Bala section.

1523 Phillips, John   To: H.T. De la Beche
Birr,
Ireland.
5 September 1843

I have been looking through Lord Rosse's telescope at the moon and planets. He has modelled a moon mountain with a curious terrace running round its crateriform summit inside the crater. My visit to Dingle was very instructive. There is a real Silurian series there. I have collected a box of fossils (20 species).

1524 Phillips, John   To: H.T. De la Beche
Bala.
13 September 1843

I have made a short reconnaissance of this region, and all is not yet clear to me. In a few days I shall have to go home to deal with British Association affairs, after which I will be available to you until Xmas. Beds here are all Lower Silurian, I believe, yet the Sections seem to back up Sedgwick's view.

1525 Phillips, John   To: H.T. De la Beche
York.
19 September 1843

I have arrived home to make arrangements for the British Association Meeting in 1844. That Meeting will be worthwhile attending, and you will be welcome to stay here. I have received a satisfactory note from Bristow, but fear his and Ramsay's estimate of length of time yet required is correct. Do you wish me to do plates, or push on to Frome?

1526 Phillips, John   To: H.T. De la Beche
York.
27 September 1843

I thought you would be satisfied with the plates. It is evident we can produce what is needed. Nomenclature is a chaos. The preparation of the pages is a work of the cabinet, not of the field, alas. The address of my bankers is on the other page.

1527 Phillips, John   To: H.T. De la Beche
York.
7 October 1843

I have overcome difficulties here, and my pony goes into stables on Monday at Gloucester. British Association and York Philosophical Society business is the reason for my tardiness. Have you seen Morris' Catalogue of 5,000 forms, and Sharpe's scheme for Devonianism in the Preface? I think the latter is erroneous.

1528 Phillips, John   To: H.T. De la Beche
Claverton Hotel,
Claverton.
12 November 1843

Hill side slips delay progress on Cornbrash and Forest Marble. I have worn out 2 pairs of shoes. It is likely I shall have to come to Town for Council of the British Association but I hope to finish this work here by 19th. Bristow is a good man who desires to know his business in depth. Ramsay must be near the end of his work in the North.

1529 Phillips, John   To: H.T. De la Beche
Claverton Hotel,
Claverton,
Near Bath.
14 November 1843

I enclose a note from Archer. Would it not be desirable to pay him off at once? Bath your sprained hand in hot or cold water. I have had difficulties with the boundaries hereabouts, and there is plenty of work for Gibbs if he can be spared.

1530 Phillips, John   To: H.T. De la Beche
Claverton Hotel.
19 November 1843

I am glad you find so much progress toward finishing the S.E. corner of the sheet. I report the N.E. corner done as far as I am concerned. Ramsay says the North is complete. Bristow is still at work. I will commence my winter stay in London if the Council Meeting is held in December.

1531 Phillips, John   To: H.T. De la Beche
Athenaeum.
30 November 1843

Rapley is satisfied with the plates. I explained we would use an artist as necessary, and left the proofs with him. Sedgwick continued his North Wales papers yesterday evening, with much credit. I mentioned Survey work. Sharpe and Murchison spoke, regretting your absence. All sorts of kind enquiries for you here.

1532 Phillips, John   To: H.T. De la Beche
St. Mary's Lodge,
York.
23 December 1843

Thank you for sending me the letters about the recommendation of the British Association on the subject of contouring the Irish maps. I found the pony well at Worcester. There is a letter from Baily expressing preference for London as a place of work. Shall I say 2 guineas per week?

1533 Phillips, John   To: H.T. De la Beche
York.
27 December 1843
As far as I can see, the course of the Irish affair is direct and forward, but I am aware that too much dependence must not be placed on sublunary affairs. I have offered Baily 2 guineas a week for 3 months.

1534 Phillips, John   To: [H.T. De la Beche]
[1843]
[Part letter only]
It is very certain that the Irish matters will be thrown into our net, and it is equally sure that I cannot work at any other matter than the Organic Remains if justice is to be done to them, to you and to myself. At the moment I am designated an 'Assistant Geologist' on the Ordnance Geological Survey. If you formally give me the charge of the Palaeontology, you will find it convenient to restore the speciality of my engagement, supposing it to endure.

1535 Phillips, John   To: H.T. De la Beche
York.
2 January 1844
I have been appointed Professor of Geology and Mineralogy in the University of Dublin. As a result, I estimate I have 8 months available for Ordnance work. I will be in Town on the 4th.
[Cited: Davies 1969, p.31]

1536 Phillips, John   To: H.T. De la Beche
1 Islington Terrace,
Kingstown.
4 January 1844
I send you my book of sketches made in 1841 between St Brides Bay and Llandovery. I will add that of other years. Re your proposal that I should undertake the Geological Survey of Yorkshire, it would be profitable to the Ordnance. Assistants would enable me to direct it from my home all the year.

1537 Phillips, John   To: H.T. De la Beche
Athenaeum.
13 January 1844
You have asked me what books I have written. My first publication was in 1819, followed by others in 1824, 1833, 1834, 1837 etc. Also the Map of the British Isles. You know how Griffith helped me on that, as did you. The Devonshire question (as you recollect) was first started, and not on a wrong scent, in my article on Geology in the Encyclopedia Metropolitana in 1833. Why do you not give Conybeare a medal?

1538 Phillips, John   To: H.T. De la Beche
London.
3 February 1844
Respecting our growing palaeontological collection, now that I am to go to Ireland, I suggest that E. Forbes be placed in charge of it. As to the field geology of Ireland, I see no difficulty. The assistants must be Irish. I would start at 3 points in southern Ireland. I would not advise a separate Museum for the Ordnance if a Museum of Economic Geology is founded.
[Cited: Davies 1969, p.31; Herries Davies 1983, p.113]

1539 Phillips, John   To: H.T. De la Beche
4 February 1844
I think our publications should be classed as follows. The laws of distribution are not to be elicited in a cabinet, and should be done by the Director or Lieutenants according to districts.

1540 Phillips, John   To: H.T. De la Beche
Link Farm,
Malvern.
1 May 1844
I send you by this post coloured sheets of South Malvern, Gloucester and Woolhope. On the Gloucester sheet I have not marked a boundary of the Lias. I suppose the Malvern traps must be red. How do you paint elvans?

1541 Phillips, John   To: H.T. De la Beche
35 Trinity College,
Dublin.
5 May 1844
My voyage from Liverpool was delightful. These ships are the way to unite the Islands. Chairs of Botany and Zoology have been added. All looks well for the progress of 'Young Ireland'.

1542 Phillips, John   To: H.T. De la Beche
35 Trinity College,
Dublin.
7 May 1844
I have seen Griffith's maps today. How valuable they are. If Lowry is pressing I could give him, for drawing, the trilobite head from Littlehope. I am glad to hear good news of sections. Are you still confined to town?

1543 Phillips, John   To: H.T. De la Beche
35 Trinity College,
Dublin.
9 May 1844
I disagree. The inferior oolite is very thick, and the line at the base of the yellow is the base of that rock. The extension of the Geological Survey into Ireland will be popular among the men of science. The occasion is excellent. I have joined the Geology Society here.

1544 Phillips, John   To: H.T. De la Beche
11 Nassau Street.
Dublin.
13 May 1844
Griffith has been called to London, where I hope he will call on you. I do not think he will finish his map in our way of work, but he is willing to let me use his documents. I have seen Larcom, but he is not a talkative person. I shall return the drawings on Wednesday.
[Cited: Herries Davies 1983, p.114]

1545 Phillips, John   To: H.T. De la Beche
11 Nassau Street,
Dublin.
14 May 1844
Your conjectures as to the real state of feeling here, and the authors of the various movements anent the Geological Survey have been prophetic. If the Government will let us have the Surveys we can gain them credit by it. Portlock was plagued and restricted here. Ireland is a land of promise. My class has grown. I wish you would put Baily on the Survey. The report on the Custom House rooms is unfavourable.
[Cited: Herries Davies 1983, p.114]

1546 Phillips, John   To: H.T. De la Beche
35 Trinity College,
Dublin.
15 May 1844
I can give the whole year to the Irish branch of the Geological Survey, by so arranging my 36 lectures at the University as to allow my supervision of the work, even in the remotest parts of the country. I await your decision.

1547 Phillips, John   To: H.T. De la Beche
18 May [1844]
I have sent a letter which you may wish to show. Hence its formality. My return from the Association appointment is more than from the Ordnance Grant. I should be glad to have the opportunity of reporting progress in the Irish Survey.
[Cited: Davies 1969, p.32]

1548 Phillips, John   To: H.T. De la Beche
Link Farm,
Malvern.
25 May 1844
I have crossed by the Mail Packet to get in a week's work here. Your letter of 22nd tells me the question of the English Survey is now settled. I am prepared to take on the Irish Survey or the Palaeontology of the British Isles. My university position will not interfere with either of these.
[Cited: Davies 1969, p.32]

1549 Phillips, John   To: H.T. De la Beche
Knightsford Bridge.
30 May 1844
I have just finished with Gibbs, who goes to Ludlow next week. He has done good work. His own map is really most creditable. When you colour the Malvern Sheet, don't forget the peculiar conglomerate I marked on my map.

1550 Phillips, John   To: H.T. De la Beche
Link Farm,
[Malvern].
31 May 1844
I have drawn all the lines on the Gloucester sheet which I know of. The

marlstone is a poser. I can draw it only at a few places, and advised Bristow to mark it where he saw it. I have no documents on the marlstone of the Bath Valleys.

1551 Phillips, John   To: H.T. De la Beche
Worcester.
2 June 1844

Can you wait till my lectures end on 25th, when I can look at the disputed boundaries near Box on my return to England from Ireland? I go tonight to Dublin.

1552 Phillips, John   To: H.T. De la Beche
Link Farm,
[Malvern].
2 June 1844

I am sorry various boundary points on the map remain perplexing. Marsh land borders are very difficult to draw with accuracy. The Box district marlstone is particularly problematic. I propose colouring the Inferior Oolite sand as both Liassic and Oolitic. The Marlstone dwindles to zero south, and reappears at Mendip.

1553 Phillips, John   To: H.T. De la Beche
Malvern.
7 June 1844

My duties in the University of Dublin, instead of opposing, almost require my connection with the Irish or English Geological Surveys. It would hardly be worthwhile to count on the two winter vacations for field work, but work can be then revised or prepared.
[Cited: Davies 1969, p.33]

1554 Phillips, John   To: H.T. De la Beche
Trinity College,
Dublin.
8 June 1844

In regard to the Memoir, there is a body of opinion that such a Survey would benefit Ireland. There is also a party who were employed on the Ordnance Memoir who have been led to expect employment on it. I propose to quit Dublin on 25 June.

1555 Phillips, John   To: H.T. De la Beche
Trinity College,
Dublin.
17 June 1844

I am sorry you are still perplexed by the Bath sheet. Marlstone is seen in small patches near Wellow. The Oolite may be a puzzle. Kane's book is rather good. I do not believe my countrymen have the least idea of the true state of Ireland.

1556 Phillips, John   To: H.T. De la Beche
Trinity College,
Dublin.
19 June 1844

I am not suprised at the result, and think it right to postpone the question of the Irish Survey. I am planning my work so as to complete my obligations to the Survey by September 30. I wish my successor well, as also the whole Survey.

1557 Phillips, John   To: H.T. De la Beche
Link Farm,
Malvern.
30 June 1844

I arrived on 28th and found your own letter as well as Ramsay's enclosing Bristow's map. There should be an inspection of junction lines. I shall examine the region east of the Abberley Hills on Monday. What are your wishes?

1558 Phillips, John   To: H.T. De la Beche
Link Farm,
Malvern.
2 July 1844

I agree with you as to the limit of an inspection of the junction of the three maps. I did not know you had modified any of Ramsay's lines. If you are in haste for this inspection, let me know. The rain has come here at last, and brown fields are turning green.

1559 Phillips, John   To: H.T. De la Beche
Malvern.
13 July 1844

I cannot regard the renewal of my engagement with the Ordnance Geological Survey as very probable, notwithstanding what you have said in your last two letters. Please give me information which will help me in preparing a Report on my activities since 1841.

1560 Phillips, John   To: H.T. De la Beche
Link Farm,
[Malvern].
16 July 1844

If you cannot tell me what arrangements you have proposed for me in connection with the Irish Survey, I must remain in ignorance. Thank you for clarifying the question of my work on the Palaeozoics. I think the proper time to meet Forbes is when his appointment commences.

1561 Phillips, John   To: H.T. De la Beche
Malvern.
24 July 1844

I return you the Board of Ordnance letter. I disagree that my nomination to the Dublin Professorship precludes service on the Geological Survey. You did not think so when in Frome you advised me to accept the Chair. As to arrangements with Forbes, the proper time is when I've completed my work.
[Cited: Herries Davies 1983, p.114]

1562 Phillips, John   To: H.T. De la Beche
Malvern.
24 August 1844

I would have written to you instantly had I known you were ignorant of Capt. James' appointment to Directorship of the Irish Survey. Forbes is being approached for Palaeontology. As far as regards the Captain, all is settled. He hammers and I lecture. You may have to defend your rate of progress. Griffith was not mentioned.
[Cited: Herries Davies 1983, p.115]

1563 Phillips, John   To: H.T. De la Beche
Link Farm,
[Malvern].
26 August 1844

Before my leaving here, I have packed up a cask of organic remains. The theodolites are also packed up in strong boxes. Gibbs has sent the main portion of the Abberley collection to town. When must we expect you in York. Your room will be ready on the 18th.

1564 Phillips, John   To: H.T. De la Beche
Llangollen.   [Aberystwyth].
21 October 1844

I am sending you sketches of the scenes we have looked at. I have seen a rock perched on a hill, doubtlessly brought by a glacier, of which your faithless mind will not admit a single splinter. Bring it by Noah's flood if you like, but bring it alone. There is nothing with it.
[Cited: McCartney 1977, p.63]

1565 Phillips, John   To: H.T. De la Beche
1 Islington Terrace,
Kingstown,
Ireland.
11 November 1844

I have taken a house, as above, for the winter. I am preparing sketches for you, and will give a short preface with a specimen description. I can see what ought to be done, plan how it should be done, and keep others in the right line for doing it.

1566 Phillips, John   To: H.T. De la Beche
1 Islington Terrace,
Kingstown.
16 December 1844

I now enclose a specimen catalogue. I have found among my papers something that belongs to you, a correction which requires a small colour change, but no alteration of lines. The climate here is much milder than at York.

1567 Phillips, John   To: H.T. De la Beche
Kingstown.
19 December 1844

I have not seen Burmeister's work, but believe it will be worthy of study. I have been unfortunate as far as Crustacea, Belemnites and Trilobites are concerned. If you have any peat near you, please send me a pinch of the clay or marl beneath it.

1568 Phillips, John   To: H.T. De la Beche
1 Islington Terrace,
Kingstown.
25 January 1845

Your letter came as a suprise. I was not aware you had put my name on the List for a Medal. When I shall look on the Medal, I shall remember through what kind hands it has passed to my own.

1569 Phillips, John   To: H.T. De la Beche
Kingstown.
6 March 1845

I am resigning my Professorship with this term, which ends in a few days. My books and collections are packed for York. Please send Sections, Fossil plates etc. to York. I have all the summer before me for Survey Work.
[Cited: Davies 1969, p.33; Oldroyd 1992, p.411]

1570 Phillips, John   To: H.T. De la Beche
St Mary's Lodge,
York.
18 March 1845

I have been arranging an office at home, with facilities for drawing etc. Your room is always ready, should you want to come. The views in front of the house are improved by Nasmyth's operations. Ramsay's parcel arrived safely.

1571 Phillips, John   To: H.T. De la Beche
St. Mary's Lodge,   Museum of Economic Geology,
York.   Craig's Court,
26 March 1845   London.

I send you sketches for the Malvern wood cuts. I intend to go there early next week. Could Aveline and Gibbs meet me at Abberley. I would take pains to teach Aveline anything needful.

1572 Phillips, John   To: H.T. De la Beche
St. Mary's Lodge,
York.
3 April 1845

Can you send me, to Gloucester, any map on which your Silurian lines are drawn? Where am I most likely to find Woolhope limestone in Mayhill? Aveline says he will be able to meet me on 14th. My friends are overjoyed I'll be working in the North.

1573 Phillips, John   To: H.T. De la Beche
Knightsford Bridge.
16 April 1845

I am pencilling some considerations on your quarter sheet, as Ramsay can find no scrap of map. We shall have finished Abberley by Tuesday. Aveline is very steady and sure of hand and head.

1574 Phillips, John   To: H.T. De la Beche
Knightsford Bridge.
18 April 1845

From this date my income from the University would have been doubled. I have reiterated my choice in favour of the Survey.
[Cited: Davies 1969, p.34; Morrell & Thackray 1981, p.443]

1575 Phillips, John   To: H.T. De la Beche
Knightsford Bridge.
18 April 1845

Herewith a coloured sketch of Aston Ingham, May Hill. I return the quarter sheet with one pencil line on it. On second thoughts, as this postman's only bag is his hat, I send you a sketch of the upper line of Wenlock.

1576 Phillips, John   To: H.T. De la Beche
[Knightsford Bridge].
18 April 1845

On my previous sketch of Aston Ingham I have marked the supposed downthrow wrongly. Herewith a second coloured sketch. It is a puzzling spot.

1577 Phillips, John   To: H.T. De la Beche
Knightsford Bridge.
19 April 1845

Herewith a sketch of a puzzling case in the Abberley Hills. Also a sketch of faults. Perhaps you know where the mystery lies. In the Malvern tract, I found small trilobites in black shale. Perhaps Reeks could see if there is potash or soda in the enclosed specimen.

1578 Phillips, John   To: H.T. De la Beche
St. Mary's Lodge,
York.
24 April 1845

I have suspended work on our section and come home to work on the Volume. I propose to put all forces on the latter, so as to produce it quickly, and thus advertise our other work. As to Yorkshire, the sooner you let me advertise my duty on it, the better.

1579 Phillips, John   To: H.T. De la Beche
[York].
6 May 1845

I am busy arranging drawings for the Book, and preparing a map of each district. I have only my work on Tortworth, and would like Baily to copy the other Silurian lines for me. I have to come to Town on 14th for a meeting of the Association Council, and wish to consult you then on progress.

1580 Phillips, John   To: H.T. De la Beche
[York].
25 May 1845

This is an excellent commencement, and I suppose we will have to refer in subsequent pages to your introductory thesis. I am glad you have begun to print. I will have a series of Malvern papers ready by 31 May. On 2 June I am going to Northampton and Cambridge.

1581 Phillips, John   To: H.T. De la Beche
York.
8 July 1845

I send today an account for £5.5.0, and also 16 drawings for the Malvern, Abberley and May Hill report. I have drawn a good many more, but these seem enough. Yesterday I looked at the new railway line, and have written to get regular authority to examine it before earth covers the cutting.

1582 Phillips, John   To: H.T. De la Beche
Rutland Arms Inn,
Bakewell.
17 July 1845

This is the devil of a country for mineral veins. I am directing my attention to the section, which is a disputed thing. But I think it is almost exactly Yorkshire over again.

1583 Phillips, John   To: H.T. De la Beche
St. Mary's Lodge,
York.
21 August 1845
[Part letter only]

After finishing my sketch of the railway sections between York and Malton, I have been in daily hopes of the dreadful weather taking up and allowing me to get away to the Derwent with some hope of seeing the meadows about it. I suppose you find the Waterfordians very like the Haverfordians, or at all events like the Pembrokians generally. As to N. Devon, I am glad to find that the view which the fossils suggested in 1840-41 is confirmed by every good reason.

1584 Phillips, John   To: H.T. De la Beche
St. Mary's Lodge,
York.
23 August 1845

I could never get specimens of Tenby goniatites which allowed of the specific characters' being determined. These might be found in masses at Whitehall Yard Museum. They are fossils not found in real Silurians, and probably nowhere go into the Magnesian Limestone. The Tenby species are the same as in Yorkshire, Derbyshire and the north of Ireland.

1585 **Phillips**, John To: H.T. De la Beche
Baslow,
Derbyshire.
31 August 1845

I have been walking the limestone edge of the west of Derbyshire, and am working up the characteristic points of the toadstones. Tomorrow I start the northern limestone boundary. There has been no rain since Tuesday morning.

1586 **Phillips**, John To: H.T. De la Beche
St. Mary's Lodge,
York.
19 September 1845

I should be glad of the assistance of James for the next quarter. I enclose a letter to him. I have been fortunate in tracing the toadstones on the hills. From what you say the first territory done in Ireland will be a productive one. Come via York to Jarrow.

1587 **Phillips**, John To: H.T. De la Beche
[30 September 1845]

Travelling charges for the Geological Survey of Yorkshire and Derbyshire July-September 1845, amount to £1.15.6.

1588 **Phillips**, John To: H.T. De la Beche
St. Mary's Lodge,
York.
31 December 1845

In looking through the manuscript I find proper subjects for wood-cuts, but will collect them all together first. The Broad and Narrow Gauge Commissioners are here. Airy has returned from his Darlington trip. The speed and load were less than on the Great Western. However, better to tide it on Pony among the Welsh Hills at 5 miles per lawful hour, than to be smashed to atoms at 57½.

1589 **Phillips**, John To: H.T. De la Beche
St. Mary's Lodge,
York.
6 January 1846

I glad you had a merry Christmas. I enclose receipt in amended form. We are to see a huge fish head from the Speeton Clay today at our Philosophical Meeting. The man who found it asks £40. I hear James has made good progress round Worksop.

1590 **Phillips**, John To: H.T. De la Beche
St. Mary's Lodge,
York.
10 January 1846

Thank you for your very full pages. Will you not accompany this with a map, like that of Devon & Cornwall? I wish 'horizontal sections' had been called 'sea-level sections'.

1591 **Phillips**, John To: H.T. De la Beche
St. Mary's [Lodge,
York].
2 February 1846

You will see by the enclosed letter that James seems not quite so ill provided as I fancied. I hope he will listen to my advice. I enclose a brief report of my progress. I think we should have Inspectors of Mines to prevent accidents.

1592 **Phillips**, John To: H.T. De la Beche
St. Mary's Lodge,
York.
5 February 1846

I have just remembered that you said I had marked conglomerate on my map at Little Malvern. This is entirely a mistake. There is no conglomerate there. I sent a batch of manuscripts to the printer yesterday.

1593 **Phillips**, John To: H.T. De la Beche
St. Mary's Lodge,
York.
23 March 1846

I am glad you approve of the commencement of Malvernia. I don't think there is likely to be much advanced that you will not agree with. Malvern is an example in the east of what you are doing in Wales.

1594 **Phillips**, John To: H.T. De la Beche
York.
2 April 1846

I am obliged to send the second copy of the cut, as the first is with Martin for correction. In 10 days I go off to Derbyshire. Thanks for the hint about the Railway. Should not the Survey be a court of appeal for speculators as well as a Government enterprise?

1595 **Phillips**, John To: H.T. De la Beche
Green Man,
Ashbourne,
Derbyshire.
14 May 1846

I have made some cuts to set James on his line about Derby. We are on New Red here. The limestone is 3½ miles off at the nearest point.

1596 **Phillips**, John To: H.T. De la Beche
18 May 1846

A day of rain, after fine weather, allows me to look through proofs which I send to be made into pages. I shall try to get to the Isaac Walton Hotel, Ilam tonight. Ashbourn Post Office remains my address though.

1597 **Phillips**, John To: H.T. De la Beche
Isaac Waltons Inn,
Ilam,
Ashbourne,
Derbyshire.
20 May 1846

The weather is very rude and stormy. This is one of the small fishing stations on the Dove but the cold weather has kept the gents away. You would be interested in the subterraneous courses of the rivers here.

1598 **Phillips**, John To: H.T. De la Beche
Isaac Walton Hotel,
Ilam,
Ashbourne,
Derbyshire.
23 May 1846

I have found several fossil specimens in the limestone here. Thank you for sending me Salter's paper. I have never had such help as this. The limestone boundary is more difficult than expected, and the strata are wonderfully crumpled. Axes not wider than Craig's Court, nor as long as one of Brougham's speeches.

1599 **Phillips**, John To: H.T. De la Beche
Walton Hotel,
Ilam,
Ashbourne.
26 May 1846

I think of leaving here on Friday morning. My next address will be Buxton. I have gathered a collection of fossils today.

1600 **Phillips**, John To: H.T. De la Beche
Isaac Walton Inn,
Ilam,
Ashbourne,
Derbyshire.
28 May 1846

James has finished in Nottingham and is on his way to Derby. I need duplicate maps on which Belper and Ashbourn are situated. Please arrange.

1601 **Phillips**, John To: H.T. De la Beche
3 Hall Bank,
Buxton.
2 June 1846

I hope now to clear up a great piece of the N.W. corner as to its detail, and have found lodgings for myself and my sister. Would you and Forbes come down to look at things here? One can radiate in all directions by coach in summer.

1602 **Phillips**, John To: H.T. De la Beche
3 Hall Bank,
Buxton.
7 June 1846

The weather is very hot. I have been examining the trap rock here. Some will say I have made these long and weary lines of trap to exalt my own credit in greenstone and amygdaloid. I hope you will defend me from this imputation. I hope you can bring Forbes with you when you come to visit us here.

1603 Phillips, John To: H.T. De la Beche
Buxton.
23 June 1846

Thanks for the maps which have arrived. The weather has turned foul, and I have completed another sheet of Type for the printers. I have found more trap in this region.

1604 Phillips, John To: H.T. De la Beche
York.
13 July 1846

After inspecting Longnor, I fell and bruised my legs. I have come home for a day, and will then visit Southampton. I don't suppose you will stay in Ireland long. I enclose the account of travelling expenses.

1605 Phillips, John To: H.T. De la Beche
Castleton,
Derbyshire.
30 July 1846

I am here for a few days to complete some fragments of work. I have several points to revisit. The Romans worked the mines here. My legs are all right again.

1606 Phillips, John To: H.T. De la Beche
Baslow,
Derbyshire.
7 August 1846

I am gradually working my way into the true coalfield, and mean to acquire the whole section before meeting you at Southampton. I find it will be quite practicable to do one general section across all Derbyshire.

1607 Phillips, John To: H.T. De la Beche
York.
17 August 1846

Thanks for the proofs which I am now revising, but I am hampered by the lack of woodcuts which were not sent by the Printer. I suppose you will be anxious to get on with the second volume.

1608 Phillips, John To: H.T. De la Beche
St. Mary's [Lodge],
York.
21 August 1846

Allow me to send these sheets to you, as 3 woodcuts done by Martin do not appear to be in the Printer's hands. Martin would soon put it right, if the Printer really has not got the blocks.

1609 Phillips, John To: H.T. De la Beche
Kidderminster.
25 September 1846

I hope to be at Ashbourn on Tuesday. Have you anything for me to say to James? I found fossils on the Worcester road. Williams is wrong about his inferences from the Uphill traps. Horner has told me of the general desire that you accept the Presidency of the Geological Society. You must not fear any misconstruction and not decline the office.

1610 Phillips, John To: H.T. De la Beche
York.
6 October 1846

I was glad to hear from you in Wales. If the result of your enquiry gives Sedgwick something below the lowest depth of Lower Silurian I shall be rejoiced. The completion of the Lower Palaeozoic Series requires a Cambrian formation below the Murchisonian groups. I suppose Ramsay will explain to James what he is to do. Charlesworth has published the first volume of his London Geological Journal. Sedgwick is in S. Wales.

1611 Phillips, John To: H.T. De la Beche
St.Mary's Lodge,
York.
4 November 1846

I have been recovering from influenza. Today I shall send up to Reeks the remaining part on Malvern. I am charmed with your progress in Ireland. Regards to Oldham. Ramsay has sent me the Lingula.

1612 Phillips, John To: H.T. De la Beche
St. Mary's Lodge,
York.
25 November 1846

I will be in Town on 2 December for the Association Council meeting. I shall bring my manuscript and drawings. Have you found some Spirifera in the Llandeilo fossils? I saw some in Bowman's collection marked as from there.

1613 Phillips, John To: H.T. De la Beche
Baslow,
Derbyshire.
1 December 1846

I have been working despite the weather. It is awkward to discriminate various formations. I shall try a little more about Chesterfield. The section across Derbyshire will be a pretty thing.

1614 Phillips, John To: H.T. De la Beche
St. Mary's [Lodge],
[York].
4 December 1846

I have finished at Chesterfield and am now working on the Memoir. It seems we must outline our new species in this volume. I shall send my drawings in a week, and those of Forbes and Salter can be added to make a grand total of novelty.

1615 Phillips, John To: H.T. De la Beche
St. Mary's [Lodge],
York.
20 December 1846

Thanks for the notice of Sedgwick's field day. A plague on all names, more particularly hard ones, and I am glad Sedgwick and Murchison do not use such mutually. Forbes tells me that Salter will work for Sedgwick for the next two months. Is it worthwhile for me to press him at all at present on my Lists? I enclose a note to Salter.
[See Phillips, J. to Salter, J.W., 20 December 1846, **1616** for enclosure].

1616 Phillips, John To: J.W. Salter
St. Mary's Lodge,
York.
20 December 1846

Forbes tells me you will be completing your work for Sedgwick during the next 2 months, and that afterwards you will be doing Welsh Palaeontology for the Survey. Baily might outline the new species you have to propose. I would inspect the results, take notes, and return to the engraver. What say you?
[Enclosed with Phillips, J. to De la Beche, H.T., 20 December 1846, **1615**].

1617 Phillips, John To: H.T. De la Beche
St. Mary's Lodge,
York.
20 January 1847

I shall be glad if you will try your hand at Celtic names, Ankerdine and Yartleton. Is not Cader a seat? I believe there is a whole shoal of Celtic names around Malvern which ought to be restored. I wish some definite decision would be made on Cambrianism, but am content to wait if Sedgwick and Murchison will let us.

1618 Phillips, John To: H.T. De la Beche
St. Mary's Lodge,
York.
11 March 1847

I left Town early to look at a point near Abberley. Let me have as much time as may be managed for revising the last sheets sent. I shall stop on my way to Swansea. Enclosed is last proof slip, with cuts and references, and also the first essays on the Palaeontology.

1619 Phillips, John To: H.T. De la Beche
Baslow,
Derbyshire.
23 May 1847

I have no doubt that the remaining type for the Memoir will reach 200 pages. It would be well to have the printers send any of the slips relating to the second portion, so as to be able to make an estimate. Space could be saved.

1620 **Phillips**, John   To: H.T. De la Beche
Baslow.
26 May 1847

Were you to come on a visit, I would show you the wonders here. It is evident that we have done two or three cuts twice over, and one has come all too late. When will you come?

1621 **Phillips**, John   To: H.T. De la Beche
Baslow.
7 June 1847

I return all the proofs for your consideration, having introduced notices of the new species which I was to describe. I think it will be best to leave these descriptions out till the end. More MS tomorrow.

1622 **Phillips**, John   To: H.T. De la Beche
Baslow.
11 June 1847

I return the proof to serve as a model for the printers to work by. I suggest the title of Palaeontographical Appendix. I have more MS ready. Be at the trouble to read my note to Salter.

1623 **Phillips**, John   To: H.T. De la Beche
Baslow.
12 June 1847

Did I tell you that when Hunt showed me the coloured Llandeilo sheet, I noticed a mistake, which should be rectified. I had not seen the coloured sheet until the other day. I go to Oxford on 14th.

1624 **Phillips**, John   To: H.T. De la Beche
Ashford but address Baslow.
1 October 1847

Congratulations on your well recovered health. I have climbed Axe Edge and Chelmorton and am still measuring toadstones. Enclosed a recipe for good health [in form of a poem].

1625 **Phillips**, John   To: H.T. De la Beche
Baslow again.
3 October 1847

I hope Salter won't drive me mad with his (unavoidable) fluctuations of opinion touching the few dubious things in my list. Still I am much obliged to him for the trouble he is taking. I have measured the toadstone here. Hopkins' brochure amounts to reasoning rightly on wrong data.

1626 **Phillips**, John   To: H.T. De la Beche
3 November 1847

I am now near the end of my Malvern work, and hope to return the remaining proofs tomorrow, followed by the Palaeontological Appendix.

1627 **Phillips**, John   To: H.T. De la Beche
[1847 FJN]
[Part of letter only]

Upon the whole I begin to think that we have had the best of the weather lately. It was unlucky that you had so much rain and that by your illness I lost the advantage of seeing more country with you. I have sent to Reeks all the quarter's papers except travelling charges.

1628 **Phillips**, John   To: H.T. De la Beche
[St. Mary's Lodge,
York].
16 January 1848

I have come home to my nurse and gruel, hoping to overcome the effects of influenza. I return all the proofs. In concluding my own particular contribution I wish to express the strong feeling of obligation to you as Director and Editor for your innumerable courtesies. If you find time do look at the bust of W. Smith. I must see if I can afford the marble for it.

1629 **Phillips**, John   To: H.T. De la Beche
26 January 1848

I return my copy of the Map. I note Lowry has left in some imperfections which if not remedied will destroy all chance of the colourer's being accurate. I return your copy for comparison.

1630 **Phillips**, John   To: H.T. De la Beche
28 January 1848
[Part of letter only]

I did not reply to your question touching my little basket of fossils intended for Edwards because I thought it certain he would send for them as he seemed so anxious about them. I have been putting the Derbyshire work a little together, but my doctors require me to be moderate in labour of all sorts except that of the jaws and legs. The whole section is drawn, except 6 miles at the end.

1631 **Phillips**, John   To: H.T. De la Beche
[St. Mary's Lodge,
York].
6 February 1848

We did talk of engraving Plate II (the long sheet of Sections) but you said it might affect the sales of the great sheets. An errata list can be added to adjust the Plates done. If you wish me to finish Plate II I will do so.

1632 **Phillips**, John   To: H.T. De la Beche
2 March 1848

I have nearly completed the section, and wish to go to Chesterfield to make sure of two or three places. I am putting a good deal of explanation on the section as it may not be long before a Memoir is ordered.

1633 **Phillips**, John   To: H.T. De la Beche
7 March 1848

I have drawn up the little report, and if I can get some official paper, will send it with this. Do you remember promising to send me the square of sheets to colour? Also to get the districts of Matlock and Tideswell on a 2 inch or larger scale?

1634 **Phillips**, John   To: H.T. De la Beche
30 March 1848

Thanks for the very complete address, and also for the map which I am now colouring. I will leave for Burton-on-Trent on Monday to do the southern boundary. I plan to finish sheets 71, 72, 81 & 82 by next March at latest.

1635 **Phillips**, John   To: H.T. De la Beche
Burton on Trent.
3 April 1848

The heat is intense for April. Where did the New Red get these pebbles from? I dare say many think the Geological Survey is too early, but if the theory and laws of geology had existed 100 years ago, they would have saved a world of blundering loss and ruin.

1636 **Phillips**, John   To: H.T. De la Beche
Post Office,
Loughborough.
27 April 1848

I gather from your letter that I coloured the Lias too heavy and blue. Today I go to Milton, and then to Bunney. I need the quarter sheet with Nottingham and Bingham on it. If at hand could you send it.

1637 **Phillips**, John   To: H.T. De la Beche
Stockport Post Office.
23 May 1848

I am now in Stockport, and will go to Macclesfield to continue the New Red boundary. I am satisfied I can correlate the Derbyshire Carboniferous groups to those of Yorkshire. The Sheffield and Manchester line crosses the Kinder axis with great boldness.

1638 **Phillips**, John   To: H.T. De la Beche
Stockport.
25 May 1848

I have not been able to map any lines here because of the extent of drift. What is your view? I presume you will give us a colour for it, or some heteromorphous dottings. Tomorrow I go to Macclesfield.

1639 **Phillips**, John   To: H.T. De la Beche
Derby.
23 June 1848

I have to go to Swansea for 4 or 5 days. I don't wonder therefore at your feeling a little weary of our Parliament of Science which can only meet in summer and interrupts business. I find my uncle's coal sections of great value here.

1640 **Phillips**, John   To: H.T. De la Beche
Ashbourne,
Derbyshire.
13 July 1848

I fear Grove might mistake your consent to give a paper to the Geological

Section for a consent to give an Evening Discourse. I've asked Percy to look into the Metallurgy of the Swansea Valley. A curious bit of country here, at the junction of New Red and Millstone grit.

1641 Phillips, John To: H.T. De la Beche
Baslow.
8 October 1848

I have finished surveying the tail end of the 30 mile section, and will start drawing when I get home. Is there any hope of your coming here? Tomorrow I go to work in the south.

1642 Phillips, John To: H.T. De la Beche
St. Mary's Lodge,
York.
29 October 1848

I took your advice and came home to work on the maps. I have to take my sister to Bridlington Quay, because of her health, and will take work with me. She thanks you for the drawings.

1643 Phillips, John To: H.T. De la Beche
St. Mary's Lodge,
York.
2 April 1849

I am leaving for the country between Chesterfield and Nottingham. I have planned 72 miles of sections. Reeks says you have inspected the bust of William Smith. I believe that only in Noble's hands could anything have been made of my poor uncle's head because he knew him at Harkness.

1644 Phillips, John To: H.T. De la Beche
York.
11 April 1849

I have just returned from Nottinghamshire, where James' lines have been inspected, partly with praise, partly with censure. I expect to be in Town on 19th, and have a question or two about the colouring of our map. Hudson seems smashed as a Chairman of Railways.

1645 Phillips, John To: H.T. De la Beche
St. Mary's Lodge,
York.
18 May 1849

I fell over a hedge yesterday, but no harm done. I have gathered a good deal of matter for the vertical section. The eastern border lines are also complete.

1646 Phillips, John To: H.T. De la Beche
Kedleston Inn,
near Derby.
26 May 1849

I have been working along the beautiful new red boundary. I recommend the Inn here, but not on or about Rent day when it is overcrowded. Red sandstone rests on Limestone shale and yields good water.

1647 Phillips, John To: H.T. De la Beche
St. Mary's Lodge,
[York].
27 June 1849

I suppose the accompanying letter will meet the case of petition for release. I will send the maps and sections to the Office, and am sorry not to be able to say I finished those four sheets myself. Another explosion! At St. Helen's.

1648 Phillips, John To: H.T. De la Beche
York.
10 October 1849

I have spent time colouring the Derbyshire sheet. There is not much work remaining to be done on it. Barker says he is satisfied there is no fault near Bakewell. Money is due to me on account of travelling charges for the June quarter.

1649 Phillips, John To: H.T. De la Beche
St. Mary's Lodge,
York.
2 December 1849

I have just returned from Newcastle, where I have seen a great deal of the mines and men. Yorkshire I will keep for winter work nearer home. I have now surveyed 2 collieries where Davy's lamp is exclusively used, others where mixed, others where no lamps used.

1650 Phillips, John To: H.T. De la Beche
St. Mary's Lodge,
York.
15 December 1849

Blackwell is here to compare notes with me. I am sure that if empowered we can do good. How to constitute the future commission will be a serious problem for you. I think the Jermyn Street mining school will be effective. There is a considerable feeling for local or district schools.

1651 Phillips, John To: H.T. De la Beche
York.
30 December 1849

Are the arrangements for the London Mining School likely to suit my friends the Viewers? They send their sons to Kings College, University College, Durham etc. How much better will our plans be?

1652 Phillips, John To: H.T. De la Beche
York.
14 January 1850

Blackwell has looked at collieries in the Newcastle district and has just left for Manchester. I think your Mining School will be well supported by the North. We agree in advising Provincial Schools. The latter should be visited by the Mining Commission.

1653 Phillips, John To: H.T. De la Beche
St. Mary's Lodge,
York.
13 March 1850

It is a long time since you have written. I have come to the conclusion that we ought to provide instruction on Mining Subjects in the districts. I am preparing the Report.

1654 [Phillips, John] To: H.T. De la Beche
St. Mary's Lodge,
York.
23 September 1850

[Part of letter only]
I still am owed £5.7.3 for travel expenses from the June quarter last year. Please send it. As regards the Geological Survey, by sending scraps of information I am in credit, but there is little relating to the series between the Millstone grit and Black shale coal to copy or extract.

1655 Phillips, John To: H.T. De la Beche
St. Mary's Lodge,
York.
27 September 1850

I have received the cheque for £5.7.3. I did not suppose you would have to alter the limestone boundary about Matlock. By all means look at the curious outline by Hopton. I must be in London in October and I can show you dips marked on my map. I have been paid off as regards Colliery investigations.

1656 Phillips, John To: H.T. De la Beche
St. Mary's Lodge,
York.
29 September 1850

Here is copy from the Home Office which seems to indicate my full discharge from the Colliery ventilation enquiry. I have not heard from Blackwell for some time. A letter from Ramsay tells me Oldham is to be married.

1657 Phillips, John To: H.T. De la Beche
22 October 1850

The Home Office wrote to me and asked me to survey a colliery which I had previously done last June and in which subsequent burnings have occurred. I asked for a fee, but they refused. I therefore have respectfully declined to perform the service.

1658 Phillips, John To: H.T. De la Beche
Magdalen Bridge,
Oxford.
1 February 1854

I will send your note to T. Wilson in Leeds who will help you over the pottery. Are there any spare copies of our Malvernia which I could give to two or three colleges here? My inaugural lecture is on Friday.

1659 Phillips, John   To: H.T. De la Beche
Magdalen Bridge,
Oxford.
8 February 1854

Thanks for the Volume and Medal. I have over 40 students, and have now delivered two lectures. There is no end of Museum work before me, health and finance permitting. When will you come to visit us?

1660 Phillips, John   To: H.T. De la Beche
Southampton.
8 September [No year]

Come to me at Mackay's, 7 Portland Terrace. Take the bedroom on the drawing room floor.

1661 Phillips, John   To:[H.T. De la Beche]
[No date] [possibly 1847]
[Part of letter only]

I hope that in respect of the tables which follow each list, there will be care taken to correct the numbers, for any omissions and additions. I dare say Mr. Salter will extend his kindness so far as to look well to this. I think you can now let it be put in sheet and page and I will not delay a single part in returning whatever you send.

1662 Phillips, John   To: [H.T. De la Beche]
[No date] [possibly 1847]
[Part of letter only]

Salter, with the help of Forbes, should still look carefully at my list of species to correct to the newest fashion genera and species and to add new genera and species in every district except Builth and N. Wales. I will then recompose my lists for the printer. As a preliminary step to hasten work I send herewith the whole of the manuscript which remains to precede the said lists.

1663 Phillips, John   To: [H.T. De la Beche]
[No date] [possibly c.1842]
[Part of letter only]

I am glad you could spare Ramsay and James for a fossilization. The findings in the Limestone of Mydrim are new to me. James knows of other points near St. Clears interesting to work, and much good will come of it. Here I am bringing the lines to bear and have nearly closed the whole boundary of the older rocks as well as traversed sufficiently the north east corner of your district. It is a pretty bit of mapping. These are those limestones more uniformly developed than Murchison's map shows.

1664 Phillips, John   To: [H.T. De la Beche]
[No date] [possibly c.1841]
[Part of letter only]

I was about to request Rees to go up to Llanboidy from St. Clears, which if you are not likely to have much more time in this Silurian region might perhaps be worth doing. I have got some more trap west of Dynevor Park and a mighty long line of jumble - a dislocation band 200 yards wide through the limestone series or north of it. James is going to Llandovery tomorrow and I am going towards Llampeter.

1665 Phillips, John   To: [H.T. De la Beche]
[No date] [possibly 1843]
[Part of letter only]

Crossing of letters is a bad contrivance. Archer's work is not very good, I think, nor indeed is it his own. But if you send the drawings tomorrow I can add two or three more for Lowry to work on, and we can see who is the best and most reasonable in charges. I have put out all the Irish fossils. They are very satisfactory. One or two errors of nomenclature probably accidental, but much proof of industry.

1666 Phillips, John   To: [H.T. De la Beche]
[No date] [possibly c.1841]
[Part of letter only]

If the thing suits you, pray name your own day. I shall in the meantime get in a good deal of what remains on the North. Hereabouts are two or three abandoned mines. James says some lead was found at Llanboidy north west of St. Clares.

1667 Phillips, John Arthur   To: [H.T. De la Beche]
Museum of Practical Geology.
27 August 1847

I thank you for the £20 sent for the purpose of paying the current expenses of the Admiralty Coals Investigation.

1668 Phillips, John Arthur   To: H.T. De la Beche
Museum of Practical Geology.
25 October 1847

I have received £10 from Cox & Co. to pay for the Admiralty experiments on coal.

1669 Phillips, Richard   To: H.T. De la Beche
Museum of Practical Geology,
6 Craig's Court,
Charing Cross.
14 May 1847

Regarding the imputation that my charge for doing analysis for Butler (3 guineas) was too high, I have ascertained the charges of Playfair, Miller and Campbell. The first would charge five guineas, and the other two named six guineas.

1670 Phillips, Richard   To: H.T. De la Beche
London.
17 September 1847

It has been proposed that I should inspect the progress and results of an experimental operation on copper smelting in Swansea. I have written to Lord Morpeth for leave of absence. Will you permit my accepting the commission?
[For reply, see De la Beche, H.T. to Phillips, J., 19 September 1847, **411**].

1671 Phillips, Richard   To: H.T. De la Beche
London.
10 August 1850

The ores sent by the Colonial Office have been analysed and the following are the results. Lead ore has no silver, but gives more than 86.6% of lead. Two specimens of copper ore are not worth analysis, but the third yields 18.54%. However, its gangue is very hard quartz.

1672 Phillips, William   To: H.T. De la Beche
Tottenham.
14 July 1825

Re your offering me the publishing of the Geology of Jamaica on the same terms as the Translations, considering the number, size, high finish etc. of the plates, sections etc., the price would have to be unusually high to cover expenses. Thus sales would be dampened. With regard to the Translations, the sale has not been large, yet my disappointment will be trifling.

1673 Phipps, Charles Beaumont   To: Gardiner
Windsor Castle.
24 December 1851

The Prince would be glad to see De la Beche and yourself on Saturday evening between 4.30 and 5.

1674 Phipps, Charles Beaumont   To: H.T. De la Beche
Windsor Castle.
2 January 1852

The Prince is glad to hear of Seymour's acquiescence in the employment of Smyth in the Superintendance of the Mining possessions of the Duke of Cornwall, and thanks you for your assistance in the matter. You are also authorised to announce that the Duke of Cornwall will give two annual exhibitions to the School of Mines at £30 each. Please draw up rules for competing for them.

1675 Phipps, Charles Beaumont   To: H.T. De la Beche
Buckingham Palace.
20 February 1852

Please send me some formal paper stating the rules under which the Prince of Wales' Exhibitions should be contended for, so that I can send an official minute for the Queen's sanction.

1676 Pickersgill, Henry William   To: H.T. De la Beche
Soho Square.
15 April 1846

Last night I proposed your being invited to our academic dinner. What is the exact way you wish to appear in our catalogue?

1677 Pickersgill, Henry William   To: H.T. De la Beche
Soho Square.
28 January 1848

I shall be most happy to place the portrait in your hands on terms much below my price, as I am anxious to place it where more care will be taken of it than what I can give in my gallery.

1678 Piddington, Henry To: H.T. De la Beche
Calcutta. Museum of Economic Geology,
2 July 1842 6 Craig's Court,
Charing Cross,
London.

Thank you for the flattering notice of my report. Lord Auckland appointed me curator of the Museum before his departure. There is so little done here. Previous work has not been published and acted on, especially Herbert's Survey for the Marquis of Hastings, and Voysey's work. I have reported on the adulteration of salt for fraudulent puposes. Mining legislation here prevents the development of mines. I am carrying out a survey on a deposit brought down by the Ganges.

1679 Piot, F. and **Meissonnier**, To: H.T. De la Beche,
Jean Louis Ernest Collenna,
Swansea. Newbridge,
20 July 1840 Glamorganshire.

Unfortunately we had already left Merthyr when your letter arrived. It would have been a great help to us when in that locality where we were given little help even by people to whom we had been recommended. We were unable to see Logan in Swansea as he was away. Mr. Vivian is not at home either, but we were given every facility for our studies in his factory. We will use your letter of introduction to Mr. Dillwyn and we shall soon leave to study the Silurian area.

1680 Piot, F. To: H.T. De la Beche
Glasgow. Ordnance Map Office,
4 October 1840 London.

When I saw you in Glasgow you offered to give me an itinerary through Devonshire that would best show the Culmiferous system. Would you write your recommendations to me at the Post Office, Ilfracomb. Can I borrow a copy of your Report at Ilfracomb or Bristol? Meissonnier has gone back to France after we visited the Highlands together.

1681 Piot, F. To: H.T. De la Beche
Barnstaple. Cardiff.
18 October 1840

I arrived at Ilfracomb on 16 October and found your letter the next morning when the post arrived. I met the Marquis of Bute in Bristol. I have already begun to follow your suggestions, but will then change them a little to meet John Taylor in Cornwall who will be there at the end of the month. I hope that the political dissension will not interfere with my travel plans. I shall be in London towards the end of November.

1682 Piot, F. To: H.T. De la Beche
Nantes. Ordnance Map Office,
20 September 1841 London.

It is with pleasure I renew my acquaintance with you, having left England without seeing you again. I have since written to Jordan, late Secretary of the Polytechnic Society of Cornwall, asking him to procure several things from you for me. He has not answered. I am working at Nantes on the Transitional formations of the Basse Loire, and the analogy between this and the Devonian has struck all geologists. I would like to purchase your map of Devon and Cornwall.

1683 Piot, F. To: H.T. De la Beche
Nantes. Ordnance Map Office,
24 October 1841 London.

Brittany, in spite of its structural resemblance to Cornwall, is very poor in metalliferous veins. However I would like to believe that in this respect its wealth is very little known. We need here some of your daring speculations to give us the courage for undertaking great enterprises, and compel us, so to speak, to know our own wealth. In accordance with your instructions, we are waiting two or three months for the map of Devonshire and Cornwall.

1684 Playfair, Lyon To: H. Hobhouse
28 Moray Place,
Edinburgh.
5 May 1844

I have been very busy canvassing in the Council Election. After the 14th, I will give my time to the completion of the Bristol and Lancashire reports. Gregory has recovered his health and will probably get the position, the best chemical appointment in the kingdom. I am pleased with the report from the Commission, especially Clark's evidence on water.

1685 Playfair, Lyon To: H.T. De la Beche
20 Cooper Street,
Manchester.
17 August 1844

Liebig's packet should be sent to his publishers who have constant opportunities of transmitting things to him. I suppose sanitary matters of Bristol are coming to an end. Please send me a copy of the form you have adopted with Registrars in saddling disease to localities, as it would assist me here.

1686 Playfair, Lyon To: H.T. De la Beche
Royal Institution,
Manchester.
23 November 1844

It is important you join us on Saturday. Lincoln wishes us to draw up a separate report on the London sewerage, in order to pass legislation forthwith. You know how this would be inexpedient, and you have more influence with Lincoln than all of us put together. I have written Chadwick and Smith, and I trust they will both be there. Let us remember the fate of the Building Bill last year.

1687 Playfair, Lyon To: H.T. De la Beche
25 May 1845 6 Craig's Court,
Charing Cross.

Analysis of artesian well water, Trafalgar Square. Besides fixed ingredients it contains 23 cu.ins. carbonic acid per gallon. It differs from Paris water by being free from salts of Potash. The Trafalgar Square water is remarkable for its large quantity of soda salt. Hardness is 5°.

1688 Playfair, Lyon To: H.T. De la Beche
Birmingham.
24 September 1845

For the last week I have been nearly idle at Riddings House. The Iron Report had to be finished, and also I was knocked up by that confounded Jarrow. I join Williams tomorrow. I hope you will be over soon, as we can go to Birmingham and Newcastle together to get information. I am doing nothing on smoke, as the Mining enquiry is sufficient for the time.

1689 Playfair, Lyon To: H.T. De la Beche
Museum.
28 August 1846

The enclosed letter reached me today, and in your absence I replied, stating that we would not want to undertake a private engagement which might later on come under public duties as result of the Commission into Private Water Bills. I have not forwarded the letter to Morpeth until I hear from you that I have done right.

1690 Playfair, Lyon To: H.T. De la Beche
London.
6 November 1846

The expenses attending fitting up the boilers for the Admiralty Coal Enquiry have so far swallowed up our funds, that without a guarantee from you as Director that any outlay beyond the vote will be made good to me, we must dismiss Wrightson, and much of the work will have to be done at the Museum. Are you in a position to afford me the necessary guarantee?

1691 Playfair, Lyon To: H.T. De la Beche
Leeds.
19 May 1850

Birmingham presses upon the Prince the view that the producer should be obliged to put his name on all articles exhibited. The Prince and the Commission wish this to be optional. A hearing will take place in June, and H.R.H. has asked me to spend the intermediate time in gathering support among the other towns for the Prince's views. I will thus be out of town this month, more than I would wish.

1692 Playfair, Lyon
[May 1850]

Playfair gave 2 lectures per week at the College for Civil Engineers, Putney till March 1850, when he resigned the lectureship. Since April 1850 he acted as Special Commissioner to the Exhibition of 1851, and has provided a substitute for duty at the Museum. However he still attends the Museum for a considerable part of the day.

1693 Playfair, Lyon   To: H.T. De la Beche
Department of Science and Art,
Marlborough House,
Pall Mall,
London.
2 January 1854

The Department is pressed to announce its intentions as to training teachers in Science. I have drawn up the enclosed draft scheme for your consideration before submitting it to Cardwell for approval. The scheme would be required to start in February. Will you ascertain the opinions of the other Professors. An early consideration of it is very important.

1694 Porter, G.R.   To: H.T. De la Beche
Board of Trade,
Whitehall.
15 February 1848

You promised in November to send me an account of our production of lead. I have not yet had it. Can I also have an account of Copper production since 1840?

1695 Porter, G.R.   To: H.T. De la Beche
Board of Trade.
8 April 1848

I send you the extract about the American copper mines from Le Play's letter, and shall be glad to hear what you have to say to it.

1696 Portlock, Joseph Ellison   To: H.T. De la Beche
Belfast.   British Association,
31 August 1839   Birmingham.
[Readdressed to:] Office of Woods & Forests, London.

I have heard only last night of your presence at the Association meeting. I greatly wish to see you. Can you come on to Belfast and pay us a visit? What do you think of Murchison's views?

1697 Portlock, Joseph Ellison   To: H.T. De la Beche
Portrush.
22 August 1841

Many thanks for remembering me with reference to Phillips' valuable work. The more I look into fossils, the more I see the difficulty of identifying forms of distant localities and the dangers of drawing too rapid conclusions from them. I was glad to see you in your rightful place as Head of the Geology section at Plymouth.

1698 Prado, Casiano de   To: [H.T. De la Beche]
Madrid.
2 July 1852

I have finally returned to Spain where I await your orders. Your good friend Pentland has recommended that I send you nickel minerals of Carratraca near Malaga and Cape Ortegal. They will be sent by boat which leaves on the 15th of the month. I have also written to Riotinto to have prepared minerals and products for you.

1699 Pratt, Samuel Pearce   To: H.T. De la Beche
27 Coleman Street.
[1849 FJN]

Herewith an abstract of my son Duncan's letter and a copy of his drawing of animal imprints on beds of sandstone in India.

1700 Pratt, Samuel Pearce   To: H.T. De la Beche
Barcelona.   Museum of Economic Geology,
2 February 1851   Jermyn Street, London.

As you intended to quote my observations respecting the recent upheaval of the granite in this part of Spain, I think it right to give you the result of more careful examinations which I have made during my present sojourn here. I find the statements which have been made, the result of which appears in that map which Sir R. Murchison exhibited at Edinburgh are quite incorrect.
[Enclosure:] The general structure of the northern part of Cataluna. [Dated 1 February 1851].

1701 Pratt [?]   To: H.T. De la Beche
27 Claremont Square.
Monday [No date]

I have just found on my table a note from Sir F. Baring, and one from yourself, informing me that I am promoted. I sincerely thank you for this further proof of the lively interest you have taken in my cause.

1702 Prestwich, Joseph Jr.   To: H.T. De la Beche
20 Mark Lane.
6 November 1852

It was not until the middle of last month that I was able to visit Paris and investigate the costs of artesian wells. I have drawn up a document which I now forward, and will call on you in 2 or 3 days.

1703 Prevost, John Lewis   To: H.T. De la Beche
Gresham Street.
26 October 1846

Favre has written an article on the Geological Survey etc. and wishes you to read it before it is printed, so that any errors might be corrected. He wishes to know if the 6 inch maps are to be coloured geologically.

1704 Purdue, John   To: [W. Buckland?]
14 Hemingford Terrace,
Islington.
[1841 FJN]

I am willing to dispose of my fossil for £4.10. If you hear of anyone wishing to purchase specimens from my large collection, please have them tell me when they will call.

1705 Pusey, Philip   To: [H.T. De la Beche]
Spring Gardens.
5 May [?]1842

Lord Spencer would like some soils analysed. If Mr. Phillips cannot manage it perhaps Dr. Playfair would.

1706 Ramsay, Andrew Crombie   To: [H.T. De la Beche]
St. David's.
25 July 1841

I enclose a sketch of what has been done. The trap rock to the south is much more perplexing than the northerly ones. Comparison with Murchison's map. The schist and slate districts form the most fertile land. Barley is the preferred crop. Potato crops best I've seen. Do you approve of my proceedings? Tremenheere is still with me. Ramsey Island is yet to be done.
[Cited: Oldroyd 1991, p.414]

1707 Ramsay, Andrew Crombie   To: H.T. De la Beche
Fishguard.
15 September 1841

The trap on the east side of Porth Lisky would be regular granite if it had mica. I saw masses of granite at Abereidy Bay. The slates there sell at £1 per thousand. A knee injury troubles me. The tide affair is getting done. I have ordered the model of Arran for the Museum.

1708 Ramsay, Andrew Crombie   To: [H.T. De la Beche]
Fishguard.
29 November 1841

Both your letters came today. Please send a few printed Survey receipts. The weather is heartrending. My yesterday's trappean theory is upset. The two bits coloured trappean (Trefgarn ridge) are not disjoined. Weather interrupted work. I expect the rock will turn out something like the large new mass I put in at St. Davids.

1709 Ramsay, Andrew Crombie   To: H.T. De la Beche
Fishguard.
5 December 1841

I will arrange for your parcel tomorrow. Enclosed is Rees' map. Please trace in markings for me. I am colouring my own work, so as to make a "triplicate" copy. I did no work yesterday due to the weather. Cutting from Examiner on Lyell at Boston. I was sorry to hear of Chantrey's death. A great national loss.
[Cited: North 1936a, pp.73-74]

1710 Ramsay, Andrew Crombie   To: H.T. De la Beche
Cardigan.
25 March 1842

My success in fossils has been beyond my anticipation. Hammering them out was extremely arduous. I have damaged my finger. Encrinites, corals, trilobites were found. Jenkins is here and very useful. I will stay on for a while. I am ruined by Doctor's expenses.

1711 Ramsay, Andrew Crombie   To: H.T. De la Beche
    Llandeilo.
    20 June 1842

The postal order and receipts have arrived. We found shells, corals and trilobites today. The rocks give an appearance exactly same as at Newcastle Emlyn. This proves that there is a difference in conditions, not age.

1712 Ramsay, Andrew Crombie   To: H.T. De la Beche
    Pumsaint.
    2 August 1842

The Theodolite etc. arrived safely today. We had a most successful day at fossils yesterday. With nine hands at work, we filled a basket. Several new univalves. All the ladies deserve an especial vote of thanks.

1713 Ramsay, Andrew Crombie   To: H.T. De la Beche
    Pumsaint.
    [4 August 1842]

18 more days are needed to complete this district. Mr Johnes took me to a fine section north of Pumsaint where I found fossil fragments. Let me know where and when you wish me to move. Playfair's note says to let him know about models if you wished any more.

1714 Ramsay, Andrew Crombie   To: H.T. De la Beche
    Pumsaint.
    5 August 1842

I have cleared up the anomaly of the southern dip of the "Caradocs". The enclosed section explains the facts of the case. The fossils are so important and I will need more than the 18 days estimated. The whole of the formation could be shown to the sea at Cardigan.

1715 Ramsay, Andrew Crombie   To: H.T. De la Beche
    Pumsaint.
    7 August 1842

My last letter contained a section showing that the Brecfa and Berris conglomerates are the same as the Taliaris ones. A Mr. Lloyd wishes to show me lead on his property. Black Jack is found with it. What is Black Jack? I have seen nothing of Murchison yet.
[Cited: North 1936a, pp.70-71]

1716 Ramsay, Andrew Crombie   To: H.T. De la Beche
    Pumsaint.
    20 August 1842

The cheque for £21 arrived. The trigonometricals are completed and the theodolite despatched. Rees has forsaken me. I shall be alone here soon. A letter from Playfair tells of disturbances there.

1717 Ramsay, Andrew Crombie   To: H.T. De la Beche
    Pumsaint.
    30 August 1842

Thanks for your letters and for keeping your present country open for me. I am not sure how much longer I shall be here. James and I have got on splendidly. I think he thoroughly understands the grand features of the country. The Johnes have left. My stony heart is only altered on the edges by Miss Johnes. With £500 a year, there is no saying what a man might do, but with £150 he must sing dumb.
[Cited: North 1936a, pp.81-82]

1718 Ramsay, Andrew Crombie   To: H.T. De la Beche
    Pumsaint,
    Llandovery.
    1 September 1842

Playfair's letter is of the right sort. The sugar affair is of great importance, and I'm sure Sir Robert will see it in the proper light. Williams is with me today. Fossils found. I like James much. I shall miss the Johnes.

1719 Ramsay, Andrew Crombie   To: H.T. De la Beche
    Pumsaint,
    Llandovery.
    10 September 1842

Things are taking a queer turn. The "Caradocs" do not stop where I first supposed. My whole train of reasoning about the country between Caermarthen and Newcastle Emlyn and perhaps to Cardigan may be affected by this affair. Phillips says it should be worked out, although outside our district. What do you think?

1720 Ramsay, Andrew Crombie   To: H.T. De la Beche
    Pumsaint,
    Llandovery.
    15 September 1842

The whole of fossiliferous nature has gone to universal and everlasting smash. I found Noeth grug fossils in a bed of conglomerate. Truth is truth in spite of theories. Berris brook = Fan = Taliaris = Pont Lodis? I am out of receipts. How shall I word James' excursions?

1721 Ramsay, Andrew Crombie   To: H.T. De la Beche
    Llandovery.
    21 September 1842

There is no doubt that the Berris brook and Pentamerus beds are the same, rolling about a little. Your letter with the plans arrived. The thing shall be done, weather permitting. I am packing the fossils today. I leave for Trecastle on Friday, doing the Carmarthen Fans from here tomorrow.

1722 Ramsay, Andrew Crombie   To: H.T. De la Beche
    Trecastle.
    27 September 1842

Fan Gehirach done yesterday satisfactorily. At Capellante today. Clouds are interfering with sights. Tomorrow to Brecon. I may be some days at Abergavenny waiting for favourable weather. I have escaped from Llandovery without a wife, but with the usual reports which always accompany me.

1723 Ramsay, Andrew Crombie   To: H.T. De la Beche
    Rhiadr [?].
    3 November 1844

The line you have drawn suits me exactly. I have drawn it on your map in red. I have observed the deceptive appearances you mention as regards cleavage and joints. The mist has been very thick and holds us up. In 8 to 10 days we should be at Aberystwyth. Ask Gibbs to chronicle dips. No cash wanted. Direct letters from Tuesday on to Devils Bridge.

1724 Ramsay, Andrew Crombie   To: H.T. De la Beche
    Dolau Cothi.
    1 December 1844

Yesterday a very good section was run. Our line is now far beyond Llanddewi. Tomorrow I will take another transverse cut, & by the end of the week shall be at the sea. I have great projects to complete before Xmas. I cannot give you an address as I am constantly on the move. I will spend Xmas here.

1725 Ramsay, Andrew Crombie   To: H.T. De la Beche
    Tregaron.
    16 December 1844

The Llanwrtyd traps are done; I also cut a long section from Llanwrtyd to Tregaron. Mist and snow make difficulties. Work for Lowry will be completed in a couple of days.

1726 Ramsay, Andrew Crombie   To: H.T. De la Beche
    Kington,
    Herefordshire.
    1 May 1845

Bristow & I arrived here yesterday and commenced work. We have been comparing maps at James' lodgings. He has worked well, and things fit well. This country is Builth over again. I have arranged another sheet of sections.

1727 Ramsay, Andrew Crombie   To: H.T. De la Beche
    Machynlleth,
    Montgomeryshire.
    5 December 1845

Your letter telling of £93 has arrived. The country north and east of Cader Idris contains traps. They do not appear on Phillips' little map, nor on Greenough's or Knipe's. We found a few graptolites near Aberystwyth.

1728 Ramsay, Andrew Crombie   To: H.T. De la Beche
    Machynlleth,
    Montgomeryshire.
    20 December 1845

I got two letters from Bristow on the Hobhouse affair. Hobhouse has made a great blunder. Bristow is always a gentleman. Aveline will have a tough job of his country. I fear the Wenlocks overlap. Why not Silurians? Henfrey was unwell.

1729 Ramsay, Andrew Crombie To: H.T. De la Beche
Dolgelli,
Merionethshire.
15 May 1846

James has written to ask about the subject you spoke of. This is the most lovely country in the universe. Its geology is stranger than fiction. Great traps etc. Selwyn has done a great deal.

1730 Ramsay, Andrew Crombie To: H.T. De la Beche
Dolgelli,
Merionethshire.
17 May 1846

Henfrey has returned. Does Williams make his approaches with authority? When will Wyley be removed from the Irish Survey? Please pay £50 into my Survey account.

1731 Ramsay, Andrew Crombie To: H.T. De la Beche
Dolgelli,
Merionethshire.
19 May 1846

Would you give the Cader Idris trap any other name than felspathic greenstone? We found a lumpy, irregular - surfaced lava all broken and tumbled of itself. One specimen has great deal of lime in its pores.

1732 Ramsay, Andrew Crombie To: H.T. De la Beche
Llanbrynmair,
Montgomeryshire.
26 May 1846

This a perplexing country, but I think we have found the key. Henfrey shows considerable work capability. He ought be given a fair trial alone. There will be room for him if James goes in September. What is your opinion?

1733 Ramsay, Andrew Crombie To: H.T. De la Beche
Dolgelli,
Merionethshire.
30 May 1846

Aveline and I leave for Church Stretton on Monday. I am bothered by the geology here. What is the fine grained specimen composed of? Is it mostly felspar? All very like Trefgarn.

1734 Ramsay, Andrew Crombie To: H.T. De la Beche
Church Stretton,
Shropshire.
6 June 1846

Forbes' account has our volume nearly ready. Should Lowry not look to the question of my plates? Bone has the proof impressions. There are many problems here concerning old physical conditions.

1735 Ramsay, Andrew Crombie To: H.T. De la Beche
Church Stretton,
Shropshire.
10 June 1846

There have been many things to reconcile here, and no end of faults. The Caradocian Sea coast is not imaginary. The limestone at the bottom of the Wenlocks must have been beachy.

1736 Ramsay, Andrew Crombie To: H.T. De la Beche
Church Stretton,
Shropshire.
12 June 1846

If there is a Cambrian system, it is assuredly here. Here we have beds without fossils, of immense thickness. On Monday we work on their relation to the Bishop's Castle Llandeilo flags. Henfrey seems working tooth and nail.

1737 Ramsay, Andrew Crombie To: H.T. De la Beche
Church Stretton,
Shropshire.
13 June 1846

£320 will be required for the present quarter. Please send it to the National and Provincial Bank at Dolgelli. The weather is extremely hot.

1738 Ramsay, Andrew Crombie To: H.T. De la Beche
Bishop's Castle,
Shropshire.
21 June 1846

We go to Chirbury on Tuesday. There is a good deal to correct. We will not be at Kington for 3 or 4 weeks. Shall I run down to settle the accounts? Trimmer could come to London.

1739 Ramsay, Andrew Crombie To: H.T. De la Beche
Bishop's Castle,
Shropshire.
21 June 1846

I am making arrangements for the accounts. I have a bad cough at nights. The work of revision goes on. Aveline's work needs explanation but not alteration. Gibbs has been overconfident, and must be curbed. I have been discussing a joint paper with Forbes. James has written twice about his "dismissal".

1740 Ramsay, Andrew Crombie To: H.T. De la Beche
Dolgelli.
29 June 1846

Henfrey has been working strenuously lately. The change of Ministry will I presume keep you longer in London than one would wish. I leave here on Wednesday.

1741 Ramsay, Andrew Crombie To: H.T. De la Beche
Dolgelli.
30 June 1846

Thanks for the additional £100. Tomorrow I start for Chirbury, having a day's work with Henfrey en route. He has been working admirably of late.

1742 Ramsay, Andrew Crombie To: H.T. De la Beche
Leintwardine.
7 July 1846

The accounts have gone up to Reeks. The balance due to you is £91.8.1. My next address will be Kington. Trimmer goes to Cardigan. Murchison's Caradoc was an antique shore. I will map these shores soon. I shall be able to map a Wales more antique than our new red sandstone Wales.

1743 Ramsay, Andrew Crombie To: H.T. De la Beche
Kington,
Herefordshire.
12 July 1846

Please sign a requisition for Sheet 40 and send it to Trimmer in Cardigan. May he measure heights with a barometer? If so, shall I order one? Travelling charges have been high this quarter. The Captain seems a good riddance. I am glad you like Oldham so much. Forbes will be with me on August 1st at Bala.

1744 Ramsay, Andrew Crombie To: H.T. De la Beche
Pen-y-bont,
Radnorshire.
17 July 1846

Aveline & I came here yesterday. I have to help Gibbs. The ground north of this is the overlapping of Wenlocks over sandstone. How Murchison and Sedgwick missed it, I don't know. We shall have to add a colour under the Silurian. Dolgelli will be my next address.

1745 Ramsay, Andrew Crombie To: H.T. De la Beche
Pen-y-bont,
Radnorshire.
20 July 1846

I will have to work here a whole week. A considerable blunder has been made on the north. Selwyn is sending my mail from Dolgelli. The loss of time annoys me considerably. I hope to be at Bala by 1st August.

1746 Ramsay, Andrew Crombie To: H.T. De la Beche
Pen-y-bont,
Radnorshire.
25 July 1846

Aveline and I have been redoing this country. As a general average, Gibbs' line is half a mile to two miles out. I am bad at scolding. Gibbs will come with me to Bala. Henfrey has difficult country at Llanbrynmair. Trimmer is progressing. He has found a shell in till at Cardigan.

1747 Ramsay, Andrew Crombie To: H.T. De la Beche
Dolgelli.
5 August 1846

Oldham is just the sort of man for us. Henfrey has resigned. It is clear he used the Survey to train himself in order to go into business with his brother on railway cuttings. All are asking for copies of the Memoirs.

1748 Ramsay, Andrew Crombie   To: H.T. De la Beche
Dolgelli.
7 August 1846
Reeks tells me you are in London. Will you come to Bala with me? Selwyn has been puzzled by the traps. Will you approve the principle that the altered country be mapped as trap? Sedgwick has been here, in my absence. There is a considerable mass of rocks here anterior to the Llandeilos.

1749 Ramsay, Andrew Crombie   To: H.T. De la Beche
Dolgelli.
8 August 1846
Where shall I send the balance due to you? I like Oldham. His notes are hearty and cordial. Fossett told me they have cut some splendid lodes. Bristow has applied for a pay increase. He deserves it.
[Cited: Herries Davies 1983, p.141; Herries Davies 1995, p.32].

1750 Ramsay, Andrew Crombie   To: H.T. De la Beche
Dolgelli.
12 August 1846
I enclose a bank order for £83.6.11. I agree we shall not miss Henfrey. I fear I must pay for copies of my own work. I hope the building matter will go on smoothly now and you can visit me here. I am sure we have another coast for the Llandeilos. Smyth's account of the Cardiganshire mines will be a prime move.

1751 Ramsay, Andrew Crombie   To: H.T. De la Beche
Dolgelli.
14 August 1846
I know Jukes from his geological papers. He is just the sort of man we need, even though his letter is injudicious. We have enough field men if all work well. Do you know what Murchison is doing with his nephew? Bath is not a good school for a Geologist.

1752 Ramsay, Andrew Crombie   To: H.T. De la Beche
Bala.
8 November 1846
The Rhiwlas limestone is providing a harvest. I will try to have a report on this country ready for Tome 2. I cannot trace the Rhiwlas in some places. Forbes and I are working together on a series of reports. He is doing splendid work.

1753 Ramsay, Andrew Crombie   To: H.T. De la Beche
Bala.
2 December 1846
May I have leave for Christmas from the middle of the month? I am glad we did not do a Memoir of this country when Forbes was here as we are not ready for it. The Rhiwlas beds are higher than they appeared to be. The ground is clear of snow now.

1754 Ramsay, Andrew Crombie   To: H.T. De la Beche
Clifton Street,
Somerset Place,
Glasgow.
18 December 1846
Thanks for the news of Bristow's increase of pay. His map is a credit to the Survey. Gordon, the Professor of Engineering here, is a capital fellow.

1755 Ramsay, Andrew Crombie   To: H.T. De la Beche
2 Clifton Street,
Somerset Place,
Glasgow.
19 December 1846
It is the intention here to memorialize Lord John Russell, for the purpose of obtaining a Museum of Economic Geology here in Glasgow, not Edinburgh. Glasgow would be the best place for the Museum. There is so much energy about the people here. All the iron masters and miners would back it. What do you think? Would you recommend a memorial? Would you pay the expenses of transport of the present collection to London?

1756 Ramsay, Andrew Crombie   To: H.T. De la Beche
Glasgow.
19 December 1846
A Mr. Crinn says that you stated that Glasgow would be preferable to Edinburgh for a Museum. Pray give me a hint how to act, and whether to encourage or discourage. I have not committed you to any opinion.

1757 Ramsay, Andrew Crombie   To: H.T. De la Beche
Cerne Abbas,
Dorsetshire.
15 April 1847
The highest beds of the Greensand are rubbly, with numerous specks of iron silicate. They have been called the junction or passage beds. Can you tell me anything about the chalky beds. They have been mapping them with Greensand. To me they should be with the chalk, both by reason of change in physical character and fossils.

1758 Ramsay, Andrew Crombie   To: H.T. De la Beche
Cerne Abbas.
26 April 1847
The weather was inclement so we have not gone to Sturminster. Tomorrow we start for Yeovil. By the end of week I hope to be in Builth. The last winter has severely tried our people. Selwyn, Gibbs & Gapper must be coddled a little.

1759 Ramsay, Andrew Crombie   To: H.T. De la Beche
Oatlands.
1 May 1847
I have been delayed here. Aveline is far from well. His eyes are still bad but improving. Before leaving I settled all Bristow's difficult points. His work is improving.

1760 Ramsay, Andrew Crombie   To: H.T. De la Beche
Builth,
Breconshire.
5 May 1847
I am about to move about quite a lot. Please forward £50 to Llandovery. Selwyn has been granted leave to visit his brothers. What goes on about the Glasgow Museum?

1761 Ramsay, Andrew Crombie   To: H.T. De la Beche
Builth,
Breconshire.
12 May 1847
Your account of young Murchison's school mishaps were amusing but annoying. He shall have a fair trial on the Survey, but I fear it will not suit his aristocratic tastes. Some of the Wenlock lines here must be redone.
[Cited: Secord 1986b, p.263]

1762 Ramsay, Andrew Crombie   To: H.T. De la Beche
Builth,
Breconshire.
29 May 1847
The plan you propose on the maps suits admirably. Despite the delay, the results here will please you. I have discovered the structure of the Carneddau, and even the pipe of the original volcano. The Wenlocks too are also understood. These results will make a prime chapter in my next paper for the Memoirs.

1763 Ramsay, Andrew Crombie   To: H.T. De la Beche
Builth,
Breconshire.
11 June 1847
I suppose Joyce has resigned because he has a good income. The position would suit me. Do you agree that I should accept the chair? It would give me a standing in the London scientific world, even if no great income could be derived from it.

1764 Ramsay, Andrew Crombie   To: H.T. De la Beche
Pen-y-bont,
Radnorshire.
4 September 1847
Forbes and I have been over the Sawdde section. He is making notes on it. Tomorrow we go to Kington. Bristow has contact with the railway engineer who offers a section. I have advised Bristow to consult me before going ahead as Ibbetson might consider it an interference with his work. My foot improves.

1765 Ramsay, Andrew Crombie   To: H.T. De la Beche
Kington,
Herefordshire.
5 September 1847
I am pleased at the antiquity of the rocks. Jukes has traced the Caradocs

close to the ashy country. I like the term Cambrian. It is as old and as good as Silurian. A letter from Trimmer informs me of his Irish absence. My foot is not strong enough for walking yet. Hence the delay in the completion of the work.

1766 Ramsay, Andrew Crombie   To: H.T. De la Beche
Ludlow.
15 September 1847

Tomorrow we move to Church Stretton. Today we have been at Clee Hills. Gibbs' work will have to be re-done. The trap rests on Coal measures and Old Red. Either we have the termination of the shales at the coal measures or the country was denuded before the trap flow. I will ask Smyth to touch it up.

1767 Ramsay, Andrew Crombie   To: H.T. De la Beche
Church Stretton,
Shropshire.
21 September 1847

The weather is very bad. As the season closes, I have thought that Jukes should do Wolverhampton, Walsall etc. during the winter, and return to the Welsh ground in summer. What is your opinion? In January I would like to bring Bristow and Selwyn to London for making clean copies and dry proofing.

1768 Ramsay, Andrew Crombie   To: H.T. De la Beche
Chirbury,
Montgomery.
10 October 1847

The weather is diabolical. I will send Jukes to do the Coal measures, with Selwyn as assistant if you approve. I am anxious to keep them contented, as I am always afraid of more lucrative offers for our best men from outside. Murchison is very silent, but we are all inclined to like him.

1769 Ramsay, Andrew Crombie   To: H.T. De la Beche
Welshpool.
19 October 1847

I have seen something like a volcanic cone or crater today. This country was done too soon, when Aveline and Selwyn were not prepared for it. Aveline now takes copious notes. In two days we go to assist Aveline and Jukes at Oswestry.

1770 Ramsay, Andrew Crombie   To: H.T. De la Beche
Pen-y-bont,
Radnorshire.
5 November 1847

The ground is intricate at this end. The traps are largely surface work. Trimmer should be restricted to the Tertiaries. He should finish to Dorchester, then do the Isle of Wight, then Salisbury to Brighton. Wales will be mapped by next Autumn. Selwyn and Jukes can stagger their localities. Do you approve of my plans?

1771 Ramsay, Andrew Crombie   To: H.T. De la Beche
Pen-y-bont,
Radnorshire.
6 November 1847

At last I've discovered the Wenlock shale over the traps and slates. Also a bed of limestone at the bottom of the shale. Here is the last piece of evidence proving the perfect overlap and disappearance of the Caradocs. Thank you for your offer of financial aid.

1772 Ramsay, Andrew Crombie   To: H.T. De la Beche
Pen-y-bont,
Radnorshire.
27 November 1847

I agree about young Murchison. I see no hope of his acquiring geological skills. If his father thinks the office of Assistant is permanent it is in contradiction to my letter, a copy of which I sent to you in Ireland.

1773 [Ramsay, Andrew Crombie]   To: [H.T. De la Beche]
[1847 FJN]
[Postscript to letter only]

News of new Museum or School of Mines is prime. When completed I hope we shall make a grand start teaching and all. Jukes and Selwyn are at Wolverhampton. I fear for young Murchison.

1774 Ramsay, Andrew Crombie   To: H.T. De la Beche
Weymouth.
26 April 1848

We have examined all the neighbouring coasts. There will be a much wider sprawl of the Purbecks to map than hitherto has been done. Bristow and Trimmer will complete the sheet. Tomorrow we will be in Chidcock.

1775 Ramsay, Andrew Crombie   To: H.T. De la Beche
Tyn-y-groes.
7 June 1848

Selwyn and I came here for a few days to satisfy his scruples. By Monday, I'll be at Llanberis, where I propose taking up quarters. There seems to be gradual passage from the Barmouth to the igneous series.

1776 Ramsay, Andrew Crombie   To: H.T. De la Beche
Port Madoc.
13 June 1848

The Harlech quarter is ready to go to Southampton. I go to Llanberis tomorrow. The Barmouth sandstones are surprising. The whole of the sandstones are intersected with traps. Sometimes cracks are filled with quartz lodes, sometimes with trap. Faults are frequent.

1777 Ramsay, Andrew Crombie   To: H.T. De la Beche
Llanberis,
Caernarvon.
18 June 1848

I was glad you had such luck in the matter of sections in Anglesea. I hope to get line between Snowdonian and Barmouth Series well to the S.W. of Llyn Cwellyn in the course of a few days. I am established at the Dolbadarn Inn.

1778 Ramsay, Andrew Crombie   To: H.T. De la Beche
Llanberis,
Caernarvon.
19 June 1848

I am glad you approve of tracing the trap dykes. I will get Selwyn to work on them when he is well up this way from Tremadoc. A note from Gapper tells of the farmer's apology. I found a big fault in Vale of Llyn Cwellyn.

1779 Ramsay, Andrew Crombie   To: H.T. De la Beche
Llanberis,
Caernarvon.
21 June 1848

Trimmer's and Gapper's affairs, I thought to send both at once. Sismonda I see tomorrow. I have found a Trilobite and a Lingula 600 feet below the boundary of the Snowdon slates and the Llanberis sandstones.

1780 Ramsay, Andrew Crombie   To: H.T. De la Beche
Llanberis,
Caernarvon.
26 June 1848

£200 will be about what is required for ordinary expenses. Sismonda and a friend here. I got on in the jabbering line much better than I expected. I think I shall be able to show that the cleavage above Nant Francon has been disturbed after it was formed by trap intrusion.

1781 Ramsay, Andrew Crombie   To: H.T. De la Beche
Llanberis.
19 October 1848

Today I took to low ground, because of the weather. I have found a mass of rock of the Barmouth series dipping easterly. The snow is melting. Forbes goes to Ireland on Tuesday with wife and maid.

1782 Ramsay, Andrew Crombie   To: H.T. De la Beche
Llanberis.
17 November 1848

In a week Jukes and Aveline will finish the Llangollen and Corwen quarters. Aveline will then inspect the cornstones. Jukes goes to Staffordshire. However it happen they cannot take long and dry proofs will be made ready in the New Museum.

1783 Ramsay, Andrew Crombie   To: H.T. De la Beche
Llanberis.
1 December 1848

I will forward your note to Nichol. I cannot spare Selwyn this winter. Gibbs' work has to be done over again. His lines are often drawn across the strike. Next summer it will be impossible to spare Selwyn out of Wales. We should have a new Oolitic hand altogether, if we have funds. Paper of Trimmer's is good. Could Woods & Forests pay for the agriculture work, and Trimmer also do the Tertiaries at the same time?

1784 Ramsay, Andrew Crombie   To: H.T. De la Beche
Llanberis.
3 December 1848

I have been working on Glyder, and in two or three days the whole will be done. Selwyn is done, yesterday. Tomorrow he will be with Talbot in Herefordshire. Jukes is in the Midlands, and anxious for a council. I will go to Wolverhampton for a couple of days, when I have finished here.

1785 Ramsay, Andrew Crombie   To: H.T. De la Beche
Capel Curig,
near Llanrwst.
30 August 1849

The weather is bad in this country. Mountain work is arduous. I will go to Birmingham for Jukes' wedding, as I am groomsman.

1786 Ramsay, Andrew Crombie   To: H.T. De la Beche
Capel Curig,
Llanrwst,
North Wales.
8 October 1849

Selwyn's address, is Pwllheli, but he is rather inaccessible. I hope Yolland has all the maps ready for me. I wish to start field-work in spring as early as possible. I will see Bristow and Trimmer in November. It will be sad if we lose any of our staff.

1787 Ramsay, Andrew Crombie   To: H.T. De la Beche
Capel Curig,
Llanrwst,
North Wales.
13 October 1849

Finding a Magistrate to certify the accounts has cost me 4 days out of my week! Bristow having accepted the new post, I feel as if I have lost a leg or an arm. If I had my own money, I would supplement what Woods and Forests pays. Selwyn will have to take over his work.

1788 Ramsay, Andrew Crombie   To: H.T. De la Beche
Capel Curig.
14 October 1849

I am glad Yolland is sending the maps. Ask Hunt to open the parcel and lay them aside for me. I hope Bristow does well in the Antipodes. It will be long before this side of the world sees his like at the Oolites. I shall be more resigned to reverses in the future.

1789 Ramsay, Andrew Crombie   To: H.T. De la Beche
Capel Curig,
Llanrwst,
North Wales.
14 October 1849

I will tell you details of Trimmer's appointment as I remember them. I was not present when the arrangement was concluded. I have abandoned the high ground here. My plans have been upset by Bristow's appointment. What are your movements?

1790 Ramsay, Andrew Crombie   To: H.T. De la Beche
Capel Curig,
Llanrwst,
North Wales.
4 November 1849

Yolland has managed the lettering of the map admirably. However, the engraver must keep all the letters within the boundaries set in the original. These maps need not go back to Southampton. I am sorry to hear of your ailment.

1791 Ramsay, Andrew Crombie   To: H.T. De la Beche
Capel Curig,
Llanrwst,
North Wales.
15 November 1849

Selwyn has made a first rate job of the Pwllheli Promontory. Black slates are unconformable to schists. The traps are all intruded. The whole country is similar to Pembrokeshire. There is no doubt about the fault west of Llwyd Mawr.

1792 Ramsay, Andrew Crombie   To: H.T. De la Beche
Capel Curig,
Llanrwst,
North Wales.
20 November 1849

The rose coloured marble has been dealt with. We could not haul up enough for a column as the inclined plane was washed away. It was too rough to take it out by boat. I have taken specimens to try polishing etc. I hope you will get my room at the New Museum set up before I arrive.

1793 Ramsay, Andrew Crombie   To: H.T. De la Beche
Trefrhiw.
30 November 1849

I am glad Bristow has a bit of grace, and we can have him do sections. Today I return to Capel Curig. I have been joining up with Aveline. This country is almost completed. The Bala beds grown trappy.

1794 Ramsay, Andrew Crombie   To: H.T. De la Beche
Caernarvon.
22 June 1850

I forgot to mention in yesterday's letter that Sir John McNeill has the levels of the Holyhead road. Could you get them from him?

1795 Ramsay, Andrew Crombie   To: H.T. De la Beche
Llanberis,
Caernarvon.
28 June 1850

I trust the estimates are passed and money in hand. £200 will do for the moment. The first day of Celtic Sectionising nearly drove me mad. An assistant refused to chain along some steep ground. I have now got a quarryman and a miner who would do anything I bid them.

1796 Ramsay, Andrew Crombie   To: H.T. De la Beche
Llanberis.
10 July 1850

The Snowdon section is completed to the sea. I have written Selwyn asking him to join up. Lord Seymour seems to have put on the screw. William sent his accounts. I had not the heart to send him your official letter. He is listed as suspended.

1797 Ramsay, Andrew Crombie   To: H.T. De la Beche
Llanberis.
18 July 1850

Herewith the accounts. I have made them out in a new fashion. I have not included my travelling bill within 15 miles, the funds not allowing the same. I shall remind you of it when you get the funds to give me a decent balance.

1798 Ramsay, Andrew Crombie   To: H.T. De la Beche
Bethesda,
Bangor.
19 September 1850

Will we meet for the Anglesea trip? Oldham is to be married on the 18th. I will go over if you do not think it will stand in the way of progress. Howell would join if a position was offered. He is very promising. I've long looked for someone like him to recommend to you.

1799 Ramsay, Andrew Crombie   To: H.T. De la Beche
Bethesda,
Bangor,
North Wales.
21 September 1850

In such a cause I must not regret Jukes' loss on this side of the water. You will find him an upright and straightforward Lieutenant. What about Anglesea? Howell would join if you offered an appointment. Touching Peach's son, I do not think we should employ him.

1800 Ramsay, Andrew Crombie   To: H.T. De la Beche
Bethesda,
Bangor.
22 September 1850

Your arrangement is better than mine. On Wednesday I will start at Dolgelli. About Hall, he is strong, but lazy, headstrong and sulky. I shall try to improve him. He asked about confirmation of his appointment, but I played for time. Bristow begins to recover.

1801 Ramsay, Andrew Crombie — To: H.T. De la Beche
Bethesda,
Bangor.
23 September 1850

Jukes is just the man to set the Castle Snobbery at defiance. When does Oldham go? I need Jukes to finish Staffordshire. Howel goes home in a day or two. I told him that if you wished him to come at once he must give up the idea of doing a quarter's drawing.

1802 Ramsay, Andrew Crombie — To: H.T. De la Beche
Tyn-y-Celyn,
Dolgelli.
29 September 1850

I am glad to hear the arrangements concerning Jukes and Oldham. It relieves my soul. Thank you for the money. Will funds admit of transference of another £100? I will write again concerning Hall. His father is a fool. Tell Smyth I got the maps.

1803 Ramsay, Andrew Crombie — To: H.T. De la Beche
Dolgelli.
1 October 1850

I enclose a document received from Aveline. There is a lie on the face of it. Will you let him know what to do? His address is Welshpool.

1804 Ramsay, Andrew Crombie — To: H.T. De la Beche
Dolgelli.
1 October 1850

I return Col. Sandham's note. Howell will do us credit. I have appointed him a station and all that sort of thing. I have arranged with Jukes to proceed to Staffordshire when his section is done. That will be in a fortnight.
[See Sandham, H. to James, H., 21 September 1850, **1868**].

1805 Ramsay, Andrew Crombie — To: H.T. De la Beche
Amlwch,
Anglesea.
14 November 1850

Thank you for the hints. I'll keep my eyes open and have no doubt you will be satisfied. I think our mode of investigation is the correct one. The rocks of Caernarvonshire are by no means so uniform as they appear. Llanerchymedd was crowded and we had to come on here.

1806 Ramsay, Andrew Crombie — To: H.T. De la Beche
Swanage,
Dorset.
27 April 1851

If the Museum opening is delayed I will push things ahead, and have asked Gibbs to come here. Forbes' observations on the Purbeck beds are remarkable. Gapper has not done good work. Lyell is not unjolly in the field though sufficiently helpless at field work. He was delighted with all he saw and will I dare say go talking about and lauding the Survey.

1807 Ramsay, Andrew Crombie — To: H.T. De la Beche
Dolgelli.
31 October 1851

It seems probable that Bristow has treated the matter of his ill-health too lightly. His cousin has informed me that they were obliged to recommend his sudden departure for the Continent. I trust this deviation from the rule will be regarded as necessity, and that a few weeks leave of absence will restore him.

1808 Ramsden, William — To: H.T. De la Beche
Wakefield. Geological Survey Office,
22 June 1847 Craig's Court,
Charing Cross,
London.

Please return to me a copy of Explanations etc. I sent to you. Some parties interested in mining operations wish to examine my principle of improvement. I have other papers in preparation on the stratification of various mines in this neighbourhood.

1809 Ransome, George — To: H.T. De la Beche
Ipswich.
12 January 1849

I enclose the arrangements for the anniversary of Ipswich Museum. Please let us know your train time and a carriage will be sent to meet you.

1810 Ransome, George — To: H.T. De la Beche
Ipswich Museum.
10 March 1849

A proof impression of Maguire's portrait of Henslow is available to subscribers of one guinea.

1811 Ransome, George — To: H.T. De la Beche
Ipswich Museum.
13 October 1849

A portrait of the Bishop of Norwich has been secured. A subscription has been opened for the purpose of placing it in the Museum. Will you add your name?

1812 Raple, William — To: H.T. De la Beche
Stationery Office.
18 May 1848

What is your opinion as to the comparative economy of tracing cotton or tracing paper, based on your practical knowledge?

1813 Raple, William — To: H.T. De la Beche
Stationery Office.
16 September 1848

I have procured the Comptroller's sanction to publish at 7/6. It will not be a very gaining price. The Bills may therefore be sent to me when the work is done, as I have authority to pass them. Please give me an Estimate for the Mine Map. I have a copy of it, and Bone has coloured it beautifully and cheaply.

1814 Reeks, Trenham
Ordnance Geological Survey.
[No date]

[Analysis of rock] from the upper part of the Wenlock Limestone, on the road from Mitchel Dean to Gloucester.

1815 Rees, Josiah Jr. — To: [H.T. De la Beche]
Milford,
South Wales.
3 May 1841

Could you let me have the new sheet containing Haverfordwest etc.? We have found fossils in sandstone close to Sowland farmhouse. We have been to Limestone at Johnston. I saw Williams yesterday, and he is to write to you concerning quarters.

1816 Rendel, James Meadows — To: H.T. De la Beche
8 Great George Street,
Westminster.
[185?]

As a favour I ask you to come to the meeting of the Engineers Institute, as a paper is to be delivered on the tract of shingle on the South and East coast, and no one knows more about it than you.

1817 Rice, Ralph — To: H.T. De la Beche
Royal Crescent, 58 Jermyn Street,
Brighton. St. James's.
28 June 1830

Dr. Pennington allowed me to go to Brighton. I have already improved in health after meeting my brother and his family. I wish you were here as you've become a favourite with them all.

1818 Rice, Ralph — To: H.T. De la Beche
At your lodgings.
[1830]

Pennington has come to see me and prescribed medicine. I wish to persuade him to let me go to Brighton tomorrow. I have completed the purchase of my house. My gratitude to you for your kindness to me at Rome and Florence.

1819 Ridsdale, G.W.H. — To: H.T. De la Beche
Graveview,
Tonbridge.
15 February 1854

Your kind letter acknowledging an old Marlow chum was most welcome. Since leaving Marlow, I served in the Navy, then took Cambridge degree and Orders, and now teach Navigation etc. Our birthdays (58th) are close to one another. I do not find mentioned in your "Geological Observer" the Alabaster Cliff at Watchet on the Bristol Channel.

1820 Rigby, William   To: J.W.D. Dundas
    Craven Hotel,
    Craven Street,
    London.
    2 July 1849
The trials on our coal have been completed, and a certified Copy of the Report has to be obtained from De la Beche. Can you procure it today or tomorrow. A large contract depends on that favourable report.
[See Dundas, J.W.D. to De la Beche, H.T., 2 July 1849, 477]

1821 Risca Works   To: H.T. De la Beche
    Newport,
    Monmouthshire.
    22 September 1842
In answer to your letter of 19 September, respecting transport charges on the Monmouthshire Canal Company's train road, they are as under.

1822 Robe, Alex W.   To: H.T. De la Beche
    16 Park Road.
    13 April 1832
The enclosed answer was received from Colby yesterday. I think he is wrong about not agreeing to the increase in price. His index of colours suggestion will allow you and Greenough to establish a system. I have sent you a fair copy of the proposals and Colby's answer.

1823 Robe, Alex W.   To: H.T. De la Beche
    16 Park Road.
    [1832]
Drummond asked me to tell you that he has heard from Kennedy that the Board have approved your proposals and ordered payment subject to Colby's very judicious conditions. This information is private. I congratulate you on your success.

1824 Robison, John   To: H.T. De la Beche
    Edinburgh.
    23 July 1842
The Committee for arranging the Agricultural Show have sent an invitation for you to attend the meeting. If you come, I hope you will stay with me. Thank you for the use you made of my note about gold coin. Gladstone proposes to look into the subject. I see the vote for the Museum was passed despite Hume's objections about the workshop for models. I have seen the young man likely to suit you for that department.

1825 Robison, John   To: H.T. De la Beche
    31 July 1842
Thank you for the offer to write to Wyon about my scheme for gold coins. The part I feel some difficulty about is the application to the edges of the coins so as to make any tampering with them as obvious as on the faces of the impressions. I regret you could not come here for the meeting. Liebig also has disappointed us.

1826 Rogers, Henry Darwin   To: H.T. De la Beche
    Philadelphia.   Geological Society,
    18 March 1834   London.
I have been engaged for the past three months in lecturing, using your work extensively. I will work in the field next summer, mostly in the coal regions of Pennsylvania. I am sending a paper through Turner to the Geological Society. I wish to avoid theorizing. A Geological Survey of Maryland was authorised this winter. I will send a report on the Geology of North America to the British Association.

1827 Rogers, Henry Darwin   To: H.T. De la Beche
    Boston.   Craig's Court,
    18 June 1849   London.
Ill health and imperative tasks have prevented my sending Geological papers. By the hands of my brother, W.B. Rogers, State Geologist of Virginia, I send a box of fossils. What specimens do you want for your Museum of Economic Geology? I have heard no news of my communication to the Geological Society.

1828 Rogers, J.   To: H.T. De la Beche
    Mawnan,   Post Office,
    Falmouth.   Helston.
    19 October 1836
As I am obliged to leave next month, I fear I will miss the pleasure of seeing you here. When will you visit this neighbourhood? There are some localities I would recommend to your attention. I am no theorist and but a very slight geologist yet my suggestions might be of help.

1829 Ross, [Lord]   To: H.T. De la Beche
    13 Connaught Place.
    3 July 1851
I believe the Museum is not open the three last days in the week. Could I have an order of admission for a party consisting of friends from Ireland who all take an interest in Practical Geology. They are obliged to leave London by an early train on Monday morning.

1830 Rosser, Highmore   To: H.T. De la Beche
    Longnor,
    Buxton.
    26 September [1848 FJN]
We have collected a goodly number of fossils, comprising many genera and species. I have found nothing in the Lower Limestone. I have found spines similar to that you bought at Ashford. We proceed to Alstonefield on Monday. My term concludes on 6 October. What of future arrangements?

1831 Rosser, Highmore   To: H.T. De la Beche
    Alstonefield,   Museum of Practical Geology,
    Derbyshire.   Jermyn Street,
    3 October 1849   St. James,
                                                             London.
I have received notice of my retention on the Survey. We commenced operations on Monday. Various fossils have been collected.

1832 Rosser, Highmore   To: H.T. De la Beche
    Alstonefield,
    near Ashbourne,
    Derbyshire.
    29 October 1849
Herewith an account of the last fortnight's activities. 2 boxes of fossils from Alstonefield have been sent to the Museum. Brassington is an unfavourable district. I have made an arrangement to collect fossils for the Bakewell Institute. I hope what I have done for the Geological Survey is satisfactory, and would welcome further employment, here or abroad.

1833 Rosser, Highmore   To: [H.T. De la Beche]
    Bakewell,
    Derbyshire.
    12 December [1850 FJN]
Adam, the proprietor of the Museum at Matlock Bath, has become bankrupt. There is a sale next week. I will get you a catalogue as soon as it is available. I have been over old ground in Derbyshire, and have collected specimens which I will send to the Museum. When is the new edition of your "Manual" coming out. Have you any employment?
[Cited: Torrens & Taylor 1990, p.176]

1834 Rosthorn, Joseph   To: [H.T. De la Beche]
    Liverpool.
    14 May 1840
Your introduction proved of great benefit to me. Buddle was most kind during my stay in Newcastle; in Staffordshire my attention was chiefly directed to the mode of working and raising coal. My own country lacks cheap fuel. Could I have your further assistance on my journey to Cornwall and Devon? Any instructions could be sent to Stutchbury. I must be in Vienna on 8th of next month.

1835 Royle, Annette Forbes   To: H.T. De la Beche
    Heathfield Lodge,
    Acton.
    22 March [1850 FJN]
Can you come to dinner on April 2nd? I will ask Professor Partridge, who is coming, to convey you, should you have no other engagement.

1836 Royle, John Forbes   To: H.T. De la Beche
    East India House.
    27 January 1849
Thank you for enquiring among the landowners of Wales for someone with a variety of mountain locales and willing to take Tibetan sheep. Talbot's domain seems suitable, and I will recommend he be requested to give them a trial. 19 left Calcutta in October, and I hope to hear of the safe arrival of some of them at least, when I will let you know what the Chairman resolves upon doing with them.

1837 Royle, John Forbes　　　　　　　　　To: H.T. De la Beche
[no date]
As the 12 April approaches, we the initiators are anxious to arrange things so that success continues to attend the establishment of the New Scientific Club. Grove, Horner, Falconer and myself would like you to take the chair. A short speech would be sufficient. Grove will give details of the proposed rules.

1838 Russell, Lord John　　　　　　　　　To: H.T. De la Beche
Downing Street.
23 June 1848
Thanks for the tables of produce of Copper mines and other works sent with your letter.

1839 Russell, Lord John　　　　　　　　　To: H.T. De la Beche
Downing Street.
2 May 1849
Thank you for the copy of the address sent.

1840 Russell, Lord John　　　　　　　　　To: H.T. De la Beche
Woburn Abbey.
13 December 1851
Thanks for the copies of the lectures forwarded with your letter of 5 December.

1841 Russia, Consul General of　　　　　　To: H.T. De la Beche
2 Winchester Buildings,
London.
2 June 1847
Herewith a parcel containing a work of Count Keyserling, which the Chief of Staff at the Imperial Mining Department at St. Petersburg has presented to you. Please acknowledge receipt to this Consulate.

1842 Sabine, Edward　　　　　　　　　　To: J.F.W. Herschel
Woolwich.
24 January 1848
Copy
Lord Northampton has announced his intention to resign the Presidency of the Royal Society. De la Beche, Grove, Faraday, and self wish to procure a limited term of office (3 years), the removal of the entertainment obligation, and lastly your acceptance of the office. I have worded this letter so as not to require a reply from you, unless it is thoroughly out of the question.
[See Herschel, J.F.W. to Sabine, E., 25 January 1848, **682**].

1843 Sabine, Edward　　　　　　　　　　To: H.T. De la Beche
Woolwich.
26 January 1848
Is this letter of Herschel final? In the line of scientific presidents, and viewing the great want of the Royal Society to be put in good working order by a man of business, Peacock seems to be the next man to Herschel.
[See Herschel, J.F.W. to Sabine, E., 25 January 1848, **682**].

1844 Sabine, Edward　　　　　　　　　　To: H.T. De la Beche
Woolwich.
20 February 1848
I am glad you are in favour of having a scientific President of the Royal Society. Regarding Council on the 24th, it does not appear to me that a resolution would be desirable, but rather an understanding. After passing our resolutions, we can send a deputation to Herschel. If he declines we shall hear his opinion. If he accepts, the business is done.

1845 Sabine, Edward　　　　　　　　　　To: Lieut. T.A.B. Spratt
Woolwich.
11 April 1848
Re your letter of last January, I have examined the records of your Magnetic Observations made in the Mediterranean in 1845 and 1846. The mere enumeration of them speaks for itself of the time and labour involved. I am glad the Admiralty take such into consideration for Promotion, and I would be glad to hear that they had done so in your case.
[See Spratt, T.A.B. to De la Beche, H.T., 19 April [1848], **1987**].

1846 Sabine, Edward　　　　　　　　　　To: H.T. De la Beche
Woolwich.
11 May 1849
I should be glad to have a few minutes conversation with you respecting the affairs of the British Association. Could Grove be present also?

1847 Sadleir, Ralph　　　　　　　　　　To: H.T. De la Beche
Committee Rooms,
Provost's House,
Trinity College,
Dublin.
29 September 1849
The Committee for raising a subscription for the widow and children of Dr. Cooke Taylor, who died of cholera, resolved to use the names of individuals having weight with public as receivers of contributions. Will you act as one?

1848 Saint-Anthoine, H.D. de　　　　　　To: H.T. De la Beche
Secretary General Institut d'Afrique,
Paris.
20 October 1840
You have been proposed as a "Membre Titulaire" of the Institute. On receiving your commitment to our aims, the Council will be glad to forward your diploma.

1849 Salmond, William　　　　　　　　　To: H.T. De la Beche
York.
21 December 1822
I have tried to get a sketch of the Whitby ichthyosaurus for you. One is published in Young's Survey of the Yorkshire coast. I have been involved in the study of Buckland's finds in Kirkdale cavern. Several foreigners of distinction have gone away perfectly convinced they are the bones of rhinoceros, elephant, hyaena and hippopotamus.

1850 Salter, John William　　　　　　　To: H.T. De la Beche
8 Pk. Pl. West.
29 January 1845
I have just been sending Sedgwick my ideas on the superposition of many doubtful parts of his country, I mean N.Wales, derived entirely from fossils and lithological evidence. I am convinced that groups of fossils of similar character are really synchronous, at least in our island. Nor do I think that Ireland is an exception, though there the series best displayed is the passage between Upper and Lower Silurians. I hope under your tuition to get a great deal of additional information.

1851 Salter, John William　　　　　　　To: H.T. De la Beche
Bridge Street,
Usk.
1 August 1846
I am leaving Usk, having been here a fortnight, and let no one say it is impossible to geologise on a honeymoon. If all ladies are like mine they are most capital hammermen and most zealous collectors. I am satisfied that the centre of the country is no lower than Wenlock shale. I have learnt all I can about the life of the time and have a large box of specimens for home. And now to Builth.
[Cited: Secord 1985, p.65]

1852 Salter, John William　　　　　　　To: H.T. De la Beche
Aberystwith.
12 September [1846 FJN]
I have just left Llandovery where we got quite overloaded with kindness from the Williams family, but that did not prevent me whisking the pony about and doing what I could to learn the country. I think I understand it and your Wenlock bay is a very pretty one, and the admiration of father Adam [Sedgwick], with whom I have worked a good deal at Llandilo and Llandovery a little. Sedgwick is now zig-zagging and verifying your sections. I long to know the result.

1853 Salter, John William　　　　　　　To: H.T. De la Beche
2a Whitehall Yard.
22 October 1847
I have not much of importance to say, but suppose you will like to know how things are getting on at home. For myself there is plenty to do, with Phillips' proofs descriptions and getting his things drawn and engraved. Hooker keeps Baily and Bone pretty closely employed so that at present Phillips plates do not get on very fast. No more Mss. from him lately. Kind regards to Oldham.

1854 Salter, John William　　　　　　　To: H.T. De la Beche
7 Rocky Park,
Tenby.
8 September 1854
Herewith a progress report on the Lower Carboniferous so far. I have

found an abundance of fossils, but nothing to separate the lower sandy beds from the Carboniferous series. It may be necessary to go to Pembroke. My belief is that you are right and that these beds are really identical to the Upper Devonian of Devon. Jukes is right too for Ireland. Huxley is hard at work. Various people wish to be remembered to you.

1855 Salter, John William   To: H.T. De la Beche
       7 Rocky Park,
       Tenby.
       15 September 1854

I am deep in Skrinkle now and most glad of your letter. I suspect however that what is the upper part of the old red sandstone (& red here) may be continued down as slaty beds in Ireland & hence both Carboniferous slate and Upper Devonian so-called be represented by those thick beds. From memory of the fossils, I can quite understand the Hook. I am sure it is Skrinkle and Caldy all over again. I came upon bivalve beds at W. Caldy very like those of Marwood and set the light-house keeper to work on them.

1856 Salter, John William   To: H.T. De la Beche
       7 Rocky Park,
       Tenby.
       3 October 1854

I have just come back from a few days run to West Angle, and I saw enough to show a great change in the character of the deposit. Neither at Caldy or Skrinkle is there any alternation of the red with the Carboniferous shales. Both finish with a sharp line and at Caldy it is a conglomerate, but some 10 or 12 feet down in the Old Red of Caldy there is a wonderful bed of Serpula and this I accidentally found at West Angle.

1857 Salter, John William   To: H.T. De la Beche
       [No date]

I was christened John William, and should the Commissioners enquire whether John William would be willing to pay a larger amount of Income Tax, pray tell them I will.

1858 Sanders, William   To: H.T. De la Beche
       Bristol.
       13 November 1840

Thank you for making me the offer re railway sections. I will accept, pending fuller information. The sum you mention will not cover expenses. I met Dr. Buckland at Glen Roy. He says £200 for 1800 miles of sections could not pay for a detailed description. Phillips' opinion is totally the reverse. Arran fossils are few. I should refrain from altering the colour which denotes Old Red Sandstone.

1859 Sanders, William   To: H.T. De la Beche
       49 Park Street,
       Bristol.
       15 March 1841

I shall be glad to know of your plans. In the meantime I am preparing the railway sections. Williams says he has found in the Cannington limestone what Sowerby calls Echini. Stutchbury will not believe it. Parker has found Goniatites at Bideford. Williams is pleased with this. Why, I know not.

1860 Sanders, William   To: [J. Phillips]
       Bristol.
       25 June 1841

I have been tracing boundary lines from Bristol northwards. About 11 miles have been completed. I will transfer all possible from my 6 inch to the one inch map. I intend to take Stutchbury to the problematic spot. He thinks Murchison was too eager in finding "Systeme Silurien" to pronounce judgement impartially. Please give any information to De la Beche.

1861 Sanders, William   To: H.T. De la Beche
       Bristol.
       28 January 1842

I have completed the task of mapping and have spent 6 months on Survey work. From the directions you gave me I presume you wish me to continue, and will make preparations accordingly. I have changed my pony for a horse and larger carriage. I will commence work again in February.

1862 Sanders, William   To: H.T. De la Beche
       49 Park Street,
       Bristol.
       21 February 1842

The information contained in your letter caused me great disappointment. I must make a material change in my plans. You led me to expect full expenses. They are £159. I would furnish particulars if it should be necessary for your application to Government.

1863 Sanders, William   To: H.T. De la Beche
       49 Park Street,
       Bristol.
       12 March 1842

My intention is to carry on the boundary lines in this district. I have parted with my horse and phaeton, and shall feel their want extremely. I will have to devote less time to the Survey. Had I received my expenses last year it would have been different.

1864 Sanders, William   To: H.T. De la Beche
       Somerset House,
       Wellington Terrace,
       Clevedon.
       25 August [1842 or 43 FJN]

I have completed the Report, but not to my satisfaction. Please correct it as you will. In reference to the Section, it ought to be appended to the report. I suggest different markings for the different formations, as it will not be in colour. I will meet you in Bristol at your convenience.

1865 Sanders, William   To: H.T. De la Beche
       Bristol.
       4 November 1843

I have been to London to give evidence with Stutchbury in the arbitration between Hemmings and the Exeter Railway Co. I saw the Museum. Herewith my long explanation of the differences which exist between Williams and myself in the mapping near Sodbury. It would be worthwhile to make the matter more clear.

1866 Sanders, William   To: H.T. De la Beche
       Bristol.
       27 April 1844

Bunt, Stutchbury and I have taken the angle of the highest point of Dundry Hill. Herewith the result. If we are correct, the Trigonometrical people are wrong. Bristow has not returned the maps. Remember I have your books, when you advise on Switzerland.

1867 [Sanders, William]
       [Bristol].
       [1847]

How do you like my section from Dundry to the Marsh? As far as we know of James' duty, he was performing it very well. 790ft. for Dundry was an error. Carpenter leaves Bristol tomorrow to take charge of the education of Lord Lovelace's children. I have heard from Bristow. Phillips has sent me a letter to Agassiz.

1868 Sandham, H.   To: James
       87 Pall Mall.
       21 September 1850

I understand that Ramsay will, in consequence of Howell's not falling short of the character I gave of him when introducing him, recommend an appointment on the Survey to De la Beche. Please forward this letter of further recommendation to De la Beche.
[See Ramsay, A.C. to De la Beche, H.T., 1 October 1850, **1804**].

1869 Schroder, Henry & Co.   To: H.T. De la Beche
       102 Leadenhall Street.   Craig's Court.
       3 March 1849

Do you wish to buy the maps referred to in d'Oesfeld's letter. If not, we are instructed to apply to you for Catalogues etc.

1870 Sedgwick, Adam   To: H.T. De la Beche
       Trinity College.   Geological Society,
       [7 March 1830]   Somerset House,
                                             London.

I take the first opportunity of performing my promise. The plate about which you enquired contains four figures. The section I have sketched ought to be lengthened by about two to one. Starling Dod is covered with blocks of Red Syenite. These blocks may be traced towards the north down the valleys into the plains of Carlisle.

1871 **Sedgwick**, Adam
Trinity College,
Cambridge.
20 April 1831
To: H.T. De la Beche
Geological Society,
Somerset House,
London.

I am confoundedly busy and not in good force, but if you are fairly at your wits end and on your last legs I would take pity on you. Indeed I had intended to draw up a short notice of the red sandstone series in the valley of the Eden as a supplement to my last papers. The President will swear when he hears of this new interruption, for he wishes me to work at the alpine paper exclusively. Perhaps I may be up at the next meeting.

1872 **Sedgwick**, Adam
Cambridge.
28 March [1832]
To: H.T. De la Beche
Athenaeum,
London.

I wrote to Lord Lansdowne and Lord Kerry as strongly as I could in recommending your proposal. I enclosed your original paper. The sum specified would hardly make the wages of a common labourer. I am expecting Conybeare and his son this evening.

1873 **Sedgwick**, Adam
Close,
Norwich.
17 December 1834
To: H.T. De la Beche
Tiverton,
Devonshire.

My knowledge of North Devon is very superficial. Weaver produced a paper on plants in the Grauwacke, but there were unbelievers in the Council of the Geological Society. Your discovery is new to English geology. Murchison is sometimes a little touchy, and Lyell extremely sensitive about his published opinions. You need not feel hurt about Austen's paper, or discussion afterwards. I do not remember your paper even being alluded to.
[Cited: North 1939b, p.1054; Rudwick 1985, pp.80, 82, 108]

1874 **Sedgwick**, Adam
Trinity College.
8 February [?1836]
To: H.T. De la Beche
Truro,
Cornwall.

Thank you for your two letters. By tracing the beds near Padstow with shells, you have done immense service, and by proving the granite to burst against the upper system of the schists you give the coup de grace to the anti-igneous blockheads. I mean to visit Cornwall as soon as I can this summer, perhaps in June. Boase has written in a very crusty temper. He knows nothing of slate cleavage. Why should he run his head against facts?
[Cited: Rudwick 1985, p.144]

1875 **Sedgwick**, Adam
Exeter.
17 October 1836
To: H.T. De la Beche
Helston.

Thank you for the use of your map. I was surprised to find so much good matter in Boase's book on Cornwall. Pray don't think our paper was an attack on you. I am certain I never heard Murchison say anything of your Devon map behind your back he would not say to your face. Errors have come from people who have copied your original. I shall rejoice at the progress of your beautiful Cornish map.
[Cited: Rudwick 1985, pp.169, 173, 174, 175]

1876 **Sedgwick**, Adam
Trinity College.
20 November 1836
To: H.T. De la Beche
Helston,
Cornwall.

I wrote to Spring Rice and told him I thought your work in Cornwall the best geological work of the kind I had ever seen since I became a geologist. I concur with some of your views, but you are mistaken in confounding culm limestones with the older limestones. You must prove the presence of greywacke over the culm beds.
[Cited: Rudwick 1985, pp.176, 189]

1877 **Sedgwick**, Adam
Trinity College.
9 May 1837
To: H.T. De la Beche
Ordnance Survey,
Falmouth.

Murchison's note-book contains all our joint remarks on the Devon problem. I have spent a miserable winter. Murchison and I agree that you must change your opinion. The culm system is not the greywacke. There are several blunders on your map. I will write again when I have more time.
[Cited: Rudwick 1985, pp.209, 210]

1878 **Sedgwick**, Adam
London.
10 July 1837
To: H.T. De la Beche
Ordnance Survey.

Hopkins wished me to forward this to you. He is anxious to work in Derbyshire. I suppose you will be glad of some local help. I am about to visit Launceston, seeing Austen and Conybeare on my return.
[See Hopkins, W. to De la Beche, H.T., 8 July 1837, **716**].
[Cited: Rudwick 1985, p.215]

1879 **Sedgwick**, Adam
Trinity College.
19 April 1838
To: H.T. De la Beche
Swansea.

I have symptoms of gout. Thank you for your two letters. I'm not surprised at what you tell me about the Devon plants. I never thought the base of the culm measures to be on a level with the base of Carboniferous series. If you can prove the culms to be below the Plymouth limestone, I will shut up shop for life.
[Cited: Rudwick 1985, pp.241-242]

1880 **Sedgwick**, Adam
Cambridge.
19 May 1841
To: H.T. De la Beche

I am amazed at Henwood's irritable letter, and this I doubt not is your feeling about it. But I am truly angry that he has quoted me, and still more sorry that I ever gave him occasion. I wish you had not put the note in your excellent and most elaborate work, because Williams' claim is downright fudge, but you had a right to quote what was printed and you had a right to suppose I know all about it. You have behaved with great forbearance in the matter. My Museum is finished, and I have the formidable task of arranging a vast collection.

1881 **Sedgwick**, Adam
Cambridge.
[?9 November 1843]
To: H.T. De la Beche

I have promised to give a lecture on Wales at our Philosophical Society on Monday, so I will want my Welsh roll of sections. Would you have the kindness to return them to the porter of the Geological Society that he may forward them to me by rail-road. It was very kind of you to undertake the reduction of my ugly sections and I am greatly obliged to you for so doing.

1882 **Sedgwick**, Adam
Athenaeum.
[?1843]
To: H.T. De la Beche

I want an hour's talk with you. I breakfast out tomorrow but I shall return early, and perhaps I could see you afterwards. Would you leave me a note here? On Wednesday week I shall have a paper on a part of North Wales. If you are to remain over that time I can see you on my return from Cambridge where I must spend the greater part of next week.

1883 **Sedgwick**, Adam
York Hotel,
Bath.
[April 1847]
To: H.T. De la Beche

I sent the corrected revise of my paper to the Geological Society yesterday. I had no spare time to call on you, though I saw Salter for a few minutes. I ought to have been here ten days since, as my medical advisers ordered me off. In three weeks I hope to return to my duties as a Member of the Council, as I wish to do. If there is any question about my paper, send it to this place. Kind regards to Forbes and other members of your staff.

1884 **Sedgwick**, Adam
Cambridge.
[May 1847]
To: H.T. De la Beche

Since my return from Bath I fully intended to be at every meeting of the Geological Society. Tomorrow I cannot come up. I am not quite well, but that would not stop me, but my nephew has been attacked by a low fever. I had removed his bed to my dining room for better air, and his mother is coming to nurse him. At present I cannot leave them. This is my apology.

1885 **Sedgwick**, Adam
Cambridge.
[June 1847]
To: H.T. De la Beche

I am truly sorry I cannot come on Wednesday. My nephew has been confined to his room, but he is much better, and I mean this morning to wrap him up in a warm covering and run him down by rail-road to

Yarmouth for the benefit of sea air. Should the motion be too much for him I can halt at Ely or Norwich. I will remain with him to the end of the week. I mean to be at Oxford.

1886 Sedgwick, Adam　　　　　　　　　　　　To: H.T. De la Beche
　　　　Cambridge.
　　　　1 November 1847

My lectures begin this morning, and as I give them five days a week (including Wednesday) I shall not be able to attend your meetings till November and a part of December are over. I want to have sent a short notice about the discovery of fossils in the Skiddaw slate of Cumberland. I will send you a short notice for the meeting on the 17th, and I will attend that day if I can. But alas! I am not to be depended on. That gouty hypochondriacal affection has got me in its grip this autumn. It is creeping over me like a moral leprosy and will be the end of me.

1887 Sedgwick, Adam　　　　　　　　　　　　To: H.T. De la Beche
　　　　Cambridge.
　　　　28 November [1847]

A gouty man, lecturing five days a week must be rather busy. This is my case. I have again written for my second batch of Skiddaw fossils, but they are not yet come. What is the cause of the delay I know not. By working hard now I shall be my own master (barring influenza which confined me 3 months last winter) after Xmas, and then I mean to be a constant attendant at your Meetings.

1888 Sedgwick, Adam　　　　　　　　　　　　To: H.T. De la Beche
　　　　Cambridge.
　　　　13 December [1847]

I fully meant to be with you on Wednesday, now that my lectures are over, but a very unexpected call to Osborne House will prevent me. I think you asked me to name someone for the consideration of the Council before they debate on the Wollaston medal. I think Forbes' Essay on glaciers stands above anything I can remember. It is a grand philosophical essay and is a decided step both in physics and geology. Were I with you, Forbes would probably have my vote.

1889 Sedgwick, Adam　　　　　　　　　　　　To: H.T. De la Beche
　　　　Cambridge.
　　　　29 December [1847]

I was last week dragged to the north by very unpleasant family business and I spent a very sad and sorrowful Xmas week. I must be in town next week to prosecute a thief. I am bound to appear on Tuesday and I may be detained by the "law's delay". I mean to be with you on Wednesday, and I will draw up a notice of two or three pages on the Skiddaw fossils. It will open a good subject for discussion. I will bring it to town on Monday.

1890 Sedgwick, Adam　　　　　　　　　　　　To: H.T. De la Beche
　　　　Cambridge.
　　　　[1847]

I am lecturing five days a week and I am sorely crippled by rheumatic gout which drives me half mad and makes me incapable of any intellectual labours. My lectures will end about the 10th of next month after which I hope to be a pretty good attender of your meetings. But I am sadly out of spirits, and hope for the future has almost fled. I have not sent my notice of the Skiddaw fossils only because the second box of them is not yet come to Cambridge. Forgive for all this stuff about my old carcass.

1891 Sedgwick, Adam　　　　　　　　　　　　To: H.T. De la Beche
　　　　14 Suffolk Street,　　　　　　　Museum of Economical Geology,
　　　　London.　　　　　　　　　　　　　　　　　　　　Craig's Court,
　　　　5 January 1848　　　　　　　　　　　　　　　　　　　Whitehall.

I am out of luck. Last week I was laid up by influenza, but I came up on Monday and brought the Skiddaw fossils with me. In fact I came up under the constraint of a sub poena from the Old Bailey. I caught a cold and today I had to reappear in Court. I must leave tomorrow, but I will leave the specimens in London, and my paper will be in good time for the next meeting.
[Letter damaged]

1892 Sedgwick, Adam　　　　　　　　　　　　To: H.T. De la Beche
　　　　Cambridge.
　　　　28 January 1848

I am still a prisoner, but at any rate you shall have my paper and I will come if I can to the meeting. The account of the Skiddaw fossils (drawn up by M'Coy) I will send today addressed to you at the Geological Society. I shall be disappointed if I can't be present. At the very latest you shall have my paper on Tuesday, i.e. a mere short notice.

1893 Sedgwick, Adam　　　　　　　　　　　　To: H.T. De la Beche
　　　　Cambridge.
　　　　[January 1848]

The morning after I wrote my former note I went to a friends house, but I had a miserable visit, being almost confined to my room. I am by no means fit to travel. The fever is all but gone, but my lungs are so tender I cannot yet endure a single gulp of cold air. Pray put my paper off for another fortnight. Can I then come on next Wednesday fortnight?

1894 Sedgwick, Adam　　　　　　　　　　　　To: H.T. De la Beche
　　　　Cambridge.
　　　　15 July 1848

I enclose a letter I have just received from Loftus of Caius College Cambridge. He used certainly to attend my lectures when he was an undergraduate seven or eight years ago. Perhaps Jukes knows him better than I do. He was full of zeal and made some progress in the study of fossils. I have no knowledge of his power as a field surveyor. I have told him there is little likelihood of his finding employment on your staff. Considering the time of year my health is unusually good.

1895 Sedgwick, Adam　　　　　　　　　　　　To: H.T. De la Beche
　　　　Scarborough.
　　　　23 December [1848]

Your last note reached me a day or two before the conclusion of my course of lectures. As soon as they were over I left Cambridge and since then have been on the ramble. You will find some account of Hailstone in the last year's address of the President of the Royal Society. It is said that he was discouraged in high quarters by persons who had no relish for geological studies. When is the next meeting?

1896 Sedgwick, Adam　　　　　　　　　　　　To: H.T. De la Beche
　　　　Cambridge.
　　　　14 March 1852

Lord Selkirk has I dare say spoken to you about a Mr. Fleming who lives close to his Lordship's seat. I promised to speak to you and say a good word for the said man. He was (and perhaps is) a schoolmaster with humble means. Fleming is no particular friend of mine whom I want to patronize, but at Lord Selkirk's request I have just stated what I believe true. I am suffering. The moment I begin any sedentary task my head gives way. I become irritable to the verge of insanity, and sometimes almost lose my memory.
[See also Fleming, E.B. to De la Beche, H.T., 8 January 1852, **541**]

1897 Sedgwick, Adam　　　　　　　　　　　　To: H.T. De la Beche
　　　　Athenaeum.
　　　　[No date]

Mr. Westall the artist is a connexion of mine, and has sent to ask me for an introduction to Mr. Hunt for the purpose of asking some questions respecting photogenic drawings. I do not personally know Mr. Hunt, but will you be kind enough to break the ice for him? I think we are on a wrong tack at the Club. Discussion of subjects should not be just after dinner. No good can come of the present plan. I go back to Cambridge this morning.

1898 Sedgwick, Adam　　　　　　　　　　　　To: H.T. De la Beche
　　　　[No date]　　　　　　　　　　　　　　　　　Geological Society,
　　　　[Watermarked 1834]　　　　　　　　　　　　　Somerset House,
　　　　　　　　　　　　　　　　　　　　　　　　　　　　　　London.

I am called away to the north of England by sudden and pressing business. I cannot therefore correct my paper, but must send it up just as I scrawled it off. You will be able to make it out, I hope, sufficiently to report on it. I hope at latest to be back in a fortnight and as soon as I can I will write over again and correct the worst parts of my P.S.

1899 Sedgwick, Adam　　　　　　　　　　　　To: H.T. De la Beche
　　　　[No date]　　　　　　　　　　　　　　　　3 Alfred [sic] Place,
　　　　　　　　　　　　　　　　　　　　　　　　　St. John's Wood.

Did you bring the maps yesterday? If not when and how am I to get them? Bring them to Lord Coles tomorrow at breakfast.

1900 Selkirk, Dunbar James, 6th Earl of　　　To: H.T. De la Beche
　　　　St. Mary's Isle.
　　　　18 December 1851

I wrote to you 17th October last but received no reply, having been applied

127

to by Fleming anent opportunities on the Geological Survey. Fleming had a school here, but now earns a pittance as tutor in a family. Is he likely to get employment he wishes on the Survey?

1901 Selkirk, Dunbar James, 6th Earl of   To: H.T. De la Beche
St. Mary's Isle.
31 December 1851
I have received your letter of 20th. I will ask Fleming to write to you about his qualifications and references. Sedgwick has already said he will speak in his favour.
[See Fleming, E.B. to De la Beche, H.T., 8 January 1852, **541** and
Sedgwick, A. to De la Beche, H.T., 14 March 1852, **1896**].

1902 Seymour, Edward Adolphus, Lord   To: H.T. De la Beche
Office of Woods.
15 March 1851
I have moved for the return respecting the Survey and Museum. My impression is that you were mistaken in your view of Sir D. Norreys' notice. I have given him a note to see the Museum as the best cure for any suspicions. Take care he is admitted when he calls.

1903 Seymour, Edward Adolphus, Lord   To: H.T. De la Beche
Office of Works.
19 January 1852
With Barry, I inspected the body found buried in the crypt of St. Stephens, and collected some dust and cloth. Would Playfair ascertain if it is embalming powder, and whether the cloth is linen or hemp?

1904 Sharpe, Daniel   To: H.T. De la Beche
12 New Ormond Street.   Post Office,
3 October 1830   Dartmouth.
[Readdressed to:] Devonport.
I have received your Pilchards which arrived in excellent condition. I shall probably go to Lisbon for 12 months to set up business contacts. This destroys my fish scheme. I will collect all the Geological information I can while at Lisbon. I am sorry to leave before the Geological Society gathering, but I wish to cross before winter sets in.

1905 Shipman, John   To: H.T. De la Beche
[Jamaica].
10 December 1824
[Receipt]
£40 received as donation to Wesleyan Mission to attend estate at Halse Hall.
[See also De la Beche, H.T. to Williams, J., 25 August 1824, **359**].

1906 Silliman, Benjamin   To: H.T. De la Beche
Yale College.   Dartmouth,
30 June 1830   Devonshire.
Thank you on behalf of myself and American Geological Society for the copies of the Tabular View of English strata. I regret I cannot again enjoy the opportunities I once had (1805-06) of studying the London Basin, the Isle of Wight etc.

1907 Simons, G.   To: H.T. De la Beche
Royal Academy,
Delft.
14 March 1851
Allow me to introduce Mr. E. van der Elst and his pupils destined for service in the Dutch colonies, who are in England for their instruction in mining affairs. I offer you also a map of the Haarlem lake, in the draining of which you are interested.

1908 Sismonda, Angelo   To: [H.T. De la Beche]
Leicester Square.
27 February 1848
Would you have the kindness to fix an hour on Monday or Tuesday when I may see you, without disturbance to your numerous and important affairs?

1909 Sismonda, Angelo   To: H.T. De la Beche
Turin.
27 October 1848
I have great pleasure in telling you that the Tertiary fossils have been sent to you. The box also contains volumes of the second series of memoirs of the Academy for your Museum, and some brochures which I ask you to forward to their appropriate destinations. Our country is in a pitiable state. Civil war is feared and even anarchy. The moderate party count on Anglo-French intervention.

1910 Sismonda, Angelo   To: H.T. De la Beche
Turin.
26 November 1848
I hope that the fossils have already arrived. I have received the parcel with the 2 copies of the 2nd volume of Memoirs which have been published under your direction. Thank you for my personal copy, and the Academy will thank you for theirs by sending you the diploma declaring you to be a member of its scientific group. The affairs of our country become more difficult as each day passes. In the middle of this political ferment I, however, remain faithful to science, but my work is hindered.
[See Turin Academy to De la Beche, H.T., 26 November 1848, **2045**].

1911 Sismonda, Angelo   To: H.T. De la Beche
Turin.
31 March 1849
The country is in a most unsettled state after the defeat by the Austrians eight days ago. I would like to know if you have received the box containing the Memoirs of the Academy, and the minerals and fossils of which I spoke in my last two letters? We have not the courage to take our work seriously here. You are fortunate in England, and I hope this fortune continues.

1912 Sismonda, Angelo   To: H.T. De la Beche
Turin.
11 November 1850
I have not replied to your letter of last February 22, because I have been awaiting the arrival of the geological maps and memoirs of which you spoke, but they have not yet arrived. Were they consigned to the Embassy in London? If so I will write for them. This autumn I made a long journey in the Savoy Alps with Studer. He wanted to verify for himself the classifications which I gave to these formations. I have heard from Collegno that all goes well with you.

1913 Sismonda, Angelo   To: H.T. De la Beche
Turin.
10 December 1850
It is so long since I have received news of you that I hasten to ask for it. Only some days ago, I received the discourse which you pronounced to the Geological Society in February 1849. There was no letter with it. Thank you for the communication there of my works on the Alps. Murchison has spoken of my labours in detail in a way most flattering to me. The good will of two great geologists such as yourself and Murchison means much to me.

1914 Sismonda, Angelo   To: H.T. De la Beche
Turin.
27 December 1850
It is only a short time since I last wrote to you, but I take the opportunity to tell you that until now the Academy has not received the sheets of the geological map of which you spoke at the beginning of this year. If they were consigned through the Legation in London, I will write to enquire about them.

1915 Sismonda, Angelo   To: H.T. De la Beche
Turin.
17 May 1851
I wish I was one of those who are such masters of their subject that I could come and pay you a visit. I would like to hear your news. The maps of which you spoke in your friendly letter of February last year have not yet arrived. If the packet was sent, then it surely has been lost. It may still be in London. I hope you have received Vol X of the Memoirs of our Academy. Vol XI has just been published and will be sent to you.

1916 Sismonda, Eugene   To: H.T. De la Beche
Secretaire,   London.
Académie Royale des Sciences,
Turin.
17 July 1848
The President thanks you for your letter of 18 May, and for the gift of the geological maps and the horizontal and vertical sections of Great Britain which you have sent. The Academy thanks you by nominating you a corresponding member. The Memoirs of the Academy will be sent to the Geological Survey Office.

1917 Sismonda, Eugene To: H.T. De la Beche
Secretaire, Craig's Court,
Académie Royale des Sciences, Charing Cross,
Turin. London.
26 October 1848
The Academy has sent you Vols. I-IX of the second series of Memoirs for the Geological Survey Office, and Vols. VII-IX for the Geological Society. In the box you will also find some fossils from Piemont which my brother the Director of the Museum has sent for the Geological Survey Office.

1918 Sismonda, Eugene To: H.T. De la Beche
Secretaire, London.
Académie Royale des Science,
Turin.
26 November 1848
I have the honour to announce to you that the physico-mathematical class of the Royal Academy of Sciences, it being proposed by Professors Ange and Eugene Sismonda, have nominated you corresponding member of the Academy. Our sincere congratulations come with the enclosed diploma. [See also Sismonda, A. to De la Beche, H.T., 26 November 1848, **1910** and Turin Academy to De la Beche, H.T., 26 November 1848, **2045**].

1919 Slaney, Robert A. To: H.T. De la Beche
University Club.
6 May 1847
I enclose a plea for the Working Classes, pointing out the necessity to improve their social condition, and the savings that would accrue from such measures. Our Health of Towns Bill ought to be carried with despatch, and will do vast good. Please put my name down as a Member of your Palaeontological Society.

1920 Smirke, Sydney To: H.T. De la Beche
24 Berkeley Square.
2 July 1846
Understanding that you and Smith have been asked to report on the stone used at the British Museum, I offer to attend any appointment you make, on behalf of my brother Sir Robert, who is unwell.

1921 Smith, Charles Harriott To: H.T. De la Beche
26 October [1830 FJN] Lockyer Street.
I return your book with thanks, having copied out of it the view of Jamaica etc. I am sorry you did not find me at home yesterday, but I was on the electioneering dinner with Sir Thomas Acland.

Smith, C.H.
16 April 1839
[see **2273**]

Smith, Joshua Toulmin To: H.T. De la Beche
8 Sarjeants Inn.
30 November 1847
[see **2274**]

1922 Smyth, Annarella To: [H.T. De la Beche]
Cardiff.
1 April 1841
We are delighted with the lithographs of the medal. Will send you copies. You asked me to remind you of procuring an official commission from your Museum for our son Warington W. to inspect machinery-making in Germany. Also what should he do at Naples etc.

1923 Smyth, Annarella To: H.T. De la Beche
Cheyne Walk,
Chelsea.
12 July [1850 FJN]
Here's the parcel to assist in illustrating the Moira Mine. Please read my husband's address to the Géographes at their Anniversary. Look at Roman Mosaic at No.11 Pall Mall.

1924 Smyth, William Henry To: [H.T. De la Beche]
Cardiff.
20 December 1840
All that I can learn of the datum of the rail-road is that it is somewhere about the sea-lock pond of the Glamorgan Canal, most likely at the spot where their Western Terminus is to be. What time will you be able to open our Cardiff Institution? I enclose a letter from Sir Francis Knowles.

1925 Smyth, William Henry To: H.T. De la Beche
3 Cheyne Walk,
Chelsea.
8 January 1850
I enclose one of three letters on the same subject. Please read it and tell me on what grounds the Commissioners were nominated. Was it by the Societies? I wish to be armed to answer enquiring friends.

1926 Smyth, William Henry To: H.T. De la Beche
Athenaeum.
15 January 1852
I have just returned from Oxford. I shall this day bring Long's retirement from the Antiquaries before Lord Mahon. Let me know how many weeks or days you can allow, for he must wind up.

1927 Smyth, Warington Wilkinson To: H.T. De la Beche
Schemnitz. Director of the
21 January 1842 Ordnance Geological Survey,
 Ordnance Map Office,
 Tower,
 London.
 Readdressed to: Southampton.
I still have not received your letter. Snow has prevented geological excursions. The richer ores of the district are now no more. There is little machinery worthy of note. Specimens of ores are available. I have made various excursions in the vicinity of Pesth. What can I do for you in this district?

1928 Smyth, Warington Wilkinson To: H.T. De la Beche
Orawitza in the Bannat. Director of the
30 July 1842 Ordnance Geological Survey,
 Ordnance Map Office,
 Tower,
 London.
 Readdressed to: Southampton.
I have received a letter authorising £20 or £30 for drawings and specimens. The only expense will be land carriage and metallic value. I have been on a week's tour of Syrmia. Herewith a catalogue of 89 specimens from Elba. The map is more difficult to send, as it is not unified. I will be delighted to receive any further commands.

1929 Smyth, Warington Wilkinson To: H.T. De la Beche
Orawitza in the Bannat. London.
21 October 1842
I have at last received your letter of 21 August. During the first month I have not found anything suitable for the Museum. I have visited various mines etc. It is now too late for the Turkey letter as I must cross the Balkan before winter commences. I will expect to hear from you at Schemnitz, to which I will return in April.

1930 Smyth, Warington Wilkinson To: H.T. De la Beche
Castlecomer.
14 October 1845
Thank you for your letter and receipt for Edwards. I have been out to the coal measures in Queen's County, and have found an interesting section. Griffith has given a section of much the same line as you selected. The sheet and reports are not yet here.

1931 Smyth, Warington Wilkinson To: H.T. De la Beche
Castlecomer.
21 October 1845
I purpose leaving for Carlow tomorrow. I suppose I can still address Wylie at Arklow. Here is a description of a dyke in Milford limestone quarry. I have laid aside some specimens. Does not this speak for a vast evolution of carbonic acid from the subjacent granite? I have seen something similar in travertine deposited by carbonic acid springs.

1932 Smyth, Warington Wilkinson To: H.T. De la Beche
Woodenbridge,
Arklow.
15 November 1845
I have only just left Tinahely for this place. It would be but a half measure to attack the Ovoca mines without some knowledge of the geology of the district. Griffith's map is insufficient. I will draw out the section roughly for your inspection. I am collecting samples. Should I send them to London? Wyley will do admirably. He is well prepared and full of zeal.

1933 Smyth, Warington Wilkinson To: H.T. De la Beche

Wooden Bridge,
Arklow.
3 December 1845

I have been tracing out the granite of Croghan and its flanks. It produces much trouble. There are slates at 750 feet so the sea must have been of tolerable depth. The mines will afford an unusual picture. Snow is preventing rapid work. I have kept a suite of specimens. My best address is Dublin.

1934 Smyth, Warington Wilkinson   To: H.T. De la Beche
Aughrim,
Co. Wicklow.
14 December 1845

I have been working here alone for the last ten days. Wyley is around Arklow. There is enough work on the mines to last through Christmas. Wyley continues to do well. I will send all the specimens together when I pass through Dublin.

1935 Smyth, Warington Wilkinson   To: H.T. De la Beche
Ovoca Hotel,
Rathdrum.
16 May 1846

I have finished at Cairn in Wexford, and am now doing surface work around these Ovoca mines. I am adding ferocious green splashes undreamt of in the Griffithian map. Wyley is at Tinahely. The Captain has left Dublin on the pier investigation. I move next to the lead mines of Glenmalure.

1936 Smyth, Warington Wilkinson   To: H.T. De la Beche
Ovoca Hotel,
Rathdrum.
31 May 1846

I have nearly completed the sheet around these mines. There is more to do than I anticipated. Flanagan was over from Rathdrum where he has had great success in fossil finding. Wyley lacks a businesslike application to one point of his work, though the Captain regards it as a serious evil. I think it a venial foible.

1937 Smyth, Warington Wilkinson   To: H.T. De la Beche
Ovoca Hotel,
Rathdrum.
8 June 1846

The heat has been intense. I am just starting for Glendalough where I will look up the lodes. My address is still Rathdrum. I have not seen Wyley again. I'm afraid there is a touch of the Hibernian irregularity about him. James returns in a week or ten days. Flanagan promises a great store of fossil localities by the time you arrive.

1938 Smyth, Warington Wilkinson   To: H.T. De la Beche
Glendalough.
28 June 1846

James tells me you're in Dublin. Today I am at Seven Churches, and cross tomorrow to Glenmalure where there is a very fair hotel at Drumgoffbridge. There I'll remain till I hear from you.

1939 Smyth, Warington Wilkinson   To: H.T. De la Beche
Rathdrum,
Wicklow.
12 August 1846

I have been for a day in Dublin, and intend to go to Ovoca tomorrow, to look at continuation of metalliferous beds towards Wicklow. The maps for Anglesea should be sent either to the office in Dublin or to Holyhead.

1940 Smyth, Warington Wilkinson   To: H.T. De la Beche
Aberystwyth.
9 September 1846

I enclose a letter from Haidinger which may interest you on the subject of trying to get up scientific societies in Vienna. He is playing a part analogous to yours. Please return the letter when writing here. I spent yesterday with John Taylor Junior with whom I was much pleased.

1941 Smyth, Warington Wilkinson   To: H.T. De la Beche
Devils Bridge.
28 September 1846

I am glad you are coming to the hills of Wales. The weather was bad and I could only do underground work. I expect it will prove another of those acute-angled junctions which seems to bring about bunches of ore. I will be detained till Friday or so. Then to Pont Erwyd.

1942 Smyth, Warington Wilkinson   To: H.T. De la Beche
Pont Erwyd.
16 October 1846

I have been double the time expected with this part. I have recorded 21 new lodes, which at some period have been worked on. Your Cornish informants have taken things for granted without looking closely into them. Cwm Ystwyth is an interesting spot. Many people enquire after you. I leave tomorrow for Goginan.

1943 Smyth, Warington Wilkinson   To: H.T. De la Beche
Birmingham.
21 December 1846

I have seen various people round by Oldbury and Dudley. A meeting of coal owners was about to write to Sir George Grey. I will take up my residence in Dudley. The Parliamentary documents have been received from Percy.

1944 Smyth, Warington Wilkinson   To: H.T. De la Beche
Chorley,
Lancashire.
[1 January 1847]

The Dudley lawyer changed his mind. I have been tracing the works in question. I am within 2 or 3 miles of Coppull. It was a small explosion, yet it killed many. Binney tells me of one that killed 12, and was not even recorded in the papers! I will look at another pit near Dudley to compare it with Oldbury.

1945 Smyth, Warington Wilkinson   To: H.T. De la Beche
Builth,
Breconshire.
[1847]

I have organized a plan of work with Ramsay. I will work on the traps tomorrow, then return to Llanidloes. Then to the mines in the north of Montgomery.

1946 Smyth, Warington Wilkinson   To: H.T. De la Beche
Llanidloes,
Montgomeryshire.
23 May 1847

I start tomorrow for Mallwyd, and expect to find the documents there. Thence to Aberystwyth. The walk from Llangurig was very "sweltry".

1947 Smyth, Warington Wilkinson   To: H.T. De la Beche
Mallwyd.
30 May 1847

I must visit the Crown Agent, Lewis Pugh, at Dolgelly. I observe I am expected to report on the Cardiganshire Crown mineral property. Letters to Aberystwyth. I have been entering the lodes of this Cowarch district, the working of which is one of the worst of imprudent and ignorant speculations I have seen.

1948 Smyth, Warington Wilkinson   To: H.T. De la Beche
Aberystwyth.
1 June 1847

I have seen Pugh at Dolgelly. Fossett unfortunately is in London at present. If the Colliery explosions pay has not come through could you let me have £20. Would it be advisable to append to the report a couple of quarter sheets with the lodes etc.?

1949 Smyth, Warington Wilkinson   To: H.T. De la Beche
Aberystwyth.
7 June 1847

Thank you for the supplies. I will send you the results of my work in 10 days. I shall have seen Pugh and extracted from him what would be needful. Please ask Hunt to copy the advertisement of Esgair y Mwyn for me. Haidinger has written thanking you for the maps etc.

1950 Smyth, Warington Wilkinson   To: H.T. De la Beche
Wellington,
Salop.
22 August 1847

Williams has recorded facts of this coal-field uncommonly well. I am leaving for Ironbridge. A local clergyman here unhesitatingly declared peat to be under the New Reds we are standing on!

1951 Smyth, Warington Wilkinson   To: H.T. De la Beche
Pontesbury,
Shrewsbury.
12 October 1847

Re your last letter enclosing Newton's on meanings of Greek phrases, his supposition concerning the first is probable. As for the second I've not heard of honey being used in the preparation of copper. I enclose Haidinger's address. The Shrewsbury sheet is more troublesome than expected.
[See Newton, C. to De la Beche, H.T., 11 July 1847, **1041**].

1952 Smyth, Warington Wilkinson   To: H.T. De la Beche
Shrewsbury.
9 November 1847

Receipt filled up as required. Please retain the £20 I owe you. Please look at the accompanying diagrams of metalliferous slates, minerals etc. I will start for the Clee hills in about a week.

1953 Smyth, Warington Wilkinson   To: H.T. De la Beche
Shrewsbury.
17 November 1847

Please pay the money you mention to my Mother. I will go to Ludlow on Saturday or Monday. I will try to send you my Cardiganshire MSS before I leave Salop.

1954 Smyth, Warington Wilkinson   To: H.T. De la Beche
Shrewsbury.
22 November 1847

I am leaving for Ludlow. Herewith the manuscripts as promised. It seems well to say a few words on the ideas of the miners. The conclusion will be a catalogue of lodes.

1955 Smyth, Warington Wilkinson   To: H.T. De la Beche
Clee Hills,
Salop.
29 November 1847

I have completed a look through this bit of coal, which turns out to be the best piece of Gibbs' work I have yet seen. I will call on Aveline on my way to Oswestry. Lewis the worker of coal here, a friend of Murchison's, is particularly obliging.

1956 Smyth, Warington Wilkinson   To: H.T. De la Beche
Ruabon,
Denbighshire.
12 January 1848

I will be in town for the Royal Hammerers' dinner. I have been working on the coalfield here, but am no further north than Ruabon. I have found beneath the lowest coal the distinctive fossils of the Lancashire and West Yorkshire coal measures.

1957 Smyth, Warington Wilkinson   To: H.T. De la Beche
Neath.
[7 April 1848]

The inquest commenced at ten, not two, so I was late. Coke waited for me, though, and on arrival I explained my role as private. The Coroner was disappointed. Struvé wishes to be remembered. I refused Glynn's assistance. I am at the "Castle".

1958 Smyth, Warington Wilkinson   To: H.T. De la Beche
Neath.
8 April 1848

The inquest has been adjourned till Monday week. The proceedings have made it evident that but for a general feeling among the magistrates, the injured parties would have had no chance of justice. The want of feeling evinced by Struvé and Randall suprised me. I have been aiding Cuthbutson the Coroner and Mr. Coke throughout. My findings will be communicated privately and I will return to London to avoid being summonsed as a witness.

1959 Smyth, Warington Wilkinson   To: H.T. De la Beche
Neath.
8 April [1848]

The Jury would want me to be sworn after visiting the mine. The ends of justice best served by informing the Coroner and Clerk of the Peace of the real state of the case. By Wednesday the owners will have altered things to a condition entirely different from what they were at time of explosion. Great neglect is already proved.

1960 Smyth, Warington Wilkinson   To: H.T. De la Beche
Amlwch.
20 June [1848]

I will fetch the compass and do what you suggest. I have been underground at the Pary's mine and seen a fine bunch of copper there. Yesterday I spent in the Llaneilian mountains. The lack of a boat prevented my going to the island off the granite-bedevilled coast towards Dulas.

1961 Smyth, Warington Wilkinson   To: H.T. De la Beche
Amlwch.
24 June 1848

I got but a poor little batch of things from Llanbabo the other day. I got in some more Granite yesterday. The Cornish captains here insisted there wasn't a morsel of granite as big as your hand. The Porphyry on Llaneilian mountain cuts a very funny little figure.

1962 Smyth, Warington Wilkinson   To: H.T. De la Beche
Llanerchymedd,
Anglesea.
13 August 1848

I examined the road Westward from here at a spot where the Granite borders on grits and conglomerates. In a quarry was a vast deposit of shells. Specimens will be sent to London in a day or two.
[Letter in mock military terms].

1963 Smyth, Warington Wilkinson   To: H.T. De la Beche
Llanfechell,
Anglesey.
10 September 1848

I fear the weather will have made your trip to Lambay Island unpleasant. I am much struck by the change to this Northern ground after the grits and black shales to the South. Is there a fault? Have we been looking at a cleavage and not a dip? Do the shales only appear to pass under green slates? A fair section will result. I will remain until further notice.

1964 Smyth, Warington Wilkinson   To: H.T. De la Beche
Llanfechell,
Anglesey.
18 September 1848

The geology of Porth Swtan is not so simple as at first sight appeared. Dips instead of being North are all to the South. The black shales dip the same way. But the fossiliferous grits and black shales overlie the green curly slates of the South, and are in turn overlaid by felspathic, talcose etc. slates.

1965 Smyth, Warington Wilkinson   To: H.T. De la Beche
Llanfechell,
Anglesey.
25 September 1848

I have been collecting specimens of Serpentine. I have an array of Elvans here, though all small. I have found a Porfido verd'antico among the dikes. Serpentine I've got in three other places. Our sections are genuine. I have been applied to by friends to report on the lode at Alston Moor. As it involves a fee, I ask your permission to agree to do the work.

1966 Smyth, Warington Wilkinson   To: H.T. De la Beche
Chester.
22 October 1848

I am now acquainted with the lead mining notions of the North. I was underground in one of Beaumont's mines, close to Sopwith's new house. I have secured a mass of vein stuff with carbonate of zinc for the Museum. I spent a couple of evenings with John Taylor Senior. I will stay here for a week.

1967 Smyth, Warington Wilkinson   To: H.T. De la Beche
Cheadle,
Staffordshire.
[1848]

I cannot return tomorrow morning as proposed. It will be the evening or so. I have evidence with respect to the Consall coal, also of use to the Survey. This is the site of the new Catholic Cathedral.

1968 Smyth, Warington Wilkinson   To: H.T. De la Beche

Beddgelert.
19 May 1849

The cycle of bad weather is interfering marginally. One or two mines here are more important than I expected. When a fossil collector gets near Snowdon again he should pick out a few good lode specimens, as there are combinations different from the South Welsh. From Tuesday to Friday will be at Ffestiniog.

1969 Smyth, Warington Wilkinson   To: H.T. De la Beche
Holywell,
Flintshire.
30 November 1849

I received a letter from Hauer yesterday. Haidinger has been instructed to draw up an outline of a proposed Geological Institute. This has been submitted to the Emperor. An Imperial decree established it on 21 November. Details given.

1970 Smyth, Warington Wilkinson   To: H.T. De la Beche
Holywell,
Flintshire.
30 December 1849

Regarding the new Museum, we ought augment our nucleus of mining implements. A series ought to be so placed that those interested can handle them etc. I have only a few more places to visit. At the end of next week, I mean to go to Prestatyn, Flintshire.

1971 Smyth, Warington Wilkinson   To: H.T. De la Beche
Ecton Mine,
Ashbourne.
18 October 1850

So many points of interest around here, I will not leave till Tuesday. The contortions of the limestone in this mountain are very striking in contrast with Derbyshire, and syn- and anti-clinals so persistent in their strike as to be counted on by the miners. There is a very good inn at Hartington.

1972 Smyth, Warington Wilkinson   To: H.T. De la Beche
Chesterfield,
Derbyshire.
20 November 1850

With the assistance of Baker's maps I have furbished up the Longstone Edge, Alport and Hubberdale districts with some additional lodes. I could get an ore-crusher for £4. I have been collecting Black shale ironstones, and enclose a list of the different beds in descending order. The box of specimens will go to the luggage train today.

1973 Smyth, Warington Wilkinson   To: H.T. De la Beche
[1850 FJN]

Engineer J.T. Woodhouse's working plans of collieries have given me a great store of facts. Phillips's coal-measure lines are drawn as a random sketch and better ignored. The New Red boundary was done with great care. Elliot has been detained as a result of a serious explosion in one of his works.

1974 Solly, Edward   To: Sir H.T. De la Beche
33 Bedford Place,   President of
Russell Square.   Geological Society.
29 November 1847

I am desired by the Widow and Administrator of my late Uncle Samuel Solly to request the return of a model or models illustrative of Saxon mines, placed in the care of the curator of the Museum of the Geological Society.

1975 Somerville, William   To: H.T. De la Beche
Bologna.   Geological Society,
February 1847   London.

Professor Bianconi of the University of Bologna has concluded that the bursting of the barrier of the Straits of Gibraltar took place in historical times. He has some questions which has no means of solving. I advised him to send 3 copies of his paper to the Admiralty, Geological Society, and Geographical Society. Soundings might be obtained from the Admiralty. You may be able to solve his doubts.

1976 Sopwith, Thomas   To: H.T. De la Beche
1 Chapel Place,
Duke Street,
Westminster.
18 March 1847

If a judicious system of inspection is established, funds may be provided by an impost. An Act in 1788 provided a tonnage rate on all coals taken down the Tyne in Keels. Local measures furnish ready means of collecting such a rate, and the practicability and prudence of these are freely admitted with reference to Bainbridge's Bill.

1977 Sopwith, Thomas   To: H.T. De la Beche
Allenheads,
Haydon Bridge.
18 March 1848

I wish to introduce to you to my friend Mr. Loftus of Newcastle. Please afford him the information he desires.

1978 Sopwith, Thomas   To: H.T. De la Beche
Allenheads,   6 Craig's Court,
Haydon Bridge.   Charing Cross.
18 March 1848

Please give an interview to Mr. Hosmer, whose invention for cleansing sewers I wish to bring to your attention. Please give my friend your favourable consideration.

1979 Southern, Henry   To: H.T. De la Beche
British Legation,
Madrid.
23 May 1845

Before I left London I forwarded a box of specimens of Coal from the Sierra Morena. I hope they are suitable for the Museum. I would like to hear about the fossils I sent to Forbes.

1980 Sowerby, James de Carle   To: H.T. De la Beche
2 Mead Place,   Lyme Regis,
Westminster Road,   Dorsetshire.
London.   Readdressed: T.H. Aveline's,
28 March 1826   Batheston,
     near Bath,
     Somerset.

I have completed the examination of the fossils you sent some time ago to my brother. I have figured several belonging to new genera or not known to English conchologists. Thanks also for the parcel left before you went from England. I will willingly make drawings of any more you send me.

1981 Sowerby, James de Carle   To: H.T. De la Beche
2 Mead Place,   Jermyn Street,
Westminster Road.   St. James's.
4 November 1829

I can name but a very few of the 18 species of Ammonites and 2 of Belemnites you have submitted to my examination. The results are remarkable and I account for the occurrence of so many new species together by supposing them to be in a formation much less known than the Oolites or Lias of England.

1982 Sowerby, James de Carle   To: H.T. De la Beche
Geological Society,   President
Somerset House.   Geologicial Society.
22 June 1847

I ask permission to take my leave in two parts, as it will be equally consistent with the business of the Geological Society and a great convenience to me.

1983 Sowerby, John Edward   To: H.T. De la Beche
3 Mead Place,
Lambeth.
16 October 1849

Enclosed for your inspection is a prospectus of a re-issue of the 2nd edition of English Botany. Should you not be provided with either edition we shall have great pleasure in adding your name to the list of subscribers.

1984 Spearman, A.G.   To: H.T. De la Beche
Treasury Chambers.   Athenaeum Club.
27 July 1838

£250 for figuring and describing the Organic Remains connected with the Ordnance Survey of Cornwall, Devon and West Somerset has been authorised by the Lords of Treasury. The money will be issued when the work is completed.

1985 **Spence**, W.   To: H.T. De la Beche
       18 Lower Seymour Street,
       Portman Square.
       13 March 1848

I enclose an address and two articles. Ransome, the Secretary of Ipswich Museum is coming to coffee, the Dean of Westminster and Sir J.P. Boileau also; will you come too?

1986 **Spencer**, Thomas   To: H.T. De la Beche
       Mount Gardens,
       Liverpool.
       18 May [no year, but post-1842]

Several persons here are trying to convince the townsfolk that water is to be found below this district. Have you ever heard of natural cavities in New Red Sandstone? I have a pamphlet coming out, and wish to check the truth of my opposition to the idea.

1987.1 **Spratt**, Thomas Abel Brimage   To: H.T. De la Beche
              Teignmouth,
              Devon.
              19 April [1848 FJN]

I enclose a copy of a letter from Col. Sabine as proof of my labours in the cause of Magnetism and other branches of science. I will bring my claims before the First Lord. If still rejected my case will be no encouragement for others to labour out of the ordinary course of duty.
[See Sabine, E. to Spratt, T.A.B., 11 April 1848, **1844**].

1987.2 **Spratt**, Thomas Abel Brimage
Abstract of Services of Lieut. Spratt R.N.
22 years in Service. 16 years attached to Archipelago Survey under Captain Graves. Recently compelled to quit for health reasons. Captain Grave's report on my work mentions capabilities, zeal etc. 4 papers on Geology published, and 3 on Geographical & Antiquarian interest. One in Philosphical Journal.

       Stanley, Edward   To: H.T. De la Beche
       23 October 1848
       see **1062**

1988 **Stephenson**, Robert
       Westminster.
       26 May 1849
[Printed] Report on the Norfolk Estuary
[See Vetch, J. to De la Beche, H.T., 17 June 1849, **2067**].

1989 **Stephenson**, Robert   To: H.T. De la Beche
       24 Great George Street,
       Westminster.
       2 July 1850

Enclosed a circular I promised should reach your hands. You probably know the author, and if you could assist him by suggestions and recommendations to parties likely to give him employment, you will oblige me.

1990 **Still**, Henry   To: H.T. De la Beche
       St. Ives.
       15 October 1837

At Falmouth I spent a few hours at the Polytechnic exhibition. Fox's experiments producing mineral veins in clay by galvanism and causing clay to become stratified had the greatest interest for me. Henwood was there and appeared to be very much annoyed because he was not allowed to show off his opposition to Fox the first day. He vows vengeance.

1991 **Still**, Henry   To: H.T. De la Beche
       Ormskirk,
       Lancashire.
       25 June 1841

I was very glad to receive yours of the 10th and to hear that you have found so many novelties. I thought you would find the lines very different along the north edge of Mr. Murchison's Silurian rocks. As the Pembroke sheets have not arrived I think it better to send the enclosed as they may be of some use to you. I have a few other sections which I collected at the pits, but they are so contradictory that they would be of no use in their present imperfect state.

1992 **Still**, Henry   To: H.T. De la Beche
       Ormskirk,
       Lancashire.
       12 July 1841

I send you sheet 40 with as much as I could make out from my notes on the old map. There is something more on my field sketches, but as it is probable that Mr. Carrington is making use of them, I do not like to apply for them. If you think it worth while to do so, you will probably find some of the lines more accurately traced on them than they could be on the old map.

1993 **Still**, Henry   To: [H.T. De la Beche]
       Llanboidy,
       St. Clears.
       30 August [1843 FJN]

I have made an addition to my collection of vegetable impressions and fresh-water shells in the carbonaceous shales, and have traced them for some miles. They shall be sent to you in about a week with as much information on the subject as the time at my disposal will permit. I shall remain here about a week and then go to Fishguard in order to pack up the fossils and send them off.

1994 **Still**, Henry
Section of Cwm Ammon Colliery on the Farm of Ystrad Ammon. The proprietor of this Colliery (Mr. Biddulph) states that the measurements in this section were taken with great accuracy.

1995 [**Still**, Henry]
       6 May 1849
Section of Colliery at Pantyffynnon by the Manager. This is a curiosity in its way, therefore I prefer sending it in its present shape, and from what I saw of the man, I think the distances are likely to be correct. H.S.

1996 **Stokes**, Charles   To: H.T. De la Beche
       4 Verulam Buildings.
       12 December 1837

Thanks for your set of inaugural lectures which I hope to profit by as soon as I can read them.

1997 **Stokes**, Charles   To: H.T. De la Beche
       Verulam Buildings.
       1 October 1840

I send copies of my paper. One is for Dillwyn. I wish to get specimens of corals with animal still alive or fresh. Can you also get me some from the West Indies?

1998 **Stokes**, George Gabriel   To: H.T. De la Beche
       5 Norris Street,
       Haymarket.
       22 July [1854 FJN]

I am sorry to have put you to so much trouble. Willis and I have agreed as to a division of territory, to allow us to put out the prospectus. I send you what I propose for myself. Willis says his will do as it stands.

1999 **Struvé**, William Price   To: H.T. De la Beche
       London.   Office of Woods
       22 January 1840   & Forests.

The emoluments etc. of the office of Keeper of Mining Records at the Museum of Economic Geology are not sufficient inducement for me to leave my position at Swansea. Thank you for the offer.

2000 **Struvé**, William Price   To: H.T. De la Beche
       Swansea.   Museum of Economic Geology,
       3 May 1849   Craig's Court,
                                                                              London.

Herewith the Cambrian newspaper from which you see that my mine ventilator works well. Should inspectors be appointed I hope you will think kindly of me. Should I be appointed I would try to discharge the duties with ability and discretion.

2001 **Stuart**, G.O.   To: H.T. De la Beche
       Literary and Historical Society,
       Quebec.
       [No date]

You have been elected an Honorary Member of the Society, and your support is solicited.

2002 Stutchbury, Samuel   To: H.T. De la Beche
Bristol.   Haverfordwest,
7 September 1841   South Wales.
Here is the information you requested: Dr. Kentish died 5th December 1832. Sanders took great pains with the sections. I have lately obtained some interesting fossils from the Oxford Clay. Herewith in sketch form my reconstruction of them. I wish to become a member of the Geological Society. Will you propose me? I'll get Conybeare, Phillips and others to sign it.

2003 Stutchbury, Samuel   To: H.T. De la Beche
7 Park Street,
Bristol.
11 September 1841
You ask my official title. It is Curator of the Philosophical & Literary Institution, Bristol. I do not feel I should trespass on Phillips' manor by doing a paper on Belemnites. Pratt's on Ammonites will be an interesting paper.

2004 Stutchbury, Samuel   To: H.T. De la Beche
Bristol Institution.
19 May 1843
Knowing your interest in the mines of the Duchy of Cornwall, herewith information on the new branch at Norton Down Pit, bearing out the assumption of a roll over at Wells Way Pit. I hope shortly to give whole of the Radstock manor both in plan and section for your Ordnance Geology records.

2005 Sweden
Svegica, Scientiarum Academia
Stockholm.
8 December 1852
Certificate of membership (foreign) of H.T. De la Beche. Signed by [W?] C. Retzius (President) and P. Wahlberg (Secretary).

2006 Sykes, William Henry   To: H.T. De la Beche
India House.
25 May 1850
Herewith the Memorandum respecting Williams. In addition to £800 he had £72 horse allowance. Through the Commissariat he had great aid with carriage, which never appeared on the books as a money charge. In all he had over £1000 per annum in advantages.

2007 Sykes, William Henry   To: H.T. De la Beche
India House.
16 July 1850
The thing is progressing without impediment at present, but do not say a word about it as we are groping in the dark. I will let you know as soon as I can.

2008 Sykes, William Henry   To: H.T. De la Beche
India House.
[?1850 FJN]
You should write a semi-official note to Capt. Shepherd explaining why a Geologist cannot be found for India, under £1000. If you propose Oldham at that sum, you might get a favourable answer sub rosa, and then you can write officially.

2009 S., Jos.   To: [?]
Dublin.
[No date]
[Part of letter only]
I chatted with Mr. O'Brien and he shortly asked me if I could tell him anything about a report that he stated he had heard from Mr. Papworth. I think viz. that Sir Henry was to take this place into his hands immediately with something more to that effect. Sullivan questioned me and spoke of a rumour that the grant for this place would certainly not pass next year, saying it was the exertions of Sir Henry to put down Kane and this Museum.

2010 Talbot, Christopher Rice Mansel   To: H.T. De la Beche
Penrice Castle.   Bridgend.
6 November 1839   Readdressed to: The Rhyddings, Swansea.
The persons most likely to give you valuable information as to the coal-measures and stratification in the vicinity of Margam are Edward Daniel, Mr. Bevan, Benjamin Daniel and Lionel Brough. Show them this letter. My gamekeeper, Issac Stubbs, will show you the park, and can tell you of many old pits and crops scarcely known to anyone else.

2011 Talbot, Christopher Rice Mansel   To: H.T. De la Beche
Margam Park.
Bridgend.
18 December 1839
I expect my cousin [William Henry Fox-Talbot] tomorrow, and would like to introduce you to each other. Take the opportunity of looking at a few more coal veins here. I will have a room for you whenever you make it convenient. I have found another old level 150 yards S of this house.
[Cited: North 1934a, p.98]

2012 Taylor, John   To: H.T. De la Beche
Chatham Place.
18 May 1832
I send you letters to a Miner, a Parson, a Banker and an Apothecary. They know most of the country you are going to explore. My son will contact you if he comes to Tavistock.
[Cited: McCartney 1977, p.41]

2013 Taylor, John   To: H.T. De la Beche
University College,
London.
3 July 1847
The Council of the University have recorded their thanks for your kindness in presiding at the distribution of prizes, and also for your excellent address.

2014 Taylor, John Jr   To: H.T. De la Beche
2 Duke Street,
Adelphi,
London.
8 March 1847
Today's Times contains another account of a colliery explosion. There is one remedy only, to sink more shafts. Are people to be killed wholesale when simple means would prevent it? Do institute enquiry upon this head.

2015 Taylor, Richard Cowling   To: H.T. De la Beche
48 South Fourth Street,   Museum of Economic Geology,
Philadelphia.   London.
[February 1842 FJN]
Respecting State Geological and other Reports, I have written to all the Governors and received annual reports from some in return. All final reports will have to be purchased. Lyell is instructing us with his lectures. His sections and diagrams are poor. I have been working on a reconnaissance map of Cuba. I have been puzzled by your white Jamaica limestone.
[Cited: Dott 1996, p.133]

2016 Taylor, Richard Cowling   To: H.T. De la Beche
48 South Fourth Street,   Museum of Economic Geology,
Philadelphia.   London.
31 March 1842
I apologise for your having to pay postage on the parcel sent. Some State reports have been sent to me. Many wish exchange arrangements. The American model would cost about $400, and I am willing to pay the expenses out of my own pocket. You would also have the Welsh model included in that price. Lyell confines his intercourse to too few people. I have a plan for working the Gold ore of Cuba.
[Cited: Dott 1996, p.133]

2017 Taylor, Richard Cowling   To: H.T. De la Beche
48 South Fourth Street,   Director,
Philadelphia.   Museum of Economic Geology,
19 May 1842   London.
I have sent various documents as per catalogue enclosed. I will have to purchase the final reports of the State Surveys. Do you wish them? Very little by way of return gifts come from Europe. Please look over the specimens I sent to the Geological Society. Myself and two friends are commencing small operations on the Gold ores of Cuba.

2018 Taylor, Richard Cowling   To: H.T. De la Beche
Philadelphia.   Museum of Economic Geology,
25 October 1842   Department of Woods and Forests, London.
I forwarded you a package last September, and also sent a letter; but

nothing has been heard. I am at a loss to account for this, as I have taken much pains to secure their safe arrival. The New York Geological Report will be a masterly affair. I sent you a statement of expenses. $18 due for carriage and postage.

2019 Taylor, Richard Cowling　　　　　　　To: H.T. De la Beche
　　　48 South Fourth Street,　　　　Museum of Economic Geology,
　　　Philadelphia.　　　　　　　　　　　　　　　Craig's Court,
　　　25 April 1843　　　　　　　　　　　　　　Charing Cross,
　　　　　　　　　　　　　　　　　　　　　　　　　　London.

It is long since I've had a letter from you. I have heard that the box sent in May 1842 has been received at the Geological Society. What do you think of my series of American coals. The distressed condition of all the States has brought Surveys to a halt.

2020 Taylor, Richard Cowling　　　　　　　To: H.T. De la Beche
　　　48 South Fourth Street,　　　　Museum of Economic Geology,
　　　Philadelphia.　　　　　　　　　　　　　　　Craig's Court,
　　　25 June 1843　　　　　　　　　　　　　　Charing Cross,
　　　　　　　　　　　　　　　　　　　　　　　　　　London.

I need your authority before making further collections for the Museum. Is there anything on Jamaica Geology in print. Have you obtained the published volumes of the New York State Survey? It is hardly selling at all. Murchison has made handsome allusion to myself in his Geological Society address.

2021 Taylor, Richard Cowling　　　　　　　To: H.T. De la Beche
　　　48 South Fourth Street,　　　　Museum of Economic Geology,
　　　Philadelphia.　　　　　　　　　　　　　　　Craig's Court,
　　　25 September 1843　　　　　　　　　　　Charing Cross,
　　　　　　　　　　　　　　　　　　　　　　　　　　London.

I have heard nothing of the money due to me from your Department. It can be paid by Bill of Exchange. My cousin John Taylor has informed me of your awareness of the amount due. I am sending you the Proceedings of the Philosophical Society of America. No notice was taken of my present to the Geological Society. Can you send me a specimen of your beautiful Geological Maps.

2022 Taylor, R. and J.E.　　　　　　　　　To: H.T. De la Beche
　　　Printing Office,
　　　Red Lion Court,
　　　Fleet Street.
　　　12 April 1847

We have 320 copies of your "Researches in Theoretical Geology" left. 1000 were printed. We cannot find the woodcuts, it being so many years since they were used. Do you wish the above copies to be sent to Craig's Court?

2023 Taylor, W.　　　　　　　　　　　　　To: H.T. De la Beche
　　　14 New Ormond Street,
　　　Queen's Square,
　　　London.
　　　7 April 1840

Am I right in believing you once said that the change from slavery to freedom in Jamaica worked well for you? And that besides, your own people remained on Halse Hall, and some from neighbouring estates removed thither? I wish to publish a statement from you, to assist in the cause of emancipation, now argued in France. I wish to assist the emancipation of the African race throughout the world.

2024 Taylor, W.　　　　　　　　　　　　　To: H.T. De la Beche
　　　14 New Ormond Street,
　　　Queen's Square,
　　　London.
　　　27 April 1840

In reply to a recent enquiry at Wesleyan headquarters. I received the enclosed copy of some resolutions passed by a committee of examination on Halse Hall. If you can give me any information on the points alluded to in a recent letter you will greatly oblige me.

2025 [Theobald, W.]　　　　　　　　　　To: H.T. De la Beche
　　　Calcutta.
　　　7 December 1848

A melancholy event obliges me to ask the favour of your acquaintance. I allude to the death of D.H. Williams which happened on 15 ult. He has left a will, and named yourself, me and Col. Forbes trustees and executors. I trust you will act in England. The cause of death was jungle fever preceded by a fall from his elephant while out on duty. I have sent an extract of details to Mrs. Williams and asked her to send it to you.
[Letter incomplete].
[Enclosure] W. Theobald to Mrs. Williams, 21 November 1848. [Details of Williams' accident and last illness].

2026 Thomas, Henry　　　　　　　　　　　To: H.T. De la Beche
　　　Sea Point House,　　　　　　　Museum of Economic Geology,
　　　Rock Island,　　　　　　　　　　　　　　Jermyn Street,
　　　[Cork].　　　　　　　　　　　　　　　　　　London.
　　　21 October 1850

At the request of Petherick I have forwarded the stone hammers discovered in Cork. While searching for copper lodes in a ravine, I found old workings which I had cleared. The hammers were found about four feet from the surface. Timber also was found. Some attribute the workings to the Danes, but it remains a conjecture.

2027 Thompson, William　　　　　　　　　To: H.T. De la Beche
　　　Donegal Square,
　　　Belfast.
　　　29 September 1846

The last volume of Geological Survey memoirs is much desired here. If Government present generally the works of the Geological Survey to any public institutions I should hope that some one of those in Belfast may be considered worthy of being the recipient of them. Our libraries here do not contain the Memoir of the Geological Survey of Great Britain, Phillips' Palaeozoic fossils, Report of Geology of Cornwall, etc. But the Report of the Geological Survey of Londonderry was presented to our Natural History and Philosophy Society by the Lord Lieutenant of Ireland.

2028 Tremenheere, George Borlase　　　　To: H.T. De la Beche
　　　3 January 1839

The Chairman and Deputy Chairman of the East India Company came to the Museum and were impressed. Royle met them there. I have put in an official application, and will let you know the result as soon as it comes out. I am working regularly with Phillips. I analysed a sample of manure used on my Uncle's farm. Worthless. My brother has been appointed Inspector of Schools.
[Cited: Stafford 1984, p.8]

2029 Tremenheere, George Borlase　　　　To: H.T. De la Beche
　　　Oriental Club,　　　　　　　　　　　　　　Swansea.
　　　Hanover Square.
　　　7 November 1839

Royle heartily approves of our notions concerning India, especially of establishing a child of your Museum. I have begun a collection of mining products and visited several mines. I am engaged in Phillips' laboratory on analyses. Buckland did not do a lecture on your map well.

2030 Tremenheere, George Borlase　　　　To: H.T. De la Beche
　　　15 George Street,
　　　Portman Square.
　　　20 May 1840

Philipps from Woods and Forests sent me several copies of Estimates. They are just what India House wanted, as the history of your Museum is there in official form. Our proposals will take some weeks to pass through India House and Board of Control. I sail on 25 July. I hope to go to Cornwall at the end of June, and even spend two days in Wales with you.

2031 Tremenheere, George Borlase　　　　To: H.T. De la Beche
　　　15 George Street,
　　　Portman Square.
　　　23 May 1840

I heard yesterday that Court and Board of the East India Company have approved our proposition that I shall take charge of specimens from the Museum, for India etc. Lord Auckland has also asked for books on Geology and Mining. My various courses continue. Lindley is a good lecturer. My uncle requested soil analysis details. Dr Grant was much pleased with your Museum. Henwood admired your model of Cornish veins.
[Cited: Stafford 1984, p.8]

2032 Tremenheere, George Borlase　　　　To: H.T. De la Beche
　　　15 George Street,
　　　Portman Square.
　　　1 June 1840

I have received an official letter from India House stating that my suggestion of the expediency of establishing Museum in India analogous to yours has been forwarded to the Government there. I am directed to take charge of collection. May I have specimens from your Museum which are of no use to you?

2033 Tremenheere, George Borlase   To: H.T. De la Beche
[?] House,
near Penzance.
1 July 1840

I have availed myself of your permission and made up a collection of specimens with the help of Phillips. I hope for others from the Survey. The Secretary of India House is prepared to accept list of books. I could not come to Wales, due to work and preparations.

2034 Tremenheere, George Borlase   To: H.T. De la Beche
Ship Lord Hungerford,   Collenna,
Spithead.   Near Cardiff,
27 July 1840   South Wales.

Thank you for all your kindness and assistance. The present Chairman and Deputy at India House seem very interested in the Museum. I have collected specimens from various sources. Here are the package and mail arrangements.

2035 Tremenheere, George Borlase   To: H.T. De la Beche
Calcutta.   Museum of Economic Geology,
5 January 1841   Craig's Court,
         Charing Cross,
         London.

The Rev. R. Everest will deliver this to you. He is one of my mentors on Indian geology. The Governor General is very well disposed towards the Museum. I go to Tennasserim to investigate the peninsula as far as Mergui. Remember me to the Llewellyns. I should be glad of more specimens. Your sword will come by the "Lord Hungerford".

2036 Trevelyan, Charles Edward   To: Earl of Lincoln
Treasury Chambers.   Chief Commissioner
16 September 1845   of Woods.
[Copy]

Re your letter of 12th inst. directions have been given by H.M. Treasury to the Paymaster of Civil Services to pay De la Beche £2000 as an advance to be repaid from the grant for Museum of Economic Geology and Geological Survey of Great Britain and Ireland.

2037 Trevelyan, Charles Edward   To: H.T. De la Beche
Treasury.
29 June 1848

The two last charges in this account are heavy, especially if Smyth is already in receipt of a Government salary. What are Smyth's emoluments as Mining Geologist to the Survey, and are there precedents for the present charges?
[For reply, see De la Beche, H.T. to Trevelyan, C.E., 30 June 1848, **414**]

2038 Trevelyan, Charles Edward   To: H.T. De la Beche
Treasury.
9 March 1850

The official answer to your application on behalf of the Palaeontographical Society was sent to you at the Geological Society on the 5th instant. I regret to say the decision was unfavourable for reasons stated in the letter.

2039 Trevelyan, Walter Calverley   To: H.T. De la Beche
Athenaeum.
7 June 1847

Do you collect specimens at the Museum? Here is some carbonate of copper from Exmouth and Devon. Should you be working in Somerset, please take up quarters with us at Nettlecombe.

      Trevelyan, Walter Calverley   To: H.T. De la Beche
      Nettlecombe,
      Taunton.
      5 September 1848
      [see **2275**]

2040 Trevelyan, Walter Calverley   To: [W. Buckland]
1 November 1849   Dean of Westminster.
[Part copy]

I think I mentioned to you my suggestion that copies of the Ordnance Map should be displayed in public situations where they would be accessible to those who have contributed to their cost but cannot afford to buy them. My cousin Sir Charles Trevelyan has learned that the Ordnance Office is prepared to recommend supplies for that purpose if demanded by Museums etc.

2041 Trevelyan, Walter Calverley   To: H.T. De la Beche
Aldborough,
Suffolk.
20 June [1852 FJN]

I forgot to mention discovery of several mining implements and artificial fuel, supposed to be Roman at Luxborough on Brendon Hill. I will try to obtain a specimen for your Museum.

2042 Trimmer, Joshua   To: H.T. De la Beche
Ringwood,
Hampshire.
7 November 1849

Thank you for your note of the 6th. Herewith a copy of the paper from the Journal of the Geological Society of Dublin, which you wished to see. The following are the discoveries I claim with respect to the erratic Tertiaries. With regard to the Wexford deposits, three distinct epochs have been confused.

2043 Trincoe, Miss   To: W. Buckland
34 Hyde Park Gardens.
12 June [1849 FJN]

Here is a valuable fossil from Whitstone pits. Can you dispose of it for a poor family?

2044 Tucker, Henry   To: H.T. De la Beche
Wakefield.
11 June 1847

I asked Lord Morpeth for employment under the Sanitary Bill. This Bill has been abandoned and a Towns Improvement Bill will soon become law. Will you intercede for me as I am anxious to obtain employment under it.

2045 Turin Academy
26 November 1848

[Certificate of membership of the Academy (Corresponding associate), of H.T. De la Beche].
[See Sismonda, A. to De la Beche, H.T., 26 November 1848, **1910** and Sismonda, E. to De la Beche, H.T., 26 November 1848, **1918**].

2046 Turner, Edward   To: H.T. De la Beche
38 Upper Gower Street.   43 Pall Mall.
21 December 1830

I wish I could answer your questions on mineral springs. I mention the ones I know of, but recommend you consult Daubeny who has made a regular study of them. Please repeat your order for analyses.

2047 Turner, Edward   To: H.T. De la Beche
38 Upper Gower Street.   10 Craven Street,
2 May 1831   Strand.

The bones of ichthyosaurus etc. all have highly crystalline texture, and are composed of bituminous matter, phospate of lime etc. The scale of Dapedium Politum has only 19% of phosphate of lime. The fossil wood from Lyme Regis was of two kinds.

2048 Turner, Edward   To: H.T. De la Beche
38 Upper Gower Street.
18 April 1832

Your friend's illness is due to a deranged state of the stomach. I should advise a change of diet. I send two prescriptions, one for antacid powder, and the other for pills. If the region of the kidney continues painful, cupping would be desirable. If convenient please have him send urine samples.

2049 Turner, Edward   To: H.T. De la Beche
University of London.
7 December 1834

You will have heard from Greenough how your Devon findings were overturned last Wednesday at the Geological Society. Murchison affirmed that your coal was from Carboniferous beds. Greenough defended you as well as he could. Lyell inclined to Murchison's view, though he spoke more guardedly. I said your only possible error was that Greywacke and

Carboniferous rocks had been confused. It will prove an important fact if you can prove fossils of Carboniferous can be found beneath Murchison's Dudley beds.
[Cited: North 1939b, pp.1053-1054; Rudwick 1985, pp.99-100]

2050 Turner, Edward   To: H.T. De la Beche
University of London.
20 January 1834
[Postmarked 20 January 1835]

Herewith a method of collecting gases and vapours. Bottles should be first filled with water and emptied just before filling with gas. They should be well sealed after collection. I was not present when Faraday alluded to Harris and the Medal. I fear I must resign Secretaryship at the Geological Society because of health. Re your Greywacke plants, does it not mean that the same flora existed before and after Old Red, and even after Murchison's upper Greywacke beds?

2051 Turner, Edward   To: H.T. De la Beche
University of London.
28 November 1835

The proper charge for a mineral analysis is 5 guineas each. We are getting on capitally at the University. The new University scheme, making us University College, is likely to be of capital use to us. The Duke is to remain President of the Royal Society for another year. Lubbock absurdly resigned as Treasurer in protest. How is the Survey going?

Turner, Edward   To: J. Lindley
University.   Horticultural Society,
11 March 1836   Regent Street.

I know of no decided factors proving that resinous wood resists decomposition less than other wood.
[See J. Lindley 1832-3, **862**].

2052 Vaughan, R.   To: [H.T. De la Beche]
1 August [1840 FJN]

Since I had the pleasure of seeing you here last year, I have found several fresh crops of coal and by the help of a young manager and one or two old miners made up a section of this country which I hope to have the pleasure of submitting to you.

2053 Verneuil, Phillippe Edouard   To: H.T. De la Beche
Poulletier de
Paris.
20 February 1848

I am taking advantage of Sismonda's visit to London to send this letter asking news of yourself and your work. He will spend two months in London and I envy his seeing your magnificent collections. I would like to come with him for it is 2 years since I was in London, and that is a long time, but family interests keep me here. When I am free in spring I will go to Spain or even Sardinia where Murchison will wait for me.

2054 Verneuil, Phillippe Edouard   To: H.T. De la Beche
Poulletier de   Museum of Economic Geology,
19 October 1850   Piccadilly,
    London.

Three months ago the Director of the geological map of Spain, M. Casiano de Prado asked me to intercede with you to obtain if possible the published works and maps of the Geological Survey under your direction. They will be of use in assisting the geological survey of Spain which has begun in the Province of Madrid.

Vernon, Robert   To: H.T.De la Beche
Westminster.
3 May 1847
[see **2276**]

2055 Vetch, James   To: [H.T. De la Beche]
8 Cliffords Inn.
22 June [1842 FJN]

I have had the pleasure of your letter of the 5th inst. by which I perceive how much Mr. J. Phillips is engaged at present but that at some future opportunity he may desire to see my Mexican specimens. You shall have one good pot of Mexican manufacture for your Museum. I regret you did not see my rock specimens from the Hebrides, Orkneys and Shetland Isles. Will they be acceptable to your Museum?

2056 Vetch, James   To: H.T. De la Beche
London.
7 July 1842

Thanks for your readiness to afford house room to my Mexican cabinet in the event of my being called suddenly from London. I have sent 2 specimens of ancient Mexican pottery to the Museum and wish they had been more worthy of a place there. You have omitted to say whether the rock specimens would be acceptable.

2057 Vetch, James   To: H.T. De la Beche
21 August 1843

I have this morning received your letter of the 20th, and a tracing in illustration respecting the treatment of Cran Burn in connection with our other operations at Windsor. I concur as near as possible in all you say. The junction you have made (and coloured green) between the Cran Burn and the new sewer a little east of Long Walk is pretty near the spot and in the portion we agreed upon.

2058 Vetch, James   To: H.T. De la Beche
13 Boxworth Grove.
24 August 1843

I was at Windsor yesterday as I had proposed and now proceed to give you an account of my mission and that you may the better understand it return your tracing with some pencillings. The proposed oval sewer from the bridge over the Cran Brook to its junction with the main sewer is exactly 22ch: 484 yds. and I am quite confirmed that such a sewer can be constructed.

2059 Vetch, James   To: H.T. De la Beche
13 Boxworth Grove.
31 August [1843 FJN]

I received your letter of 29th yesterday and wrote by the same evening to Capt. Tucker stating our wants and hope it will prove of some use, but I propose with your approbation to go to Windsor on Saturday to look at the diversion of the Cran Burn across Street Road and the Long Walk to get the exact distance.

2060 Vetch, James   To: H.T. De la Beche
8 September 1843

I have been hard at work for you all this week designing and with respect to the water works find that taking the height to be raised at 90ft, the friction on 1782 feet of horizontal pipe 10 inch diameter will be equivalent to making the height 950 feet. The tanks will occupy a very considerable space. I think I have arranged the distribution of them on the plan pretty well and have sketched a design of the construction.

2061 Vetch, James   To: H.T. De la Beche
Boxworth Grove.
21 September 1843

I am sorry you found so much more matter turning up for the Report than you anticipated. I have been over all the estimates again and have added some and reduced other items and the total for the Windsor works comes to £20,561.0.3. This includes 1/10th for contingencies. I am quite aware that none of the matters spoken of in this letter can be discussed at present, but I merely write to notice them so that when the time comes they may be considered.

2062 Vetch, James   To: H.T. De la Beche
33 Northumberland Street.
16 November [?1843]

I find the main drain of the Government Establishment at Windsor independent of the town no easy matter. It would be desirable to keep the door open for the town's people to join when it might be granted as a matter of favour, and it would also be desirable to stick to the Report as close as possible.

2063 Vetch, James   To: H.T. De la Beche
33 Northumberland Street.
18 December [?1843]

We are quite becalmed in the Windsor business, not a breath of air. In the vicinity of London we ought to have the deepest well or pit sunk that the habitable world can boast of, that we may see the geology foot by foot and the temperature also. And possibly we may come upon a good supply of hot water for salubrification of this great city for sanitary and cleansing purposes and to pay costs in some degree.

2064 [**Vetch**, James]
[No date, c. 1843]
Estimated compensation for feeder and main sewer [at Windsor]. Total £594.0.0.

2065 [**Vetch**, James]
[No date, c. 1843]
Rough estimate of First class branch sewers and sewer proceeding from main sewer in Long Walk direct to George IV gate [Windsor] Total £6,019.5.0. Rough estimate canal or feeder £1900.

2066 [**Vetch**, James]
[No date, c. 1843]
Statement of losses incurred by the Royal Sappers and Miners employed on the Windsor Survey who have been separated from their families. Total £33.10.6.

2067 **Vetch**, James To: H.T. De la Beche
Finchley Road.
17 June 1849
I left at your office my suggestions as to tide guages. I ought to have mentioned that the mean low water proposed as a zero point should be I think, thus at the entrance to the London Dock, and a name written with an iron pen on the granite rock at the Dock entrance.
[See Stephenson, R., 26 May 1849, **1988** for enclosure]

2068 **Vetch**, James To: H.T. De la Beche
36 Great Ormond Street.
11 June [No year]
I send you a copy of our Report on the Colchester Navigation Bill by which you will perceive the peril into which the sanitary state of Towns and the future drainage of the County may be placed and without check but for the Admiralty stepping in upon a question of navigation, and as one of the Health of Towns Commissioners I am desirous of bringing the subject under your notice.

2069 **Victoria**
1 May 1848
Warrant granting to Sir Henry Thomas De la Beche, Knight, the diginity rank and privileges of an ordinary Member of the Civil division of the third class or Companion of the Most Honorable Order of the Bath. (Signed) Albert, Great Master.
[See also Woods, A.W. to De la Beche, H.T., 2 May 1848, **2162**].

2070 **Victoria** To: H.T. De la Beche
Buckingham Palace. 6 Craigs Court.
[May] [No year]
The Lord Chamberlain is commanded by the Queen to invite Sir H.T. De la Beche to a ball on Friday 11 June at 9.30 p.m.

2071 **Villeneuve**, Vte. de To: M. D'Auderie
Prefecture des Bouches-du-Rhone, Prefet du Var
Marseille. à Draguignau.
25 October 1828
Allow me to recommend to you M. De la Beche the learned English mineralogist who wishes to work in your district. Affording him the means to become known to the local authorities and knowledgeable people of the country would oblige Baron Cuvier who has recommended De la Beche to me.

2072 **Villiers**, André-Jean-Francois-Marie To: H.T. De la Beche
Brochant de Athenaeum Club,
Rue St. Dominique, London.
Paris.
2 April 1832
From the moment the first edition of your Manual of geology appeared last year, I have been so satisfied with it that I have very much wanted to make a French translation of it. The second edition has convinced my publisher that this should be done. I have four translators, students of mine, ready to work on it. I was completely unaware that Ajasson de Grandsange wishes also to make a translation of the work. I have no other interest than that of procuring for France an excellent textbook of geology.

2073 **Villiers**, André-Jean-Francois-Marie To: H.T. De la Beche
Brochant de Athenaeum Club,
Paris. London.
26 April 1832
I appreciate very much the obliging manner in which you have received my plan for publishing a French translation of your Geological Manual and of the offer you make me of helping with your advice and cooperation. Ajasson de Grandsange has completely given up the translation of your Manual. He would have made a bad job of it, as he is by no means a geologist.

Vivian, Henry Hussey To: H.T. De la Beche
Hafod Works.
6 December 1849
[see **2277**]

2074 **Vivian**, Richard Hussey To: H.T. De la Beche
Master General of the Ordnance.
[c. 1835]
[March Route for De la Beche as Geologist to the Ordnance Trigonometrical Survey of Great Britain]. It is to be understood that where he finds it necessary to demand a billet he is himself to pay the customary charges to prevent their being entered in the parish or public accounts.

2075 **Voltz**, L. To: H.T. De la Beche
Paris. Athenaeum Club,
7 February 1832 London.
I have just learned that you are shortly to print the 2nd edition of your excellent Manual of Geology. As the work of M. Thirria on the Jurassic rocks of the Haute Saône contains several mistakes, and as you have copied in your table of Jurassic fossils the information given by Thirria, I feel it my duty to offer the necessary corrections as well as several other important corrections and additions.

2076 **Vyvyan**, Richard Rawlinson To: H.T. De la Beche
10 January 1842
I send you the first two volumes of my work. The notice to the reader explains my method of treating the subject. The introduction to the first volume is an attempt to account for the progressive system in the abstract, and the remainder of the first volume is devoted to the development of such an idea in reference to numbers. The introduction to the second volume explains my mode of considering Nature as one great system of progressive development.

2077 **Wagner**, Rudolfe To: H.T. De la Beche
Munich. London.
21 July 1828
I have delayed writing in order to give you satisfactory news about our common goal, but I am still no further advanced. Our government here, as elsewhere, is short of money; the foundation of our new university has thrown us into debt. Funds destined for the advancement of natural science are already exhausted for several years in advance. In spite of the black outlook I do not yet despair. I have requested Cuvier to recommend me to your government.

2078 **Walker**, Charles V. To: H.T. De la Beche
1 Albert Terrace,
Westbourne Green.
3 October 1844
I send herewith a collection of electrotype productions. I have also sent the mould of the tazza as an illustration of the process by which it was obtained. The two large heads will be ready in a few days. I must make a fresh mould of the outside of the tazza. I enclose the accounts for materials etc. which I have from time to time required. These include simply the actual sums expended, without a single sous to myself. This I must leave entirely to you.

2079 **Walker**, William To: H.T. De la Beche
Plymouth Dockyard. Ordnance Geological Survey,
28 January 1838 Swansea.
In answer to your letter of 23rd, herewith is the information you wish on tides. The range of tides varies from local causes, and augments as we ascend the Bristol Channel. True sea level is of vast practical utility in scientific enquiries. A good tide guage often indicates change before a barometer. The different specific gravities of fresh and salt water effect important changes.

Waring, Henry Franks To: H.T. De la Beche
Lyme.
6 April 1847
[see **2278**]

2080 Watson, M.L.   To: [H.T. De la Beche]
     Middlesex Hospital.
     Wednesday 2 [March or November] 1836

I am very happy to find, through the kindness of Mr. Kerr that you have it in your power to afford me great assistance in the execution of my granite tomb. From the ground to the top of the tomb will measure 7 feet 6 inches. The extreme base will be 12 feet in length and 7 feet 6 inches in width. This will give you some idea of the whole.
[See also Philipps, T.W. to De la Beche, H.T., **1204**]

2081 Way, Albert   To: H.T. De la Beche
     131 Piccadilly.
     23 February 1844

I have heard by accident that you have collected, or are desirous of collecting specimens of glass, and of fictile manufacture, for some purposes of a public nature. I may be able to be of assistance to you as Director of the Society of Antiquaries, since I have devoted a lot of attention to these matters.

2082 Way, Albert   To: H.T. De la Beche
     131 Piccadilly.
     [February 1844 FJN]

I have got some characteristic specimens. The great thing is to make the great features by instructive examples, and fill up the spaces afterwards, as occasion occurs. I think I shall be able at once to help you with some contributions. Good or rare pieces might be obtained as a temporary exhibition or deposited in the collection like the Portland Vase, to be withdrawn if required.

2083 Way, Albert   To: H.T. De la Beche
     Alderley Park,
     Congleton.
     9 March 1844

I have not been able yet to draw out as I promised the sum total of our account hitherto. On the day I left you I directed the 3 Maiolica dishes to be brought to you as you desired, and hope they came safe the morning after. They are truly incomparable. I closed the bargain for the enamel I named to you at the price to which you consented £10.

2084 Way, Albert   To: H.T. De la Beche
     12 Rutland Gate,
     Knightsbridge.
     [1847 FJN]

I am sorry to leave Town without seeing you. I sent you a dish from Lechmere the Worcester banker. Herty of Malborough Street has some enamels you ought to see. He has three pieces of pottery he wishes to present to the Museum. The seals I showed you, which were his, are spurious except three.

2085 Weaver, Thomas   To: [H.T. De la Beche]
     2 July [1830]

I should be glad if you gave me letters of introduction to Cuvier, the two Brongniarts, Brochant de Villiers and to whomsoever else you please as most likely to facilitate the objects I have in view, viz. an opportunity of seeing all that is most worthy of notice in the renowned metropolis of France. Is von Buch likely to be now in Paris?

2086 Weaver, Thomas   To: H.T. De la Beche
     Woodlands,
     near Wrington,
     Somerset.
     23 March 1831

Bearing in mind your projected Manual of Geology I send you the annexed list of organic remains. I have never seen any animal remains in the old red sandstone of the British geologists. Would you communicate to me the list observed by Hannah in the transition limestone of Plymouth? What are the species of trilobites found by Radley in Devon? And by you around Babicomb?
[Cited: Rudwick 1985, p.79]

2087 Weaver, Thomas   To: H.T. De la Beche
     Woodlands,   Geological Society,
     near Wrington,   Somerset House,
     Somerset.   Strand,
     12 April 1831   London.

I am happy to hear your Manual advances with rapidity. It would be satisfactory to me to have my paper on the South of Ireland included in your forthcoming half-volume. In a conference with Murchison, Greenough and Lonsdale it was agreed that my paper "Metalliferous Relations" should occupy 30 pages of the Transactions.
[Cited: Rudwick 1985, p.79]

2088 Weaver, Thomas   To: H.T. De la Beche
     Woodlands,   Athenaeum Club,
     near Wrington,   London.
     Somerset.
     23 April 1831

Herewith replies to your queries in your letter of yesterday. The Black rock of Cork is transition. The Queen's County limestone is decidedly carboniferous. I shall always be happy to answer any of your enquiries, and look forward to your account of the Plymouth transition.
[Cited: Rudwick 1985, p.79]

2089 Weaver, Thomas   To: H.T. De la Beche
     Geologicial Society,
     London.
     19 July 1831

Your note respecting Lyell's Geology and Beaumont's observations was brought to me this morning. The messenger boy assured me you were still at 10 Craven Street, but the servant told me you had left town. I am sorry to have missed you.

2090 Weaver, Thomas   To: H.T. De la Beche
     16 Stafford Row,
     Pimlico.
     4 June 1852

Since I had last the pleasure of meeting you at the Museum of Practical Geology, I have got up from the country a host of my books, in order that I might select therefrom such subjects as might be appropriate to the service of your noble institution. Let me know how you wish them delivered. Would you present to me your published volumes of the Government Geological Survey?

2091 West, T.   To: H.T. De la Beche
     Archbishop's Palace,
     Dublin.
     31 January 1850

In the absence of the Rev. Ralph Sadleir who has hitherto corresponded respecting the subscription for the family of Dr. William Cooke Taylor your letter to him has been handed over to me as one of his Co-Trustees & I beg to say I will be happy to take charge of your subscription if forwarded to my address.

2092 Westminster, Lady   To: H.T. De la Beche
     Motcombe House,
     Shaftesbury.
     6 February 1851

In consequence of Faraday's direction a box of specimens has been sent to the Museum of Geology. They were found at about 27 feet, when sinking for a well at Motcombe.

2093 Weston, Charles Henry   To: H.T. De la Beche
     5 Lion Place,
     Bath.
     26 April 1848

As some of the members doubted the existence of Hastings stone in the locality of the Ridgway cutting near Weymouth. I have been over various other sites and the cutting again. My conviction was confirmed and I will embody my impressions in a supplementary paper for a future date.

2094 Whitworth, Joseph   To: H.T. De la Beche
     Manchester.
     19 April 1845

In reply to your note received this morning, I enclose the account of the experiments on cast iron. Colebrookdale is the strongest and also, considering its strength, a remarkably good running iron.

2095 Whyte, C.J.   To: H.T. De la Beche
     17 Gough Square,
     Fleet Street.
     19 August [1847 FJN]

I have arrived in London on business connected with Col. Durham's

Chancery Suit, and write this in the event of not seeing you at Craig's Court. Can you procure appointments for my sons? The eldest is now 16, and as a parent I am anxious for their advancement.

2096 Wickenden, Joseph   To: H.T. De la Beche
Birmingham Philosophical Institution.   Swansea,
30 October 1839   South Wales.
Your permission is asked for the Institution to enrol you as an Honorary Member. I have spoken to Russell, and he requested me to inform you that your chains are in progress.

2097 Williams, David   To: H.T. De la Beche
Bleadon W. Cross,   Tavistock.
Somerset.
21 November 1837
I am very glad you have reviewed the precincts and aid me with your strong arm in dissipating the "Murchisoni-Sedgwickean" phantom, that fine metaphysical non-entity of the anticlinal line. If "a zealous collector" of facts as well as fossils should simply show these matters will it not quietly but gravely impeach their credibility as to other parts of their Devon mythology? They will gain no kudos by their flourish in the west.

2098 Williams, David   To: H.T. De la Beche
Bleadon W. Cross,   Ordnance Map Office,
Somerset.   Tower,
20 February 1838   London.
I wrote to Professor Phillips a few days ago to know when he wished to have such fossils from Devon and Cornwall as I had that I might forward them in time. I am ashamed to say I have to unpack, reduce and label them, a task I anticipate with anything but pleasure, but tomorrow I will set about it. Can I furnish you with anything for your report? I am convinced of the truth of your views on Devon culm, and believe Murchison, Lyell and Sedgwick will in the end agree with you.
[Cited: Rudwick 1985, p.235]

2099 Williams, David   To: H.T. De la Beche
Bleadon W. Cross,   Ordnance Map Office,
Somerset.   Tower,
16 March 1838   London.
I have at last packed up a box which I purpose forwarding to Bristol this evening and thence by coach to Town tomorrow. I left a box with Lonsdale in May last containing trilobites, fossil wood and many of my best organics from Devon. I hope you will be able to procure these from him, as they were from my private collection. Phillips has another box.

2100 Williams, David   To: H.T. De la Beche
Bleadon W. Cross,   Swansea.
Somerset.
24 December 1838
I would gladly avail myself of any mode you may suggest to avoid the trouble and expense of packing and repacking and the to and fro carriage of the fossils from York to London. I think my North Devon ones are labelled as such, and as my main object in forwarding them is to establish the identity of many species with those of the South, my object I think might be answered if Professor Phillips would favour me with a communication before the meeting how far in his judgment such identity exists.

2101 Williams, David Hiram   To: H.T. De la Beche
Pontypool.
21 April 1840
In reply to your request I have brought my work down the Taff Valley. My section embraces the whole of the strata from the Old Red Sandstone down to the turnpike. I have no doubt you will work between the Taff and Rhumney Vallies. There are five large faults between them. I hope you will excuse my suggesting a district for Rees, and advising on yours.
[Cited: North 1936a, p.67]

2102 Williams, David Hiram   To: H.T. De la Beche
Cardiff.
19 July 1840
I have extended the greater part of the lower measures to join the Taff river, and there found two anticlinal lines crossing the limestone E of the Taff extending so far down as the 4th milestone on the road from Cardiff to Newbridge. I think the south junction extends much further south. I shall not return to Cardiff before Saturday evening, so as to work on as fast as I can. I shall be able to accomplish this side of the Taff river within the time.

2103 Williams, David Hiram   To: H.T. De la Beche
Caermarthen.
16 December 1840
In reply to yours of yesterday, the error west of Pentyrch must have been made in consequence of information given me by Mr. Jenkins. I am very glad you have detected it. I enclose a list of veins as given me by Mr. Jenkins. I think we should wait until Mr. Russell's pits are sunk at Risca before attempting to ascertain the exact number.
[Cited: North 1936a, p.67]

2104 Williams, David Hiram   To: H.T. De la Beche
Caermarthen.
17 December 1840
While out on duty this morning I lost 7 portfolio squares of the Caermarthen map. I returned and had them cried by the town crier and asked at all the houses on my route. I fear they will be used as pictures on some cottage wall. Please send me sheet 41 and I will re-do the work without waste of time. Should I hear of the lost maps I will inform you immediately.
[Cited: North 1936a, p.68]

2105 Williams, David Hiram   To: H.T. De la Beche
Caermarthen.
29 December 1840
By this day's mail I send the book and organic remains packed up in a hamper. I am glad your calculations of heights agree so well with mine. I make the height of Twym Barllwym as 1446 feet. Lieut. Williams of the Ordnance Trigonometrical Survey estimates it at 1160 feet.
[Cited: North 1936a, pp.87-88]

2106 [**Williams**, David Hiram]
[c. 1840]
[Section of Measures of Upper Ebbw Vale]

2107 [**Williams**, David Hiram]
[c. 1840]
[Section of Cwm Telery Shaft belonging to Thomas Brown; Section of Measures from Abersychan to the Cwm pits in the Ebbw Fach, Monmouthshire; Section of Coal Measures sunk through at Risca, Monmouthshire, by John Russell].

2108 [**Williams**, David Hiram]
[c. 1840]
[Section of Coal Measures at Cwm Celyn and Blaina; Section of Cwm Celyn and Blaina Globe pit; Section of Cwm Celyn and Blaina Deep pit].

2109 Williams, David Hiram   To: H.T. De la Beche
Haverfordwest.
4 May 1841
You will be glad to hear I have discovered at Broadhaven a bed of limestone in the coal measures containing fossils like those of another place. Would it not be a fair conclusion to come to that they are the same measures. I enclose a drawing of the 5 figures of fossils.
[Cited: North 1936a, p.69]

2110 Williams, David Hiram   To: H.T. De la Beche
Coleford.
18 August 1841
I do not think much trouble would be required to make Bristow understand our mode of proceeding. I should not feel any inconvenience by allowing Mr. Bristow to accompany me. Should you decide to send him here, the sooner he comes the better for him as I shall have finished the mining part of this district in a month or so.
[Cited: North 1936a, p.72]

2111 Williams, David Hiram   To: H.T. De la Beche
Coleford.
12 September 1841
The enclosed is my work at Landshipping which you will find is laid down carefully and according to the best evidence of the old men at Col. Owen's collieries. Since I have been here I have been hard at work. There are 140 shafts in the forest, but not a single section has been kept. I am the more surprised the late Mining Commission did not recommend the practice. The importance of this has already been felt by a great many persons here who have been tempted to sink by some sharks here.

2112 Williams, David Hiram   To: H.T. De la Beche
Coleford.
25 September 1841

I trust you will find the enclosed Survey Account correct. I have now to report my progress here. The Forest of Dean coalfield is finished. I have more confidence in my work here than in Pembroke. There will be some difficulty in making a contour survey of the forest, as it is so covered by wood. The quantity of the coal worked in the coalfield is very small, considering how long it has been under operations.
[Cited: North 1936a, p.68]

2113 Williams, David Hiram   To: H.T. De la Beche
Coleford.
30 September 1841

After reading your note I think Chepstow would be the best place for me to go next as the mountain limestone chain from the Forest extends that way south and there is also about 600 acres of coal measures at the Chase 19 miles from here on the Chepstow road, so that on looking at the map I find there is about 60 square miles of country round Chepstow and north of the Severn I could do very well. I should then be able to extend all the work within a drive of Gloucester.

2114 Williams, David Hiram   To: H.T. De la Beche
Coleford.
10 October 1841

My labours in this neighbourhood are coming to a rapid termination. I have finished a section from NE to SW 12 miles long and another on the north crop 4 miles long. I have also commenced another section from E to W which will be 9 miles long. I conceive this will be quite enough to show everything connected with this coal district. Stokes of Tenby begs me to remind you about the copy of the longitudinal section on the large scale. He is now going to sink in one of the troughs near the sea, and is now convinced there is no coal of any value under my fossiliferous bed, the discovery of which on the western side he conceives to be of great importance in classing the measures, and of course of value to every person connected with mining.

2115 Williams, David Hiram   To: H.T. De la Beche
Chepstow.
19 October 1841

I arrived here last night from Coleford and will commence my labours tomorrow morning. During the last part of my stay at the Forest of Dean I have paid some attention to the structure of the great sandstone series at the bottom of that basin. I have detected Stigmaria ficoides, Sigillaria, boulders of coal and trap boulders occasionally. This resembles so closely the character of the Townhill Sandstone, that I shall I think be right by calling it the great Pennant rock. The great coal and iron deposit of South Wales is entirely wanting.
[Cited: North 1936a, pp.69-70]

2116 Williams, David Hiram   To: H.T. De la Beche
Chepstow.
1 November 1841

On extending my work to Shirenewton I happened to fall in with some men boring a deep hole for coal near that place. The bore-hole happens to be 4 chains north of the carboniferous limestone and of course they are just on the upper beds of the Old Red Sandstone. They have also sunk a shaft between 200 and 300 feet and have built the yard to receive the coal. Do you think I had better inform Mr. Hallis that he is deceived by these rogues and will be a ruined man without a question of doubt if he does not discontinue?
[Cited: North 1936a, p.68]

2117 Williams, David Hiram   To: H.T. De la Beche
Chepstow.
4 November 1841

I wish to have some information from you about a vein in the Maesteg section. On looking over Mr. Logan's book very carefully I observe a rock called or supposed to be Falsecockshot. This appears at the bottom of the lane and is seen near the wooden bridge. I have compared all the sections and achieved some results. I will wait for your opinion before I shall be able to close the north and south sections.

2118 Williams, David Hiram   To: H.T. De la Beche
Chepstow.
13 November 1841

I do not think from your replies of 11th and 12th, that I have made myself clear about the Maesteg section. I have not done anything further about Mr. Hallis's speculation. While out at work I met a gentleman when crossing his property and he asked me if I had been at Shirenewton and what I thought about the trial for coal. I told him my view of the affair which he believed. Mr. Hallis must have now spent nearly double the sum I mentioned previously.

2119 Williams, David Hiram   To: H.T. De la Beche
Chepstow.
20 December 1841

I have received from my brother the long-expected section. It embraces 3,591 feet of ground, with 102 feet of coal. With the assistance of this section you will be able to calculate the value of the Llangeinor property. As soon as I have received your instructions where to send this section, it will be forwarded without any delay.

2120 Williams, David Hiram   To: H.T. De la Beche
Chepstow.
28 December 1841

I am glad the section arrived safely. I shall ask my brother some questions about the bottom part of his section. I think he was not able to draw it on the scale of 30 feet to the inch as he intended doing. I hope you find the enclosed account for the quarter correct. I am getting on well with this district. There is a good deal of magnesian limestone here. I have been told Mr. Hallis has stopped boring at Shirenewton, and think he learned of my opinion indirectly.

2121 Williams, David Hiram   To: H.T. De la Beche
Lower Redland.
21 April 1842

As I shall very soon want to work on the sheet south of Bristol, it will be necessary I should have two red brown wet proofs, one for field work and the other as duplicates, and one dry proof. You will also have the goodness to send me some drawing paper and pencils, as I am quite out of them.

2122 Williams, David Hiram   To: H.T. De la Beche
Lower Redland.
10 July 1842

I have seen Mr. Rees and gave him the map he requires, and we have arranged to communicate with each other so as to enable us to join our work. He left yesterday for Wales. Rees and myself will work the whole district from the Mendips to Dundry hill working eastward. I hope this arrangement meets with your approval, and I will proceed with all speed to its accomplishment.

2123 Williams, David Hiram   To: H.T. De la Beche
Lower Redland.
26 July 1842

You will be delighted to hear I have found a bed of underclay with Stigmaria ficoides on the top of which I have 1 foot of carbonaceous shale coal. This discovery will be interesting to show that conditions favourable for the deposition of coal commenced at a much earlier period here than was originally supposed. You will also be glad to hear the result of my measurements of the carboniferous limestone of this neighbourhood which I send herein.

2124 Williams, David Hiram   To: H.T. De la Beche
Yatton.
21 August 1842

The second half of the £10 note has not arrived. Someone may have detained it and I must enquire at the Bristol Post Office. On the other hand you may still have it. You will be glad to hear I am pushing on with my field-work. All the Nailsea coal measures have not been sunk through.

2125 Williams, David Hiram   To: H.T. De la Beche
Yatton.
13 September 1842

On extending my work through Wrington manor I fell in with some mines belonging to Mrs. Avelin, on the examination of which I find all the faults are filled with iron, manganese and lead. A large subsidence has been of assistance, and a cutting made. It may not promise something good, and I have persuaded Mrs. Avelin to do nothing more for the present. Perhaps it would be better to leave the extension of the cutting until you come over.

2126 Williams, David Hiram   To: H.T. De la Beche
Yatton.
27 September 1842

After some trouble I have succeeded in getting something like a true statement of the coal trade at Bridgwater. The total quantity of coal taken into that port would not exceed 10,200 tons per month. The annual total would not exceed 132,000 tons. The Newport and Cardiff best coals are considered best by trades people as being more durable. The forest coal burns away much quicker on account of its being more bituminous, but the richer people like it best for drawing room fires.

2127 Williams, David Hiram   To: H.T. De la Beche
Yatton.
2 October 1842

The cost of coal at Merthyr during a period of 7 years was as follows, including royalty, agency, hauling etc. I enclose a receipt for the conveyance of self from Yatton to Nailsea. By using the railway I have considerably augmented my return of work during the last eight weeks, which you will find on a reference to the map to be above 30 square miles.

2128 Williams, David Hiram   To: H.T. De la Beche
Wells.
10 November 1842

Yesterday I arrived here in the afternoon. When you forward the Forest of Dean documents, please send some drawing pencils. A trial for coal has lately been made in the neighbourhood of Banwell. I obtained a section from the prospector, who asked me the cost of setting up an engine and sinking the shaft. I estimate £2,400, too much to risk under the circumstances.

2129 Williams, David Hiram   To: H.T. De la Beche
Wells.
4 December 1842

In obedience to your instructions I am taking all care to show the lower shales of the Carboniferous limestone when possible to do so. From the continued wet weather I have been prevented from doing as much as I could on the higher ground. I have begun a new drawing of the Forest of Dean west-east section, but do not have the originals with me. You will be glad to hear I have laid the foundation of a large columnar drawing of the coal measures of Gloucestershire.

2130 Williams, David Hiram   To: H.T. De la Beche
Frome.
24 May 1843

The section of the Carboniferous limestone along the lower Avon is written in detail in the large section book, forwarded many months ago. Having some time ago discovered a very large fault on the south side of the Mendips I have again found it in the coal measures near Coleford from which place it appears to take a direct course for Radstock. I would like to see the sections of that locality which Mr. Stuckley has. Would you have the goodness to send me a few receipt forms?

2131 Williams, David Hiram   To: H.T. De la Beche
Frome.
29 May 1843

With as little delay as possible I forward you the outlines required. The Downside rock I am inclined to consider a variety of Lias, as I have seen a similar deposit in the Lias by Shepton Mallet. The two small patches left uncoloured on the map I think are trap. With regard to the Harptree grit, I have found fossils in both the Lias and Oolite very much resembling those found in the above rock.

2132 Williams, David Hiram   To: H.T. De la Beche
[postmarked Wrexham].   Museum of Economic Geology,
19 February 1844   Craig's Court,
                                                    London.

[Section of Measures at Stratton Common belonging to the Duchy of Cornwall; Section of beds seen in Col. Horner's Wood in Mills Park; Section of Quarry; Section of Measures at Radstock; Section of Colliery in Radstock; Section of Coal Seams at Holcombe Old Works].

2133 Williams, David Hiram   To: H.T. De la Beche
Llangollen.
15 August 1844

I am at a loss what division to make on the map without your consent, as it would hardly be right for me to make minor divisions of the upper and lower rocks of this district, where there are none in Nature. The existence of organic remains here depend on the development of arenaceous conditions, so that considerable difficulty will be found in identifying these rocks as Silurian, as the physical conditions differ considerably from the typical districts where the order was first established.

2134 Williams, David Hiram   To: H.T. De la Beche
Oswestry.
2 November 1844

As things here have turned out just as I thought they would, it is necessary to inform you that in another week my labours in the coal measures, gravel and new red will be completed. Agreeable to your orders I have confirmed my labours within the coal-fields taking in of course all the gravel and new red coming within my reach. You will like the delineation of the great gravel drift, and when complete will be found to extend from the Mersey to the Severn. On about Thursday next I shall go down to Shrewsbury to look for quarters.

2135 Williams, David Hiram   To: H.T. De la Beche
Watling Street.
1 May 1845

I rejoice to hear that the application which you have so kindly made on my behalf is now likely to be soon granted. This addition of £54.7.6 to my present pay will be received by me with gratitude and will stimulate me to fresh exertions.

2136 Williams, David Hiram   To: H.T. De la Beche
Watling Street.
8 June 1845

On finding the appointment in connection with our Indian affairs, and that a connection will be kept up between it and our Survey, I will accept the appointment under the conditions stated in your letters. You may depend that I shall not omit to do all in my power to exert myself on a mission of so much consequence as the Survey of the Indian coalfields. I hope it will eventually turn out both in an economic point of view and a branch of the Geological Survey of Great Britain. My wife will regret the separation, but she is too good to object to the idea as it will do much for our little ones.
[Cited: Stafford 1989, p.113]

2137 Williams, David Hiram   To: H.T. De la Beche
Camp [?] Near Bagnah,
India.
2 February 1846 [sic]
[probably 2 February 1847 from internal evidence]

I am now making a map of the Damoodah coalfield. The more I see of it the more I am convinced that the measures are developed on a large scale. From my present knowledge it contains 200 feet of coal which so far as thickness goes it contains more fuel than any coalfield in Great Britain or America. I have an assistant who is employed in preparing a base map. This part of India was never regularly surveyed.

2138 Williams, David Hiram   To: H.T. De la Beche
Spence's Hotel,
Calcutta.
18 February 1846

I arrived here too late by two months to take an establishment to the field. I have proposed to the Deputy Governor of Bengal to proceed alone to the Damoodah coal deposits. This trip will only be a preliminary step before taking the field with tents and elephants. After having looked round the district I shall be better able to state to the Government what I shall absolutely want. You will be applied to for an assistant. He ought to have £450 to £500 per annum.

2139 Williams, David Hiram   To: H.T. De la Beche
Ranugange,
India.
4 April 1846

I am now fairly within the great coalfield of the Damoodah and Adji rivers. In one locality I found 12 seams of coal, equal to 48 feet. The measures are nearly horizontal and associated with grey sandstones so like our Pennant grit at home that I can hardly fancy I am so far from you. I am making a large collection of fossil plants. Oh, by Jupiter, you will be delighted to hear that these coals have underclays with impressions very like Stigmaria.

2140 Williams, David Hiram   To: H.T. De la Beche
Ranugange,
India.
30 April 1846

With this I have enclosed a bill of exchange for £50 which I will trouble you to send to my wife. You will be glad to hear that this coalfield is likely to turn out a large quantity of coal. Mr. Sim's wife here has died of

cholera, as also a Mr. Playfair (brother to our Dr. Playfair I am told). You will be so good not to communicate the death of Mrs. Sims to anyone who would be likely to communicate it to my wife. It is advisable that the person who is appointed to join me should be in the best of health.
[Cited: Stafford 1984, p.10]

2141 Williams, David Hiram To: H.T. De la Beche
Ranugange.
1 June 1846

With this I send you a set of bills of exchange for £50 which you will be pleased to send to my wife. Cholera is raging at Calcutta to a fearful extent. It is my intention not to return there until the rain has set in, and all cooled down. The Government are anxious to know to what extent this coalfield is developed. I have made a collection of sandstones. You will be suprised when they come to hand to find them so like the sandstone of the English coal measures.

2142 Williams, David Hiram To: H.T. De la Beche
Ranugange.
1 June 1846

I have just heard from home that my youngest daughter died on 26 March last. I am much grieved. As my wife in her trouble will probably apply to you for your advice and assistance to put herself and the children in mourning, I ask you to pay it and I will repay you.

2143 Williams, David Hiram To: H.T. De la Beche
Damoodah River.
26 August 1846

With this I have much pleasure to send you a copy of my report to the Government of Bengal. Would you believe I was asked for a geological map of the Damoodah coalfield. I told the Deputy Governor it was quite impossible to furnish a geological map without a ground plan of the country.
[Enclosure] Copy letter D.H. Williams to F.J. Halliday, 10 August 1846. Summary of my observations during my late journey through the North Western Provinces of Bengal and more especially on the Damoodah and Adji great coalfield.
[Cited: Stafford 1984, pp.8, 10]

2144 Williams, David Hiram To: [H.T. De la Beche]
Damoodah.
[c. 1846]

Damoodah coal-field, India. Degrees of temperature at the surface and in the mine. Depth of shaft 90 feet. [Two pages of tables] [Enclosed also sketch of underclays and coal plant]

2145 Williams, David Hiram To: H.T. De la Beche
Damoodah Valley.
28 April 1847

Since I last wrote to you I have been sinking shafts and making excavations in this neighbourhood in the lower sandstone and shales in which I found 17 seams of coal amounting in all to more than 100 feet of coal. Two of the beds have turned out to be coking coals. I am sending a report of my findings to the Government of Bengal, and enclose a copy.
[Enclosure] D.H. Williams to C. Beadon, 28 April 1847. I have the honour to acquaint you for the information of the Deputy Governor that the coal discovered in shaft 13 of my excavations in the Damoodah coalfield which is 9 ft. thick has turned out to be a coking coal.

2146 Williams, David Hiram To: H.T. De la Beche
Loong,
Bengal.
1 August 1847

You will I am satisfied be glad to learn the result of my late suvey in the Damoodah coal-field. The measures are developed on a grand scale, and separable into four marked divisions. The whole mass is 3,110 feet. The great south fault must be very considerable. Next October I expect to go to Central India to report on the gold mines lately discovered and brought to the notice of the Deputy Governor of Bengal.

2147 Williams, David Hiram To: H.T. De la Beche
Spence's Hotel,
Calcutta.
22 December 1847

It is with much pleasure I have to inform you that the Survey of the Damoodah Valley is completed and the maps and sections were submitted to the Deputy Governor of Bengal on the 9th of this month, the day on which I arrived in the capital. I had as you are previously aware the laborious task of making a new geographical survey; this I must tell you was accomplished in 6 months, and the area surveyed was 1150 square miles of which the coalfield occupies 804 square miles. The sections exhibited 50 beds of coal having a total thickness of 353 feet. I appoint you executor and trustee to my will.

2148 Williams, David Hiram To: H.T. De la Beche
Camp Harancebagh.
25 June 1848

I have just returned from the field to take up my quarters here during the rains. I have been searching for coal since I last wrote to you. The sandstone in the mountains resembles the sandstone of the coal measures, but I have not found the least indication of coal, and it is my conviction that the sandstones do not belong to the coal measures.

2149 Williams, H.F. To: Secretary of the
Constantinople. Geological Society
20 November 1848

Since my arrival in the Turkish capital to take charge as Commissioner for the new boundary line between Turkey and Persia, Angus, the Naturalist, has resigned. I have asked the Secretary of State for a substitute, but would like someone with geological ability, and have been referred to you. £100 a year is the salary, plus all expenses.

2150 Williams, Wadham P. To: [H.T. De la Beche]
Bleadon Rectory.
28 October 1850

It is impossible to describe by letter my father's collection. There are some unique Saurians from the Lias of the neighbourhood. Also cave bones from this neighbourhood, lion, tiger, hyena, buffalo, etc., and a splendid femur of a Megatherium. The Great Western railway is convenient for transport. Could Woodward come down for a day or so if the Society wished to purchase the collection.

2151 Williams, Wilb. T. To: H.T. De la Beche
Penryn,
Cornwall.
15 December 1849

I enclose a letter of introduction from Enys. What is your opinion of two specimens of stone, to be used for the repair of streets and roads in London? Sir Charles Lemon offered me an introduction to you, but I thought one would be sufficient.

2152 Willis, Robert To: H.T. De la Beche
Cambridge.
5 March 1854

Thank you for your kind offices which will save me a world of trouble. I am glad Forbes is to be in the party, and wish you were also. I have seen Stokes, and have no doubt that a proper application to him will be accepted. I'm sure you and all our party at Jermyn Street will rejoice at this.

2153 Willis, Robert To: H.T. De la Beche
Cambridge.
22 July 1854

No doubt our new professor has seen you and given you his own syllabus. I think I had better leave mine in the same state as before. As for the rest of the Prospectus, I see nothing to change except the parts already revised. I will be in town on 1st August, when I hope to see you.

2154 Wilson, Charles Heath To: H.T. De la Beche
6 Duke Street,
Portland Place.
7 December 1847

My connection with Somerset House is at an end, and I am anxious to provide for your door. I have found an artist for the purpose. He will model it for £50. I would stipulate a time. Stevens was 9 years in Italy, and the most skilful ornamentist I know. I will give him your drawings if you approve.

2155 Wilson, Charles Heath To: [H.T. De la Beche]
Paris.
[No date]

I have some difficulty in choosing for you. I am to have 2 specimens of free stone polished to represent marble. I have bought some items from the exhibition. May I buy you a superb candelabrum 5 feet high, 250 francs, in iron, modern, very cheap and exquisite specimen of workmanship? Please pay £60 to my credit through Coutts. Please answer instantly or I may leave.

2156 Wilson, John To: H.T. De la Beche
Royal Agricultural College.
18 May [1847 FJN]

The first instalment of the Report has been sent by Great Western Railway. It contains the marrow, being particulars of the experiments on the evaporative powers of the coals tested. The remaining part of the Report will be ready in a day or two, comprising drawings to scale of the boiler house and apparatus, and a tabular presentation of the results.
[On Museum of Economic Geology stationery].

2157 Wilton, Edward To: [H.T. De la Beche]
West Lavington,
Devizes,
Wiltshire.
11 September 1835

I discovered in Steeple Ashton an escutcheon with the Arms of Beach and Timms family on it. Your father frequently visited the burial place of the family. It is now in possession of a carpenter of Steeple Ashton, and I could obtain it for you if you wished.

2158 Wilton, Edward To: H.T. De la Beche
West Lavington Grammar School.
4 April 1851

I have lately found an allusion to the Beach family in Aubrey's Natural History or Wiltshire. It speaks of marchasite in the grounds of Mr. Thomas Beach of which Prince Rupert made trial, but without effect. Should you be visiting this district I would gladly show you the memorials of the Beach family.

2159 Wilton, Edward To: [?L.L. Dillwyn]
West Lavington,
Devizes.
23 October 1863

Many years ago I met Mr Gowland and from him gathered considerable information about the Beach family, his first wife being sister to Lt Col De la Beche, son of Thomas and Helen Beach. About 65 years ago Col De la Beche was at Ashton and removed some painted glass from the window. This is no doubt the glass Mrs Dillwyn refers to.

2160 Wood, C. To: H.T. De la Beche
D.S.
14 February 1852

Thank you for sending the lectures. I wish I had time to come and hear some of them.

2161 Wood, V. To: H.T. De la Beche
Hilton Hall,
Trade Houses.
7 December 1849

I have not lost sight of the subject of your letter but have been quietly talking it over with several of our coal owners. We shall have a General Meeting of the Trade the latter end of this month when I purpose bringing it before them, and will expect get a memorial signed by all the Trade here. At least I shall do all I can to accomplish it.

2162 Woods, Albert W. To: H.T. De la Beche
Herald's College,
Doctors Commons.
2 May 1848

Her Majesty having appointed you C.B., I am commanded by H.R.H. Prince Albert, Great Master of the Order, to transmit to you the enclosed Warrant.
[See Victoria 1 May 1848, **2069** for warrant].

2163 Woods [?] To: H.T. De la Beche
[Watermark 1836] [1841 FJN] Swansea.

I fear you will think that I have forgotten my promise of attempting a sketch of the fossil tree. It has however been long in readiness and I have been only waiting for an opportunity to send it over. I also venture to beg your acceptance of two other drawings of specimens.

2164 Wyon, Leonard Charles To: H.T. De la Beche
22 Bloomfield Road,
Maida Hill.
21 April 1853

I have the dies of your Medal which appear to be in a perfect state, and will call on you at your convenience. I have changed residence from the Mint, as new regulations do not allow residence in the Mint for those officers who undertake private commissions.

2165 Wyon, William To: H.T. De la Beche
H.M. Mint.
29 October 1842

The Golden Head remains in my possession, but I will send it to your daughter in a day or two. Regarding Robinson's project of preventing sweating of gold coin, I do not think coins are sweated to any extent, and the plan itself would be impracticable. Allan Cunningham has had a third attack of paralysis. When do you return to London?

2166 W., H. To: H.T. De la Beche
24 June 1850

I quite forgot that I am to be at Cambridge today. I will be at your Palace next week, and will give you a day's notice.

2167 Yolland, William To: H.T. De la Beche
Ordnance Map Office,
Southampton.
30 September 1843

I have caused one sketching case and strap, and one 2ft steel straight edge to be forwarded to you in compliance with your request of 23rd inst. Please transmit to this office unserviceable stores. Your balance of serviceable and unserviceable stores is not in order.

2168 Yolland, William To: H.T. De la Beche
Ordnance Map Office,
Southampton.
8 June 1847

What do you want done to the transfers to stone of Sheet 58 and the portion of sheet 40? I sent you the impression you required some time since.

2169 Yolland, William To: H.T. De la Beche
Ordnance Map Office,
Southampton.
21 May 1849

In reply to Hunt's letter of 19th, copies of Sheet 37 charged to Stationery Office were supplied in August and October.

2170 Yolland, William To: H.T. De la Beche
Ordnance Map Office,
Southampton.
26 October 1849

I have sent you an impression from the Geological plates, that you may see the style in which the writing is engraved. Let me know if you wish it altered in any way.

2171 Yolland, William To: H.T. De la Beche
Ordnance Map Office,
Southampton.
23 May 1851

It is unfortunate that in some way or other, an error appears to have crept in with reference to your understanding with Mr. Cornewall Lewis as to what portion of the one-inch map should be now engraved. It must be done as cheaply as possible. You must select some spot other than Dublin and its vicinity. When we get an order to proceed I will let you know, and in the meantime you may turn over in your mind what other locality will best suit you.

2172 Yolland, William To: H.T. De la Beche
Ordnance Map Office,
Southampton.
12 December 1851

I return to Mr. Cornewall Lewis' note respecting the one inch map of Ireland, and when we get the instruction from the Treasury through Pall Mall, I will take care that the proper sum is named for the two sheets to be inserted in the estimates, if that is the extent on which we are to commence.

Unidentified authors

2173 Unidentified To: H.T. De la Beche
28 February 1837   Athenaeum.
[Appointment at the Museum of Economic Geology].

2174 Unidentified To: H.T. De la Beche
[Ordnance Office].
1 May 1841
The accompanying letters speak for themselves. Have the kindness to return them. You will observe that the opinion I have expressed to Mr. Henwood is just that which is to be found in the last paragraph of your letter of the 28 April which I yesterday received. Nothing can be more satisfactory than the statement you make in that letter and the testimony afforded by the Proceedings of the Geological Society 1840 to the merits of your Geological Survey of Cornwall and Devon.

2175 D...[?], John To: H.T. De la Beche
House of Commons.
19 July 1846
Can you call on Lord [?] tomorrow at the House of Lords.

2176 Unidentified To: H.T. De la Beche
Louvain.   Geological Society,
9 August 1848   London.
Having finished the first part of my synoptic table of fossil and living species of the family 'Arcacées', I send a copy of it to the Society. Would you also forward the other copies to those to whom they are addressed?

2177 Unidentified To: H.T. De la Beche
18 January 1849
Thanks for your report. Parker shall present it this evening.

2178 Unidentified [possibly William Branwhite Clarke]
[New South Wales].   To: W.W. Smyth
15 August 1851
You will have heard and no doubt with surprise of the auriferous character of our interior near Bathurst and at Moreton Bay 500 miles to the north, so that the intervening country, which is mostly of the same geological character may be supposed to be more or less rich. It is generally found in the alluvial deposit on the banks of mountain streams in the form of scale gold.

2179 Unidentified [Labouchere, Henry] To: H.T. De la Beche
Board of Trade.
23 December 1850
A memorial has been presented to Lord John Russell signed by the principal persons connected with the mining interest in South Wales asking for the encouragement of the Government in the improvement of the education of Mining Engineers. As your thoughts must have been much directed to this subject, what is your opinion on the subject and the best mode of bringing it about. It seems to me that the idea can only be carried out in conjunction with the exertions of the Mining Interest itself.
[Same author as **2180**].

2180 Unidentified [Labouchere, Henry] To: H.T. De la Beche
Board of Trade.
11 July 1852
Thank you for the papers which you have sent me respecting the School of Mines.
[Same author as **2179**].

Anonymous authors

2181 Anon.
1 November 1823
Section of the veins of coal belonging to Trecastle Colliery in the Parish of Lanharry in the County of Glamorgan. N.B. These veins were surveyed by Messrs. Martin and Davis in 1816.

2182 Anon.
19 April 1829
[Note in French]

2183 Anon.
[Watermark 1829]
[Notes from Jones' *History of Brecknockshire* re Clydach Ironworks, Beaufort Works, Brecknock and Monmouthshire Canal and Llanelly Works, Brecknockshire on prices of coal, mode of working, transport].

2184 Anon.
[Watermark 1832]
[Pencil sketch of an owl with a rat and a stanza of verse about the nightingale]

2185 Anon.
[Watermark 1833]
[?referee's comments on a MS of over 270 pages]

2186 Anon.
[Watermark 1834]
[?referee's comments on De la Beche's *How to observe*, 1835]

2187 Anon.
1833-4
Account of strata bored through at Pyethorn in the Township of Wilpshire near the Blackburn & Whalley Road on an Estate belonging to the Right Honourable Lord de Tabley.

2188 Anon.
[post-1837]
Reports House of Commons from 1715 to 1801 16 vols
General Index 1801 to 32
General Index 32 to 44
Catalogue of Reports from 1696 to 1834
Catalogue of Reports to 1837

2189 Anon.
[Watermark 1838]
[Plan for establishing a mining college at Truro]

2190 Anon.
[?c. 1840]
A section of the Coal and Iron Mine under Cafen Foce in the Parishes of Haleston and Tythegeson; Section at Cwm Risca and Tondu.

2191 Anon.
[Watermark 1841]
Dundry Church to Bristol and Exeter Road
[survey measurements]

2192 Anon.
[?c. 1841]
Preselly Top on Maenclochog Church
Carn Llanllawer on top of Carnyngly
[survey measurements]

2193 Anon.
[Watermark 1842]
[Notes on burial; several addresses]

2194 Anon.
15 November 1843
[Survey measurements, Somerset]

2195 Anon.
[c. 1844]
[List of causes of deaths in Bristol parishes 1842 and 1843]

2196 Anon.
1846
[List of reports on e.g. Fine Arts, Tidal Harbours, Drainage of Ireland etc.]

2197 Anon. To: R.T. Abraham
Exeter.   Heavitree,
4 February 1847   Exeter.
Re Penn Recca quarry, Staverton, Devon. I.B. White and Son are making an adit to drain it. They came unexpectedly on greenstone, as shown in the plan. The men get £17 per fathom for tunnelling the greenstone.

2198 Anon. To: H.T. De la Beche
Killaloe Marble Works,
Dublin.
1 August 1848
Please excuse the delay in providing the marbles which you ordered, but

there has been difficulty in obtaining the green marble. All will be ready soon.

2199 Anon.
[?c. 1849]
It seems in contemplation to convey the sewerage of London down to the Essex and Kent marshes. No means can be devised for scouring such an extent with little or no descent. A sediment will soon form at the bottom. It will be a melancholy reflection after expending some couple of millions if London should prove in fifty years to be in a worse state than ever. One million pounds of manure could be recovered annually.

2200 Anon.
Van Dieman's Land.
18 October 1850
This magnificent tree was discovered on the Estate of Richard Backer Esq. of Macquarrie Plains. It was 12 feet high and embedded in lava.

2201 Anon.
Mumbles Lights.
15 January 1842
The rise and fall of the tide at Mumbles Head is 36 feet.

2202 Anon.   To: H.T. De la Beche
3 South Parade,
York.
30 January 1852
There are only four persons on the Survey who can sketch hills, having been instructed in the art by Mr. Dawson. Of those now remaining I am the senior having received my appointment in 1828. Mr. Hall received his the following year and the other two - Messrs. Giles and Durrant joined a few years after us. If others attempt to draw the hills without proper instruction it would prove a failure.

2203 Anon.
[no date]
[Notes on Jamaica]

2204 Anon.
[no date]
Presented by Raleigh Mansel Esq., Heathfield, Swansea.

2205 Anon.
[no date]
Specific Gravity.

2206 Anon.
[no date]
M. E. de Verneuil
53 Rue de la Madeleine
Paris.

2207 Anon.
[no date]
[Map of Ireland]

2208 Anon.
[no date]
Narrative of A Voyage to the Southern Atlantic Ocean in the years 1828-29-30, performed in HMS Chanticleer under the command of the late Captain Henry Foster FRS. By W.H.B. Webster, Surgeon of the Sloop.
[above title only]

2209 Anon.
[no date]
[Notes on fossils]

2210 Anon.
[no date]
[List of fossils]

2211 Anon.
[no date]
[Chemical analysis]

2212 Anon.
[no date]
[Table: pressure and temperatures]

2213 Anon.
[no date]
[Curriculum vitae] I have knowledge of Chemistry, Botany and Physiology, Geology and Agriculture.

2214 Anon.
[no date]
[Names and addresses of engineers]

2215 Anon.
[no date]
[Survey measurements South Pembrokeshire]

2216 Anon.
[no date]
[Notes on pay and expenses]

2217 Anon.
[no date]
[Notes on metals]

2218 Anon.
[no date]
[Notes on olive and carob trees]

2219 Anon.
[no date]
[Unfinished pencil tracing of the Forest of Dean area]

2220 Anon.
[watermark 1834]
[Manuscript copy: tithe for Bishop of Exeter]

2221 Anon.
[watermark 1833]
[Manuscript copy: Parliament Roll of the 50th year of Edward III, No. 130 on Stannaries of Devon]

2222 Anon.
[watermark 1837]
[Manuscript copy: stannaries of Devon and Cornwall]

2223 Anon.
[watermark 1837]
[Manuscript copy: Patent Roll of the 50th year of Edward III p.1 m.2, appointment of stampers of tin in Devon]

2224 Anon.
[watermark 1836]
[Manuscript copy: Parliament Roll of the 8th year of Edward II. ml5. n.ll5, Petition from the poor men of Devon that the Tinners do maliciously subvert, dig, destroy and waste the arable lands, meadows, groves, houses and gardens of the same men]

2225 Anon.
[watermark 1836]
[Manuscript copy: Charter Roll of the 36th year of Henry III m.18, Tinners of Cornwall and Devon]

2226 Anon.
[watermark 1836]
[Manuscript copy: Patent Roll of 35th year of Edward [?]m.43, Tinners of Cornwall and Devon]

2227 Anon.
[watermark 1836]
[Manuscript copy: Parliament Roll of 8th year of Edward II, Tinners of Devonshire]

2228 Anon.
[watermark 1837]
[Manuscript copy: Patent Roll of 5th year of Henry IV p.1 m.19, Confirmation for the Tinners of Devonshire]

Miscellaneous printed items

2229
[watermark 1823]
Index to the Ordnance Survey of Great Britain
Sold by James Gardner, 163 Regent Street, London. Indicates the number of Plates published, engraving for publication, or in progress.

2230 *Illustrated London News*
8 April 1848
[report] The Museum of Economic Geology
[illustration] Piccadilly front of the Museum of Economic Geology

2231 *Illustrated London News*
24 May 1851
[illustration] Opening of the Museum of Practical Geology - The Great Hall.
[report] The Museum of Practical Geology.

2232
[c. 1843-1847]
A New Plan of Bristol, Clifton and the Hot-Wells.
Published by J. Chilcott, Wine Street, Bristol.
[Some areas marked] Blue for fever, Red for cholera.

2233
[no date]
Chart of the World illustrative of the impolicy of slavery.
Published by J. Cross, 18 Holborn.

2234
[no date]
[view of] Wycombe Church, Bucks: previous to 1273

2235
[no date]
[Pages 137-140; page of errata; and plates 38, 40, 41, 42, 43, 43A and 44 on the dentition of kangaroos and wombats. (in French)]

2236
[no date]
[Plate (Tav. I) comprising 11 figures of elephant bones]
Lit. Targioni. A. Servantoni del. Tofani dis.

2237
[post 1845]
[Half of printed page containing details of publication dates and prices of *Transactions of the Linnean Society* up to 1837 and annotated up to 1845]

Addenda

2238 Aveline, [?] Elizabeth　　　　　　　　　To: H.T. De la Beche
[?1848]
Thank you for sending me the "portrait charmant". What you may have lost in beauty since the portrait of your youth you have gained in expression. You look like a wise man with some cares and much thought. It is in itself a beautiful engraving doing great credit to Walker.

2239 Barr　　　　　　　　　　　　　　　　To: H.T. De la Beche
[no date]
The Board of National Education are fitting up a fishing school at Galway (in the Claddagh). The Board have asked me to write a book for schools in seaports. I was glad to meet your wishes about the Transactions of the Royal Irish Academy for your Library. Hope you have not put me out with Sir Robert Kane. Thought him odd last time we met.

2240 Booker, Thomas W.
[c. 1837]
Notes on experiments by D.W. Linden April 28 1760 on the water in Taffs-well in Glamorganshire.

2241 Booker, Thomas W.
24 June 1839
Section of strata from Abersychan to the Cwm Pits in the Ebbw Fach. Copied from a paper lent me by Edward Jones Rudry June 24th 1839.

2242 Booker, Thomas W.
24 June 1839
An account of the minerals on the Waunfawr Coalery south of Risca. Copied June 24th 1839 from paper lent me by Edward Jones Rudry.

2243 Booker, Thomas W
24 June 1839
Section of levels at Rudry Works. Copied June 24th 1839 from paper lent me by Edward Jones of Rudry.

2244 Coffin, Walter　　　　　　　　　　　　To: H.T. De la Beche
Llandaff.
25 September 1844
As soon as I can go to Cardiff I will send you the Taff Vale Railway Act. In the meantime you may like to know the tolls on iron and coal. We are sinking the second pit. We have gone through the Abergorki, Upper Four Feet and Six Feet.

2245 Conybeare, William Daniel
[c. 1824]
[3 pages of manuscript on geology]

2246 De la Beche, Henry Thomas
15 July 1819
[Passport used on tour of France, Switzerland, Italy etc. 1819-20 issued by V. de Latour-Mauborg, French Embassy, London].

2247 De la Beche, Henry Thomas
[watermark 1820]
[2 pages of notes on geology of Italy]

2248 De la Beche, Henry Thomas
[watermark 1828]
[1 page of notes on social animals].

2249 De la Beche, Henry Thomas
[watermark 1829]
[Notes "On the origin of the thermal springs" by Daubeny, published by De la Beche in *A Geological Manual* 3rd edition p. 606].

2250 De la Beche, Henry Thomas
[watermark 1830]
Notes from Mantell's *Geology of Sussex* 1822.

2251 De la Beche, Henry Thomas
[watermark 1830]
Notes from Fitton *Annals of Philosophy*, New Series Vol 8 November 1824.

2252 De la Beche, Henry Thomas
[watermarks 1830 & 1831]
[9 pages of manuscript:] It may at first sight appear somewhat strange to call the attention of naval men to geology.

2253 De la Beche, Henry Thomas
[watermark 1831]
Animal substances
Most frequent ultimate elements of animal matter.
Oxygen, Hydrogen, Carbon, Nitrogen, Sulphur and Phosphous.

2254 De la Beche, Henry Thomas
[watermark ?1833]
[Memo: compass bearings in Cornwall and note on a section near Polperro].

2255 De la Beche, Henry Thomas
[post-1836]
[5 pages on notes on water temperatures and thermal springs].

2256 De la Beche, Henry Thomas
[post-1836]
[2 pages of notes on phosphate of lime in bones, scales and teeth; and temperature at depth below Paris].

2257 De la Beche, Henry Thomas
Liliput.
March 1838
[10 pages of] Extracts from Washington Irving's "Sketch Book".

2258 De la Beche, Henry Thomas
[?c. 1848]
[4 pages of notes on] Streets, Courts and Alleys [of Bristol].

2259 De la Beche, Henry Thomas
[1848]
That the Wollaston Medal for the year 1848 be presented to Dr Buckland for the many eminent services he has rendered to Geological Science and for the numerous and brilliant discoveries which he has made known to the World through this Society.

2260 De la Beche, Henry Thomas   To: [?A.C. Ramsey]
[watermark 1849]
[?part of draft letter] You may, if you like, state to Gapper that he has his increase of pay from the 1st, and I will write you the regular official tomorrow.

2261 De la Beche, Henry Thomas
[no date]
[1 page] Note from Sowerby's *Mineral Conchology* Vol I 1813.

2262 De la Beche, Henry Thomas
[no date]
References to Webster's Publications 1814 to 1825.

2263 De la Beche, Henry Thomas
[no date]
Note from Conybeare and Phillips Outlines of the geology of England and Wales (1822) [before the appearance of Mantell's *Geology of Sussex*].

2264 De la Beche, Henry Thomas
[no date]
Notes on Lake Superior from Captain Bayfield
Transactions of the Literary and Historical Society of Quebec Vol 1.

2265 De la Beche, Henry Thomas
[no date]
[1 page of notes on colours for landscape paintings].

2266 [?]James, Henry
28 October 1842
Coloured sketch vertical sections Woolhope, Malvern and May Hill.

2267 Lemon, Charles   To: H.T. De la Beche
Carslow.
4 January [1852 FJN]
I have been in daily expectation of a summons to London to one of my sisters, who is in an alarming state of health, and I intended there to have thanked you for the publications which you have sent me at different times. But there appears to be a respite in her case, and perhaps I shall not go up yet. I do not wish to defer longer thanking you for the account of your proceedings which I delight to follow. You have begun prosperously and I hope will continue to have the wind in your sails.

2268 Lemon, Charles   To: H.T. De la Beche
Carslow.
10 October [no year]
I have been so busy with the Polytechnic and have now the geological in hand, for which I have two papers besides my address, that I have not had time to thank you for your very kind assistance.

2269 Macfadyen, James,
[watermark 1821]
Description of three species of the Genus Ficus, Natives of Jamaica.

2270 Philipps, Trenham W.
1837
Statement of monies owing.

2271 Phillips, John
1838
Notes on some Devonshire fossils 1838 (6 pages). Notes on metals (1 page). Memoranda on bearings in Cornwall (1 page).

2272 Phillips, John   To: H.T. De la Beche
Athenaeum.
23 July 1843
[Part of letter only]
You ask what you can do to relieve my beastly trials in Whitehall Yard - Give me my drawers! A hundred and twenty drawers! Colby has got a notion that in your evidence you condemned the book by Captain Portlock. I told Colby that he must have been misinformed. There is certainly much praise due to Portlock for his endeavour to do justice to his Palaeontological Subjects.

2273 Smith, C.H.
16 April 1839
Section of the beds cut in sinking to the Three Feet Vein at the Double Pit above Tyr Bach Llansamlet; section of beds occurring in the Clydach Valley as observed on the west side of the railroad to Shieks Collieries in the 5 feet vein of coal between the Lower Forge and Pont y Lon.

2274 Smith, Joshua Toulmin   To: H.T. De la Beche
8 Serjeants Inn.
30 November 1847
From a short paper which I took the liberty of enclosing to you a few days ago you would perceive that I am now engaged in a careful examination of all the Ventriculidea. In the Museum of the Geological Society is a specimen in which the structure happens to be well preserved. May I make a section of it? It is already broken into pieces.

2275 Trevelyan, Walter Calverley   To: H.T. De la Beche
Nettlecombe,
Taunton.
5 September 1848
Having lately been looking at the limestones here, I find several spots where they appear and which are not indicated in the Geological Survey as shown in the tracing enclosed. The series of Portugese building stones I promised you is nearly complete. Do you wish for any blocks of alabaster from Watchet?

2276 Vernon, Robert   To: H.T. De la Beche
Westminster.
3 May 1847
Robert Vernon of No. 50 Pall Mall and Ardington, Berks has long employed his great wealth in the patronage of the fine arts and has collected at enormous expense the best specimens of the English School. He is now building the steeple at Ardington Church at his own cost. When his collection goes to the nation with his name, he asks that a name might be accompanied with an accessory that would show his sovereign's approbation of his conduct.

2277 Vivian, Henry Hussey   To: H.T. De la Beche
Hafod Works.
6 December 1849
We have all the materials prepared for making you a small collection of copper products but I find it difficult to arrange them to my own satisfaction. I thought of getting a stand made with divisions for each process showing the gradual advancement of the metal and the resulting products of each process, at the same time placing a few nice specimens alongside the ordinary rough products.

2278 Waring, Henry Franks   To: H.T. De la Beche
Lyme.
6 April 1847
I am in communication with the Department of Woods and Forests respecting a lease of stone ledges on this coast, and understand that you are appointed to survey the adjacent coast at my expense. When you are at Lyme, please come to dinner with our mutual friends Captain and Mrs. Benett.
[Enclosed with Benett, C.C. to De la Beche, H.T., 8 April 1847, **79**].
[See De la Beche, H.T. to Waring, H.F., 9 April 1847, **410** for reply].

Miscellaneous items

2279 Medal
Obverse: Bust facing right, 'H.T. DE LA BECHE' vertically behind.Neck inscribed 'W. WYON R.A.'
Reverse: Two crosssed hammers in centre of beech wreath. 'SIS MEMOR USQUE ME1' above. Inscribed 'L.C. WYON' below.
Edge: Engraved '   , JULY 1893'. Name of recipient removed. Awarded at Royal School of Mines to Samuel Warren Price.
AE, diameter: 45mm, edge thickness: 4mm, weight: 51.57g.
[Previous accession no. 84.52G.]

2280 Medal
Description as **2279** but edge thickness 5mm and weight 66.97g.
Formerly in the possession of James Holmes, First Office Keeper at the Museum of Practical Geology, Jermyn Street, London.
[Previous accession no. 37.166]

2281 Medal
1841
Plaster moulds and casts.
Obverse: As **2279**. Reverse: Palm trees, 'REWARD FOR GOOD CONDUCT' above. 'HALSE HALL/JAMAICA' below in exergue.
Diameter: 45mm
See: Brown, L., *A catalogue of British historical medals 1837-1901*, No. 2002.
[Previous accession no. 38.767]

2282 3ft radius quadrant
c.1820
Brass, on heavy wooden tripod. Inscribed Troughton. F.J. North in the preface to his unpublished biography of De la Beche refers to 'the discovery, amongst the collections of the Cardiff Municipal Museum . . . of a surveyor's quadrant with the name H.T. De la Beche engraved on its metal scale'.
[Cited and figured: Holbrook 1992, p.111 and fig.117, p.69]

2283 Bust of Henry Thomas De la Beche
c.1845
Miniature (22cm high) reproduction in plaster of a bronze bust by E.H. Baily, R.A.
[Presented by Miss Dorothea and Miss Violet Ramsay of Porthmadog]
[Figured: North 1936a, plate 2]

PERSONAL NAMES

Aberdeen, George Hamilton Gordon (1784-1860), 4th Earl of 192, 221
Abernethy, Dr 849
Abraham, Mr [of Montreal] 917
Acland, Thomas Dyke (1787-1871) 1-4, 483, 1921
Acraman, Daniel Wade (c.1775-1847)
    picture collection of 304
Adam, Sir Frederick (1781-1853) 363
Adam, William (c.1794-1873)
    sale of Matlock Bath Museum belonging to 1833
Adams, Mikhail Ivanovich
    mammoth hair found in Siberia by, 343, 346
Adare, Edward Richard Windam Wyndham-Quin, Viscount (1812-1871) 5-7, 757, 1376
Agassiz, Louis Jean Rodolphe (1807-1873) 8, 134, 311, 621, 891, 905, 952, 1037, 1071, 1548, 1598, 1867
    & glacial theory 189, 1507
    & British Association Fossil Fish Report 1150
Ainsworth, William 865
Airy, George Biddell (1801-1892) 496, 497, 1053, 1588, 1922, 2051
Ajasson de Grandsagne
    & French edition of De la Beche's *Manual* 2072, 2073
Albert, H.R.H. Prince (1819-1861) 9, 10, 591, 605, 617, 630-633, 968-973, 1161, 1691, 2069
    meeting with De la Beche 252, 253, 1673
    on use of sewage 629
Alder, Mr 800
Allen, Mr 963, 964
Alves, Mr [of Jamaica] 352
Amouroux, J.V.F.
    collection of 350
Anderton, Mr 1684
Angus, Mr
    resignation from post of Naturalist on Turkish-Persian Boundary Commission 2149
Anning, Charles Churchill 11
Anning, Joseph (1796-1849) 11
Anning, Mrs Mary (c.1764-1842)
    plesiosaur in possession of 299
Anning, Mary (1799-1847) 182, 977
    biographical notes on 11
    sale of ichthyosaur by 180
    ichthyosaur in possession of 297
    discovery of complete plesiosaur by 302
    on *Hybodus* 952
Anning, Richard (c.1766-1810) 11
Anson, Col. 650
Ansted, David Thomas (1814-1880) 558, 732
Ap John, James 779, 1555
Arago, Dominique François Jean (1786-1853) 1180
Arbuthnot, George (1802-1865) 13
Arbuthnot, General Robert (1773-1853) 1183
Archer, Mr 1497, 1521, 1529, 1530, 1665
Argyll, George Douglas Campbell (1823-1900), 8th Duke of 14-16, 1183
Arkwright, Richard (1732-1792) 850
Armstrong, William George (1810-1900)
    elected Fellow of the Royal Society 743
Armstrong, Mr [of Bristol] 246
Arnott, Dr 233
Arnoux, G.[of Toulouse] 17
    letter of introduction of 1195
Artis, Edmund Tyrell (1789-1847)
    biographical notes on 18
    fossil collection of 18
Artis, Elizabeth 18
Ashworth, Lady 559
Ashworth, Miss
    engagement to Edward Forbes 559
Askham, William 2022

Askin, Mr 1943
Atherstone, W. Guybone 19
Atkinson, Charles C. 2013
Atkinson, Mr [of Jamaica] 352, 920
Atkinson, Mr 2114
Attwood, Melville (1812-1898) 1966
Aubison de Voisins, Jean Francois d' (1769-1841) 349
    publications 500
Aubrey, John (1626-1697)
    Natural History of Wiltshire 2158
Auckland, George Eden, Earl of (1784-1849) 20-23, 612, 680, 681, 683, 706, 1678, 2031, 2033-2035
    death of 713
Auerbach, F. 1002
Auld, Robert 2055, 2056
Auriol, John Lewis 298, 349
Auriol, Mrs Anna 346, 349
Austen, Robert Alfred Cloyne (1808-1884) 24-41, 561, 589, 941, 945, 1403, 1873, 1878, 2098
    as potential Geological Society President 584
    fossil collection of 37, 1334, 1335, 1342, 1358, 1363
    & Devon Controversy 621
    publications of 621
Austen, Mrs 559, 589
Austin, Mr 239, 240
Aveline, Miss Barbara (?1778-1846) 339
Aveline, [?]Elizabeth (c.1779-1856) 2238
Aveline, Henry Thomas 42
Aveline, Thomas Huddle (c.1777-1839) 1980, 2087, 2088
    in Militia 2088
Aveline, William Huddle (d.1839) 43, 341, 349, 360, 373, 533
    as Elizabeth (Bessie) De la Beche's grandfather 339, 373
Aveline, William Talbot (1822-1903) 42, 44, 45, 119, 120, 561, 583, 588, 794, 1008, 1368, 1378, 1380, 1388, 1389, 1397, 1399, 1402, 1407, 1408, 1412, 1430, 1437, 1441, 1442, 1444-1447, 1450-1452, 1460, 1463, 1474, 1543, 1556, 1571, 1572, 1728, 1730, 1732, 1733, 1739, 1742-1744, 1746, 1752, 1768, 1769, 1782-1784, 1789, 1793, 1796, 1797, 1803, 1955, 2125
    laudatory reference to 1573
    health of 1759, 1761
Aveline, Mrs 1759, 2125
    mines of 2125
Avory, Mr 594
Aylmer, George William 898

Babbage, Charles (1792-1871) 46, 302, 718, 890
    drawings of 46
Babington, William
    biographical notes on 47
Back, Mr 1020
Backer, Richard 2200
Baddle, Mr 1275, 1282
Bagot, Sir Charles (1781-1843) 878-880
Bagot, Richard (1782-1854) 878, 880
Bailey, Mr 608
Bailey, Dr 301
Bailey, Sir T. Lawrance? 301
Bailey, Mr [?of Bristol] 1867
Bailliere 84
Baily, Edward Hodges (1788-1867) 421
    bust of De la Beche by 2283
Baily, Francis (1774-1844) 898, 914
Baily, William Hellier (1819-1888) 543, 544, 560, 567, 569, 581, 588, 703, 709, 1545, 1555, 1562, 1579, 1615, 1616, 1797, 1853
    fossil drawings of 407, 542, 545, 552, 1536, 1542, 1740, 1742
    salary of 1532, 1533
Baily, Mr 1154
Bain, Andrew Geddes (1797-1864) 19, 585
Bainbridge, William (1812-1860) 48, 1976

Baker, John (fl.1814-1850) 49
    offer of sale of specimens collected by 49
Baker, Mr 568, 961
Baker, Samuel [building contractor British Museum] 1920
Baker, Mr [of Bridgwater] 2041
Bakewell, Mr 915
Balfour, Dr Edward 736
Ball, Robert (1802-1857) 50-52
Ball, Thomas 558, 756, 1749
Ball, Mr [shell collector] 800
Ball, Mrs 564
Balmat, Jacques (1762-1834)[alpine guide] 344, 346
Balmat, Pierre (Jr. & Snr.) [alpine guides] 344, 346
Bancroft, George (1800-1891) 53-56
Bancroft, Dr [of Jamaica] 920
Bandiner, Mr
    collection of china of 454
Bandinck, James [of Foreign Office] 1979
Bang, Mr 598
Bankhart, Mr
    patent on copper smelting 1670
Banks, Lady Dorothea (1758-1828) 150
Banks, Sir Joseph (1743-1820) 148, 150, 342, 1332, 1537
    health of 153
    resignation as President, Royal Society 153
Barber, Mr [shell collector] 800
Bardwell, William 56
Barff, John 57
Baring, Francis Thornhill (1796-1866) 58-61, 387, 662, 665, 1201,
        1202, 1205-1209, 1214, 1215, 1217, 1220, 1228, 1230, 1243,
        1244, 1256, 1259-1262, 1264-1266, 1268, 1274, 1300, 1701
Barker, James 62, 63, 739, 1648
    geological maps of Derbyshire 1972
Barlow, John 64
Barnes [of East India Company] 896
Barnett, Mr 110, 1458, 1461
Barnett, Hoare & Co [bankers of 62 Lombard St.] 1489, 1491, 1526,
        1533, 1582, 1589, 1610, 1634
Barr, [?Frederick] 2239
    considered as Keeper of Mining Records 1249, 1250, 1254
Barrande, Joachim (1799-1883)
    laudatory reference to 1022
    publications 639, 1022
        Système Silurien... de la Bohême 642
Barratt, Capt. 946
Barrer, Mr
    on origin of red snow 343
Barrington, Shute (1734-1826) Bishop of Durham 170
Barrow, Mr 149
Barry, Charles (1795-1860) 65, 66, 240, 979, 982, 1242-1244, 1246,
        1255, 1262-1264, 1266, 1267, 1300, 1903
    criticism of 1264
    health of 1303
Barry, Sir J. 665
Barton, Sir John 2165
Barton, Mr 2165
Barton, Mr 171
Bates, Mr
    British Association Railway Section Committee 189
Bath and Wells, Bishop of
    notes on Mendip Mines from Book of 852
Bathhurst, Henry (1762-1834) 3rd Earl 151, 152, 157, 1255
Bayfield, Captain 2264
Bayley, Sir Henry 123
Bayley, William Butterworth (1782-1860) 2028, 2030
Baylis, B. 67, 68
Beach, Helen *née* Hynes (d.1771) 2159
Beach, Janet 456
Beach, John Hynes (c.1766-1803) 2159
Beach, Robert (d.1673) 2159
Beach, Thomas (d.1801) 456, 2158, 2159
Beach, William 456
Beach family 2157-2159
Beadon, Cecil 2145
Beaufort, Sir Francis (1774-1857) 69, 70, 542, 1987
Beaufort, Henry Somerset (1792-1853), 7th Duke of 342

Beaumont, John (d.1731) 2089
Beaumont, Thomas Wentworth (1792-1848)
    lead mine owner 1966
Beaumont, Mrs
    water divining by 150
Beazley, Mr 39
Becher, A. B. 69, 71, 72
Beck, Dr [of Copenhagen] 891
    criticism of Deshayes 891
Becke, Henry (1751-1837) 147, 162, 302
Becker
Beddes, Joseph 73
Bedford, E. 70
Bedlington, William 74
Belcher, Captain Edward (1799-1877) 75
    to survey Pacific coast 622
Bell, George Gray 76
Bell, Thomas (1792-1880) 77
    & Royal Society 911, 1104, 1186, 1194
    proposed as Secretary, Royal Society 913
    reptile monograph 94
Bellamy, J. C. [of Plymouth] 1158
Belleroche, F. A. 488
Belsey, Mr 237
Belshaw, Thomas 78
Benett, Miss A. M. 91
Benett, Capt. Charles Cowper (c.1788-1878) 79, 80, 696, 2278
Benett, Etheldred (1776-1845)
    sale of collection of 91
Benett, Philip Bowles (1835-1863) 80
Benett, Sarah 80, 81
Bennett, Mr 945
Bennett, Mr [?Dudley Coal Owner] 1943
Bennett, Squire 1728
Benson, Starling (1808-1879) 456, 869
Benton, Mr 977
Berkeley, Craven
    duel with Capt. Boldero 1272
Berthelot, Sabine
    map of Canary Islands by 501
Bessel, Friedrich Wilhelm (1784-1846) 1440
Bethune, Captain
    expedition to Borneo 542
Beudant, Francois Sulpice (1787-1852)
    on geology of Hungary 1927
Bevan, Evan
    surveyor at Neath colliery explosion inquest 1958
Bevan, Mr
    Cwm Avon Works 2010
Bianconi, Giacomo 82-84, 1975
Biddulph, Mr
    Cwm Aman colliery owner 1994
Bilton, Mr 1335, 1337
Binks, Mine Captain [of Ponterwyd] 1942
Binney, Edward William (1812-1881) 1944
Biram, Ben 85
Birch, Samuel 86
Birch, Col. Thomas James Bosvile (c.1768-1829)
    collection of 341
    crustacean fossils of 169, 170
    ichthyosaur in possession of 699
    plesiosaur in possession of 300
Bird, Dr [of Swansea] 456
Birmingham, Mr 2
Bischof, Karl Gustav Christoph (1792-1870)
    publications, *Lehrbuch der chemischen physicalischen geologie*
        (1848) reviewed 932
Blackwell, J. Kenyon 87-89, 1650, 1652, 1656
    report on colliery explosions 98, 538
Blackwell, Thomas E. 90
Blaine, H. & R. S. 32
Blauner, Mr
    as Swiss natural history dealer 202
Blayds, Elizabeth 91
Blayds, Mr 91
Blissett, Joseph
    mineral collection offered to Bristol Institution 302, 304

Bl[oc?]he, Francois [fossil collector, N. France]
   De la Beche purchase of fossils from 350
Blore, Edward (1787-1879) 1266
Boase, Dr Henry Samuel (1799-1883)
   criticism of 897, 1874
   publications, [?book on Cornwall] 1875
Boileau, Sir John Peter (1794-1869)
   vice president of Ipswich Museum 1985
Boldero, Captain H.G. [MP] 399, 1534
   duel with Craven Berkeley 1272
Bone, Charles Richard (1808-1875) 703, 738, 1096, 1106, 1734, 1853
   fossil drawings of 407, 551, 560, 566, 567, 705, 709
   laudatory references to 705, 1813
Bone, Henry Pierce (1779-1855)
   engraving of portrait of De la Beche by 415
Bonner, Col. 2034
Bontell, Charles 92
Booker, Thomas W. 2240-2243
Booth, Prof. 2017
Boromeo, Count Vitello
   mineral collection of 363
Bostock, John (1773-1846)
   nominated President, Geological Society 175
Boswell, Mr 1213, 1251
Botfield, Mr
   offered ichthyosaur by Mary Anning 180
Boucher, Mr [of Jamaica]
   Jamaican estates of 352
Boué, Ami (1794-1881) 93, 497
   Murchison's criticism of 1009
   English translation of *Memoir on the Geology of Germany* by 423
   on geology of Italy 332
Boulanger, Mr
   translation of De la Beche's publications by 500, 502
Boulton, Matthew (1728-1809) 1825
Bowdich, Thomas Edward (1791-1824) 343, 346
Bowerbank, James Scott (1797-1877) 94-96, 550, 573, 578
Bowman, Mr 1612
Bradley, W.H. 97
Braikenridge, George Weare (1775-1856)
   collection of 300
   ichthyosaur in possession of 296, 297
Brand, H. 98-100, 732
Brand, Mr
   on origin of red snow 149
Brande, William Thomas (1788-1866) 101, 102
   publications 101, 102
   & geological education of R.I. Murchison 1020
Brandis, Dr 337
Bray, Capt. 945
Brayley, Edward William (1802-1870) 103
   application for Chair at University College, London 103
Bree, C. N. 95
Breislak, Scipione (1748-1826) 348
   mineral collection of 348, 363
Breton, Lieut. 712
Breuner, Count 158
   continental tour of 153, 155
Brewster, David (1781-1868) 104
Brickenden, Richard Thomas William Lambart (1809-1900) 105
Brickwell, William 106-107
Briggs, George D. 108
Bright, Benjamin Heywood (1787-1843) 109-112, 1441
   collection of 1442, 1457, 1458
Bright, Henry (1784-1869) 113-116
Bright, Richard (1754-1840) 301, 302, 304
   collection of 109, 300, 1458
Bright, Dr Richard (1789-1858) 686
Bright, Sarah 112
Bright, Mr 2111
Bristol, Dean of [Becke, Henry (1751-1837)] 147, 162, 302
Bristow, Henry William (1817-1889) 118-128, 572, 574, 575, 582, 583,
     585, 588, 591, 794, 1503-1505, 1513, 1515, 1516, 1525, 1530,
     1550, 1552, 1555, 1557, 1726, 1728, 1759, 1764, 1767, 1770,
     1774, 1786, 1793, 1797, 1800, 1866, 1867, 2110
   application to New South Wales Geological Survey 125
   appointment to New South Wales Geological Survey 126,
     1787-1789
   health of 1807
   laudatory reference to 1528
   pay increase for 1749, 1754
   recommendation for New South Wales Geological Survey 124
Bristow, Whiston 129
   biographical notes on 129
Bristow, Mrs [of Broxmore Park, cousin to H. W. Bristow] 1807
Broadhurst, Mr [of Jamaica] 352
Brocchi, Giovanni Battista (1772-1826) 146
Brochant de Villiers, André-Jean-Francois-Marie (1772-1840) See Villiers,
   André-Jean-Francois-Marie Brochant de
Brock, Sir Michael [at Tenby] 456
Broderip, William John (1789-l859) 115, 130, 182, 307, 620
Brodie, Benjamin Collins (1783-1862)
   & Royal Society 911
Brongniart, Adolphe Theodore (1801-1876) 32, 131, 132, 903, 2085
   publications, *Prodrome d'une histoire des végétaux fossiles* (1828)
     1329
Brongniart, Alexandre (1770-1847) 17, 132, 133, 343, 346, 496, 534,
     1195, 1548, 2081, 2085
   collection of 343, 346
   on geology of Monte Ferrato, near Prato, Italy 363
Bronn, Heinrich Georg (1800-1862) 134
   publications, *Index Palaeontologicus* 134
     Lethaea Geonosticus 134
Brooke, Sir James (1803-1868) 992, 993
Brooker, Mr 925
Brookes, Joshua (1761-1833) 302
Brough, Lionel [of ?Oakwood] 2010
Brougham, Henry Peter, Baron Brougham and Vaux (1778-1868) 622,
     657, 834, 1598
Brown, John (1780-1859) 135, 136
Brown, Robert (1773-1858) 137, 138, 141, 149, 452, 568, 570, 686,
     906, 908, 912, 913, 1058
   in Paris 2053
Brown, Thomas
   Cwmtillery coal mine 2107
Brown, Dr 867
Brown, 1323
Browne, Mr 302
Brunel, Isambard Kingdom (1806-1859)
   British Association Railway Section Committee 189, 397, 1375,
     1376, 1858, 1859
Bryce, James Junior (1806-1877) 139, 1264
   application for Chair of Geology, Queen's College, Belfast 139
Buccleuch, Walter Francis Scott (1806-1884), 5th Duke of 140
   as head of Metropolitan Sanitary Commission 615
Buch, Baron Christian Leopold von (1779-1853) 141, 177, 503, 645,
     1022, 1408, 1898, 2077, 2085
   award of Wollaston Medal to 141
   in Oxford 145
   map of Lake of Lugano presented to Bristol Institution 977
   on limestones at Nice 364
   publications, on brachiopods 141
   publications, on Canary Islands 501
   specimens of, in Neuchâtel Museum 347
   tour with Elie de Beaumont 491
   theory of volcanoes of 141
Buch 143
Buckingham, Richard Temple Nugent Brydges Chandos Grenville
   (1776-1839), 1st Duke of 301, 302
   Hebridean tour with Buckland 304
Buckland, Francis Trevelyan (1826-1880) 179, 180, 307
   breeching of 183
   piddling cartoon of, by De la Beche 1031
Buckland, Mary née Morland (1797-1857) 142, 143, 164
   health of 176, 178
Buckland, William (1784-1856) 8, 43, 107, 110, 130, 131, 144-200,
     204, 231, 256, 280, 302, 306-312, 341, 342, 358, 364, 379,
     387, 396, 457, 470, 473, 496, 535, 536, 554, 621, 690, 703,
     718, 719, 797, 876-878, 880, 899, 903, 965, 1006, 1009, 1018,
     1037, 1058, 1063, 1152, 1167, 1274, 1300, 1301, 1375, 1376,
     1389, 1395, 1514, 1531, 1704, 1819, 1874, 1875, 2029, 2040,
     2042, 2043, 2268
   acquisition of specimens 148, 151, 152

advice sought 256, 315, 1037
& Glacial Theory 189, 537
& Devon Controversy 623
& geological education of Murchison 1020
& Lyme Regis landslip 1339
appointment as Professor of Geology at Oxford 148
appointment to Canonry of Christ Church, Oxford 173
arrested in Italy 146
as candidate for President, Corpus Christi College 167
award of Royal Society Copley Medal 165
award of Wollaston Medal 2259
coach accident 160
collection of 301
comments on De la Beche's Nice paper 362, 364, 620
continental tour, 1816. 146
continental tour, 1820. 153, 155
criticism of Geological Survey plan to publish *British Organic Remains* 1490
display cases of 1217
elected Honorary Member, Cambridge Philosophical Society 152
elected to British Museum committee 1883
epitaph, "Mourn, Ammonites, Mourn" 155, 156
fieldwork in Dorset 153, 157, 177
gas shares of 142, 455
health of 142, 143, 151, 152, 455, 1152, 1489, 1490
hyaena den theory of 162, 165, 172
in Glen Roy 1858
in lithograph *A coprolitic Vision* 432
Kirkdale Cave 1849
lectures of 157, 167, 176, 1709
map of Glamorgan 200
opposition to Swansea British Association meeting 1020
plan for a geological map of Devon 147
proposed visit to Scotland 158, 159
publications 151, 152
    on agriculture 194, 196
    on Alpine limestones 156, 157
    Bridgewater Treatise (1836) 309, 340
    Excavation of Valleys by Diluvial Action (1824) 163
    Geology of Weymouth (1836) 156, 157, 179, 180, 184, 186-188, 891, 1009
    On Ichthyopodolites or petrified trackways of ambulatory fishes... (1843) 191
    Reliquiae Diluvianae (1823) 165, 167, 168, 170-172, 182, 301
    Secondary Formations between Nice & the Col di Tende, (1835) 177, 364
    Vindiciae Geologicae 153, 154, 204
social invitation to 1171
on trace fossils 191
tour of Hebrides 304
vice president, Ipswich Museum 1985
visit to Derbyshire 150
visit to Devon 149, 152, 157, 160, 163
visit to South Wales 146, 147, 150, 151, 158, 160
visit to Somerset 152
Buddle, Thomas 1834
    section of Newcastle Coalfield 182, 322
Bull
    presentation of model dredge to Museum of Practical Geology 579
Bullock, William
    ichthyosaur in possession of 296, 298
Bumper, Mr [Bookseller] 376, 529
Bunbury, Charles James Fox (1809-1886) 701, 725
    fossil plant collection of 904
    proposed as Foreign Secretary, Geological Society 904
Bunbury, Edward Herbert (1811-1895) 862, 578
Bunsen, Baron Christian Karl Josias von (1791-1860) 141, 201-203
Bunt, Thomas Gamlen 1866, 1867
Burge, Mr 114
    as Attorney General of Jamaica 352
Burgess, Thomas (1756-1837), Bishop of St David's 204
Burgliesh, Lord
    as Ambassador to Florence 363
Burgoyne, Sir John Fox (1782-1871) 205, 206, 237
Burke, Mr 760, 761, 779
    Attorney General [? of Ireland] 1170, 1316

Burlington, William Cavendish, 2nd Earl of (1808-1891)
    suggested as President of Royal Society 906
Burmeister, Hermann (1807-1892)
    publications, *Die Organisation der Trilobiten* (1843) 1567
Burney, Dr
    on July 1819 comet 343
Burton, Mr [?of St Leonards] (d.1837)
    funeral of 623
Busk 1854
Busry, Mnr 474
Bute, John Crichton Stuart, 2nd Marquess of (1793-1848) 1681
Butler, G. [of Ordnance] 207, 1876
Butler, Mr 1669
Byham, R. 208-215, 261, 276, 379
Byng, Mr 238

Caernarvon, Henry John Herbert (1800-1849), 3rd Earl 151
Caldcleugh, Mr
    paper by 622
Calding, Mr
    letter of introduction of 597, 598
Callender, William R. 216
Calton, Mrs 309
Cambridge, Adolphus Frederick (1774-1850), Duke of 471
Cameron, Capt. 650, 815, 818, 922, 1084, 1094, 1101
Campbell, Mr [of University College, London] 1669
Candolle, Augustin Pyrame de (1778-1841) 217, 218, 342, 344, 346
Canning, Charles John Canning (1812-1862), Earl 219, 220, 402, 405, 766, 768, 770, 775, 859, 1317
    on emancipation of slaves 357
Canning, Sir Stratford (1786-1880) 1289
Canterbury, Archbishop of 162
Capodistrias, Mr 217
Cardwell, Edward (1813-1886) 593, 1693
Carey & Lea [publishers of Philadelphia] 1826
Carlisle, George William Howard (1802-1864), 7th Earl of [Viscount Morpeth] 224-229, 417, 713, 812, 1103, 1324
    [See also Morpeth]
Carpenter, Mr 302
Carpenter, Dr 1854
    at St Thomas's Hospital 1867
    leaving Bristol 1867
    tutoring post of 1867
Carrington, Frederick A. 229
Carrington, Mr 1992
Carwen
    coal mines of 1295
Cary 1434, 1444-1446, 1448, 1464
Castile, Mr
    Jamaican estate of 352
Castlereagh, Robert Stewart (1769-1822), Viscount 151, 152
Cawdor, John Frederick Campbell (1790-1860), 1st Earl of 230, 947, 1020, 1385, 1386
Cawdor, Dowager Lady 185
Chadwick, Edwin (1800-1890) 231-243, 483, 1322, 1686
Chalmers, J. H. 708
Chambers, Richard 1985
    Scottish raised beaches compared with R. Seine 980
Chamisso, A. de 331
Chance, James 1194
Chandelon, Y. 244
Chandos, Lord 194
Chaning Pearce, Joseph see Pearce, Joseph Chaning
Chantrey, Francis Legatt (1781-1841) 173, 245, 2165
    cast of Coal Measures plant by 164
    cast of Plesiosaur jaw by 163
    death of 1709
    modelling plesiosaur 302
Chantrey, Lady 1025
Chapman, S.J. 246, 247
Charles, G.S. 248
Charles, Mr 2
Charlesworth, Edward (1813-1893) 1150
    as possible successor to W. Londsdale at Geological Society 1469
    London Geological Journal of 1610
Charleton, Dr E. [of Newcastle] 322

Charlton, Isaac 249, 1042
Charlton, William Henry [of Hesleyside] 322
Charters, S. 250
    continental tour of 250
Chatterton, Mr 302
Chawner, Mr [?Office of Works Architect] 1217, 1249, 1250, 1253, 1254, 1260, 1268
    building stones presented to Museum of Economic Geology by 1266
Cheffling, Mr [?Chefflings]
    account of Great Western Railway 1308, 1310
Child, James Mark [of Begelly House]
    collieries of, in Pembrokeshire 351
Childs, Edward 809
Christie, Samuel Hunter (1784-1865) 682, 1842
Church, John [Comptroller of HMSO] 1233
Clarendon, George William Frederick Villiers (1800-1870), 4th Earl
    1094
Clark, George Thomas (1811-1898) 190
Clark, J.A. 251-253
Clark, Sir James (1788-1870) 241, 728, 906
Clark [?of Sandgate, Kent]
    collection of Gault fossils presented to Museum of Practical Geology 580
Clark, Mr
    shell collection of 800
Clark, Dr 1684
Clarke, Rev. William Branwhite (1798-1878) 691
Clarke, Mr
    British Association Railway Section Committee 189
Clarke, Mr 468
Clawes, Thomas 254
Clibbono, Edward 255
Clift, Mr [of Birmingham Gas Works] 291
Clift, William (1775-1849) 176, 699, 1006
    quarrel with Everard Home 302
Clowes, Messrs William & Sons [Printers] 684, 935, 1578
Clutterbuck, J.W. 256
Coates, Thomas 257
Cock, Valentine B. (d.c.1824) 352
Cockerell, Charles Robert (1788-1863) 301, 304
Cocks & Biddulph 1321, 1323, 1668, 1730, 1739, 1750, 1760, 1780, 1796, 1800
Coffin, Walter (1785-1867) 258, 2244
Coke, H. S. [Town Clerk of Neath 1839-66] 1957, 1958
Colby, Thomas Frederick (1784-1852) 207, 210-213, 215, 232, 259-290, 370, 376, 378-380, 388, 529, 546, 547, 623, 624, 717, 757, 949, 950, 1157, 1214, 1216, 1229, 1303, 1306, 1308, 1462, 1508, 1532, 1564, 1823, 2074, 2272
    & Irish Geological Survey 1562
    & system of colours on geological maps 1822
    criticism of 837-839
    Geological Society of Dublin 1157
    Geological Society of Dublin Presidential address of 268
    on scale of Ireland maps 775
    plan to publish Ordnance Survey map of Bristol 302
Colding, A. 291
Cole, Sir Christopher (1770-1836) 145, 146, 149-151, 153, 154, 159, 165, 166, 172
    donation of natural history specimens to W. Buckland 152
    portrait of 152
Cole, Henry 292
Cole, John 121
Cole, Lady Mary Lucy (1776-1855) 145-154, 157, 162, 164-168, 171, 172
    portrait of 152
Cole, Lord [?] 1899
Colenso, [?John William (1814-1883)] 1270, 1279
Coleridge 301, 302
Collegno, H. de 503, 1025, 1028, 1912
Collingwood, G.V. Graham 293
Colquhoun, [?]J.H. 294
Condamine, Rev. Henry Malcolm de la (1823-1854) 295
    publications 295
Congreve, Miss 299
    ichthyosaur in possession of 297
Conrad, Timothy Abbott (1803-1877) 8
Constantine, Nikolaevich, Grand Duke (1827-1892) 894

Conway, Mr [near Pontypool]
    fossil collection of 1360, 1389
Conybeare, William Daniel (1787-1857) 146, 147, 162, 163, 296-316, 354-358, 473, 496, 499, 505, 624, 690, 889, 1417, 1746, 1849, 1875, 1878, 2002, 2245
    & Lyme Regis landslip 1339
    as Vicar of Axminster 179
    attitude to Glacial Theory 310, 311
    continental tour 146
    criticism of 534
    election to Institut de France, Academie Royale des Sciences 306
    geological map of Bristol 302, 1860
    health of 977
    laudatory reference to 1537
    lecture on plesiosaur discovery 302
    lectures 33, 314
    modifications to King Coal's Levee 150
    nomination for presidency of Section C, British Association 1015
    on stratigraphic names 532
    poem about Buckland by 158
    preaching at Oxford 302
    publication of 152, 153, 308
    publications, *Outlines of the Geology of England & Wales* proposed 2nd edition 316, 1760, 2263
    response to G. Penn's criticism of Buckland 301, 355
    views on church reform 303
    views on slavery 302, 303
    visit to Gower caves 164
    visit to Madeira 313
    visit to Sedgwick in Cambridge 1872, 1877
    visited by W. Buckland 152, 157
    withdrawal from Bristol Institution 305
    work on Sanskrit 306
Conybeare, John Josias (1779-1824) 302
    death of 305
Conybeare, [sister of W.D. Conybeare] 306
Cook, I. 121
Cook, Dr 747
Cooke, George (d.1842) [? of Tortworth] 1475
    fossil collection of 748
Cooke, John (d.1823) death of 167
Cooke, Mr 522
Cooke Taylor, Dr (d.1849) [of Trinity Coll, Dublin]
    death of 1847
    subscription for family 1847
Cooper, Sir A. 160
Copeland, Aderman 1155
    pottery business of 461
Coplestone, Edward (1776-1849) 306
    review of Buckland's *Reliquiae Diluvianae* 171
Cord 904
Corda, August Carl Joseph (1809-1849)
    Murchison's criticism of 1022
Cordier, Pierre Louis Antoine (1777-1861) 317, 325, 1022
Cornish, Hubert
    views of Devon & Dorset coast 163
Cotton, Joseph [of ?Isle of Wight] 583, 585
Couch, Mr 946
Coulomb, Charles Augustin de (1736-1806) 657
Coulon, Mr [of Neuchâtel] 347
Coutts [Bank] 1309, 1311, 2155
Cox, William 44
Cox, Mr 1291
Cox, Mrs (née Poole) 1856
Coxworthy, Franklin 318
Craig, Mr
    railway sections of 1376
Crawshay, William (1788-1867) 342
Crighton, Dr Alexander (1763-1856) 149
Crinn, Mr [chemist, of Glasgow] 1756
Croker, John Wilson (1780-1857) 319
Cromwell, Oliver (1599-1658)
    skull of, in University Museum, Oxford 342
Crook, Mr 301
Cubitt, [?William (1785-1861)] 610, 665
Cumberland, George Senior (1752-1848) 302, 320

collection of 300, 341, 342
  ichthyosaur belonging to 296
Cumberland, George Junior 320
  collection of 320
Cuming, Hugh (1791-1865) 554, 572
Cumming, Rev. Joseph George (1812-1868) 321, 573
Cunnington, William (1813-1906) 1148
Cunningham, Allan (1784-1842)
  health of 2165
Currie, Robert 322
Curtis, Mr
  pension of 1469
Cuthbutson, Mr
  coroner at Neath colliery explosion inquest 1957, 1958
Cuvier, F. 346
Cuvier, Georges Jean Leopold Chretien Frederic Dagobert (1769-1832) 171, 204, 296, 298, 303, 306, 323-326, 343, 346, 363, 796, 897, 1366, 2071, 2077
  Maastricht mosasaur of 349
  meeting with De la Beche 343, 346
  on Buckland's hyaena den theory 165
  plesiosaur jaw cast for 163
  publications, *Recherches sur les Ossemens Fossiles* 350
  receipt of specimens from W. Buckland 157
  request for drawings from De la Beche 161

Dakin, Mr 735
Dalgleish, Mr 1213, 1239
Dalhousie, James Andrew Broun Ramsay (1812-1860), 10th Earl 713
Dalman, Johan Wilhelm (1787-1828) 1333
Damon, Robert (1814-1889)
  collection of 573
Daniel, Benjamin
  Messrs O'Neil's works 2010
Daniel, Edward [of Taibach] 2010
Daniell, John Frederic (1790-1845) 389, 911, 1244, 1262, 1267
Danvers, F. Dawes 327
D'Archive 586
Darlington, John 328
Darwin, Charles Robert (1809-1882) 329-331
  advice on Jamaica animal breeds sought by 329
  convinced by Glacial Theory 537
  objections to R. A. C. Austen's coral reef theory 28, 31
  on origin of coral reefs 331
  publications, [coral reefs] 331
  publications of 622, 1043, 1044
Daubeny, Charles Giles Bridle (1795-1867) 231, 332, 333, 496, 2046
  publications, on the geology of Sicily, *Edinb. Phil. Jl* (1825) 332
D'Auderie 2071
Davey, Mr 941
Davidson, Thomas (1817-1885) 566, 2053
Davidson, Messrs & Co 466
Davies, William 119, 121
Davis, Mr 2181
Davy, Edmund (1806-1885) 304
Davy, Sir Humphry (1778-1829) 165, 301, 302, 946
  on copper sheathing for ships 302
Daw, Robert [of Cardiff] 1924
Dawes, John J. 334, 2034
Dawes, M. 198
Dawson, Mr 2202
Day, M.J. 335
Day, William 336
Day, Mr 570
Day & Haghe 1137
Dean, Mr [shell collector] 800
De Caumont, Mr 32
Dechen, Ernst Heinrich Karl von (1800-1889) 337, 504, 1537
  geological map of France by 504
  publications, 337
Deck, Isaiah (1792-1853) 676
Defrance, Jacques Louis Marie (1758-1850) 32
De Frank 1329
de Groot, Mr 1907
De Koninck, Laurent Guillaume (1809-1887) 556

De la Beche, Bessie
  see De la Beche, Elizabeth Junior
De la Beche, Elizabeth Junior (Bessie) (1819-1866) 1, 7, 26, 28, 32, 41, 80, 81, 110, 338-340, 349, 373, 696, 1219
  see also Dillwyn, Elizabeth (Bessie),
  birth certificate of 338
  birth of 345
  engagement 29
  extract from register of births pertaining to entry concerning 443
  marriage of 832, 948
De la Beche, Elizabeth, Senior (d.1833)
  ?as Elizabeth De la Beche's grandmother 339, 373
De la Beche, Henry Thomas (1796-1855) 1-7, 9-56, 58-73, 75-106, 109-143, 156, 161, 163, 169, 170, 173, 174, 175, 177, 178, 180-184, 186-199, 201-203, 205, 206, 208-210, 214, 217-255, 257-260, 262, 264-268, 270-275, 277-279, 281-284, 287-292, 294-299, 301-305, 312-315, 317-322, 324-337, 339-478, 480-493, 495-547, 549-577, 579-612, 614-618, 620-626, 630-671, 673-692, 694-727, 729-787, 789-799, 801-808, 810-824, 826-861, 863-974, 976-1008, 1010-1034, 1036, 1038-1057, 1059-1112, 1114-1607, 1609-1683, 1685-1691, 1693-1703, 1705-1813, 1815-1841, 1843, 1845-1987, 1989-1993, 1996-2039, 2041, 2042, 2044-2105, 2109-2148, 2150-2158, 2160-2172, 2244, 2246-2265, 2267, 2268, 2272, 2274-2278
  advice, assistance and information sought from 80, 97, 109, 113, 116, 135, 206, 232, 235, 315, 329, 533, 542, 605, 610, 612, 614, 643, 651, 652, 655, 661, 666, 674, 680, 681, 687, 694, 700, 746, 827, 828, 857, 870, 881, 896, 978, 983, 984, 989, 1016, 1041, 1053, 1134, 1179, 1197, 1254, 1269, 1281, 1283, 1331, 1812, 1847, 1986, 1987, 2015, 2020, 2023, 2095, 2151
  appointments
    certificate of employment on Ordnance Survey 370
    proposed position at British Museum 199
    nominated Director, Museum of Economic Geology 1209, 1214, 1215, 1217
    intended resignation of 623
  boat at Lyme (illn.) 341
  cartoons and lithographs
    Awful Changes (1830) 179, 367 (illn.)
    Duria Antiquior (1830) 180, 182, 368
    Duria Antiquior Germany parody 183
    Frank Buckland 'piddling' 1031
    suggestions for Quaternary, Tertiary and Carboniferous reconstructions 183
    The Irregularities of Sol visited upon his System (1841) 395, 396
    A Coprolitic Vision 432
  collection of 297, 300, 341, 431
    purchase of fossil crocodile 161
    purchase of specimen at sale of Bullock's Museum 342
    plesiosaur specimen of 163, 169, 170
    ichthyosaur skull donated to Cuvier 325
    ichthyosaurs in possession of 297, 300, 342
    request for specimens from De la Beche 442, 444
    return of specimens to 138
    specimens figured in Sowerby's *Mineral Conchology* 1980
  Devon Controversy 378, 621-624, 837-839, 897, 902, 1012, 1013, 1226, 1332, 1873, 1875-1877, 1879, 1880, 2049
  donation to Jamaican Wesleyan mission 1905
  drawings of Alpine fossils 156
  family and personal life
    Beach family 2158
    Beach family coat of arms 2157
    birth certificate of daughter 338
    birth of daughter 338, 345, 443
    bust of 939, 2283
    cameo of 363
    Christmas greetings to family by 419
    health of 117, 448, 556, 591, 816, 830, 1112, 1150, 1386, 1387, 1460, 1461, 1464, 1529, 1530, 1624, 1627, 1686, 1695, 1790
    investments of 801-803
    medals struck for 688, 1401, 1659, 2164, 2279-2281
    order of spectacles from Dollands 374
    parody newspaper, "The Mining Chronicle" by 377

155

passport of 343, 346
photographs of (illn.) 422
photographs of family of (illn.) 422
portrait 2279
portrait (illn.) 415
shares of 801-803
wine bill of 254
geological map of Normandy by 532
geological map of Jamaica by 355
Geological Survey
    authority to appoint assistants 215
    commencement of Cornwall Survey 261
    expenses on Cornwall Survey 269, 379, 529
    report on progress of Cornwall Survey by 263, 276, 277
    Devon survey supported by Sedgwick 1872
    discovery of Carboniferous plants in Devon grauwacke 1330, 2049
    reaction to criticism of Devon mapping 378
    march route of 2074
    week's work on Devon Survey 373
    geological map of Devon 497-499, 502
    salary & expenses of De la Beche 211, 212, 214, 276-282, 285, 286, 376, 379, 397, 399, 529
honours
    knighthood 58, 1422, 1423
    Companion of the Bath 709, 2069, 2162
    nominated Chevalier de l'Ordre de Leopold 853
    award of 1851 Great Exhibition medal 9
introduction sought 202, 686
investigations of Colliery explosions by 414, 2044
invitation to publish by Society for the Diffusion of Useful Knowledge 257
invitation to Sèvres 133
invited to join Metropolitan Sanitary Commission 615
Jamaica property 260, 301-305, 352, 354, 355, 357-359, 688, 689, 959, 1905, 2023, 2024
jetty at Lyme by (illn.) 341
laudatory references to 280, 473, 645, 972, 1299
    work on geology of France praised 473
lecture on 1851 Great Exhibition 10, 999
letter of introduction 323, 324, 363, 2071
meeting with Prince Albert 252, 253
meeting sought with 203
MS notes, Contributions towards a theory of the Earth's structure 427
MS notes, Humbug 424
MS translation of Boué's Geology of Germany 423
offer of accommodation to 248
offer to chair Royal Society meeting 412
opinions and views
    attitudes to slaves 352, 354, 355, 2023
    criticised for his attitude to Mantell's work 534
    criticism of British politicians 347
    criticism of English abroad 344, 346, 363
    criticism of monks 363
    on economic advantages of geological maps 372
    on ichthyosaur research 341, 342
    on Military College, Great Marlow 342
    on need for a general view in geology 363
    on origin of earthquakes 352, 357
    on origin of flint 343
    on religion/religious ceremony 341-344, 346, 348, 349, 363
plan for emancipating the Negroes 369
position sought with 1200
proposal for a College of Mines 385, 386
publications
    Annals of Philosophy 307
    A selection of the geological memoirs... in the Annales des Mines (1824) 304, 355
    Report on the coals suited to the steam navy (1848-51) 294, 524, 1194
    Admiralty Manual of Scientific Enquiry (1849) 680, 681, 683, 684
    A Geological Manual (1831) 39, 93, 104, 183, 326, 339, 428, 470, 493, 495-497, 499-501, 534, 690, 796, 846, 896, 948, 974, 1001, 1036, 1833, 2075, 2086, 2087
    A Geological Manual American edition 1826
    A Geological Manual French editions (1831) 496-502, 855, 1011, 2072, 2073
    A Tabular View (1827) 133, 489, 495, 796, 1906, 2086
    Notice on the Diluvium of Jamaica (1825) 173
    Geological Notes (1830) 12, 533
    Remarks on the geology of the south coast of England from Bridport ... to Babbacombe Bay (1822) 341
    On the geology of Southern Pembrokeshire (1826) 173, 949, 1009
    Remarks on the geology of Jamaica (1827) 332, 531, 1672
    On the geology of ... Weymouth (1836) 156, 157, 179, 180, 184, 186-188, 891, 1009
    Geology of the South Coast of England (1824) 163
    On the Geology of the Coast of France (1824) 350, 473
    How to Observe (1835, 1837) 135, 136, 502, 834-836, 846, 930, 1204, 1224
    On the geology of ... La Spezia (1833) 93, 365, 366, 490
    On the geology of the environs of Nice (1829) 130, 361, 362, 364, 375, 490, 620
    On the geology of Nice [abstract] (1829) 619
    On the geology of Tor & Babbacombe Bays (1827) 130, 360, 620
    On the Formation of Rocks of South Wales and Southwestern England (1846) 408, 409
    On the temperature and depth of Lake Geneva (1827) 849
    Parliament Building stones report 528
    Report on geology of Cornwall, Devon & West Somerset (1839) 34, 35, 37, 266, 272, 276, 281, 283, 284, 378-380, 387, 390-394, 434, 504, 624, 949, 957, 1085, 1156, 1157, 1680, 1880, 1984
    Researches in Theoretical Geology (1834) 428, 500-502, 621, 846, 890, 897, 921, 1013, 1826, 2022
    Researches in Theoretical Geology American edition 692
    Researches in Theoretical Geology French translation 500-502
    Report on the state of Bristol and other large towns (1845) 524
    Sections and Views (1830) 12, 141
    The Geological Observer (1851) 77, 335, 442, 508, 524, 644, 799, 918, 930, 1031, 1046, 1819
    The Geological Observer 2nd Edition (1853) 632, 645
    ichthyosaur paper (1821) 161, 296, 300
quadrant 2282
reference or introduction sought from 103, 139, 644, 805, 876-878, 1159, 1160
responsibilities of 225
request for autograph of 216
request for commission to 250
School of Mines inaugural lecture 63, 685, 856, 986
servants, Martin, biographical details of 343
    Henri and Self (illn.) 363
Smoke enquiry 1308
social invitation from 53, 54
social invitation to 137, 241, 410, 656, 810, 842, 844, 1170, 1171, 1173, 1174, 1177, 1676, 1835, 2070
society memberships and positions
    Secretary of Geological Society 175
    Corresponding Associate of Turin Academy 1910, 1918
    President, Section C, British Association 1016
    Honorary Member, Birmingham Philosophical Institution 2096
    Honorary Member, Quebec Literary and Historical Society 2001
    Honorary Member, Société de Physique et d'Histoire Naturelle, Geneva 606
    Member, Geological Society of France 493
    Member, Imperial Society of Naturalists, Moscow 1002, 1003
    Member, Société Météorologique de France 603
    Member, Swedish Academy of Science 2005
    President of Geological Society 554, 723-725, 904, 910, 994, 1320, 1609
    Geological Society Anniversary Addresses 969, 972, 1145, 1148, 1161, 1180
    Vice President, Geological Society of Ireland 51, 756
    Corresponding Associate of Turin Academy 2045

Corresponding Member, Academy of Natural Sciences of
    Philadelphia 1001
Corresponding Member, French Academy of Sciences 475
Corresponding Member, Institut de France 602
Membre Titulaire, Institute d'Afrique 1848
Fellow of Royal Society 342
Royal Society 1842, 1843, 1845
invited to chair new Scientific club 1837
University College, London, De la Beche presenting prizes at 741,
    2013
visits and tours
    ascent of Vesuvius 363
    at St Austell 340
    continental tour (1819) 343-347
    continental tour (1828) 217, 363, 426
    expedition to Mont Blanc 344, 346
    temperature and depth, Lake of Geneva 345, 346, 849
    intended visit to Ionian Islands 363
    Italy (1828-29) 363, 490
    Jamaica 352-358
    Lyme Regis, for foreshore report 79, 410, 2278
    Minehead 1
    Paris 133
    sailing Dover to Calais 343, 346
De la Beche, Letitia (*née* Whyte) (1801-1866) 338, 342, 344, 346
De la Beche, Rosalie Torre (1834-1858) 441, 448, 449, 454, 465, 807,
    1922
    education of 1047, 1048
    photographs of (illn.) 422
De la Beche, Thomas (d.1801) 2157, 2159
Delesse, Achille Ernest Oscar Joseph (1817-1881) 442
    publications of 442
De Lor, Philippe Francois 443
De Luc, Jean André (1727-1817)
    collection of, with nephew 344, 346
De Luc, André [nephew of the above]
    collection of 344, 346
Denham, Capt. Henry (1800-1887)
    sailing to Pacific 656
Denison, William Thomas (1804-1871) 444
Denny, Henry (1803-1871) 91
Denton, Bailey 1294
Denton, Mr [shell collector] 800
Derby, Edward Smith Stanley (1775-1851), 13th Earl of 1058
De Rottermond, E.S.
    appointment on Canadian Geological Survey 885
    salary of 885
De Saussure, Horace Benedict (1740-1799) 148, 1366, 1378, 1434
    publications, *Voyages dans les Alpes* 344, 346
Deshayes, Gerard Paul (1796-1875) 93
    award of Wollaston Fund to 900
    Beck's criticism of 891
Desor, Pierre Jean Edouard (1811-1882) 8
de Stail, Madam [Chateau at Coppet, Lake of Geneva] 345, 346
De Verneuil, Philippe Edouard Poulletier (1805-1873) 442, 566, 1022,
    1026, 1027, 1180, 1181, 1420, 1422, 1436
    fossil collection 38
    on Devonian fossils 1403
    publications, with Murchison, Geological Structure of Russia (1841)
        1402
    visit to Vienna 639
    visit to Montreal 887
Deville, Charles Sainte-Claire (1814-1876) 508
Diaz, Mr 496
Dick 1273, 1292
Dickie, George (1812-1882) 95
Dickinson, J.D. 445, 446
Dieterie 447
Dilke, Charles Wentworth (1789-1864) 552
Dillwyn, Elizabeth (Bessie) (1819-1866) 441, 448-452, 455, 456, 459,
    464, 465, 560, 599, 660, 810, 841, 1443, 1444, 1517, 2159,
    2165
    See also De la Beche, Elizabeth Junior
    children of 422
    health of 456, 457, 668, 1123, 1124, 1449
    laudatory reference to 561
    photographs of (illn.) 422
Dillwyn, Elizabeth Amy (1845-1935) 450, 465
    photographs of 422

Dillwyn, Fanny Llewelyn (b.1810) 456
Dillwyn, Henry De la Beche (1843-1892) 450, 451, 465
    birth of 191
    photograph of (illn.) 422
Dillwyn, Lewis Llewelyn (1814-1892) 450-458, 460, 461, 465, 802,
    875, 1102, 1441, 1443, 1451, 1479, 1482, 1959, 2159
    photographs of (illn.) 422
    pottery 461
    Tenby friends of 456
    visit to Radnorshire 807
Dillwyn, Lewis Weston (1778-1855) 149, 152, 165, 166, 172, 456-464,
    560, 800, 872, 874, 970, 993, 1020, 1469, 1679, 1959, 1997
    fossil collection of 458
    health of 458, 459, 463
    plant collection of 840
    publications
        Coleopterous insects found in the neighbourhood of Swansea
            (1829) 949
        Contributions towards a History of Swansea (1840) 456, 457,
            459, 460
        Materials for a Fauna and Flora of Swansea (1848) 463
Dillwyn, Mary (b.1816) 456, 464
    health of 457
    photographs of De la Beche and family by 422
Dillwyn, Mary De la Beche (Minnie) (1839-1922) 450
    photographs of (illn.) 422
Dillwyn, Miss [?Mary] 441
Diney, Mr 1281
Dinkel, Joseph (1805-1891) 95, 905, 1333
Dixon, Mrs A.M. 466
Dixon, Francis Graham 467, 468
    employment sought by 466-468
Dixon, Colonel Matthew [of Royal Engineers] 468
Dixon, Frederick (1799 -1849) 578
    collection of 573
Dobson, I.
    book work on Iron 2017, 2018
Dolland, Messrs
    De la Beche's order for spectacles from 374
Donaldson, Thomas L. 469
Donegue [Geological Survey, Ireland] 1088
Donkin, Bryan (1768-1855) 470
Donne, Mr [of Lyme Regis] 897
Downes, Mr 304
Downie 1727
Dree, Marquis de
    collections of 484
    health of 484
Drummond, Capt. 1214, 1240, 1678, 1823
Dublin, Richard Whately (1787-1863), Archbishop of 1847
Dudley, John William Ward (1781-1833), 1st Earl
    Jamaican estate of 352
Duffield, R. Dawson 471, 472
Dufrenoy, Ours Pierre Armand Petit (1792-1857) 34, 473-476, 500, 503,
    504, 1180-1182, 1198, 1408, 1683, 2053
    award of Wollaston medal to 505, 506
    elected Honorary Member Geological Society 473
    tour with Elie de Beaumont 492
Dumas 885
Dumont, Andre-Hubert (1809-1857) 586
Duncan, Philip Bury (1772-1863) 148, 535
    poem about Buckland by 158
    sketch prompted by King Coal's levee 150, 151
    theory on origin of red snow 148
Duncan, Mrs I. 307
Duncannon, John William Ponsonby (1781-1847), 1st Baron 1201,
    1203, 1208, 1209, 1211, 1213-1217, 1219-1221, 1229, 1232,
    1233, 1242, 1247-1255, 1257-1260, 1262, 1264, 1268
    death of brother (Sir Frederick Cavendish Ponsonby) 1221
    health of 1202
Dundas, James Whitley Deans (1785-1862) 477, 1820
Dundonald, Thomas Cochrane (1775-1860), 10th Earl 478
Du Noyer, George Victor (1817-1869) 814, 815, 833, 1105, 1110, 1123,
    1124, 1128, 1133
    appointment of, as assistant, Geological Survey of Ireland 1085
    letter of introduction of 1064

salary of 1109
visiting card of 1064
Dunraven, Windam Henry Quin, 2nd Earl of (1782-1850)
  estates of 7
Dunstonville, Lord de 850
Durham, Edward Maltby, Bishop of 322
Durham, Shute Barrington, Bishop of 170
Durham, John George Lampton, 1st Earl of (1792-1840) 660
Durham, Col. 2095
Durrant 2202
Dyer, Mr 687
Dyer, Dr
  collection of 300

Eastlake, Charles Lock (1793-1865) 480, 481
Eastlake
  publications of 622
Eastwick, Mr 742
Eaton, Prof. Amos (1776-1842) 695
Ebel, Johann Gottfried (1764-1830) 344, 346
Ebelmen, J. 447, 482, 1193
Ebington, Viscount 242, 483
Eddy, Capt. 514, 515
Eddy, Mine Captain 1942
Eden, Rodney [of ?Swansea]
  visit to Norway 456
Eden, Mr (of Bristol?) 302
Edge, Mr 661
Edgeworth, Miss 341
Edmond 880
  offered post as Keeper of Mining Records 871, 872
Edmonds 1261, 1264, 1265
Edwards, Henri Milne (1800-1885) 96, 484-486, 571, 572
  publications, Monograph, British Fossil Corals 486
Edwards, Lewis 1930
Edwards, William (1719-1789) 342
Edwards, Frederick Erasmus (1799-1875)
  collection of Tertiary fossils 569
Edwards 1369
Egerton, Sir Philip de Malpas Gray (1806-1881) 8, 95, 96, 487, 573,
    623, 898, 899, 1272
  publications
    Figures & descriptions of British Organic Remains 571
    on *Pterichthys, QJGS*, (1848) 1043
  specimens from, for Museum of Economic Geology 1236
Eichthall, Charles d' 488
Eldon, John Scott (1751-1838), 1st Earl of 148
Elie de Beaumont, Jean Baptiste Armand Louis Léonce (1798-1874) 34,
    132, 489-510, 845, 855, 1034, 1180, 1408, 1683, 1700, 2072
  Alpine journeys of 490, 502, 503
  as originator of "Hercynian" 185
  attack on 797
  award of Wollaston Medal to 505, 506
  elected Honorary Member Geological Society 473
  elevation theory of 491, 493, 494, 495
  geological map of France by 473, 492, 502, 503, 505
  on Auvergne 309
  on elevation craters 501, 897
  on limestones at Nice 364
  publications of 508, 509
  tour with Dufrenoy 492
  tour with von Buch 491
Eliot, Lord 661, 662
Ellicomb, G. C. 213, 379
Elliott, Mr [of Monkwearmouth colliery] 1973
Ellis, Henry (1777-1869) 1058
Elst, E. van der (1826-1897)
  letter of introduction of 1907
Emlyn, Lord 1385
Enard, Mon. [of Geneva] 344, 346
Englefield, Sir Henry Charles (1752-1822) 147, 341, 1221
Enniskillen, William Willoughby Cole, 3rd Earl (1807-1886) 8, 95,
    1053, 1411, 1546
  collection of 51, 558, 1075
Enys, John Samuel (1797-1872) 511-523, 672, 806, 2151
Erman, Georg Adolf (1806-1877)
  at British Association meeting (1842) 1440

Evans, D. Th. 524
Evans, Mr 1186, 1194
Everest, Rev. R. 2035
Exchequer, Chancellor of
  (1840 January) 327
  (?1851 December) 856
Exeter, Rector of 150, 154
Eyde 95
Eynard [of Geneva] 363

Fairbairn, Mr 693
Fairholme, G. 309
Falconer, Hugh (1808-1865) 1021, 1023, 1837
  appointed as Botanist, India 2138
Fanshaw, Captain William
  repairs to Lyme Cobb by 341
Fanshawe, Edward 207, 215, 276, 379
Faraday, Michael (1791-1867) 65, 525-527, 530, 638, 682, 850, 1037,
    1254, 2092
  & Royal Society 906, 1053, 1842, 2050
Farey, John (1766-1826) 309, 310
Faubin, Le Comte de [Director General of Museum at Paris] 343
Favre, Jean Alphonse (1815-1890) 191, 1514
  article on Geological Survey for Bibliotheque Universelle 1703
Fawcett, Helen 51
Fawcett, Joshua 528
Fawcett, Rev. Mr 1287
Featherstonhaugh, George William (1780-1866) 1037
Fenton, John [of Fishguard] 954
Fenton, Rev. 109
Ferguson, Robert
  defence of De la Beche in House of Commons 624
Ferrier, J. 261, 282, 287, 529
Fischer, J.C. 530
Fischer de Waldheim, Johann Gotthelf Friedrich (1771-1853) 1002
Fisher, Mr 302
Fitton, William Henry (1780-1861) 47, 141, 166, 302, 314, 413, 473,
    496, 531-539, 575, 612, 622, 876, 1009, 1434, 2017, 2018
  & Devon Controversy 1012
  & geological education of R.I. Murchison 1020
  criticism of Conybeare & Phillips 534
  criticism of De la Beche 534
  criticism of Greenough's map 534
  criticism W. Smith 534
  criticism of T. Webster 534
  involvement with Society for the Diffusion of Useful Knowledge 257
  Mantell praised by 534
Fitzwilliam, Charles William Wentworth (1786-1857), 3rd Earl 540, 625
Flanagan, James (d.1859) 557, 558, 593, 755, 766, 772, 1073, 1079,
    1084, 1086, 1088, 1097
  laudatory references to 751, 1081
  salary of 751, 1081
Fleetwood, Mr
  specimens from, for Museum of Economic Geology 1231
Fleming, E. B. 541
  employment of Geological Survey sought by 541, 1895, 1900, 1901
Fleming, Rev. J. 173
Fleming, Dr 342
Fleming, Dr [shell collector] 800
Fleming, Mr 306
Fletcher, Alexander & Co 2034
Floud, Mr
  search for coal by 28
Forbes, Edward (1815-1854) 19, 50, 51, 84, 314, 321, 403, 407, 458,
    484, 526, 539, 542-593, 609, 633, 709, 723, 725, 737, 738,
    742, 756, 761, 762, 764, 769, 771, 811, 889, 1018, 1021, 1022,
    1025, 1030, 1062, 1072, 1077, 1086, 1103, 1163, 1164, 1189,
    1303, 1601, 1614-1616, 1621, 1628, 1662, 1727, 1734, 1739,
    1743, 1746, 1752, 1753, 1764, 1766, 1767, 1769, 1774, 1789,
    1801, 1806, 1853, 1979, 1883, 1966, 1987, 2152
  & BAAS, Swansea meeting 1640
  collection of 554
  criticism of Jukes 5837
  criticism of F. McCoy 558
  criticism of John Phillips 544, 545
  criticism of Sedgwick 555

engagement of 559
  health of 544, 577, 743
  inheritance of land by 564
  in Ireland 1079-1081, 1090, 1102, 1781
  invitation to lecture at Royal Institution 545
  laudatory reference to 1602
  lectures of 592
  move from Edinburgh to London 544
  position with Geological Survey 1559-1562, 1565
  proposed for post at Museum of Economic Geology 1538
  publications, *Figures & descriptions of British Organic remains* (1849) 552, 565, 566, 569, 575, 578
  publications, Geological Survey Memoirs 937
  recommended for Wollaston Medal 1888
  reluctance to become Geological Society President 584, 589
  resignation as Geological Society Curator & Librarian 542, 543, 547
  visit of Thomas Colby to 546, 547
  wedding of 561, 562
Forbes, Mrs Edward 560-562, 564, 575, 1102
  health of 576, 579
Forbes, James David (1809-1868) 8, 330, 594-596, 1030
  proposal for Geological Survey Museum in Scotland 596
  publications, [on glaciers] 1888
Forbes, Col. [of Engineers in India] 2025, 2138
Forbes, Dr [of Jamaica] 352
Forchhammer, Johann Georg (1794-1865) 523, 597, 598
Forster, Frank 322
Forster, W. I. [of Tynemouth] 322
Fortescue, Lord 741
Fortifications, Inspector-General of 207, 211, 213, 215, 269, 276-278, 280, 281, 286, 289, 379, 649
Fossett, Mr 1748, 1749, 1948, 1953
Fox, Robert Were (1789-1877) 513-516, 523, 599-601, 804, 942, 944, 1214
  Dipping Needle Deflector 288, 381-383, 388, 600
  electrical experiments with clay 599-601
  experiments producing mineral veins in clay 1990
  parody on magnetism experiments of 377
  parody on temperature in mines experiments of 377, 599-601
  publications, mineral veins 946
Fox, Col./Gen. 838, 839, 842
Fox-Talbot, William Henry
  see Talbot, William Henry Fox
Fox, Henderson & Co [Birmingham] 1494
France d'Houdetot, Le Compte 344, 346
Francis, George Grant (1814-1882) 626
Francis, Mine Captain [?Matthew Henry (1810-1869)] 1942
Frank, Mr 1284
Franklin, John (1786-1847) 157, 302
Franklin 1020
Frere, Mr [of Llanelly] 2183
Fressange, Mr
  metal castings by 421
  trial of 421
[?]Fridau, Baron, Francis of Stiria
  expedition to Ceylon 644
Fripp, Charles Bowles (d.1849) 304
Fry, Peter Wickens
  photographs of Dillwyn family by 422

Gall, Dr 343
Galloway, T. 570
Gapper, Dr 301, 303
Gapper, J.C. 123, 572-574, 577, 1778-1799
  criticism of 1806
  health of 1758
  pay 2260
Garby, [?]John 604
Gardener, Mr 123, 180, 182, 188
Gardiner, J.R. 605, 1316, 1673
Gardner, James 2, 378, 1420, 1877
Garnett, Jeremiah (1793-1870) 693
Geach, Mr [of St Austell] 943
Geoffroy Saint-Hilaire, Etienne (1772-1844) 343, 346
George IV (1762-1830) 148
  coronation of 158

George, William 45
Gell, Phillip 170
Gerville 1329
Gelling, Mr [clergyman] 1264
Gibbs, Richard (d.1878) 62, 120, 555, 563, 582, 583, 585, 588-592, 607-609, 1505, 1510, 1513, 1515, 1529, 1540, 1549, 1563, 1571, 1723, 1739, 1744, 1752, 1791, 1799, 1800, 1806, 1832, 1955
  criticism of mapping of 1746, 1766, 1783
  health of 1758
  pay of 1509
Gibbs, Mrs 1391
Gibson 585
Giesecke, Sir Charles Lewis 306
Gilbert, Davies (1767-1839) 511, 512, 517, 522, 658, 914
  and 1820 Presidency of Royal Society 153
Gilbertson, William (1789-1845)
  collection of Carboniferous Limestone fossils 1356
Gilby, Dr [of Clifton] 342
Giles 2202
Gladstone, William Ewart (1809-1898) 1272, 1824
Gloucester, John Plumptre, Dean of 147
Glover, Thomas 701
Goadby 554
Godwin, George (1815-1888) 610
Godwin-Austen, Robert Alfred Cloyne (1808-1884)
  see Austen, Robert Alfred Cloyne
Goldfuss, Georg August (1782-1848) 892, 1332, 1351, 1405, 1826
Gontard, Mr [proprietor of Baths at St Gervais] 344, 346
Gorden, Mr 738
Gordon, Hubert 1268
Gordon, Dr [of Jamaica] 352
Gordon, Prof. Lewis D.B. 1754
Gore, Hon. C. 1256, 1300, 1313
Gore, Mr 91
Gosnor, Dr
  on Canadian geology 882
Goss, Mr [of Leeds] 530
Gossett, W. Driscoll 611
Gourlie, William (1815-1856) 1755
Gower, Abel Lewis (d.1849) 873
Gower, J.R. 612
Gowland, Thomas 466, 2159
Gowland, Janet 2159
Gradby, Henry 613
Graham, George 614
Graham, Sir James Robert Gordon (1792-1861) 1309
  proposing mining explosion enquiry 1308
Graham, Professor Thomas (1805-1869) 103, 1684
Grant, Dr 2031
Grant, Mr [?of Neath] 1959
Granville, Granville George Leveson-Gower (1815-1891), 2nd Earl 616-618, 918
Granville, Augustus Bozzi (1783-1872) 898
Graves, Capt. 1987
Gray, John Edward (1800-1875)
  anti-Royal Society attitude of 570, 1058, 1148
  burning his paper on Chitons 570
Green, John 51
Green, Miss
  tavern of, in Jamaica 352
Green, Mr
  Jamaican estate of 352
Green, Lieut. [of Navy] 657
Greenhill 1294
Greenough, George Bellas (1778-1855) 35, 47, 103, 175, 188, 309, 322, 396, 473, 496, 537, 544, 619-625, 834, 835, 837, 843, 898, 901, 909, 912, 1018, 1229, 1873, 2087
  amendments to De la Beche's Tor & Babbacombe Bays Paper 360
  & Devon Controversy 621-623, 624, 838, 897, 902, 2049
  & index of colours on geological maps 1822
  continental tour of 153
  criticism by Lyell 897
  criticism of 652
  criticism of Macculloch by 623
  criticism of map of 534, 1437
  defence of De la Beche 624, 838

geological map of 145, 153, 179, 492, 624, 625, 1727, 2088
notes on De la Beche's Nice Paper 362
preparation of second edition of geological map of 622, 624
publications, *A critical examination of the first principles of Geology*
criticism of 153
support of De la Beche 621, 623, 2049
Greenwich, Mr 315
Greere, Francis 626
Gregory, William (1803-1858)
appointment as Professor of Chemistry, Edinburgh 1684
Greig, Charles 627, 628
Grenville, William Wyndham Grenville (1759-1834), Lord 148, 150, 151, 154, 161, 162, 302
health of 172
Grenville, Lady
interest in botany of 150
Gresswell, Rev. [or Greswell] [of Tortworth]
fossil collection of 747, 748
Grey, Charles (1804-1894) 629-633
Grey, Henry George (1802-1894), 3rd Earl 612, 636, 637, 674, 1027, 1267
Grey, Sir George (1799-1882) 59-61, 98, 100, 238, 240, 414, 488, 634, 635, 1656, 1943
Griffith, Richard John (1784-1878) 756, 766, 773, 776, 1022, 1111, 1407, 1461, 1521, 1537, 1543, 1554, 1562, 1930, 1932, 2042
criticism of 763, 1070, 1071, 1443
geological map of Ireland by 306, 756, 1070, 1071, 1084, 1522, 1542, 1544
on Devon fossils 1583
Griffith, Mr
British Association Railway Section Committee 189
Groombridge, Mr 305
Grotius 204
Grove, William Robert (1811-1896) 568, 570, 638, 682, 906, 907, 911, 1020, 1104, 1194, 1837, 1842, 1845, 1846
& BAAS meeting, Swansea 1640
nominated Secretary, Royal Society 570
Guidoni, Girolamo 293, 363, 2182
Guilding, Mr
appointment as ship's naturalist 148
Gulpeth, Mr
collections of 51
Gundry, Mr 123
Gutch, John Wheeley Gough (1809-1862)
visit to Norway 456
Gutch, John Matthew (1776-1861) 302
Gwynne, Mr [of Kilkiffeth Farm, Pontfaen] 954
Gwynne, Mr [of Llandwith, Builth] 342
Gyde 1024

[?]Hackey the Engineer] 515
Haidinger, Wilhelm Karl von (1795-1871) 639-645, 1940, 1949, 1969, 2053
geological map of Austria by 1022, 1928
title of 1951
Hailstone, Rev. John (1759-1847) 1895
Hailstone, Rev. J. [of Bottisham, nephew of J. Hailstone (1759-1847)] 1895
Hale, Mr 28
Haley, [Murchison's servant] 1022
laudatory reference to 1025
Hall, Sir Benjamin (1802-1867) 646, 2115
mine at Abercarn 201
Hall, Elias (1764-1853) 956
Hall, Henry 647
Hall, James (1811-1898) 8
publications, On... soft parts of Orthoceras, *QJGS*, (1849) 1043
Hall, Jos. 648
Hall, Col. Lewis Alexander 237, 649, 650, 792, 1084, 1098, 1101, 1108, 1868
Hall, Miss Polly
lodgings of, in Kingston 352
Hall, Mr [of Kings College, London] 1332, 1333
Hall [employed on Geological Survey] 1797, 1798, 1802, 2202
character of 1800
Halliday, F. J. 2143

Hallis, Mr [of Shirenewton House]
boring for coal by 2116, 2118, 2120
Hallowes, Mr 1750
Halse, Mr 1990
Hambro & Son 598
Hamel, Joseph von (1788-1862) 401, 651
at British Association Meeting 1842. 1440
Hamilton, Sir Charles (1767-1849) 896
Hamilton, Charles William 652, 1541, 1543, 1544
attacks on R. Griffith 756, 772
criticism of 781
Hamilton, William John (1805-1867) 141, 622, 653, 898, 901, 1027, 1922
as potential Geological Society President 584
paper on Asia Minor 864
Hamilton, William Rowan (1805-1865) 1058, 1069, 1316
Hamilton, Mr 152
Hamilton, Dr 115
Hamilton 1408
Hammond, E. 654
Hammond, Mr 363
Harcourt, Rev. William Venables Vernon (1789-1871) 655, 1294, 1384
Harding, Major William 2097
fossil collection of 1334, 1342, 1352, 1376, 1469
Harding, Mr
specimens from, for Museum of Economic Geology 1231
Hardy, Thomas Duffus (1804-1878) 391, 394
Hardwick, P. 656
Harford, John Scandrett (1788-1866) 301, 302, 304
Harkness, Robert (1816-1878) 541
Harness, Mr 2012
Harris, William Snow (1791-1867) 657-668, 915
award of Royal Society Copley Medal to 658, 1006, 2050
dispute with Navy over lightning conductors 657, 659, 661-665, 952
electrical experiments of 657, 658
increased pension sought by 665
meteorological observations by 659, 660
son of 661
Harris, Mr 123
Harvey, E. 108
Harvey, George 669, 670
Harvey, W. 671
Hasker, Mr [shipowner of Penryn] 519
Haskings, Elizabeth 11
Haskings, John 11
Haskings, R. 672
Hastings, Lady Barbara (1810-1858), Marchioness of 561, 578
collection of Tertiary fossils 569, 1151
crocodile from Isle of Wight 1144
Hastings, R.G. 673
Hastings, Marquis of 1678
Hauer, Franz von (1822-1899) 640-645, 1022, 1969, 2053
publications of 639, 1949
Haughton, Rev. Samuel (1821-1897)
paper on Carboniferous Limestone of Menai Straits 593
Hawes, Benjamin (1797-1862) 127, 612, 636, 674
Hawker, Rev. Peter (c. 1773-1833)
collection of 342
Hawkin, Mr 306
Hawkins, Edward (1780-1867)
publications, *Archaeological Journal* (1847) vol 4 1041
Hawkins, Thomas (1810-1889)
publications, *Memoirs of Ichthyosauri & Plesiosauri* (1834) 309
Hawkins, Mr 1895
Hawksley, Mr 291
Hawle, Ignaz (1783-?1868)
criticism of, by Murchison 1022
Hayward, A. 675
Heath, Messrs & Co 488
Heber, [?]Reginald 158, 159
Hebert, Edmond (1812-1890)
paper in *Bull. Geol. Soc. Fr.* on Isle of Wight 583, 588
Heer, Oswald (1809-1883) 642
Hemmings 1865
Henderson, Captain Alexander 708
triangulation of coast under 269

Henfrey, Arthur (1819-1859) 464, 584, 589, 1730, 1733, 1736, 1738, 1741, 1746, 1751
   health of 1728
   laudatory reference to 1732, 1740
   pay of 1740
   resignation of 1747, 1748, 1750
Henland, Mr 306
Henley, Henry Hoste (1766-1833) 11
Hennah, Rev. Richard (1766-1846) 2086
   collection of 373, 1334, 1338
Hennett, Mr 919
Henslow, Rev. John Stephens (1796-1881) 676, 677
   portrait of, Ipswich Museum 1810
Henwood, Mr G 515, 946, 1157, 1365, 1880, 2029, 2031, 2174
   criticism of 1990
   criticism of R W Fox 601, 942, 1990
Henwood, William Jory (1805-1875) 678
Herbert, Mr 1243, 1249, 1253, 1260, 1266-1268
Herbert, Capt. (d.c.1828)
   & geological survey of Himalayas 1678
Hereford, John Merewether, Dean of (1797-1850) 1291
Herschel, John Frederick William (1792-1871) 35, 302, 496, 497, 638, 679-685, 911, 1056
   as editor of Admiralty Manual of Scientific Instructions 680, 681, 683, 684
   as reviewer of Whewell's Inductive Sciences 957
   proposed for Presidency of Royal Society 1842, 1844
   refusal of Presidency of Royal Society 682, 906, 907, 1058, 1843
Herty, Mr
   collection of antiquities 2084
Hey, Rev. 2157
Hibbert, George (1757-1837) 686, 687
Hibbert, W.J. 688
Hibbert and Co. 689
Hibbert, Mr ?George [of Jamaica] 352
Hibbert, G W ?& S 959, 1200
Hichens, Mr 5
Higgins, Mr [of Bristol]
   fossil insects from Cotham Marble in possession of 1610
Higgon, Mr 953
Hilbert, Mr 113
Hill, Rowland (1795-1879) 661
Hill, Mr [?of Ordnance Survey] 940, 941
Hinch, Mr 1243, 1266, 1268
Hitchcock, Edward (1793-1864) 690-692, 1826, 2017
Hobhouse, H. 1684, 1728
Hodges, Dr 81
Hodges & Smith, Messrs 1813
Hodgkinson, Eaton (1789-1861) 693
Hoeninghaus, Friedrich Wilhelm (1770-1854) 183
Hoffschmidt, C. d' [Belgian Minister of foreign affairs] 853
Holdsworth, A.H. 694
Holdsworth, Mr 179
Holl, Harvey Buchanan (1820-1886) 695
   collection 1334
Holland, George 349, 696
Holland, Henry Richard Vassall Fox (1773-1840), 3rd Baron 327
Holland and Sons 697
Hollins, Peter (1800-1886) 698
   & statue of Robert Peel 698, 1197
Hollingsworth, Mr
   Jamaican estates of 352
Holmes, James [Museum porter] 1312, 737, 1628, 1630, 1760, 1889
Holmes, Robert 80
Home, Sir Everard (1756-1832) 148, 297, 298, 302, 699
   on Ichthyosaurus 341, 342, 699
   publications, *Phil. Trans.* (1818) 341
   reaction to Conybeare & De la Beche's ichthyosaur paper 161
Homfray, John 456
Honey, Mr 148
Hooker, Joseph Dalton (1817-1911) 552, 560, 584, 585, 589, 700-712, 935, 1628, 1739, 1853, 1883
   Antarctic voyage of 702
   Borneo visit of 706, 713
   Egypt visit of 707
   health of father 701
   income of 702, 704
   Indian journey of 703, 705, 706, 709, 713
   introducing Prime Minister of Nepal 710
   Nepal visit of 710
   position with Geological Survey 702
   publications of 704, 705
   publications, *Mem. Geol. Surv.* 937
   Tibet visit of 713
Hooker, William Jackson (1785-1865) 19, 198, 686, 713, 714
   health of 701
Hope, Mr 1293
Hopkins, Evan 715
Hopkins, William (1793-1866) 585, 716-721, 892, 944, 1625, 1885
   & geological survey of Derbyshire 1878
   award of Wollaston Medal to 721
   on rotation of earth 720
   proposal for a geological map of Derbyshire 278, 309, 716, 717
   recommended by De la Beche for Geological Survey of Derbyshire 380
   seeking scientific post on Railway Board 718, 719
Hopkinson, Mr [retired banker of Regent St., friend of Greenough] 1228
   offer of specimens to Museum of Economic Geology 1229, 1232
Horne, Sir A. 343
Horner, Leonard (1785-1864) 722-727, 737, 904, 905, 1219, 1414, 1609, 1837
   Geological Society Anniversary Address 722
   publications, On the mineralogy of the Malvern Hills *Trans. Geol. Soc. Lond.*, (1811) 1416
Horner, Col. 2132
Hornes, Moritz (1815-1868) 640-644, 1969
Hoskins, J. Elliott 728
Hosmer, Mr 729
   and sewer cleaning 729, 1978
Howard, Mr 1303-1305
Howell, Henry Hyatt (1834-1915) 730, 731
   appointment to Geological Survey 730, 1800-1802, 1804
   attachment to Geological Survey sought for 732, 1868
   laudatory reference to 1798, 1799, 1868
   vacation work with Ordnance Survey 1868
Howell, Thomas 732
Howley, William [Bishop of London] (1766-1848) 154
Howse, Richard 95
   shell collection of 800
Hudson, James 629, 1274
Hudson, George (1800-1871) 1644
Hugi, Franz Joseph (1796-1855) 897
Hull, Edward (1829-1917) 731
Hullmandel, Charles Joseph (1789-1850) 1138
   lithographs by 431
Humboldt, Freidrich Willhelm Heinrich Alexander von (1769-1859) 151, 201, 302, 363, 495, 524, 908, 1898
   portrait of 968
Hume, Joseph (1777-1855) 113, 733-736, 1824, 2030
   criticism of Museum of Economic Geology 1272, 1273
Humphreys, J. D.
   shell collection of 800
Humphreys, [?Charles] 304, 1303
Hunt, Robert (1807-1887) 561, 568, 709, 737-740, 819, 989, 1101, 1111, 1163, 1469, 1615, 1623, 1636, 1641, 1642, 1670, 1727, 1743, 1788, 1790, 1796, 1799, 1897, 1949, 2139, 2169
   health of 739
   Poetry of Science 971
   salary 1312
Hutton, James (1726-1797) 141
Hutton, Robert (1785-1870) 51, 306, 564, 624, 661, 741, 779, 901
Hutton, William (1797-1860)
   biographical details of 322
   collection of 322
   publications, *The fossil flora of Great Britain* 322
   subscription in aid of 322
Huxley, Thomas Henry (1825-1895) 1854
Hyams, Mr 2034

Ibbetson, Levett Landon Boscawen (d.1869) 31, 544, 575, 742-745, 1764
   salary as surveyor of railway sections 745
Ilchester, 3rd Earl of (1787-1858) 153

Ilchester, 4th Earl of
    see Strangways, William Thomas Horner
Illingworth, William (1764-1845)
    translation of Wm De Breos charter to Swansea 459, 460
Ingham, Robert [of Westoe] 322
Inglis, Robert Harry (1786-1855) 1058
Inglis, Robin H. 746
Irving, Washington (1783-1859)
    extracts from *The sketch book* 2257
Irving, General 541

Jacobi, Moritz Hermann von (1801-1874)
    at British Association Meeting (1842) 1440
Jackson, Thomas 65
Jackson, Mr [of Bristol Observatory] 342
Jacquemont, Victor (1801-1832)
    in India 495
    publication on Himalayas 1853
Jaeger, [?George Friedrich (1785-1866)] 489
James, Henry (1803-1887) 402, 546, 550, 552, 747-792, 808, 1079,
        1084, 1313, 1314, 1318, 1459, 1461, 1468, 1469, 1475, 1493,
        1508, 1717, 1718, 1720, 1726, 1729, 1732, 1749, 1751, 1868,
        1932, 1936, 1937, 1938, 2266
    appointed Director Geological Survey, Ireland 1562
    appointment as Director-General, Ordnance Survey 791
    assistance sought on piers 766, 770, 772, 773, 776, 779, 780, 808,
        859, 1935
    birth of son of 755
    criticism of Griffith 763
    De la Beche's criticism of 402, 403
    health of 780
    laudatory reference to 1457
    resignation of 780-782, 1743
James, Trevor Evans 751, 793-795, 873, 1356-1360, 1367-1369, 1377,
        1378, 1380-1383, 1386-1390, 1394, 1396-1398, 1400, 1402,
        1405, 1407-1409, 1412, 1415, 1417, 1424-1426, 1430-1432,
        1435, 1437, 1441-1446, 1450-1454, 1460, 1463, 1464, 1466,
        1474, 1478, 1491, 1503, 1506, 1510, 1512, 1513, 1516, 1586,
        1589, 1591, 1595, 1600, 1605, 1609, 1610, 1644, 1663, 1664,
        1666, 1729, 1866, 1867
    health of 1379, 1381, 1382
    termination of employment on Geological Survey 404, 406, 795,
        1739, 1743
James, Mr [of Robeston Wathen]
    sale of goods of 1379
Jameson, Robert (1774-1854) 246, 796-799, 1022, 1030
Jameson 332
Jeffrey, Mr 297
Jeffreys, John Gwyn (1809-1885) 800
    collection of 572-575, 800
    sale of collection of 800
Jelly, Rev. Henry (b.1801)
    collection of 1503
Jenkin, Alfred 601
Jenkin, Charles 945
Jenkins, Sir Richard (1785-1853) 2028
Jenkins, Mr [?of Pentyrch] 2103
Jenkins 1710
Jenkyns, John 801-803
Jenner, Dr 301
Jenner, Mr 840
Jennings, Frank [of Cork] 1583
Jesse, Mr
    coal specimen for Museum of Economic Geology 1295
John, Lord 1016
Johnes, Miss 1714, 1717
Johnes, [?John] 1713
Johnes, Mrs 1712, 1717, 1718, 1724
Johns, Charles Alexander (1811-1874) 804-806
Johnson, G. 1976
Johnson, James (c.1764-1844)
    collection of 296, 297, 299, 300, 320, 342
    ichthyosaurs in possession of 296, 297, 299, 303, 342
Johnson, Prof.
    British Association Railway Section Committee 189

Johnston, Mr [contractor on Plymouth breakwater]
    dispute with Government on work on Plymouth breakwater 373
Johnston, Dr [shell collector] 800
Johnson & Co [of Hatton Garden] 1671
Jones, Calvert Richard (1804-1877) 807
Jones, Edward [of Rudry] 2241-2243
Jones, Col. Harry David (1791-1866) 766, 768, 770, 773, 776, 808, 859
Jones, Thomas Rupert (1819-1911)
    proposed as Assistant Secretary Geological Society 915, 916
    Monographs of 96
Jones, Theophilus (1758-1812)
    extracts from *History of the County of Brecknock* 2183
Jones, David & Co. [bankers, Llandovery] 1405, 1760
Jones, Lloyd & Co. [bankers, London] 1405, 1760
Jones, Mr [of Cardigan] 1404
Jones, Mr [of Swansea] 919
Jones, Rev. R. 809
Jones-Bateman, M. 810
Jordan, Thomas B. (1807-1890) 194, 397, [?]513, 599, 600, 1274,
    1275, 1284, 1293, 1301, 1469, 1517, 1665, 1682, 1865
Jovet, Monsieur 1923
Joyce, Mr 91
Joyce
    resignation [from Chair of Geology at University College London]
        1763
Jukes, Joseph Beete (1811-1869) 551, 554, 561, 563, 566, 571, 582,
        583, 791, 811-820, 833, 1043, 1046, 1109, 1765, 1767-1770,
        1773, 1782, 1784, 1793, 1797, 1800, 1802, 1804, 1894
    application for Chair at Trinity College Dublin 579
    appointment as Local Director, Geological Survey of Ireland 813,
        1129, 1799, 1801
    Australian visit on HMS Fly 812
    beard of 572
    criticism of 583
    criticism of D. Sharpe 811
    debts of 812
    employment on Geological Survey 1751
    marriage of 812, 1785
    paper on geology of Australia 811
    salary increase 997
    salary 223
    sketch of cystoid by (illn.) 563

Kaltner, Mr 1010
Kane, Sir Robert John (1809-1890) 193, 403, 738, 758, 760, 761, 764,
        767, 768, 770, 772, 774, 776, 779, 782, 814, 815, 821-831,
        1046, 1067-1070, 1073-1075, 1086, 1087, 1106, 1133, 1144,
        1310, 1312, 1314, 1316, 1318, 1323, 1555, 2009, 2239
    criticism of 557, 593, 1066
Kapodistrias, ?Avgoustinos (1778-1857) 217
Karck, Theodore 1869
Kater, Captain Henry (1777-1835) 301, 302, 914
Kaulbach, Wilhelm von (1805-1874)
    method of Fresco painting of 924
Kay, Dr W. [of Clifton] 627, 628
Kay-Shuttleworth, James Phillips (1804-1877) 805
Keilhau, Baltazar Mathias (1797-1858)
    paper on Norwegian gneiss 727
Kelly, Sir Fitzroy (1796-1880)
    in Italy 256
Kendall, Prof. 2020
Kendle, John 832
Kennedy, John Studdert (c.1832-1856) 593, 833
Kennedy 1823
Kenner, G. 981
Kennison, Mr
    see Kynaston, Cabot
Kenrick, Mr [of Pontypool] 1707
Kentish, Dr [of Bristol Institution] (d.1832)
    death of 2002
Kenwood, Mr 1215
Kenyon, Mr 535
Ker, Charles Henry Bellenden (?1785-1871) 623, 834-842
Kerr, Mr 939, 2080
Kerry, Lord 260, 1872
Keyserling, Alexander Andreevich, Count von (1815-1891) 1420, 1443,
    1449, 1456, 1461, 1841

at British Association Meeting (1842) 1440
distribution of publications of 1022
Kidd, John (1775-1851)
publications, Mineralogy of St Davids, (*Trans. Geol. Soc. Lond.* Vol 2, 79-93, 1814) 351
Killaly, Richard G. 843
Kindle, Mr [landowner, Fowey Consols area] 2029
King, Captain Arthur Phillip (1738-1814)
King, William (1809-1886) 95, 96
King, Mr
shell collection of 800
Kington, V.F. 842
Kirby, [?William (1759-1850)]
president of Ipswich Museum 1985
Klaproth, [?Martin Heinrich (1743-1817)]
analysis of limestone by 317
Kleville, C.L. 845
Knight, Rev. Hey [of Newton Nottage] 147
at Paviland 171
Knight, Charles (1791-1873) 834, 836, 846, 898, 1204
Knipe, Mr 1278
geological map of 1727
Knowles, Sir Francis 1924
Koch, Albert C. (d.1867)
sale of collection of Missouri vertebrates of 536
Konig, Charles (1774-1851) 1058, 1208, 1213, 1214, 1217
Konigsbraun, Baron
as landscape painter on Austrian Ceylon expedition 644
Kreeft, Mr 530
Krigstatscher, Jos. 847
Kynaston, Cabot (1793-1866) 1854

Labouchere, Henry (1798-1869) 2179, 2180
Lambert, Mr 342
botanical specimens from 164
Lamorte, Mr 32
Lamouroux, Jean Vincent Felix (1779-1825)
fossil collection of 350
Landale, D. 848
Landsborough, Dr [of Saltcoats] 541
Landseer, Edwin Henry (1802-1873) 2276
Lane, Mr [Bronze manufacturer of Birmingham] 1185
Langham, Mr [Assistant Deputy Surveyor Forest of Dean] [of Coleford] 1282
Lansdowne, Henry Petty-Fitzmaurice (1780-1863), 3rd Marquis of 152, 227, 260, 1056, 1872
Lansdowne, Lady
visit to Vesuvius 146
Larcom, Thomas Aiskew (1801-1879) 747, 749, 753, 766, 773, 776, 1532, 1544
appointed Commissioner of Public Works, Ireland 1068
Lardner, Dr 2016
Lascelles, Hon. W.S. 57, 1808
Laugier
limestone anaylsis by 317
Laurence, Oswald
position sought on Geological Survey 1154
Laus, Mr 1010
Lavier, Mr 1215
Lawley, Sir Francis 1278
Lawrence [?of Bristol] 1867
Laws, Mr 1976
Layard, Mr 865
Leach, William Elford (1790-1836) 162, 343, 346, 355, 849
De la Beche's letter of introduction from 343, 346
naming of *Dapedium politum* 342
on naming *Proteosaurus Ichthyosaurus* 342
Leaford, Lord 114
Lechmere, Mr [banker of Worcester] 2084
Lee, John Edward (1808-1887)
fossil collection of 1334, 1352, 1358, 1360, 1389
Lee 227
Leeds, Duke of 1255
Lees, Mr [of Ordnance Survey] 212, 285
Le Hante, Capt. 2042
Lemon, Lady Charlotte (d.1826) 151, 153, 160, 166

Lemon, Sir Charles (1784-1868) 111, 151, 153, 160, 171, 195, 389, 513, 601, 604, 850-852, 1272, 1293, 2151, 2267, 2268
& plan for School of Mines 1333
bust of 939
health of 166
Leo XII (1760-1829), Pope
death of 363
Leopold I (1790-1865) King of the Belgians 853
LePlay, Pierre Guillaume Frederic (1806-1882) 854, 1695
Levrault and Co. 855
Lewin, Miss
water divining by 150
Lewis, David 1405
Lewis, George Cornewall (1806-1863) 856, 2171, 2172
Lewis, Rev. Thomas Taylor (1801-1858) 1468
Lewis, Mr [coal worker/?owner, Clee Hills] 1955
Lewis, Mr 119
Liebig, Justus von (1803-1873) 1685, 1825
analysis of coprolites by 655
on analysis of air 233
Lilford, Lord 1183
Lincoln, Henry Pelham Fiennes Pelham Clinton, Earl of (1811-64) 766, 767, 770, 774, 808, 821, 857-861, 1270-1274, 1276-1281, 1283-1285, 1288, 1289, 1291, 1292, 1294, 1295, 1297-1300, 1303-1315, 1686, 2036
at Nottinghamshire Quarter Sessions 1303
death of brother 1286
[see also Newcastle under Lyme, 5th Duke of]
Linden, D.W. 2240
Lindley, John (1799-1865) 32, 862, 890, 902, 2031
identifying Coal Measure plants for De la Beche 890
palaeobotany lecture notes 862
Lindsay 558
Linné, Carl von (1707-1778) 302
criticism of 152, 154
Liverpool, Robert Banks Jenkinson (1770-1828), 2nd Earl of 148
Liverpool, Charles Cecil Cope (1784-1851), 3rd Earl of 64
Llewelyn, John Dillwyn (1810-1882) 456, 461, 841
Llewellyn, Mr 957
Llewyer 1741
Lloyd, Humphrey (1808-1881) 52
Lloyd, John Augustus (1800-1854) 496
Lloyd, Mrs [of Fishguard] 954
Lloyd, Mr 1399, 1405, 1424
Lloyd family
working Gilfach lead mine, 1435
Lloyd, Mr [of Carmarthenshire]
lead on property of 1715
Loftus, William Kennet, (?1821-1858) 863-867
letter of introduction 1977
position sought on Geological Survey 1894
on Turkish Persian Boundary commission 863-867
Logan, Edmond 876, 885
Logan, William Edmond (1798-1875) 476, 868-888, 903, 917, 1036, 1152, 1387, 1414, 1436, 1438, 1496, 1521, 1679, 2117
application for post of Provincial Geologist, Canada 876-878
appointment of, as Provincial Geologist, Canada 881
assisting Greenough with 2nd Edition of map 625
British Association Railway Section Committee 189
criticism of 695
departure for Quebec 456
fieldwork in Canada 883-885
geological map of South Wales 869, 872, 873, 876, 878-880, 882, 2104
?offered post on Indian Geological Survey 885
visit to Canada 873-875, 879, 880, 882, 883
visit to Pyrenees 868
London, William Howley (1766-1848), Bishop of 154
Lones [the engraver] 1221
Long
retirement of from Society of Antiquaries 1926
Longman (Publishers) 101, 102, 150, 312, 314, 737, 889, 1747, 1760
Lonsdale, William (1794-1871) 26, 29, 34-36, 177, 182, 183, 188, 747, 749, 890-892, 903, 1331, 1403, 1443, 1474, 1475, 1479, 1503, 1528, 1530, 1991, 2086-2089, 2099
Chalk fossils of 891
health of 1469

laudatory reference to 1506
resignation of, as Assistant Secretary & Librarian, Geological Society 1469
subscription raised for 535-537
Loraman, Mr
exhibition of pictures by, at Bristol Institution 304
Loscombe, Mr
shell collection of 800
Louvay, Mr [of Bayeux] 350
Lovelace, D 37
Lovelace, Lord
Dr Carpenter as tutor to family of 1867
Lowry, Joseph Wilson (1803-1879) 1098, 1281, 1505, 1521, 1530, 1629, 1632, 1641, 1665, 1725, 1734
drawings of 544, 552, 566, 567, 571, 572, 575, 704, 759, 1330, 1495, 1519, 1524, 1525, 1531. 1542
Lubbock, John William (1803-1865) 720, 893, 914, 1056
proposed as Fellow of Geological Society 909
resignation as Treasurer, Royal Society 2051
Lucas, J. 164, 166, 171, 172
Luckington, Sir James [Chairman of East India Company] 1836
Lutfield, Mr 1222
Lutke, F. 894
Luuyt, Paul L.
letter of introduction to De la Beche 507
Lycett, J. (d.1882) 743
Lyell, Charles (1797-1875) 8, 32, 39, 54, 84, 94, 101, 141, 170, 175, 306, 496, 512, 524, 526, 534, 570, 586, 622, 718, 723, 725, 736, 895-918, 947, 1045, 1058, 1397, 1398, 1520, 1521, 1806, 2051
& Devon Controversy 621, 623, 837, 897, 902, 1012, 2049, 2098
& Royal Society 906-908, 911-914, 1058
as President, Geological Society 897-902
as Secretary, Geological Society 531
attitude towards De la Beche 1873
award of Royal Society Medal to 621
criticism of 2016
criticism of Boase 897
criticism of Greenough 897
criticism of illustrations used by 2015
death of mother 916
health of father 915
idea of gradual changes 1031
in New York 875
in support of Murchison 621
invited by Forbes to Isle of Wight 591
lecturing in United States 1709, 2015, 2016
on Canadian geology 882
on Reform Bill 895
on *Stigmaria* 876
publications, *Manual of Elementary Geology* 917, 918
publications, *Principles of Geology* 309, 497, 1873, 2089
reaction to criticism 1873
Lyon, J.W. 919
Lyons, Mr [shell collector] 800

McBridrow 552
McCallum, Alexander [fossil collector, Girvan] 1032
McCarsland, Mr 306
McClellan [display case maker] 1217, 1218
McClelland, John [in India] 2138
McCoy, Frederick (c.1823-1899) 578, 763, 766, 767, 769, 772, 787, 937, 1071, 1072, 1079, 1443, 1833, 1883, 1892, 1893, 1936
class size of 1046
criticism of mapping of 558, 1083, 1084
feud with T. Oldham 784
health of 756, 761
laudatory references to 784
McCulloch, 1352, 1370, 1491, 1531, 1578
Macculloch, John (1773-1835)
criticism of 623, 1877
geological map of 158, 271
Macdonald, Major 559
Macfadyen, James 920, 2269
recommended for post of Curator, Jamaica Botanic Garden 686
MacFadyan, Mr 173

MacGillivray, John (1822-1867) 552, 560, 1165
MacGillivray, William (1796-1852) 800
Mackenzie, Sir George Stuart (1780-1848) 921
on Tertiary volcanic rocks of NE Atlantic 921
Macglashan, Dr [of Jamaica] 352
McKinnon, Mr 688, 689
McKinson, George 938
McLauchlan, Henry (1791-1881) 265, 266, 270, 283, 289, 310, 622, 625, 939-958, 1392, 1875, 1991
geological observations of 939-954, 958
salary of, on Geological Survey of Cornwall 376, 529
MacLeod and Co. [of Liverpool] 194
McMahon, T. 959
McManus 238
McMurray, Amelia 960
MacNeill, Sir John Benjamin (1793-1880) 1085, 1794
Madden, Dr [of Penicuik] 233
Magistrini 83
Maguire, Hugh J. 922, 923
letter of introduction of 827
Ordnance Survey work of 922
request for employment on Geological Survey 922
Maguire, T.H.
portrait of J.S. Henslow 1810
Mahon, Philip Henry Stanhope (1805-1875) 5th Earl 1926
Malacarne, Prof. [of Milan] 363
Malcolm, Admiral Sir Pulteney (1768-1838)
fleet of 363
Mallet, Robert (1810-1881) 756, 924, 1079
Maltby, Edward, (1770-1859), Bishop of Durham 322
Managhini 83
Manderson 1118
Mann, Mr [of Jamaica] 352
Mantell, Gideon Algernon (1790-1852) 512, 575, 925-931
and belemnites 1330, 1148
and Owen 568, 589, 1148
death of 583, 584
laudatory references to 534, 583, 584
publications
Geological excursions round the Isle of Wight (1847) 927
Illustrations of the Geology of Sussex (1827) 534
Maxillary and dental organs of *Iguanodon* 931
Memoirs on Belemnites 931
specimens exhibited at Geological Society by 925, 929
Mantell, Walter Baldock Durant (1820-1895) 926, 928, 930
Marcus, Otto 932
Marlborough, George Spencer (1766-1840), 5th Duke of 342
Marryat, Joseph 454
Marsh, Dudley
materials for Great Exhibition, 1851. 78
Marten 1282
Martin, John 933, 934
Martin, L.C. [Printer] 935, 936, 1594, 1607, 1608, 1614, 1620
Martin, Peter John (1786-1860) 532
publications, *Geological Memoir on part of W. Sussex*, 1828. 534
Martin, Mr [of South Wales] 2181
Marton, Henry [of Biddick Hall] 322
Masurnath, Mr
on *Stigmaria* in seatearths 695
Mathens, Mr 1943
Mather, Mr 739
Matson, Col. 781
Mattick, Robert [carpenter, Steeple Ashton] 2157
Mawe, [?John (1766-1829)] 147
Mayer, Dr 849
Mayman, Mr 1259, 1261
Mayne, Mr 238, 240
Meade, Thomas 961
Medlicott, Henry Benedict (1829-1905) 592, 593, 962-964
appointment at Roorkee College 962-964
salary of, on Irish Geological Survey 1117
Medlicott, Joseph G. (c.1825-1866) 814, 815, 1075, 1082, 1087, 1089, 1107, 1110, 1117, 1127
Medlicott, Samuel (b.c.1832) 1109
Meismer, Prof. [of Bern] 348
collection of 348

Meissonier, Jean Louis Ernest (1815-1891) 1679, 1680
Melbourne, William Lamb (1779-1848), 2nd Viscount 58
Melville, J. 965
Meredith, Augusta
 marriage to J.B. Jukes 812
Meredith, James 121
Merewether, John, (1797-1850) Dean of Hereford 1291, 1292
Messengers, Messr
 materials for Great Exhibition (1851) 78
Metcalfe, Mr [shell collector] 800
Metternich, Prince Klemens Wenzel Nepomuk Lothar von (1773-1859) 1949
Meyendorff, [?]Alexandre von 966, 967
Meyer, V. 968-973
Meyer 1567
Michell, Mr 1207
Michelotti, Jean 974
Miles, Philip John (1773-1845) 304
Millbourn, William 975
Miller, Mr C. 1352
Miller, Hugh (1802-1856) 1163
Miller, J. 976
Miller, Johann Samuel (c.1779-1830) 302, 303, 531, 977, 1349
 cataloguing W. Buckland's collection 168, 169, 301
 collection of 300, 1330
 proposed appointment at British Museum 162
Miller, Mr/Dr
 geology lectures of 920
 museum of, at Kingston 352
Miller, William Allen (1817-1870) 1669
Milligan, Dr
 on coalfields of Tasmania 444
Milne, A. 727, 978, 979, 1208, 1216, 1275, 1276, 1285, 1298, 1300, 1309
Milne, David 980
Milne, John 981
Minto, Gilbert Elliott, 2nd Earl of (1782-1859) 661-663, 667
Minton, Herbert (1793-1858) 1195
Mitchell, James W. 983, 984
Moggridge, Matthew 985
 plants for Ramsay from 449
Monkhouse, W. [lithographer, York] 1370, 1610
Montague [shell collector] 800
Montmollin, Auguste de (1808-1898) 347, 1024
Monteagle, Lord 986
 see Spring-Rice, Thomas Moore 727
Moore, F.W. [booksellers, Philadelphia] 2019
Moore, John Carrick (1804-1898) 788, 987
Moore, William 988
Moore, Mr
 metal castings by 421
 trial of 421
Morgan, George [of Carmarthenshire]
 copper ore on property of 626
Morley, Edmund Parker (1810-1864), 2nd Earl 989-991
Morley, J. 992, 993
Morlot, Adolf von (1820-1867)
 map of Austrian Alps by 1022
Morpeth, George William Frederick Howard (1802-1864), Lord 411, 568, 704-706, 786, 982, 994-998, 1318-1320, 1322, 1670, 1689, 2044
 [see also Carlisle]
Morrell, George F. 999
Morris, John (1810-1886) 95, 413, 571, 578, 742, 915, 1414
 paper on *Nautilus* (*QJGS*, Vol 4, p193, 1848) 539
 publications, *Catalogue of British Fossils* 1527, 1956
Morrison, Capt. 1290
 Forest of Dean to Gloucester railway proposed by 1281
Mortimer, Mr 952
Morton, Owen 1000
Morton, Samuel George (1799-1851) 1001
 publications, *Synopsis of Organic Remains* 1001
Moseley 1004, 2267
Mott, Mr 44, 45
Mountenry, Barclay de 1005
Mudge, Robert 1006

Mudge, Richard Zachariah (1790-1854) 208, 370, 658, 660, 892, 939, 1204, 1223
Mudge, William (1762-1820) 342
Mudie, R. 1007
Mulcaster, Sir Frederick William (1772-1846) 207, 211, 213, 215, 263, 269, 276-8, 280, 281, 286, 289, 379
Mulgrave, Mr 1216
Murchison, Kenneth 1008
Murchison, Roderick Impey (1792-1871) 34, 38, 39, 118, 119, 141, 175, 176, 183, 185, 322, 473, 496, 499, 535, 537, 552, 555, 584, 596, 624, 642, 654, 718, 723, 725, 727, 737, 762, 852, 879, 884, 898, 901, 916, 939, 947-949, 966, 967, 975, 1009-1034, 1043, 1046, 1146, 1163, 1184, 1331, 1334, 1338, 1353, 1358, 1360, 1365, 1386, 1389, 1399, 1409, 1420, 1433, 1436-1439, 1443, 1447, 1449, 1450, 1456, 1461, 1468, 1469, 1471, 1492, 1520, 1521, 1531, 1537, 1607, 1615, 1617, 1618, 1663, 1696, 1700, 1706, 1715, 1736, 1742, 1744, 1751, 1765, 1851, 1860, 1898, 1913, 1955, 1991, 2017, 2020, 2053, 2075, 2087
 & Devon Controversy 378, 621-623, 837, 902, 1012, 1013, 1873, 1875, 1877, 1879, 1880, 2049, 2097, 2098
 & erratics 2042
 & Old Red Sandstone of Russia 1348
 attitude towards De la Beche 1873, 1875
 autobiographical details of 1020
 British Association Railway Section Committee 189
 continental tours 1022, 1025
 criticism of 310
 criticism of A. Boué by 1009
 criticism of Geological Society Presidential Address of 1452
 criticism of Hawle & Corda's work on trilobites from Prague 1022
 criticism of map of Carmarthen area of 1378, 1382, 1383, 1417
 criticism of map of Malverns of 1426, 1470
 criticism of Phillips "Palaeozoic fossils of Cornwall" by 1397
 criticism of Samarsky by 1017
 criticism of Sedgwick by 1019
 death of brother 1034
 early fieldwork of 1020
 geological education of 1020
 offer of specimens to Museum of Practical Geology 1029
 on Cambrian-Silurian boundary 1018, 1019
 on Devon geology 26, 29, 31, 35, 36
 on geology of Alps 1026, 1028
 on geology of Arran 1858
 on gradual changes 1031
 on Indian geology 97
 on Old Red Sandstone 1011, 1348
 on Russia 1016
 on stratigraphic nomenclature 1011, 1014
 publications,
  Geological sketch of... Sussex... *Trans. Geol. Soc. Lond.* 2nd Ser. Vol.2 (1825) 1020
  on geology of Cheltenham *Proc. Geol. Soc.* Vol.1 (1832) 1417
  On the geological structure of the Alps (1849) *QJGS* 1026-1028
  Siluria (1854) 1033, 1034
  Silurian System (1839) 950-952, 958, 1103, 1405
  Ueber die Silurischen Gesteine Bohmens, nebst einigen Bemerkungen ueber die devonische Gebilde in Mähren, *Leonhard v. Bronn, N. Jahrbuch* (1848) 1022
  with De Verneuil, Geological Structure of Russia (1841) 1402
 visit to Vienna 639, 1022
Murchison, Roderick M. 1022, 1035
 attached to Geological Survey 1008, 1761, 1768, 1772, 1773
 college report of 1035
 school 1751
Murchison, Mrs [Lady] Charlotte (*née* Hugonin) (1788-1869) 1020
Murphy, Thomas 755, 772
Murphy (d.1836)
 death of 901, 939
Murray, Alexander (1810-1884) 879, 883, 885, 887, 1036, 1454, 1455, 1460, 1461, 1463
 salary of 885
Murray, Sir George (1772-1846) 399, 1300
 reference for W.E. Logan 876, 878
Murray, Lady Louisa
 death of 878
Murray, Peter 1037
Murray [publishers] 684
Mushet, [?David (d.?1859)] 2115

165

Napier, Charles James (1782-1853) 1021, 1023, 1038
  resignation from government of Sind 1038
Napier, Mr 1023
Napoleon III (Louis Napoleon) (1808-1873) 585
Nasmyth, James (1808-1890) 1039
Nasmyth, Sir John 1570
Naumann, Carl Friedrich (1797-1873) 1025
Nazimoff, Vladimir 1002
Necker de Saussure, Louis Albert (1786-1861)
  Chateau at Coppet, Lake Geneva 345-347
Necker, Madame 347
Neill 727
Nelson, Capt.
  offer of coral to Geological Society 987
Nesbit, John Collis (1819-1862) 1043
Nesti, Prof. [of Florence] 620
  duties of, at museum in Florence 363
  on geology and fossils of district around Florence 363
Newbold, Captain Thomas John (1807-1850) 865, 866
Newcastle-under-Lyme, Henry Pelham Fiennes Pelham Clinton
  (1811-1864), 5th Duke of 1040, 1255
  [see also Lincoln]
Newton, Charles Thomas (1816-1894) 1041, 1951
Newton, Sir Isaac (1642-1727) 171, 204
Nichol, Mr 1358
Nichol 1783
Nicholas I of Russia (1796-1855) 651
Nicholl, Sir John (1797-1853) 158, 159
Nicholl, Miss 158
Nichols, Charles 1042
Nicol, Dr D. 453
Nicol, James (1810-1879) 539, 727, 811, 829, 915, 928, 1016,
  1042-1046, 1143, 1982
  appointment of, as Prof. of Geology, Queens College, Cork 1045
  resignation of, from Geological Society 1045
Nicol, William (c.1768-1851) 703
Noaks, E. 1047, 1048
Noble, Matthew (1816-1876) 1643, 1648
Noel, Lady
  water divining by 150
Norrey, Stephen MP 1049
Norrey, Sir D. 1902
Norris, J.P. 1050
Northampton, Spencer Joshua Alwyne Compton, 2nd Marquis of
  (1790-1851) 561, 718, 901, 1051-1060
  British Association Railway Section Committee 189
  retiral as President of Royal Society 682, 1842
  visit to Museum of Economic Geology 2029
Northumberland, Hugh Percy (1785-1847), 3rd Duke of 150
Northumberland, Algernon Percy (1792-1865), 4th Duke of 1056, 1061,
  1976
Norton, Mr 8
Norwich, Bishop of (1837) 1062, 1229
  portrait of, Ipswich Museum (1849) 1811
Notter, Richard [of Kilmoe, Co. Cork] 2026
Nuttall, Thomas (1786-1859) 1826

O'Brien, William Smith (1803-1864) 1770
O'Brien, Mr [builder, Dublin] 2009
O'Connor, Feargus Edward (1794-1855) 1770
Oesfeld, M. [of Berlin] 1869
O'Farrall, Mr [Secretary of Admiralty] 661
Odlham, Charles Aemilius (?1831-1869) 1065, 1107
Oldham, N.A. 1063
Oldham, Thomas (1816-1878) 51, 52, 255, 557, 558, 561, 562, 567,
  579, 783, 824, 833, 962, 963, 1064-1133, 1611, 1743, 1747,
  1785, 1802, 1852, 1853
  appointment as Local Director of Irish Geological Survey 784, 1065
  appointment to Geological Survey of India 1115, 1116, 1119,
    1121-1123, 1133, 2008
  asked to be Section C Secretary, BAAS 1077
  elected FRS 1098
  feud with F. McCoy 784
  injured in coach accident 1071, 1072
  laudatory references to 552, 1749
  lectures of 1077
  marriage of 1656, 1798, 1961
  resignation from Geological Survey Ireland 1131, 1801
  salary & expenses of, on Irish Survey 1115, 1121
  salary on Indian Survey 1116, 1122
O'Neil, Messrs
  Works near Pyle 2010
Ora, William MP 322
Orbigny, Alcide Charles Victor Dessalines d' (1802-1857) 539, 1567
  publications, 508, 509
Ord, Major 1134
Ordnance Surveyor General 210, 212
Ordnance, Master General of 214, 215, 269, 379, 401, 838, 1179, 1822
Organ, Mr 739
O'Riley, Mr 993
Ormerod, George Wareing (1810-1891)
  publications, Saltfield of Cheshire, QJGS Vol 4 (1848) 1135
Ormerod, Mr
  poem on death of 148
Ottley, Mr
  fossil collection 1352
Owen, Captain 71
Owen, Col. 1992
Owen, Col. [of Landshipping]
  collieries of 2111
Owen, Richard (1804-1892) 8, 19, 93, 195, 987, 1071, 1136-1153,
  1166, 1403
  & BAAS British Fossil Reptiles 1136-1141, 1150
  & BAAS Report on British Fossil Mammals 1139
  and Mantell 568, 584, 926, 1148
  & Royal Society 911, 912
  as potential Geological Society President 584, 589
  holiday with Buckland 1152
  proposed as Honorary Palaeontologist to Geological Survey 405
  publications, Eocene mammals from Hordle, QJGS Vol.4 (1848)
    1143, 1148
  publications, Fossil reptiles from New Jersey, QJGS Vol.5 (1849)
    1146
Oxford, Richard Bagot, Bishop of 878, 880

Paccard, Dr Michel-Gabriel (1757-1827)
  collection of 344, 346
  description of 344, 346
Page, Dr [of Westminster]
  disposal of Bandiner's China collection by 454
Page, Mr [of Bristol]
  collection of 300
Page, Thomas (1803-1877) 1293, 1295
Paget, Edward (1775-1849) 108
Paley, Dr 204
Pallas, Peter Simon (1741-1811)
  discovery of Siberian meteorite 148
Palmer, Dr [of Jamaica] 352
Papworth, Edgar George (1809-1866) 1154, 1241
Papworth, George (1781-1855) 758, 760, 2009
Parish, Sir Woodbine (1796-1882) 898, 901
Parker, Mr
  Jamaican estate of 352
Parker, Mr [of Exeter] 1859
Parker, Mr 1243
Parrot 594
Parry, Capt. William Edward (1790-1855) 157, 302, 542
Partridge, Prof. 1835
Pasely, Mr 826
Pasini, Lodovico (1804-1870) 1184
Pasley, Lady 451
Pattison, S.R. [of Launceston] 1332,
  collections 1334, 1352, 1337
Paul [Quaker Cornish copper mine director] 373
Payne 1292
Peach, Charles William (1800-1886) 195, 198, 851, 1156-1165, 1583,
  1990
  collection of 1158, 1161
  pay, pension and conditions of service as Coastguard 1158
  post sought as Customs Comptroller, Fowey 1160
  post sought as fossil collector 1157
  post sought as Principal Coast Officer, Spalding 1159

Peach, William (b.1833)
  post sought as fossil collector on Geological Survey 1164, 1799
Peacock, Rev. George (1791-1858)
  suggested as Royal Society President 1843
Pearce, James 1166
Pearce, Joseph Chaning (1811-1847) 91, 1148, 1253
  biographical notes on 1166
  collection of 1166, 1503
Peel, Sir Robert (1788-1850) 43, 48, 148, 192, 194, 195, 197, 199, 397, 535, 536, 617, 655, 835, 908, 1056, 1058, 1063, 1273, 1274, 1278, 1294, 1295, 1297, 1300, 1303, 1305, 1306, 1384, 1389, 1718
  as Irish Secretary 1170
  statue in Birmingham 698, 1197
Pellatt, Apsley (1791-1863) 740, 1178
Pemberton & Co. [of Dublin] 1069
Pendawes, Mr 1207
Penn, Granville (1761-1844) 301
  criticism of 355
  opposition to Buckland's *Reliquiae Diluvianae* 171
Pennant [shell collector] 800
Pennethorne, James 1179, 1297, 1300, 1317, 1325
Pennington, Dr 1817, 1818
Penny, James 756, 772, 775-777, 786, 828, 1075, 1933, 2009
  expenses of 785, 1067, 1071, 1072
Penny, Mr [son of James Penny]
  employment sought by 828
Penrose & Evans
  as proprietors of Eskyn (Eaglesbush) Colliery 1958, 1959
Pentland, Joseph Barclay (1797-1873) 84, 161, 176, 299, 364, 489, 506, 1180-1184, 1698
Percy, John (1817-1889) 17, 63, 291, 421, 698, 743, 831, 1185-1197, 1326, 1819, 1943, 1944
  & BAAS meeting, Swansea 1640
  and photography 1196, 1197
  & statue of Robert Peel 1197
  appointment as lecturer, School of Mines 1187
  experiments on water on stone 1197
  marriage contract 1187
  metallurgical work of 1188, 1189
Perkins, Mr 607
Pernolet 1198
Petherick, Captain John 1199, 1207, 1934, 2026
Pethoud, H. 1200
Petrie, Henry (1768-1842) 392, 393
Peynhausen 1537
Philipps, Trenham W. 402, 405, 417, 420, 666-668, 704, 705, 713, 768, 1201-1327, 2270
  health of 1322
Phillips, Ann (1803-1862) 112, 1374, 1409
  discovery of fossils in Llandovery Conglomerate in Malvern Hills 111, 112, 1450
Phillips, Edward [of Fishguard] 954
Phillips, H. 121
Phillips, John (1800-1874) 15, 27-30, 32-39, 52, 62, 95, 98, 102, 110-112, 131, 141, 196, 258, 322, 400, 458, 484, 538, 542-547, 550, 560, 563, 568, 623, 655, 702, 723, 725, 740, 747, 751, 754, 770, 786, 876, 883-885, 903, 914, 944, 956, 961, 967, 1022, 1030, 1077, 1158, 1328-1666, 1678, 1697, 1706, 1719, 1766, 1767, 1833, 1851, 1853, 1856, 1860, 1863, 1865, 1867, 1893, 1993, 2002, 2003, 2055, 2098-2100, 2112, 2123, 2271, 2272
  appointments
    appointed Professor at Dublin 1535, 1541, 1561
    appointment to Chair at Dublin 400
    as Survey's Chief Assistant Geologist 1397, 400
    employment of on Geological Survey 387, 397, 400, 1337, 1341, 1347, 1353-1355, 1384, 1389, 1390, 1534, 1559-1561, 1565, 1574
    post at King's College, London 1337, 1353, 1537
    post at Royal Institution 1353
    post at York 1337, 1353
    post with British Association 1349, 1353, 1535, 1547
    resignation from chair at Dublin 1569
    resignation from King's College, London 1341
    resignation from Yorkshire Museum 1341, 1349, 1408
  salary of, at Dublin 1574
  salary on Geological Survey 1390
  & BAAS *Catalogue... of fossils of Great Britain* 1330
  & Geological Survey of Ireland 1554, 1556, 1559-1562
  & Geological Survey of Yorkshire 1536, 1578, 1587
  award of Geological Society Wollaston Medal to 1568
  bank account 1462, 1489, 1526, 1533
  British Association Railway Section Committee 189, 1858
  criticised by Edward Forbes 544
  criticism of Murchison's map of Carmarthen area 1378, 1382, 1383, 1417
  criticism of Murchison's map of Malverns 1426, 1470
  education of 1415
  elected to Athenaeum 1411, 1412
  financial situation of 1342
  fossil drawings of 407, 1549
  geological map of 1727
  health of 1069, 1406, 1604, 1605, 1611, 1628, 1630
  lectures 1342, 1349, 1415, 1534, 1649, 2272
  on belemnites 1330, 1331
  on fossils of Cornwall & Devon 1334, 1335, 1346
  on income tax 1414
  Oxford class of 1659
  plan for Geological Survey of Ireland 1538, 1543, 1548
  proposal that Phillips describe Cornwall Survey fossils 387
  proposal that Phillips describe South Wales fossils 1349
  proposed plan of work for Geological Survey 1373
  publications
    list of publications 1537
    monograph on belemnites 1567
    monograph on crusteacea 1567
    A Guide to Geology (1834) 1330, 1537
    Geology of Yorkshire (1829) 1329, 1537
    On geology of Havre, *Phil. Mag.* (1830) 1328, 1329
    Palaeozoic Fossils of Cornwall.. (1841) 1340, 1342, 1343, 1346-1350, 1352, 1354, 1358, 1362, 1363, 1366, 1368-1371, 1397
  receipt of De la Beche's personal medal 1659
  time available to work on Geological Survey of Ireland 1546, 1548, 1553
  tour of Devon & Cornwall 1335-1337
Phillips, John Arthur (1822-1887) 1667-1668
Phillips, Richard (1778-1851) 103, 219, 233, 397, 655, 1183, 1243, 1251, 1254, 1264-1268, 1270, 1274, 1281, 1283, 1284, 1288, 1289, 1292, 1296-1298, 1302, 1308, 1319, 1322, 1352, 1435, 1469, 1669-1671, 1705, 1820, 2028-2031, 2033-2035
  analysis of clay by 989, 990
  analysis of ore by 1671
  criticism of 1262, 1287, 1296, 1301, 1303, 1306
  health of 1314, 1671
  inspection of Swansea copper smelting 411, 1670
  laboratory 870, 1246, 1249, 1250, 1253, 1260-1262, 1273, 1301
  lectures of 1302, 1303
  salary of 398, 1249, 1256, 1257, 1275, 1278, 1312
  work of 398
Phillips, William 1672
Phillips, William [draper of Fishguard] 954
Phillips, W.L. 121
  criticism of 534
Phipps Charles Beaumont (1801-1866) 1273, 1294, 1673-1675
Phyffer, General
  model of Alps by 348
Pickering, Dr 108
Pickersgill, Henry William (1782-1875) 1676, 1677
Pictet, Marc Auguste (1752-1825) 344, 346-348
Piddington, Henry 1678
Pigott, Capt. 2035
Pilla, Leopoldo (1805-1848) 83, 1025
Pinck [the Brewer, Bath] 246
Pinney, Mr 302
Piot, F. 1679-1683
Pirey 1688
Piron 306
Pitts, Mr
  culm mine of 832
Pius IX (1792-1878), Pope 1025

167

Playfair, James
    death of 2141
Playfair, John (1748-1819) 141, 921
    visit to Vesuvius 146
Playfair, Lyon (1818-1898) 14, 78, 195, 196, 222, 291, 481, 488, 538,
    544, 568, 591, 630, 631, 709, 747, 916, 1273, 1278, 1286,
    1289, 1309-1311, 1313, 1315, 1319, 1322, 1326, 1511, 1669,
    1684-1693, 1705, 1713, 1716, 1718, 1721, 1726, 1738, 1749,
    1903
    & mines explosions enquiry 1308
    death of father 2137
    death of son 2140, 2141
    health of 1686
    invited to be Great Exhibition travelling Commissioner 617, 1692
    lectureship at Putney College for Civil Engineers 1692
Plumptre, Dr John (1753-1825), Dean of Gloucester 147
Plumptre, Mr [A?] [son of above] 147
Ponsonby, Sir Frederick Cavendish (1783-1837)
    death of 1221
Ponsonby, Lord 1922
Poole, Miss 1856
Porter 570
Porter, Mr [shell collector] 800
Pourtales, Count 347
Porter, Sir Robert Ker (1777-1842) 173
Portland, Margaret Cavendish (1714-1785) Duchess of 1255
    shell collection of 800
Portlock, Joseph Ellison (1794-1864) 749, 751-753, 771, 818, 1064,
    1079, 1493, 1508, 1545, 1696, 1697, 2272
    criticism of 751, 754
    fossil drawings of 407, 786
    publications, Report on the geology of Londonderry (1843) 1085
    sailing for Corfu 751
Potter 1391
Potts, Dr 892
Powell, Mr [?of Swansea] 456
Powell, Col.
    mines of 1748
Prado, Casiano de 1698
Pratt, Duncan [of 22nd Bengal Infantry] 1699
Pratt, Samuel Pearce (1789-1863) 109, 544, 1699, 1700
    collection of 1503
    paper on ammonites 2003
Pratt[?] 1701
Prestwich, Joseph (1812-1896) 581, 586, 592, 1702
    as potential Geological Society President 584
Prevost, John Lewis (1796-1852) 1703
Prèvost, Louis Constant (1787-1856) 304, 1031
Price, Sir Ralph 293
Price, Sir Rose
    Jamaican estate of 352
Price, Samuel Warren
    De la Beche medal awarded to 2279
Prichard, James Cowles (1786-1848) 302, 977
Prideaux, Prof. 389
Prince Regent (George IV) 148
Princess, James 1678
Protheroe, Mr 1284
    finances of 1281
Pugh, Lewis [Crown Agent, Dolgellau] 1947-1949
Purcell 1854, 1855
Purdue, John 1704
    collections of 551, 555, 1704
    echinoid for sale by 1704
Pusey, Philip (1799-1855) 397, 1273, 1705
Pyrie, Mr [of Lyme] 696

Quetelet, Lambert Adolphe Jacques (1791-1874) 1487

Radley, Mr [of Newton Bushel, Devon] 2086
Radnor, Earl of 228
Ramond
    quoted on history 341
Ramsay, Andrew Crombie (1814-1891) 118, 119, 128, 235, 418, 449,
    545, 547, 552, 554-556, 558, 562, 563, 567, 568, 575, 709,
    792, 814, 816, 817, 833, 917, 918, 1008, 1024, 1036, 1043,
    1073, 1103, 1164, 1343, 1362-1365, 1367, 1382, 1390, 1400,
    1408, 1412, 1419, 1424, 1425, 1429, 1432, 1439, 1446, 1448,
    1449, 1454, 1460, 1464, 1469, 1471, 1472, 1478, 1486, 1498,
    1500, 1503, 1504, 1509, 1512, 1513, 1515, 1516, 1525, 1528,
    1530, 1550, 1552, 1557, 1558, 1570, 1572, 1573, 1610, 1611,
    1656, 1663, 1684, 1685, 1689, 1706-1807, 1852-1854, 1868,
    1942, 1945
    appointment to Geological Survey 1016
    accidents to 1377, 1408, 1417, 1710
    health of 560, 1407, 1706, 1707, 1739, 1764, 1765, 1798, 1800
    offered Chair of Geology at University College, London 1763
Ramsay, Capt. William [RN] 848
Ramsden, William [of Wakefield] 57, 995 , 1808
Randall, Mr
    solicitor at Neath Colliery explosion inquest 1958
Rankin, Thomas [of Bristol Institution] 2002
Ransome, E. 1985
Ransome, George [of Ipswich Museum] 1809-1811
Raple, William 1812-1813
Rapley, Mr 544, 1369, 1370, 1529, 1531, 1734
Rashleigh, Sir Colman 941
Rashleigh, Philip (1729-1811)
    mineral collection of 941
Raspe, Rudolph Erich (1737-1794)
    on geology of Pembrokeshire 947, 949
Rassam, Hormuzd (1826-1910)
    Nineveh excavations of 867
Raw, Mine Captain James [of Cwmystwyth] 1942
Rawlinson, Henry Creswicke (1810-1895) 865, 867
Rawlinson, C. 915
Rawson, Rawson W. 879, 880, 886
Reed, Mr 511
Reeks, Trenham (1823-1879) 89, 246, 571, 576, 701, 707, 709, 1073,
    1075, 1081, 1083, 1093, 1100, 1106, 1109, 1112, 1119, 1120,
    1124, 1126, 1130, 1268, 1272, 1277, 1281, 1284, 1287, 1296,
    1303, 1310, 1577, 1611, 1627, 1643, 1667, 1725, 1727, 1735,
    1739, 1740, 1742, 1743, 1748, 1775, 1780, 1786, 1787, 1796,
    1797, 1814, 1944, 1955, 2139
    analysis of rock by 1814
    health of 1283, 1301, 1306
    salary 1275, 1288, 1312
Rees, George [of Risca Works] 1821
Rees, Josiah (1808-1873) 873, 1242, 1368, 1378, 1388, 1389, 1394,
    1397-1399, 1401, 1402, 1404, 1406, 1407, 1412, 1413, 1421,
    1424, 1430, 1431, 1435-1437, 1439, 1452-1457, 1460, 1461,
    1463, 1464, 1466, 1469, 1664, 1666, 1709, 1716, 1814, 2030,
    2101, 2122
Reeve, Messrs [publishers] 462
Rehausen, Baron 1032
Reid, Dr David.Boswell (1805-1863) 1214
Renard 1002
Rendal [?Rendel] 1283
Rendel, James Meadows (1799-1856) 373, 1816
Reniby, Mr 2
Rennie, Sir John (1794-1874) 933
Rennie [?George] 682, 1842
Renoir, Mr 189
Respighi, Prof. Lorenzo (1824-1889)
    visit to London of 84
Retzius, ?W.C. 2005
Reynolds, Mr 301
Rice, Sir Ralph 363
    health of 1817, 1818
Rice [of Brighton] 1817
Richards, Capt. John 515
Richards [the Engineer] 513
Richardson
    materials for Great Exhibition (1851) 78
Richardson, Mr
    surveyor at Neath Colliery explosion inquest 1958
Richardson 1020
Richardson, Mr 1415
Richardson, Sir John (1787-1805)
    expedition (1848) 1845
    Polar Dinner 137

Richmond, Charles Gordon Lennox (1791-1860), 5th Duke of
    Lord Lieutenant of Ireland 1170
Ridsdale, G.W.H.
    biographical details of 1819
Rigby, William 477, 1820
Rigby & Hancock 1820
Rigny, V. de 323
Riguy, Admiral de 363
Riley, Dr Henry (1797-1848) 903, 918
Riocreux 2081
Rippingille, Edward Villiers (1798-1859) 304
Robe, Alex. W. 210, 266, 273, 274, 284, 289, 380, 383, 892, 948, 949,
    1240, 1331, 1822, 1823
Roberts, George (1803-1860) 11, 988
    book on Lyme 696
Roberts, Mr [of Swansea] 452
Robinson, Sir John 1039, 2165
Robison, John 1824, 1825
Roe, Mr 232
Rogers, E. 87
Rogers, Henry Darwin (1808-1866) 1146, 1148, 1517, 1826, 1827
    health of 1827
    paper on Alps 1827
    on Appalachian geology 1026
    Report on geology of New Jersey 695
    visit planned to BAAS York meeting 1501
Rogers, J. 1828
Rogers, William Barton (1804-1882) 8, 695, 1827
Rogers, Messrs 547
Roget, Peter Mark (1779-1869) 908, 1056
    resignation of, as Secretary Royal Society 1842
Romer, Carl Ferdinand von (1818-1891) 134
Rootsey, Samuel 304
Roscoe, Mr 301
Rose, Mrs 150
Ross, Captain
    as Commissioner of Plymouth Dock Yard 373
    inspection of Dartmoor Prison by 373
    Polar Dinner 137
Ross, Sir James Clark (1800-1862)
    expedition (1848) 1845
    in South Polar Sea 662
Ross, Sir John (1777-1856)
    North West Passage Expedition (1818) 148
Ross, Lord 1829
Rosse, William Parsons, 3rd Earl (1800-1867)
    telescope of 1523
Rosser, Highmore 1830-1833
    employment on Geological Survey 62, 1830-1832
    employment sought by 1832
Rosthorn, Joseph 1834
Rowe, Mr
    culm mine of 832
Rowbotham, Mr
    specimens from, for Museum of Economic Geology 1231
Royer 343, 346
Royle, Annette Forbes 1835
Royle, John Forbes (1799-1858) 1836, 1837, 2028, 2029
Rubridge, Dr 19
Rule, Captain 614
Rumfort, Countess of 347
Rupley, Mr 1367
Russell, John
    pits at Risca 2103, 2107
Russell, Ino. [of Risca Works] 1821
Russell, Lord John (1792-1878) 665, 688, 718, 719, 1838-1840, 2179
    installed as Rector, Glasgow University 1755
Russell, Mr [of Birmingham] 2096
Ruthven, John
    graptolites in Skiddaw Slate discovered by 1025, 1886
Ryan, Sir Edward (1793-1875) 1678

Sabine, Colonel Edward (1788-1883) 554, 638, 682, 907, 911, 914,
    1058, 1842-1846, 1987
Sadleir, Rev. Ralph 1847, 2091
Saint-Anthoine, H. D. de 1848

St David's, Bishop of 204, 1020
St David's, Dean of 889
Sale, Captain William 352, 353
Salmond, William 1849
Salter, John William (1820-1869) 541, 551, 555, 558, 562, 573, 763,
    769, 1598, 1613-1615, 1621, 1622, 1625, 1641, 1661, 1662,
    1740, 1742, 1752, 1797, 1799, 1801, 1850-1857, 1883, 1887
    and income tax 1857
    Geological Survey pay of 1739
    honeymoon of 1851
    report on private collections of Silurian fossils by 560
Samarsky
    criticism of, by Murchison 1017
Sanders, William (1799-1875) 302, 1337, 1338, 1358-1360, 1362,
    1367, 1368, 1411, 1412, 1415, 2002, 2123, ?1457, 1503, 1504,
    1506, 1507, 1509, 1511, 1609, 1858-1867
    & British Association Railway Section Committee 189, 1858, 1859
    collection 1335, 1336
    plan for geological map of Bristol 1858
    working for Geological Survey 1858-1865
Sandham, Col. H. 1804, 1868
Sandon, Lord 1272
Sardinia, King of
    criticism of 345-347
Saul, Mr
    offered ichthyosaur by Mary Anning 180
Saulini, T. 363
Savage, Miss Sarah 339
Saxby, Stephen M. 95, 539
Saxony, King of 1177
Saxton [?Joseph 1799-1873] 658
Sayer, Mr 302
Scharnhorst, General de
    map collection of 203
Schmarda, Ludwig Karl (1819-1908)
    on Austrian expedition to Ceylon 644
Schonbein, Christian Friedrich (1799-1868)
    at British Association Meeting (1842) 1440
Schroder, Henry & Co. 1869
Schrötter, Prof. Anton von (1802-1875) 641
Schweitzer, Baron de/von 348, 349
Scott, Mr
    botanical specimens from Australia & New Zealand 158
Scott, M. 74
Scott, Mr (Lieut.) [of Navy] 657
Scott, Sir Walter (1771-1832) 148, 309, 2165
Scouler, John (1804-1871) 652, 2042
Scrope, George Poullett (1797-1876) 178
    criticism of 307, 308
    criticism of, as Secretary, Geological Society 175
    publications 306
Sedgwick, Adam (1785-1873) 26, 31-35, 47, 111, 141, 173, 180, 185,
    280, 301, 314, 379, 387, 496, 541, 578, 621, 680, 681, 718,
    899, 901, 947-951, 1009, 1022, 1163, 1331, 1333, 1338, 1365,
    1385, 1396, 1407, 1424, 1434, 1449, 1450, 1461, 1471, 1522,
    1524, 1615-1617, 1744, 1748, 1752, 1765, 1850, 1853,
    1870-1899, 1900, 1901, 2075
    & Cambrian/Silurian boundary 791, 1018, 1019
    & Devon Controversy 273, 378, 622-624, 902, 1332, 1873,
    1875-1877, 1879, 1880, 2097, 2098
    & stratigraphic nomenclature 1014
    at Cheltenham 1445
    collection 1334
    criticised by Murchison 1019
    criticism of 555
    health of 623, 624, 1025, 1445, 1489, 1877, 1879, 1883, 1884,
    1886, 1887, 1890, 1891, 1893, 1894, 1896
    in Devon 15, 28-30, 36
    in Wales 30, 1610, 1852
    invited to Osborne House 1888
    lecture to Cambridge Philosophical Society 1881
    lectures at Cambridge 1886
    on Purbeck and Portland Beds 532
    paper on North Wales read to Geological Society 1531, 1882
    paper on Skiddaw fossils 1890-1893
    prebendaryship at Norwich 890, 1873
    publications 1870, 1871, 1898

publications, *A synopsis of the classification of the British Palaeozoic Rocks* (1855) 555
    refuses Bishopric of Norwich 1226
    recommendation of Hopkins to map Derbyshire 278, 380
    support for De la Beche and Devon Geological Survey 1872
    visit to Cornwall 1874, 1875
    visit to Gower caves 164
Sedgwick, Richard [nephew of Adam]
    health of 1884, 1885
Selkirk, Dunbar James, 6th Earl of 541, 1896, 1900, 1901
Selwyn, Alfred Richard Cecil (1824-1902) 566, 1728, 1729, 1745, 1747, 1748, 1760, 1767-1770, 1773, 1775, 1776, 1778, 1782-1784, 1786, 1787, 1789, 1791, 1796-1800
    health of 1758
    salary increase 997
Sergeant, Mr [of Treasury] 1292
Setter [of Ashford]
    collection of 1830
Severs, Captain 352
Seymour, Edward Adolphus (1775-1855), 11th Duke of Somerset 524, 650, 666-668, 1032, 1049, 1119, 1131, 1326, 1674, 1796, 1902, 1903
Sharpe, Daniel (1806-1856) 95, 578, 1471, 1492, 1531, 1765, 1904
    as potential Geological Society President 584
    critisicm of 811, 1527
    "fish scheme" of 1904
    intention to go to Lisbon 1904
    publications, on slaty cleavage, *QJGS*, (1849) 1043
Shepard, Prof.
    fossil collection of 1345
Shepherd, Forest
    criticism of 887
Shepherd, Capt. 2008
Shipman, John 1905
Short, Mrs 174
Shrewsbury, Lord 363
Shuttleworth, Philip Nicholas (1782-1842) 151, 152
Silliman, Benjamin (1779-1864) 1906
Simms 108
Simms, Mr 376, 529
Simons, G. 1907
Simpson, Sir George (1792-1860) 885
Simpson, Mr [De la Beche's agent in Jamaica] 352, 353, 920
Simpson, Mr [the Engineer] 1243
Simms, Frederick
    death of wife 2140, 2141
Sismonda, Angelo (1807-1878) 1908-1915, 1917
Sismonda, Eugene (1815-1870) 1916-1918
Sismonda 1779, 1780, 2053
Skinner, Mr
    collection of Coal Measures plants given to Bristol Institution 301
Slaney, Robert Aglionby (1792-1862) 234, 1919
Slattery, Mr 242
Smeaton, John (1724-1792)
    construction of Eddystone lighthouse 372
Smedley, Edward (?1788-1836) 306
Smirke, Sir Robert (1780-1867) 1293
    health of 1920
Smirke, Sydney (1798-1877) 1920
Smith, Charles [of Swansea]
    visit to Norway 456
Smith, Charles Harriot (1792-1864) 610, 1920, 1921
    on Parliamentary Building Stone Commission 979, 1266
Smith, C.H. 2273
Smith, D. Southwood [?Thomas Southwood] 466
Smith, James (1789-1850) [of Deanston] 232
    land drainage by 194, 195
Smith, Sir James 464
Smith, Sir James Edward (1759-1828) 686
Smith, Joshua Toulmin (1816-1869) 2274
Smith, Thomas Southwood (1788-1861) 238, 240, 1686
Smith, William (1769-1839) 246, 979, 1476, 1550, 1552, 1585
    1815 map 1332
    & Parliamentary Building Stone enquiry 65, 1242, 1243, 1255
    bust of 1628, 1643, 1648
    coal sections of 1639
    collection of 1537
    criticism of 534
    debts of 1348
    pension 1352
    widow of 1342
Smith, Dr [of Jamaica] 352
Smith [the Mason] 1264, 1266
Smith, Mr [of Welsh Court] 747
Smith, Mr [of Grand Junction Canal Company] 417
Smith 302
Smyth, Annarella 1922, 1923
Smyth, William Henry (1788-1865) 75, 395, 679, 908, 1923-1926
    daughter of 396
    in Cardiff 396
Smyth, Warington Wilkinson (1817-1890) 568, 604, 639, 640, 642, 709, 739, 764, 769, 770, 772, 865-867, 1071, 1076, 1077, 1196, 1197, 1289, 1327, 1335, 1740, 1750, 1766, 1800, 1802, 1922, 1927-1973, 2178
    colliery explosion investigations by 1943, 1957-1959
    continental tour 1927-1929
    geological locality in ?Carpathians (illn.) 1929
    on mines of Wales 1940-1942
    on mines of Wicklow and Wexford 420
    paper on mines (?of Turkey) 864
    payment for investigation of colliery explosions by 414, 1948, 1949
    salary of, as Survey Mining Geologist 414, 2037
    to superintend mines of Duke of Cornwall 1674
Solly, Edward (1819-1886) 1974
Solly, Samuel (1805-1871) 1974
Somerville, William (1771-1860) 1975
Somerville, Mary (1780-1872) 82, 83
Sopwith, Thomas (1803-1879) 109, 110, 1275, 1282, 1799, 1966, 1976-1978, 2016, 2113
    British Association Railway Section Committee 189
    map of Forest of Dean 1280
    model of Forest of Dean 2115
    suggested as Mineral Surveyor, Duchy of Cornwall 978
Southern, Miss 451
Southey, Robert (1774-1843) 154, 301, 302
Sowerby, George Brettingham
    proposed publication on British Palaeontology 1610
Sowerby, George Brettingham (1812-1884) 95
Sowerby, James de Carle (1787-1871) 578, 1982
    laudatory reference to 1850
    publications, *Mineral Conchology* 1980, 2261
Sowerby, James (1757-1822)
    Mineralogy 147
    English Botany, 2nd Edition 1983
Sowerby, John Edward (1825-1870) 1983
Sowerby 34, 39, 141, 149, 355, 413, 458, 800, 849, 1859, 1980
Sparling, W. 467, 468
Spencer, Lord 1253, 1256, 1258, 1705
Spratt, Thomas Abel Brimage (1811-1888)
    biographical details of 1987
    magnetic observations of 1845, 1987
Spring-Rice, Thomas (1790-1866) [Lord Monteagle] 837, 940, 1217, 1220, 1258, 1876, 2029
    elevation to peerage 1265
Squire, Mr [of Ware, Quaker] 809
Standidge, Mr 275
Stanger, Dr 585
Stanhope, Lady Hester Lucy (1776-1839)
    fossil fish collection of 146
Stanley, Edward (1779-1849) 1062, 1229
    portrait of, in Ipswich Museum 1811
Stanley, Captain Owen (1811-1850)
    expedition of 552, 560, 1062
Stanley, Lord 221
Staunton, Mr
    building stone of, from Limerick 66, 1246
Steer, Mr 179
Stephens, A.J.
    & Royal Society 912
Stephenson, Sir Benjamin (d.1839)
    death of 1256
Stephenson, Robert (1803-1859) 189, 195, 242, 1988, 1989

Stevens, Alfred (1818-1875) 2154
Stewart, Robert 1268
Stewart, Mr
    coal pit of 848
Still, Henry 265, 266, 270, 283, 941, 945, 948, 949, 953, 1392, 1990-1995
Stock, Dr 1038
Stokes, Charles (1783-1853) 156, 163, 182, 342, 457, 578, 1991, 1996, 1997
    & Devon Controversy 1012
Stokes, George Gabriel (1819-1903) 1998
Stokes, Captain John Lort (1812-1885)
    maritime survey of Australia and New Zealand 612
Stokes, Mr [of Hean Castle, Tenby] 2114
Stokes, Mr 2152
Strangways, William Thomas Horner Fox (1795-1865) [4th Earl of Ilchester] 154, 155, 170
    at Paviland Cave 172
    continental travel of 146, 158, 168, 171
    fieldwork in Dorset 153
    geological observations in Turkey 622
    geology examination of 145
    in Russia 159, 160
    on Russian geology 157
    proposed as Geological Society Vice President 152
    publications of 150, 160
    return from Russia 151
    specimens sent to W. Buckland by 148, 151, 152
    studentship at Christ Church 152
    work in Russia 149
Stratford, William Samuel (1791-1853) 669
Strickland, Hugh Edwin (1811-1853) 622, 731, 744, 1443, 1447, 1470
Struvé, William Price 1999, 2000
    & Neath Colliery explosion inquest 1957-1959
    assisting W.E. Logan on S. Wales map 878, 882
    declines post of Keeper of Mining Records 1999
    mine ventilation of 2000
    recommended for post of Keeper of Mining Records 872
Strzelecki, Paul Edmund (1797-1873)
    knighthood 1027
    on geology of Tasmania 444
Stuart, G.O. 2001
Stubbs, Isaac [gamekeeper, Margam] 2010
Stuckley, Mr 2130
Studer, Bernhard Rudolf (1794-1887) 189, 330, 1912
Sturgeon, Mr 952
Stutchbury, Samuel (1798-1859) 109, 903, 918, 1360, 1834, 1859, 1860, 1865-1867, 2002-2004
    membership of Linnean Society 2003
    membership of Royal Academy of Turin 2003
    membership of Royal Geological Society of Cornwall 2003
    membership of Yorkshire Philosophical Society 2003
    wishes to join Geological Society 2002
    suicide of brother 1858
Styan, Harriet 339
Suess, Eduard (1831-1914) 645
Sullivan [of Dublin] 2009
Sumner, John Bird (1780-1862) 154, 302, 306
Sussex, Augustus Frederick (1773-1843), Duke of
    as President, Royal Society 2051
Swinfield, Mr 1240, 1245
Sykes, William Henry (1790-1872) 570, 898, 911, 1054, 2006-2008
Sylvester 1204, 1213, 1216, 1223, 1224
Symonds, Sir William (1782-1856) 661

Talbot, Charlotte Louisa (1800-1880) 167
    health of 172
    marriage of 307
Talbot, Christopher Rice Mansel (1803-1890) 154, 155, 157, 165, 965, 1836, 2010, 2011
    at Oriel College, Oxford 153
    examination of 172
Talbot, Jane Harriet (1796-1874) 145, 146, 148-152, 155, 157, 158, 162
Talbot, Mary Theresa (1795-1861) 147, 157, 158, 162, 166, 167
Talbot, Misses 145, 146, 152, 153
Talbot, Lord 1170

Talbot, William Henry Fox (1800-1877) 1469, 2011
Tankerville, Charles Bennet, 4th Earl of
    collection of 148
Taylor, Hugh [of Earsdon] 322
Taylor, John (1779-1863) 614, 898, 1941, 1954, 2012, 2013
    mining interests in Devon & Cornwall 373
    plan to form Cornish Mining School 302
    suggested as Mineral Surveyor for Duchy of Cornwall 978
Taylor, John Junior 1940, 2012
Taylor, Mrs John 2012
Taylor, Richard (1781-1858) & J.E.[Printers] 186, 1279, 1287, 1336, 1406, 1408, 1450, 1459, 1499, 1630, 2020, 2022
Taylor, Richard Cowling (1789-1851) 2015-2021
    models of 2015, 2016
    plan to work Cuban gold 2016, 2017
    publications 2015
Taylor, W. 2023, 2024
Taylor, Dr William Cooke (1800-1849)
    subscription for family of 2091
Taylor, Mr 1223
Taylor & Walton, Messrs [publishers, Upper Gower Street] 1685
Tcherikoff, Col. 867
Tennant 578
Thennett, R.I. 690
Theobold, W. 2025
Thirlwall, Connop (1797-1875), Bishop of St David's 1020
Thirria, Charles Edouard (1796-1868)
    work on Jurassic criticised 2075
Thomas, Alfred
    map of Pembrokeshire by 1009
Thomas, Henry 2026
Thomas, Horace 121
Thomas, Richard 121
Thomas, Mr [of Welfield, Builth] 342
    trilobites given to De la Beche by 342
Thomas, Mr [of Pencerrig, Builth] 342
Thompson, Mr [shell collector] 800
Thompson, Dr [?editor of Annals of Science] 849
Thompson, William 2027
Thornley, I.D. [Customs, Fowey]
    death of 1160
Thorpe, Archdeacon
    as candidate for Mastership at Trinity College, Cambridge 958
Tindall, Edward [Craig's Court Housekeeper] 1213, 1217, 1222, 1225, 1230, 1231, 1238, 1239, 1264, 1300, 1301
    salary 1312
Tomlinson, Mr 664
Torrens, Henry [Secretary Asiatic Society, Bengal] 1678
Tothill, Mr 740
Traherne, John Montgomery (1788-1860) 456
    marriage of 307
Tranter, Mr 254
Treattel & Wurth [booksellers] 796
Treffry, Austen 1161
Tremenheere, George Borlase (1809-1896) 1050, 1380, 1678, 2028-2035, 2141
    certificate in chemistry 2030, 2031
    character of 1706
    plan for India geological museum 2028
    publications 1678
    with Ramsay in S.W. Wales 1706
Tremenheere, Hugh Seymour (1804-1893) 2034
    appointed Inspector of Schools [in Wales, 1839] 2028
Trevelyan, Sir Charles Edward (1807-1856) 414, 2036-2038, 2040
Trevelyan, Sir John [of Nettlecombe] 1819
Trevelyan, Walter Calverley (1797-1879) 144, 150, 159, 322, 2039-2041, 2275
Trevethan, Mine Captain 1942
Trevithick, Richard (1771-1833) 518
Trimmer, Joshua (1795-1857) 1738, 1739, 1742, 1743, 1746, 1751, 1765, 1770, 1774, 1779, 1786, 2042
    agricultural knowledge of 418
    appointment to Geological Survey 1789
    on diluvial deposits of Caernarvonshire 181
    on erratics 2042
    proposed resignation of, from Geological Survey 418

publications 2042
work on New Forest 418, 1783
Trincoe, Miss 2043
Trist, Mr 949
Troost, Gerard (1776-1850) 2017
on geology of Tennessee 695
Troughton & Simms 1464, 1743, 2282
Tucker, Henry 2044
Tucker, Capt. 1294, 1868
Tuckey 1268
Turle, Miss 339
Turner, Edward (1798-1837) 657, 658, 897, 1228, 1826, 2046-2051
& Devon Controversy 621, 2049, 2050
charges for mineral analysis 2051
resigning as Secretary of Geological Society 2050
Turner, Richard 613
Turner, Mr 644
Turner, Mr [of Sunderland] 172
Turpin, Mr [of Jamaica] 352

Underwood, Mr 493
Upton, Mr [?son of Lord Templeton] 344, 346
Urquhart, [?David (1805-1877)]
poem on death of Mr Ormerod 148

Valenciennes, A. 1182
Van Amburgh 1455
Vansittart, Mr 162
Vaughan, R. 2052
Verneuil, Philippe Edouard Poulletier de (1805-1873) 2053, 2054
Vernon, Robert (1774-1849) 2276
Vernon, William 301
Vernon, Mr [Lord Lincoln's Secretary] 766
Vetch, Captain James (1789-1869) 232, 242, 243, 858, 2055-2068
Vicary, Capt. 1021, 1023, 1038
Vickerman, Miss 451, 465
Victoria, Queen 2069, 2070
visit to Ireland rumoured 1100
visit to Parliament, 1848. 240
Villeneuve, Vte. de 2071
Villiers, Andre-Jean-Francois-Marie Brochant de (1772-1840) 306, 497, 498, 500, 501, 503, 2072, 2073, 2085
Vivian, [?Charles Crespigny (1808-86)]
proposal for a Cornish Mining School 389
Vivian, Henry Hussey (1821-1894) 2277
Vivian, John Henry (1785-1855) 154, 461, 513, 1016, 1639, 1679
Vivian, Richard Hussey (1775-1842) 839, 1211, 2074
Vivian 363
Viviani, Prof. [University of Genoa] 347
Voltz, L. 2075
Voysey, Henry Wesley (c.1791-1824) 97, 1678
Vyvyan, Richard Rawlinson (1800-1879) 737, 2076

Wagner, Rudolf 2077
publications 2077
Wahlberg, P. 2005
Waldegrade, Lord
collieries of, at Radstock 2004
Wake, Mr [comptroller, Stationary Office, Dublin] 1127, 1128
Walker, Charles Vincent (1812-1882) 2078
Walker, William (1791-1867)
engraving of portrait of De la Beche 415, 2238
Walker, William 2079
Walker, Prof. 2020
Walker 590
Waller, Mr [the Engineer] 1242
Wallich, Nathaniel (1786-1854)
drawings of plants from India 164
visit to India 164
Walligen [?] 582
Wallin [?] 587
Walters, D. 109-111
Walton, ?William [of 17 Grosvenor Place, Bath]
collection of 1503
offer to purchase collection of T. Cooke 748
Warburton, Henry (1784-1858) 542-544, 623, 624
defence of De la Beche in House of Commons 624

Ward, Captain 97
Ward, Mr
specimens from, to Museum of Economic Geology 1236
Waring, Henry Franks (1806-1874) 79, 410, 2278
Warren, Prof J.C. [MD of 176, Regent Street, London] 692
Washbourne, Dr Oakley 301
Waterhouse, George Robert (1810-1888) 578
Watson, [sculptor of bust of De la Beche; of Penrhyn] 939, 1224
Watson, Musgrave Lewthwaite (1804-1847) 1204
tomb of (illn.) 2080
Watt, James (1736-1819) 850
Way, Albert (1805-1874) 2081-2084
Weale, Mr 693
Weaver, Thomas (1773-1855) 652, 902, 1936, 2085-2090
& Devon Controversy 1013, 2098
presentation of early geological books to Museum of Economic Geology 2090
publications of 152, 621, 2086, 2087
suppression of paper by Geological Society 1873
visit to Paris 2085
Webb, Philip Barker (1793-1854) 170
map of Canary Islands 501
Webber, John (?1750-1793) [draughtsman to Captain Cook] 348
Webster, Thomas (1772-1844) 145, 175, 473, 496, 531, 575
criticism of, by Fitton 534
publications 2262
on freshwater formations of Isle of Wight 534
drawing plesiosaur 302
Webster, Mrs 156
Weld, Charles Richard (1813-1869) 412
Wellington, Capt. 2067
Wellington, Arthur Wellesley, 1st Duke of (1769-1852)
funeral of 585
Werner, Abraham Gottlob (1749-1817) 146, 306, 897, 1434
West, T. 2091
Westall, William (1781-1850) 1897
Westminster, Lady 2092
Westminster, Lord 1303
Weston, Charles Henry 2093
publications, on geology of Ridgway QJGS (1848) 1043, 1890, 2093
Weyman, William 121
Whately, Richard (1787-1863)
as author of epitaph, "Mourn, Ammonites, Mourn" 155, 1847
Wheatstone, Charles (1802-1875) 195, 1469
Wheatstone [?Daniel] 1244, 1262, 1267
Whewell, William (1794-1866) 280, 387, 496, 497, 623, 624, 718, 955-957, 1563, 2098
as candidate for Mastership at Trinity College, Cambridge 958
proposed as President, Geological Society 901
tidal theory of 952
White, Mr 1222
specimens from, for Museum of Economic Geology 1225
White, I.B. & Son 2197
Whitehurst 309
Whittacre, Dr 152
Whitworth, Joseph 2094
Whyte, C.J. 2095
Whyte, Letitia [Mrs H.T. De la Beche] (1801-1866) 342, 344, 346
birth certificate of daughter 338
Wickenden, Joseph 2096
[?]Wicksteed [the Engineer] 513-515
Wilberforce, William (1759-1833) 113, 302
Wilcox, Dr C.
Purbeck fish of 571
Wiley & Putnam 2019
Wilkie, Mr 173
Wilkin, Mr [Cornish miner in India] 1678
Wilkinson, Charles Hunnings (1763-1850)
collection of 342
ichthyosaur in possession of 342
Wilks, Mr 309
William [assistant on Geological Survey] 1752, 1760, 1799
health of 1796
Williams, Rev. David (1792-1850) 1342, 1348, 2097-2100
& Devon Controversy 35, 952, 1875, 1880
fossil collection 624, 1332, 1334, 2041, 2099, 2100
fossil collection offered to Geological Society 2150

Williams, David Hiram (d.1848) 561, 562, 873, 1115, 1383, 1395, 1400, 1410, 1443, 1446, 1450, 1451, 1453, 1455, 1457, 1458, 1503, 1507, 1530, 1609, 1688, 1718, 1719, 1730, 1815, 1821, 1859, 1861, 1865, 1950, 2006, 2030, 2101-2148
   accounts of 1460
   appointment in India accepted by 2136
   brother 2119, 2120
   death of 446, 2025
   death of daughter 2142
   loss of field maps by 2104
   need for assistant on Indian Survey 2138, 2141
   request to East India Company for assistance for widow of 416
   search for coal in India by 416, 445, 446, 479, 2025, 2136-2141, 2143-2148
   salary and expenses of 479, 1116, 2135
   terms of engagement by East India Company 479
Williams, Lt. [of Ordnance Survey] 671, 2105
Williams Deacons & Co 1264
Williams, Fanny 81
Williams, H.F. 2149
Williams, The Hon. John 359
Williams, Paul 517, 520
Williams, Wadham P. 2150
Williams, Wilb. T. 2151
Williams, Dr 146, 158, 1760
Williams, Mr [connoisseur of Plymouth China] 991
Williams, Col. 864, 866, 867
Williams [collector on Geological Survey] 1405, 1436
Williams, Miss 1712
Williams family Llandovery
   (see also Ramsay) 1852
Williamson, Sir G. 707
Willis, Robert (1800-1875) 1871, 1998, 2152, 2153
Willson, Walter Lindsay (d.1878) 756, 761, 770, 772, 776, 814, 1066, 1084, 1086, 1102, 1105, 1107, 1110, 1123, 1124, 1133
Wilson, Charles Heath (1809-1882) 2154, 2155
Wilson, John 2156
Wilson, Mr [?Professor] 530
Wilson, Lady (Miss Jane Weller)
   collection of 159
Wilson, T. [?potter of Leeds] 1658
Wilson, Mr
   British Association Railway Section Committee 189
Wilson (the Pedestrian)
   at Weymouth 341
Wilson, Mr [Museum of Economic Geology laboratory scientific assistant] 1690
Wilton, Edward 2157-2159
Winfield, R.W.
   materials for Great Exhibition, 1851. 78
Winstanley, T.W. [?hotelier, Manchester] 1440
Wise, Mr 993
Witcomb, Mr 1533
Witham, Rev. Thomas [of Larbington Hall] 322
Witte, Madame
   model of moon in possession of 679

Wollaston, William Hyde (1766-1828) 165, 167, 171, 302, 356, 850
   on analysis of limestone 332
   on origin of red snow 149
   on vision 302
Wood, C. 2160
Wood, Enoch & Son
   potteries of 461
Wood, Nicholas [of Hetton Hall] 322, 938, 1649, 1652
Wood, Searles Valentine (1798-1880) 95
Wood, V. 292, 1265, 2161
Wood, Mr 1976
Woodcock, Mr 169, 170
Woodhouse, John Thomas (1809-1878) 616
   colliery plans by 1973
Woods, Albert W. 2162
Woods[?] 2163
Woodward, Samuel P. (1821-1865) 1443, 2150
Woodward 578, 1892, 2150
Woolcott [casemaker, Museum of Economic Geology] 1225, 1226, 1236, 1239
Wordsworth, Dr Christopher (1774-1846)
   resignation as Master of Trinity College, Cambridge 958
Worsley, Thomas (1797-1885) 1871
Wortley, Mr 1241
Wreford, William 1160
Wright, Romley 310
Wright, Mr
   shell collection of 800
Wrightson, Mr [Museum of Economic Geology laboratory scientific assistant] 1690
Wrottesley, John, 2nd Baron (1798-1867) 572, 1056
Wyley, Andrew (c.1820-1885) 761, 766, 767, 769, 770, 772, 776, 814, 815, 1071, 1095, 1128, 1133, 1931, 1935-1937
   criticism of 1102, 1109
   health of 1107
   laudatory reference to 1932, 1934
   mapping of, criticised 1097, 1109
Wyon, Leonard Charles (1826-1891) 2164
Wyon, William (1795-1851) 1825, 2165
   medal for De la Beche by 688, 1401

Yarrel, Mr 946
Yates, James [of London] 1393
Yates, Rev. James (1789-1871) 898
Yates, Mr 337
   & Devon Controversy 621
   paper on Diluvium 797
Yeates 1088
Yolland, Capt. William (1810-1885) 239, 287, 922, 1449, 1786, 1788, 1790, 2167-2172
Young, Rev. George (1777-1848)
   publications, *A geological survey of the Yorkshire Coast* (1822) 1849

Zurlaufen, General 348

PLACE NAMES

Aachen [Aix-la-Chapelle], Germany
    description of 349
    Cathedral 349
    mineral springs at 2046
Aarberg, Switzerland
    fossil tortoises from 348
Aare, River, Switzerland 348
Abberley Hills, Worcestershire 1452, 1540, 1548, 1553, 1554, 1557, 1558, 1563, 1571, 1573, 1578, 1618
    Memoir 1578-1581, 1611, 1631
    trilobite from 1536
    sketch geology map and section of (illn.) 1577
    Welsh placenames near 1617
Abbeville, France 343, 346
    Cathedral 343, 346
Abbotsbury, Dorset 151-153, 155, 157
    tides at 341
Abbotskerswell, Devon, geology of 341
Abbots Leigh, Somerset
    Leigh Woods 342
    geology of 342
    section to Cooks Folly (illn.) 342
Aber, Caernarvonshire 1793, 1796
Aberaeron, Cardiganshire 1852
Abercarn, Monmouthshire 87
    Sir Benjamin Hall's mine at 201
Aberdare, Glamorgan 2101
    explosion at Lletty Shenkin (1849) 98
Aberdaron, Caernarvonshire 1786
Aberdeen, Aberdeenshire 95
Aberdeenshire
    Aberdeen 95
    Peterhead 1163, 1164
Abereiddi Bay, Pembrokeshire
    fossils from 950, 951
    geology of 351, 948, 949, 1707
    slate quarries 1707
Abergarw, Glamorgan 2117
Abergavenny, Monmouthshire 342, 1721, 1722
    view near (illn.) 342
Abergele, Denbighshire 725, 1942
Aberglasney, Carmarthenshire 1623
Abergoleu, Carmarthenshire 1426
Abergorlech, Carmarthenshire 1720
Abergwili, Carmarthenshire 204, 394, 1400, 1424, 1431, 1666
    Pant-y-Glien Quarry 626
    fossils from 1425
    geology of 1401, 1435
Abergwenlais, Carmarthenshire 1426
Abermarlais, Carmarthenshire 1426
Abermawr, Pembrokeshire 1709
Abernant, Carmarthenshire 1397, 1426
    section (illn.) 1394
    geology of 1429, 1439
Aberpergwm, Glamorgan 2101
Abersychan, Monmouthshire
    Coal Measures at 2107
Aberystwyth, Cardiganshire 877, 1564, 1723, 1746, 1791, 1852, 1940, 1946-1949
    graptolites 1727
Accrington, Lancashire 953-958
Acquapendente, Italy
    geology of 363
Acton, London
    Heathfield House 1835
Admiralty 20-23, 69, 70, 72, 659, 662, 663, 665, 1845, 2177
Adriatic, shells of 458
Aiguebelle, France
    geology of 347

Ailsa Craig, Ayrshire, curling stones from 1032
Aix-la-Chapelle, Germany see Aachen
Aix les Bains, France
    geology of 347
    mineral springs at 347
Albany, United States 883, 2019
    Institute 2016
Aldeburgh, Suffolk 2041
Alderley, Cheshire 2083
Aldershot, Hampshire
    establishment of barracks at 206
Alessandria, Italy 347
Alfrick, Worcestershire 1540
Allahbad, India 2138
Allenheads, Northumberland 1977, 1978
Almondsbury, Gloucestershire 1867
Alnwick Castle, Northumberland 150, 1061
Alport, Derbyshire 1972
Alps 344-347, 1912
    red snow 148
    W. Buckland on limestones of 156, 157
    fossils of 156
    fold, Savoie Alps (illn.) 347
    limestones of 156, 157, 364
    geology of 491, 640, 1022
    R. I. Murchison on geology of 1026
Alstonefield, Staffordshire 608, 1830-1832
Alston Moor, Cumberland
    lead mine 1965, 1966
Alum Bay, Isle of Wight 582
    coloured sands of 581
    sellers of coloured sands at 581
Alverdiscott, Devon
    lead mine near 624
Alveston, Gloucestershire, Ship Inn 1860
Ameglia, Italy
    geology of 363
Amherst, United States 690-692
Amlwch, Anglesey 1805, 1960, 1961
Amsterdam, Netherlands 349
    description of 349
    Naval Dockyard 349
Ancona, Italy 363
    geology of 363
Andeer, Switzerland 363
Andernach, Germany 349
    millstones from 349
Anglesey, 1777, 1787, 1798-1800, 1805, 1939
    Amlwch 1805, 1960, 1961
    Beaumaris 2042
    Bryngwran 1964, 1965
    Carmel Head 1964
    Dulas 1960
    Four Mile Bridge 1963
    Holyhead 1128, 1522, 1523, 1794, 1939, 1963
    Llanbabo 1960
    Llanerchymedd 1805, 1961, 1962
    Llanfachraeth 1964
    Llanfairynghornwy 1962
    Llanfechell 1963-1965
    Llangefni 1964, 1965
    Parys Mountain 1960, 1961
    Porth Swtan 1964
    geology of 1961-1965
    sketch section (illn.) 1963
Annecy, France 347
    geology of 344, 346
Antarctic 702
Antigua 352, 353

Antilles 202, 304
Antrim, Ireland
    Chalk erratics from 2042
Antwerpen, Belgium 349
    Cathedral 349
    Museum 349
    description of 349
Aosta, Italy
    De la Beche on town and people of 344, 346
    Roman remains at 344, 346
Appalachian Mountains, United States
    geology of 883, 1026
Appennines, Italy 1025
    geology of 332, 347, 348, 363
Apperley, Gloucestershire
    geology of 1470
Ararat, Mount, Turkey 865, 867
Arctic, birds of 148
    exploration of 148, 302
Ardington, Oxfordshire
    Church 2276
Arendal, Norway
    analysis of feldspar from 17
Arenig, Merionethshire 1752
Argenteuil, France
    fossil elephant from 343, 346
Argyllshire
    Jura 70
    Staffa 1680
Arklow, Ireland 762, 1128, 1931-1934
    erratics 2042
    geology of 1096
Armagh, Ireland
    fossil fish from 558, 1079, 1080
Arno, River 347, 363
    fossil hippopotamus from valley of 343
    fossil vertebrates from valley of 363
Arquata Scrivia, Italy
    geology of 363
Arran, Bute 541, 1721, 1725
    Blackland Point 1858
    Brodick 1858
    geology of 1858
    model of, for Museum of Economic Geology 1707
Arromanches, France
    geology of 350
    section (illn.) 350
Arthurstown, Ireland 755, 1124
Arve, River, France 344, 346
Ashbourne, Derbyshire 1596-1600, 1609, 1640, 1831, 1832
    Green Man 1595
    geology of 1971
    sketch geology map of 1640
Ashburton, Devon 29-31, 1332, 1336, 1337, 1681
    fossils from 1338
Ashby de la Zouch, Leicestershire 1635, 1636, 1639
Ashford, Derbyshire 647, 1587, 1624, 1625, 1830
    marble works at 982
Ashford, Ireland 1071, 1097
Ashleworth, Gloucestershire 1500
Ashley, Wiltshire
    geology of 1550, 1552
Asia Minor 1922
    Strickland and Hamilton in 622
Asti, Italy
    mammoth teeth from 347
Aston Ingham, Gloucestershire
    sketch geology map of (illn.) 1575, 1576
Atherfield, Isle of Wight 539
Atlantic Ocean
    crossing by De la Beche 352, 353
    speed of crossing by W. E. Logan 874
Auerbach, Germany
    geology of 960
Aurangabad, India 97
Australia 674, 706, 732, 1029

Bathurst 2178
Sydney 977
Tasmania 444, 712
New South Wales Geological Survey 124-127, 1787-1789
seeds from 157, 158
mineral wealth of 444
fossils from 554
bone cave fossils from 796, 797
maritime survey of 612
coal in 612, 674
gold 646, 2178
lectures on gold for emigrants to 646
J.B. Jukes' paper on geology of 811
shell specimens for Museum of Practical Geology 1062
Austria
    Hallstatt 1022
    Graz 1022
    Innsbruck 1022
    Linz 1927
    Mauthausen 1927
    Salzburg 156, 1022
    Steiermark 146
    Tyrol 141, 153, 156, 172, 348
    Vienna 146, 153, 363, 639-645, 1022, 1282, 1834, 1927, 1928, 1949
    maps of 363, 643, 645, 1022
    proposal for geological survey of 640, 643, 1928
    Geological Institute formed 1969
    1848 rebellion 642
    Imperial Geological Institute 644
    Geological Survey of 644, 645, 739
Auvergne, France 309
Auvernier, Switzerland 347
Auxerre, France 146
    Cathedral 343
    fossils from 343
Avallon, France 343, 346
Avoca, Ireland 1934, 1936, 1937, 1939
    Vale of 1096
    Mines 1077, 1932, 1935
Avon, River 1550
    Carboniferous Limestone fossils of 1372
    valley of 308
Awre, Gloucestershire
    geology of 1502
Axe, River 3
Axminster, Devon 43, 151, 163, 164, 179, 188, 311, 312, 341
Axmouth, Devon 284, 410
    Head, geology of 341
Aymestrey, Herefordshire 119, 1354, 1468
Ayrshire
    Ailsa Craig 1032
    Girvan 1032
    Saltcoats 541

Babbacombe, Devon
    De la Beche's MS on geology of 360
    geology of 621, 1338
    Marble 341
    trilobites from 2086
Babylon, Iraq 865-867
Bacharach, Germany 349
Bad Kissingen, Germany 905
Badminton, Gloucestershire 1515, 1516
Baffin Bay, Canada 148
    fossil plants from 862
    origin of red snow at 148, 343
Baghdad, Iraq 865-867
    deaths from fever (1849) 865
Baggy Point, Devon 1336
    fossils 1340, 1346
    geology of 31, 1337
Bagni di Lucca, Italy
    geology of 363
    section (illn.) 363
    warm spring at 363

Bagshot Heath, Surrey 308
Bahamas, coral reef specimen from 987
Bakewell, Derbyshire 63, 159, 982, 1587, 1648, 1833, 1972
    Institute 62, 1832
    mineral veins near 1582
    Rutland Arms Inn 15827
Bakhtiari, Iran 867
Bala, Merionethshire 554, 555, 563, 1376, 1432, 1449, 1461, 1471,
    1522, 1524, 1525, 1739, 1743, 1745, 1746, 1750
    difficulties of mapping around 1748, 1752
    fossils of 555, 609, 1371, 1375, 1492, 1524
    fossils of, comparison with Ireland 557
    geology of 1752, 1753
Balbriggan, Ireland
    geology of 1107
Balleroy, France
    geology of 350
Ballinagar, Ireland 1097
Ballycotton Bay, Ireland
    comparison of rocks of, with Cornwall & S Wales 779
Ballymanus Bridge, Ireland 1097
Baltic Sea 157
    Silurian of islands of 1018
Baltinglass, Ireland 1066, 1086
    geology of 1086
Bampton, Devon 1876
    geology of 1340
Banc-y-felin, Carmarthenshire 1382
Baneh, Iran 867
Bangor, Caernarvonshire 802, 1522, 1523, 1784, 1789, 1793, 1802
Banwell, Somerset
    trial for coal near 2128
Barcelona, Spain
    coloured sketch section of 1700
    geology of 1700
Bareuth, Germany see Baruth
Barlow, Derbyshire 411
Barmouth, Merionethshire 1775, 1949
Barnsley, Yorkshire
    colliery explosion investigation (1847 or 6) 2044
    mining disaster, (1849) 76
Barnstaple, Devon 25, 657, 1336, 1337, 1681, 1875, 1876
    fossils 1332
    geology of 1877
Barristown, Ireland 772
Baruth, Germany
    bone caves near 146
Basel, Switzerland 348
    Cathedral 348
    description of 348
    Library 348
Basingstoke, Hampshire 744
Baslow, Derbyshire 1606, 1613, 1619, 1621-1625, 1654
    sketch, fault in limestone near (illn.) 1585
    Wheatsheaf Inn 1620, 1641
Bath, Somerset 91, 109, 110, 155, 157, 307, 458, 748, 1008, 1020,
    1033, 1166, 1409, 1515, 1552, 1554, 1556, 1558, 1699, 1884,
    2093
    Claverton Manor 1180
    coach to Bristol 342
    collections in 1503
    De la Beche's comments on 342
    fossils from 342
    geological mapping of 1512, 1525, 1528-1530, 1552, 1555
    ichthyosaurs from 342
    Montague House 1166
    plan for natural history museum, botanic garden, and Professorship of
        Natural History at 342
    school 1751
    waters of 246
    York Hotel 1883
Batheaston, Somerset 246, 655, 1980
Bathford, Somerset
    geology of 1550, 1552, 1566
Bathurst, Australia 2178
Bathurst, Canada 883

Battersea, London 106, 107
Bavaria, Germany 488
Bayeux, France 169, 350
    gravel pits near 350
    quarries near 350
Beachy Head, Sussex
    geology of 511, 512
Beaminster, Dorset 179
Beaufort, France 344, 346
Beaulieu-fawr, Carmarthenshire 1401
Beaumaris, Anglesey
    erratics 2042
Beaune, France 343, 346
Beauvais, France 343, 346
Beckington, Somerset 961
Beddgelert, Caernarvonshire 1775, 1796
    mines near 1968
Bedfordshire
    Woburn 1840
Bedwas, Monmouthshire 2102
Beer, Devon
    Head 341
    Head (illn.) 433
    section (illn.) 433
Begelly, Pembrokeshire
    geology of 351
    J.M. Child's collieries at 351
Belfast, Ireland 1696, 2027
    Lough 306
    Natural History and Philosophy Society 2027
    Queen's College 139, 1111
Belgium
    Antwerpen 349
    Brussels 146, 349, 853
    Gent 349
    Leuven 349
    Liège 146, 244
    Mechelen 349
    Namur 146
    Ostend 343, 488
    Spa 250
    St Truiden 349
    Tienen 349
    Tongeren 349
    geology of 349, 1487
    geology of, compared with S W England 1181
    King Leopold of 853
    Tertiary fossils compared with Isle of Wight 591
Bellagio, Italy 363
    section, Bellagio Point to Monte San Primo (illn.) 363
Bellano, Italy
    geology of 363
    sections near (illn.) 363
Bellegarde, France 344, 346
    Perte du Rhone 344, 346
Belper, Derbyshire 1600
Bembridge, Isle of Wight
    geology of 584, 585, 590
Bengal, India 2138, 2143, 2146
    Asiatic Society of 1678, 2141
    Geological & Mineralogical Museum of Asiatic Society of 1678,
        2141
Benson, Oxfordshire 342
Benton, Pembrokeshire
    Castle 351
    trap rocks of 351
Berkeley, Gloucestershire 1469, 1512, 1513, 1515
    canal 1550
Berkshire
    Maidenhead 150, 308, 342
    Reading 308, 744
    Salthill 342
    Sandhurst 41
    Slough 744
    Windsor 744, 1294
    Windsor Castle 19, 219, 630, 633, 971, 1673, 1674, 2057-2066
Berlin, Germany 141, 146, 1869

Museum of Art 924
Royal Academy 201
Bern, Switzerland 146, 348, 1866
   description of 348
   geology of 1028
   natural history dealer of 202
   Natural History Museum 348
Berry Head, Devon
   geology of 341
Berwickshire
   Cockburnspath 792
   map of 792
Berwyn Mountains 1432, 1461, 1492
   geology of 1371, 1524
Bethesda, Caernarvonshire 1796, 1798-1801
   Penrhyn Quarry, erratics in 2042
Betsy, Wheal, Devon 373
Betws-y-Coed, Caernarvonshire 1784, 1799, 1968
Bex, Switzerland
   geology of 344, 346
   gypsum works at 344, 346
Biarritz, France
   nummulitic limestone 8
Bickington, Devon
   fossils from 1346
   geology of 1338
Bideford, Devon 34, 378, 621, 1013, 1336, 1337, 1681
   coal plants from 1331, 1877
   Culm mines near 832
   goniatites from 1859
Bincombe, Dorset 1764
Bingen, Germany 349
Bingham, Nottinghamshire 1636
Birkenhead, Cheshire 922
Birmingham, Warwickshire 78, 291, 334, 698, 743, 744, 812, 975,
      1185-1197, 1345, 1499, 1688, 1691, 1943, 2034
   British Association meeting (1839) 35, 36, 1261, 1264, 1334, 1696
   British Association meeting (1849) 464, 812, 1030, 1109, 1192,
      1629, 1785
   bronze manufacturing 1185
   Committee of Birmingham Peel Testimonial 698
   Edgbaston 1134
   glass works 1185, 1186, 1188
   Museum 701, 1356
   Philosophical Institution 2096
   statue of Sir Robert Peel 698, 1197
Birr, Ireland 1522, 1523
Bishops and Clerks, Pembrokeshire
   geology of 948
Bishop's Castle, Shropshire 607. 1738, 1739, 1771
   geology of 1735, 1736, 1769
Bishop's Lydeard, Somerset 2
Bishopsteignton, Devon
   geology of 36, 341
Bitlis, Turkey 867
Blackburn, Lancashire
   levelling of 922
Blackdown Hills, Devon 28, 543, 544, 961
   belemnites 1331
   fossils of 458, 542
Black Forest, Germany see Schwarzwald
Black Head, Cornwall 944
   graptolite from 1163
Blackheath, London
   H.M. De La Condamine's geological map of 295
Blackmoor Vale 3
Black Mountain, Carmarthenshire 1721
Black Ven, Dorset 341
Blaendyffryn, Carmarthenshire
   geology of 1437, 1438
Blaina, Monmouthshire
   Coal Measures at 2108
   Cwm-celyn 2108, 2127
Blanc, Mont 343-348
   De la Beche's expedition to 344, 346
   geology of 344, 346

   Glacier de Bionnassay 344, 346
   Glacier de Boisson (illn.) 344, 346
   new route of ascent 344, 346
   relief model of 499
Bleadon, Somerset 2041, 2097-2100
   Rectory 2150
Blenheim Palace, Oxfordshire
   description of 342
Blue Anchor, Somerset
   "Alabaster Cliff" at 1819
   geology of 1819
   Rhaetic Bone Bed at 179
Boconnoc, Cornwall 946
Bodinnick, Cornwall
   fish beds at 1161
Bodmin, Cornwall 943
Bohemia, Germany 146, 645
Bolca, Monte, Italy 343
   fossil fish remains 146
Bologna, Italy 82, 83, 84, 348
   University of 1975
Bolsena, Lago di, Italy
   geology of 363
Bolsover, Derbyshire 1624
   building stone from 1246, 1255
Bolt Head, Devon 1876
Bolton, Lancashire
   Coal Measures plants from 890
   fossil tree in situ from 1612
Bolton Abbey, Yorkshire
   seeds from 162
Bolzano, Italy 1022
Bombay, India 97, 865
Bonassola, Italy
   geology of 363
Bondy, France
   fossil elephant from 343, 346
Bonn, Germany 146, 250, 337, 349, 678
   University of 932
Bonneville, France 344-346
Bonsall, Derbyshire 1655
Boppard, Germany 349
Bordeaux, France
   caves near 176
Borghetto di Vara, Italy
   geology of 363
Bormio, Italy 1022
Borneo 992
   Captain Bethune's expedition to 542
   coal from 1315
   J.D. Hooker's visit to 706, 713
Boscastle, Cornwall 1875
Boston, United States 8, 690, 879, 880, 1826, 1827
   Lyell lecturing in 1709
Bottisham, Cambridgeshire 1895
Boulogne, France 38, 343, 346, 845, 1334
   description of 343, 346
Boulston, Pembrokeshire
   geology of 351
Bournemouth, Dorset 561
Bovey Tracey, Devon
   tourmaline from 147
Box, Wiltshire 1558
   geology of 1550-1552, 1557
Bracciano, Lago di, Italy 363
Braco, Italy
   section near (illn.) 363
Bradford, Yorkshire 528
Bradford-on-Avon, Wiltshire 1166, 1503, 1552
Brading, Isle of Wight 585
   Harbour 583, 587
Brampton Bryan, Herefordshire
   geology of 118
Bran, Afon 1430, 1437
Branscombe, Devon
   cliffs (illn.) 341

geology of 341
    section (illn.) 433
Brassington, Derbyshire 608, 1832
Brasted, Kent 533
Bratislava, Czechoslovakia 146
Braunton, Devon 1336
Brawdy, Pembrokeshire
    geology of 351
Bray, Ireland 1087
Brechfa, Carmarthenshire 1383, 1397, 1426, 1717, 1720
    geology of 1385, 1394, 1401, 1712, 1714, 1715
    sketch section (illn.) 1714
Brecon, Breconshire 1360, 1426, 1442, 1445, 1449, 1721
    anticline N of 1421
    Beacons 1413, 1722
    Beacons, geology of 342
    Beacons, tilestones 342, 1348
    canal 342
    copper mine near 342
    fossils 1348
    geological survey sheet 1409-1411
    Old Red Sandstone 1481
    Ordnance map of 1388
    section to Newport (illn.) 342
    sketch map (illn.) 1410
Breconshire
    Brecon 342, 1348, 1360, 1388, 1409-1411, 1413, 1421, 1426,
        1442, 1445, 1449, 1481, 1721, 1722
    Llanwrtyd Wells 342, 457, 1724, 1725, 1852
    Trecastle 1400, 1405, 1407, 1408, 1426, 1721, 1722
    Ynyscedwyn 455
    Ystradgynlais 65
Breda, Netherlands 349
Bredon Hill, Worcestershire 1434, 1452
    geology of 731
Breidden Hills, Montgomeryshire 1950
Brendon Hills, Somerset 2041
Brentor, Devon 25, 1875
Breslau see Wroclaw, Poland
Brevent, Mont, France
    De la Beche's ascent of 344, 346
    geology of 344, 346
Bricquebec, France
    geology of 350
Bridgend, Glamorgan 80, 102, 872, 952, 965, 2010, 2011
    lead in Lias of 176
    mineral veins around 7
Bridgnorth, Shropshire 2088
Bridgwater, Somerset 284, 315, 2041
    Canal 2126
    coal imports 2126
    Dock Company 2126
    Lias of 176
Bridgwater Level, Somerset 3
Bridlington, Yorkshire 1642
Bridport, Dorset 123, 156
    belemnites 1331
    geology of 341
Brienzer See, Switzerland 348
    geology of 348
Brighton, Sussex 470, 748, 1027, 1281, 1303, 1770, 1818
    Royal Crescent 1817
Bristol 113, 151, 168, 175, 191, 246, 296, 297, 299-302, 320, 463,
    505, 573, 623, 627, 628, 700, 1020, 1272, 1273, 1335, 1336,
    1358-1360, 1368, 1370, 1371, 1398, 1401, 1445, 1506, 1519,
    1680, 1681, 1834, 1858-1867
    Brandon Hill 351
    brass manufacture 2035
    Brislington 163, 303-305, 307, 354-358, 1861
    British Association meeting (1836) 35, 266, 378, 622, 658, 902,
        1015, 1875
    building stones 342
    Clifton 156, 169, 170, 246, 298, 332, 342, 351-353, 531, 627, 628,
        1372-1374, 1409, 1849
    Clifton, geology of 341, 342, 655
    coach to Bath 342

fever cases in 627, 628
fossil fish of, compared with Ireland 558
fossil insects in Cotham Marble from 1610
geological map of 1860-1865
geological map of coalfield 1308, 1310
Geological Survey in 2121-2129
Geological Survey sheet 1411
geology around 903
George & Railway Hotel 742
Hotwells 342, 794
Infirmary 627, 628
Institution 109, 301-305, 342, 355, 357, 977, 1389, 2002-2004
Metropolitan Sanitary Commission 1684-1686
Mr Hennett's (Brunel's) Yard 919
Observatory 342, 1866
Ordnance Survey map of 302, 1866
Pennant Quarry at Fishponds 742
Philosophical Society 301-304
Redland 2121-2123
Sander's plan for geological map of 1858
Seamills 342
section, Cooks Folly to Abbotsleigh (illn.) 342
Stapleton, geology of 342, 1861
steamer 1369
steamers to Cardiff from 455
steamers to Swansea from 455, 463
streets. courts and alleys of 2258
to Gloucester railway 742, 747, 748
Triassic vertebrates from 903, 918
White Lion 175
Bristol Channel 342, 352, 353, 696, 1819
    tides of 2079
Britanny, France
    druidical remains in 302
    geology of 1682, 1683
Briton Ferry, Glamorgan
    geology of 879
Brixton, London 715
Broad Haven, Pembrokeshire
    geology of 351, 2109
Broadstairs, Kent 343
Brockenhurst, Hampshire 1151
Bromley, London 295
Brompton, London 610, 1005
Brook, Isle of Wight 581
Bruchsal, Germany 349
Brue, River 3, 794
Brushford, Devon
    fossils of 1346
Brussels, Belgium 146, 349, 853
Bruton, Somerset
    geology of 1550
Brynberian, Pembrokeshire 954
Bryngwran, Anglesey 1965
    Madoc Inn 1964
Buallt, Carmarthenshire 1426, 1431, 1432, 1437, 1439, 1440, 1663
    fossils from 1433
Buckfastleigh, Devon 29
Buckinghamshire
    Dropmore House 161, 172
    Great Marlow 342, 1819
Budapest, Hungary 1927
Budleigh Salterton, Devon
    geology of 341
Buet, Mont, France 344-346
Builth Wells, Radnorshire 121, 342, 401, 1413, 1449, 1562, 1662, 1726,
    1729, 1744, 1758-1761, 1763, 1768, 1770, 1771, 1851, 1852,
    1945, 1946
    Carneddau, geology of 342, 1762
    De la Beche's comments on 342
    fossils from 342
    section, Carneddau (illn.) 342
    trap rock at 342
    trilobites from 336, 342
    trilobites from (illn.) 431
Bulley, Gloucestershire 1470, 1471, 1478, 1500

Bullslaughter Bay, Pembrokeshire
  geology of 351
Bunny, Nottinghamshire 1636
Bures, Essex/Suffolk
  fossils from 471, 472
Burford, Oxfordshire 342
Burgundy, France
  wine from 343, 346
Burma
  Maulamyaing 2035
  Mergui 2035
  Tenasserim 1678, 2035
Burrington, Somerset
  cave deposits at 42
Burton, Pembrokeshire
  geology of 351
Burton-upon-Trent, Staffordshire 1634, 1635
Bute
  Arran 541, 1707, 1721, 1725, 1858
Buttermere, Cumberland 1870
Buxted, Sussex 64
Buxton, Derbyshire 1599-1603, 1625, 1830, 1832
  [?Ashwood Dale] (illn.) 1165
  baths 1601
Bwlch Capel, Carmarthenshire 1394, 1400
  graptolites from slate quarry at 1425
Bwlch Gwyn lead mine, Cardiganshire 1941

Cader Idris, Merionethshire
  geology of 1727, 1731
Cadgwith, Cornwall
  coloured section near (illn.) 1990
Caen, France 350
  crocodile/plesiosaur from 163
  Museum 350
  Stone 350, 610
Caeo, Carmarthenshire 1403
Caerau, Carmarthenshire 1426
Caer Bwdy Bay, Pembrokeshire
  geology of 351
Caer Caradoc, Shropshire 1744
Caerleon, Monmouthshire 1358, 1389
Caernarfon, Caernarvonshire 1794
  J. Trimmer's paper on Diluvium of 181
  sketch section, Cardiff to Caernarfon (illn.) 1492
Caernarvonshire 449
  Aber 1793, 1796
  Aberdaron 1786
  Bangor 802, 1522, 1523, 1784, 1789, 1793, 1802
  Beddgelert 1775, 1796, 1968
  Bethesda 1796, 1798-1801, 2042
  Betws-y-Coed 1784, 1799, 1968
  Caernarfon 181, 1492, 1794
  Capel Curig 1767, 1785-1793, 1796
  Carnedd Llewelyn 1789
  Clynnog-fawr 1789
  Conwy 725
  Cwellyn, Llyn 1777, 1778
  Dinorwic 1781
  Dolwyddelan 1796
  Glyder Fawr 1777, 1784
  Llanberis 562, 1775, 1776-1784, 1792, 1795-1797
  Llanllyfni 1968
  Llŷn 1791, 1792
  Menai Strait 181, 496, 593, 1492
  Moel Siabod 1796
  Moel Tryfan 181, 496, 2042
  Mynydd Mawr 1791
  Nant Ffrancon 1779, 1780
  Penmachno 1968
  Penrhyn du 1791
  Pentir 1781
  Peris, Llyn 1791
  Porthmadog 1775, 1776
  Pwllheli 1767, 1768, 1786, 1791
  Snowdon 181, 436, 559, 1461, 1777, 1778, 1781, 1792, 1795, 1796, 1968
  Trefriw 1789, 1793
  Tremadog 1610, 1611, 1778
  Ty'n y Groes 1775
  Ysbyty Ifan 811
Caerphilly, Glamorgan 342
  Castle, description of 342
  Castle (illn.) 342
  coal crop workings 342
  De la Beche's comments on 342
  Mountain 342
Cagliari, Sardinia
  R. Wagner's paper on bone breccia at 2077
Cahore Point, Ireland
  geology of 1084
Cairo, Egypt
  geology of 707
Caistor, Lincolnshire
  frit from 676
Calais, France 146
  description of 343, 346
  lighthouse 343
Calcutta, India 164, 709, 710, 963, 1836, 2025, 2032, 2034, 2035, 2137, 2140, 2141, 2146
  geology of 709
  Museum of Economic Geology 1678, 2028-2035
  Oriental Bank 2146
  Spence's Hotel 2138, 2147
Caldey Island, Pembrokeshire 1343
  fossils of 1372
  geology of 351, 1344, 1358, 1363, 1583, 1854-1856
Caledfwlch, Carmarthenshire 1426
California, United States
  mercury mines 13
Callington, Cornwall 379
Cam, Gloucestershire 747
Camborne, Cornwall 942
Cambridge 624, 676, 716-721, 1578, 1580, 1872, 1875, 1882, 1885-1890, 1892, 1893, 1895-1897, 1998, 2152, 2153
  British Association Meeting (1833) 1826
  Caius College 1894
  Downing College 1871
  Philosophical Society 152, 952, 1022, 1881
  St Peter's College 380
  Trinity College 1870, 1871, 1874, 1876, 1877, 1879
  Trinity College, Mastership of 958, 1009
  University Commission 685
  Woodwardian Museum 1022, 1029, 1880, 1895
Cambridgeshire
  Bottisham 1895
  Cambridge 152, 380, 624, 676, 685, 716-721, 952, 958, 1009, 1022, 1029, 1578, 1580, 1826, 1870-1872, 1874-1877, 1879-1882, 1885-1890, 1892-1897, 1998, 2152, 2153
  Comberton 676
  Ely 1885
  Peterborough 18, 172, 1536
Camelford, Cornwall 210
Camerton, Somerset
  geology of 1550, 1552, 1555
  sketch section near (illn.) 1555
Canada 873, 879, 880, 885
  Baffin Bay 148, 343, 862
  Baie des Chaleurs 883, 885
  Bathurst 883
  Bay of Fundy 903
  Cape Breton Island 885
  Cape Turnagain 302
  Coppermine River 157, 302
  Edmonton 885
  Gaspé 882, 883, 885, 888
  Halifax 874-876, 880, 882
  Hudson Bay 885, 887
  Icy Cape 302
  Joggins 882-885

Mackenzie River 302
Melville Island 311
Montreal 874, 880, 882-888, 917
New Brunswick 883
Nova Scotia 695, 875, 876, 883, 885
Ottawa 885, 887
Port Daniel 883
Quebec 456, 874, 2001
River/Gulf of St Lawrence 883, 885, 887, 888
Saskatchewan 885
Temiscamingue, Lac 886
 coal 874, 875
 dates of sailings from Liverpool to 880
 fossil tortoise from 917
 Geological Survey of 875, 885
 geology of, compared with New York 884, 888
 W.E. Logan in 456, 873-875, 879, 880, 882-885
Canary Islands 141, 202
 Berthelot's map of 501
 von Buch's paper on 141, 501
Canaston Bridge, Pembrokeshire
 geology of 351
Canaston, Pembrokeshire 1485
Canterbury, Kent
 Cathedral 343
Cape Breton Island, Canada
 coalfield 885
Capel Curig, Caernarvonshire 1767, 1785-1793, 1796
Capri, Italy 363
 Naples boats with Capri in the distance (illn.) 363
 panorama, Vesuvius to Capri (illn.) 363
Capua, Italy 363
Cardiff, Glamorgan 108, 176, 189, 306-309, 396, 671, 715, 958, 1348, 1357, 1358, 1681, 1922, 2102
 coal exports 2126
 Collenna 873, 1679, 2034, 2052, 2102
 Institution 1924
 sketch section, Cardiff to Caernarfon (illn.) 1492
Cardigan, Cardiganshire 873, 876, 947, 948, 958, 1375, 1386, 1387, 1397, 1404, 1407-1409, 1710, 1714, 1719, 1742, 1743
 flint gravel near 877
 geology of 1419, 1428
 till 1746
Cardiganshire
 Aberaeron 1852
 Aberystwyth 877, 1564, 1723, 1727, 1746, 1791, 1852, 1940, 1946-1949
 Bwlch Gwyn 1941
 Cardigan 873, 876, 877, 947, 948, 958, 1375, 1386, 1387, 1397, 1404, 1407-1409, 1419, 1428, 1710, 1714, 1719, 1742, 1743, 1746
 Cwmsymlog 1942
 Cwmystwyth 1723, 1941, 1942, 1952
 Devil's Bridge 1723, 1941, 1948
 Esgairmwyn 1948, 1949
 Goginan 1940, 1942, 1952
 Lampeter 1394, 1395, 1428, 1433, 1435, 1438, 1486, 1664, 1719
 Llanddewi Brefi 1724
 Llanfair Clydogau 1719, 1940, 1947
 Nant y Creiau 1941, 1952
 Pennar 1947
 Ponterwyd 1941, 1942, 1948
 Trefilan 1724
 Tregaron 1724, 1725
 Tynfron 1952
 Ystum Tuen 1941
Cardington, Shropshire
 damage to trees at 45
Carentan, France
 geology of 350
Carew, Pembrokeshire 456
 geology of 351
Carinthis 146
Carlisle, Cumberland 595, 1870
Carlow, Ireland 558, 771, 772, 1101, 1102, 1109, 1931, 1933
 geological map of 1102, 1105, 1119

maps of 757
1" map of 791
Carmarthen, Carmarthenshire 336, 1378, 1379, 1381-1383, 1389, 1390, 1392-1394, 1397, 1400, 1403, 1424, 1426, 1432, 1434, 1437, 1712, 1719, 2103-2105
 Bay 873
 coach to Llanmadog 1399
 coloured sketch map of geology to SE of (illn.) 1425
 copper ore near 626
 fossils 1419
 geological map of 814, 1431
 geology of 793, 1395, 1401, 1425, 1428, 1429, 1431, 1466
 Ivy Bush Inn 1382
 lead mines SE of 1435, 1666
 town crier 2104
Carmarthenshire
 Aberglasney 1623
 Abergoleu 1426
 Abergorlech 1720
 Abergwili 204, 626, 1394, 1400, 1401, 1424, 1425, 1431, 1435, 1666
 Abergwenlais 1426
 Abermarlais 1426
 Abernant 1394, 1397, 1426, 1429, 1439
 Banc-y-felin 1382
 Beaulieu-fawr 1401
 Black Mountain 1721
 Blaendyffryn 1437, 1438
 Brechfa 1383, 1385, 1394, 1397, 1401, 1426, 1712, 1714, 1715, 1717, 1720
 Buallt 1426, 1431-1433, 1437, 1439, 1440, 1663
 Bwlch Capel 1394, 1400, 1425
 Caeo 1403
 Caerau 1426
 Caledfwlch 1426
 Carmarthen 336, 626, 793, 814, 873, 1378-1383, 1388-1390, 1392-1395, 1397, 1399-1401, 1403, 1419, 1424-1426, 1428, 1429, 1431, 1432, 1434, 1435, 1437, 1466, 1666, 1712, 1719, 2103-2105
 Carn Coch 1396, 1438, 1440
 Carreg Cennen Castle 1387
 Castell Cogan 1381, 1432, 1435
 Castell Pigyn 1426
 Clogyfran 1380, 1406, 1407, 1429, 1431, 1485, 1487
 Clynderwen 1397, 1401, 1664
 Court Henry 1426
 Coygan 1381, 1382, 1431
 Craig Twrch 1719, 1724
 Creigiau Ladis 1718, 1719
 Crug 1400, 1439
 Cwm Aman 1994
 Cwmfelin Boeth 1991
 Cynwyl 1383
 Cynwyl Elfed 1394, 1400, 1419, 1426
 Cystanog 1435, 1666
 Dinas Fawr 351
 Dolaucothi 1724, 1725
 Dryslwyn Castle 1431
 Dynevor 1431, 1664
 Eglwys Fair 1377
 Felin nant yr Eglwys 1426
 Ffosddu 1394, 1400, 1426
 Ffynnon 1377
 Ffrwd 1426
 Foel Cywyn 1426
 Gilfach 1435
 Godor 1401
 Goetre-uchaf 1400
 Golden Grove 1431, 1432, 1438-1440, 1623
 Henllan Amgoed 1991
 Kidwelly 625, 869, 878
 Laugharne 351, 1378, 1381, 1432
 Llanarthney 1383, 1388, 1395, 1431, 1432
 Llanboidy 954, 1392, 1397, 1664, 1666, 1993
 Llanddowror 793, 1378, 1380, 1382, 1383, 1431

Llandeilo 157, 626, 1336, 1360, 1382-1386, 1388-1390, 1392, 1394-1401, 1403, 1429, 1431, 1433, 1437-1439, 1486, 1623, 1666, 1711, 1714, 1852
Llandeilo Abercywyn 1431, 1432
Llandovery 1360, 1388, 1389, 1396-1402, 1404, 1405, 1407-1410, 1416, 1424, 1426, 1431, 1435, 1442, 1445, 1446, 1448, 1536, 1663, 1664, 1714, 1716, 1718-1722, 1760, 1852
Llandybie 1387
Llanelli 878, 2126
Llanfihangel-uwch-Gwili 1397, 1426
Llanfynydd 1426
Llangadog 336, 1388, 1389, 1399, 1401, 1402, 1408, 1431, 1433, 1435, 1437, 1438, 1441, 1852
Llangathen 1426
Llanglydwen 954, 1392, 1397
Llangunnor 1395, 1425
Llangynog 1424, 1431
Llansawel 1388, 1397, 1426, 1433
Llansteffan 1380, 1381, 1431, 1432
Llanwrda 1426
Maesgwrda 1381, 1382
Mallaen, Mynydd 1718, 1719
Meidrim 1382, 1393, 1394, 1397, 1400, 1426, 1429, 1431-1433, 1437, 1439, 1440, 1663
Melin Ricket 1394, 1400, 1426, 1431
Merthyr 1394, 1425
Middleton Hall 1431, 1437
Mynydd Mallaen 1718, 1719
Nant y Caws 1428, 1431
Nant y Mwyn 1666
Newcastle Emlyn 958, 1378, 1388, 1394, 1397, 1401, 1711, 1719
Newchurch 1394, 1397
New Inn 954
Ogofau 1403, 1424, 1435, 1440, 1666
Pantyffynnon 1995
Pant y Glien 626
Pembrey 878
Penddaulwyn 1431
Pendine 351, 1378, 1485
Pen y Gaer 1426
Pen y Mulfre 1382
Pen y Parc 1426
Pen y raen 1426
Pigyn Shon-Nicholas 1720
Pistyllgwyn 1400, 1426
Plas y Golomen 1394, 1426
Pontarllechau 1435
Pont Gwladys 1439, 1440
Pumsaint 1402, 1426, 1433, 1449, 1712-1721
St Clears 793, 1377, 1378, 1381, 1383, 1392-1394. 1397, 1400, 1407, 1426, 1429, 1431, 1432, 1437, 1440, 1486, 1663, 1664, 1666, 1712, 1993
Taliaris 1397, 1401, 1419, 1426, 1429, 1432, 1433, 1439, 1714, 1715, 1717
Tal-Sarn 1724
Tavernspite 1378
Troedrhiw 1426
Typica 1431
Wernddu 1426
Ynys Brechfa 1426
Carmel Head, Anglesey 1964
Carn Coch, Carmarthenshire
   geology of 1438, 1440
   section 1396
Carnedd Llewelyn, Caernarvonshire 1789
Carningli, Mynydd 957
Carnmenellis, Cornwall 382, 383
Carouge, Switzerland 345, 346
Carpathians 146, 147
   geology of 1028
Carrara, Italy 363, 490
   geology of 363
   Marble quarries at 363
   section from Marble Quarries to Lavenza (illn.) 363
   section from Marble Quarries, westward (illn.) 363
   section, top of pass between Carrara and Massa (illn.) 363

sections near Carrara (illn.) 363
Carreg Cennen Castle, Carmarthenshire
   geology of 1387
Cassel, France 250
Castellammare, Italy 363
   description of 363
Castell Cogan, Carmarthenshire 1381, 1432
   geology of 1435
Castell Pigyn, Carmarthenshire 1426
Castle Ashby, Northamptonshire 1052, 1053, 1055, 1056, 1183
Castlebythe, Pembrokeshire 954
Castlecomer, Ireland 1931
   coalfield 1930
Castle Howard, Yorkshire 14
Castle Malgwyn, Pembrokeshire 873
Castlemartin, Pembrokeshire
   geology of 351
Castleton, Derbyshire 608
   Roman mining at 1605
   sketch section at (illn.) 1605
Castletown, Isle of Man
   King William's College 321
[?]Castletown [Bearhaven], Ireland 780, 782
Cauville, France
   section (illn.) 350
Cefn Cave, Denbighshire 2042
Cefn Cribwr, Glamorgan 7
Cenis, Mont, France 155
   Col du 347, 348
Cerne Abbas, Dorset 122, 1758
   geology of 1757
Ceylon
   Adam's Peak 1134
   Austrian expedition to 644
   Matale 1134
   Matara 1134
   Nuwara Eliya 1134
   Pidurutalagala 1134
   geological specimens from 1134
Chagny, France 343, 346
Chaleurs, Baie des, Canada 885
   geology of 883
Chalon sur Saône, France 343, 346
Chambery, France
   geology of 347
Chamonix, France 344, 346, 348
   section of valley of (illn.) 348
Champlain, Lake, United States, 883
Chard, Somerset 3, 2126
Charfield, Gloucestershire 749
   section of Bristol to Gloucester railway near (illn.) 748
Charlestown, Cornwall 940
   fossils from 1156
   mines at 945
Charmouth, Dorset 188, 410
   fossils 1330
   geology of 341
Charton Bay, Devon
   galena at 179
Cheadle, Staffordshire 1967
Chelmorton, Derbyshire 1624
Chelmsford, Essex
   Chronicle 472
Cheltenham, Gloucestershire 1445, 1450
   coal supply to 1281
Chepstow, Monmouthshire 173, 1388, 2113-2120
   tides at, during Great Storm of 26 November 1703. 2079
Cherbourg, France
   geology of 350
   visit of R.A.C. Austen 32
Cheshire
   Alderley 2083
   Birkenhead 922
   Chester 68, 739, 1375, 1564, 1966
   Congleton 2083
   Davenham Hall 750

Place Names   *Cheshire - Cornwall*

    Macclesfield 1637, 1638
    Macclesfield Forest 1135
    Middlewich 1135
    Runcorn 2042
    Tarporley 487
    Warrington 701
    salt 1135
Chesil Beach, Dorset 313
    pebble sizes of 341
Chester, Cheshire 739, 1375, 1564, 1966
    sewerage of 68
Chesterfield, Derbyshire 1613, 1614, 1632, 1643, 1800, 1972
    Angel Inn 1620
Chesterford, Essex
    frit from 676
Chevroux, Switzerland 347
Chew, River
    geology of 307
Chew Magna, Somerset 2131
Chewton Mendip, Somerset
    Lias of 176
Chiasso, Switzerland 363
Chiavari, Italy 363
    geology of 363
    panorama, Portofina and (illn.) 363
Chiavenna, Italy
    geology of 363
Chideock, Dorset 123, 1774
Chile, earthquake 2051
Chiltern Hills, drift deposits of 310
China 1286
Chippenham, Wiltshire 90
Chipping Camden, Gloucestershire 731
Chirbury, Shropshire 1738, 1740, 1741, 1768
Chorley, Lancashire 1944
    Alison Hall 328
Chudleigh, Devon 29-31, 1335-1338, 1681, 1879
Chur, Switzerland
    geology of 363
Church Stretton, Shropshire 563, 1733, 1734, 1737, 1766, 1767, 1771, 1796
    damage to trees at 44, 45
    geology of 1735, 1736
Cilciffeth, Mynydd, Pembrokeshire 957
Cirencester, Gloucestershire 342
    Training College 806
Civita Castellana, Italy 363
Clarbeston, Pembrokeshire 1991
    sketch section, from Preseli through (illn.) 1993
Claverton, Somerset 1528-1530
    Claverton Hotel 1528-1530
Clee Hill, Shropshire 310, 1452, 1952, 1955
    geology of 1766
    sketch section of (illn.) 1766
Cleenish, Ireland 558
Clevedon, Somerset 1863, 1864
    geology of 342
Cleveland, United States
    coalfield 885
Cliff End, Isle of Wight
    geology of 582, 586, 587
Clogyfran, Carmarthenshire
    geology of 1380, 1406, 1407, 1429, 1431, 1485, 1487
Clonakilty, Ireland
    baryte from mine at 780
Clonmel, Ireland 1125
Clun, Shropshire 121
    geology of 119
Cluses, France
    geology of 348
Clynderwen, Carmarthenshire
    geology of 1397, 1401, 1664
Clynnog-fawr, Caernarvonshire 1789
Clytha, Monmouthshire 1358, 1359
Coalbrookdale, Shropshire
    coalfield 2088

    iron 2094
Coalpit Heath, Gloucestershire
    miners pay at 2127
Coblentz, Germany see Koblenz
Coburg, Germany 146
Cockburnspath, Berwickshire 792
Colchester, Essex 135, 136, 471, 472
    Navigation Bill 2068
Coleford, Gloucestershire 1282, 2110-2115
Coleford, Somerset 2130
Collenna, Glamorgan 873, 1679, 2034, 2052, 2102
Collingwood, Kent 679-684, 1058
Collonges, France 344, 346
Colmar, France
    coal from 348
Cologne, Germany 146, 250, 349
    Cathedral 349
    fossil elephant from 343
Colwall, Herefordshire 1442, 1452, 1479
Colwell Bay, Isle of Wight 590
Combe Martin, Devon 1875
    fossils from 1346, 1408
    geology of 1336
Comberton, Cambridgeshire
    frit from Roman Villa at 676
Combs, Suffolk
    boring at 677
Comeragh Mountains, Ireland 1126
Como, Italy
    description of 363
    fossils from 363
    geology of 363
    panoramic section from, to Monte Francesca (illn.) 363
    Port of (illn.) 363
    section from, to Lago di Lugano (illn.) 363
Como, Lago di, Italy 363
    boat and oar (illn.) 363
    Between Onno and Lecco (illn.) 363
    iron mines near 363
    panorama (illn.) 363
    Rezzonico (illn.) 363
    view from Villa Pliniana (illn.) 363
Condé sur Noireau, France
    geology of 350
Congleton, Cheshire 2083
Connecticut, United States
    geology of 695
Connemara, Ireland 652
Consall, Staffordshire 1967
Constantinople (Istanbul), Turkey 153, 154, 864, 1922, 2149
    geology of 622
Conwy, Caernarvonshire 725
Copenhagen, Denmark 597, 891
    University Geological Museum 597
    waterworks at 291
Coppet, Switzerland 344-347
Coppull, Lancashire
    colliery explosion at 1944
Cordoba, Spain
    coalfield to W of 1979
Coregna, Italy
    section from La Spezia to (illn.) 363
    sections (illn.) 363
Corfu, Greece 363
Cork, Ireland 652, 776, 777, 779, 814, 816, 1101, 1522, 2026
    British Association meeting (1843) 903, 1520, 1521
    geological map 1114
    geology of 1583
    Queens College 829, 1045, 1046, 1111
Cornwall 27, 195, 273, 830, 897, 1020
    Black Head 944, 1163
    Boconnoc 946
    Bodinnick 1161
    Bodmin 943
    Boscastle 1875
    Cadgwith 1990

Callington 379
Camborne 942
Camelford 210
Cape Cornwall 2079
Carnmenellis 382, 383
Charlestown 940, 945, 1156
Crowan 943
Dodman Point 939
Dolcoath Mine 160, 264
Duchy of 341, 515, 605, 857, 978, 1279, 1674, 2004, 2132
Enys 513-516, 519, 520, 522, 672
Falmouth 340, 378-380, 383, 513, 521-523, 599, 600, 623, 717, 779, 804, 837, 838, 939, 940, 942-944, 946, 1161, 1225, 1828, 1877, 1990, 2012, 2034
Flushing 520
Fowey 198, 434, 513, 515, 668, 941, 942, 1156, 1160-1163, 1346, 1876, 2029
Fowey Consols (illn.) 434, 513
Gerrans Bay 940, 1157, 1161
Gorran Haven 946, 1156-1159, 1161, 1163, 1337, 1990
Greeb Point 939
Gribbin Head 946
Gurnards Head 382
Helston 43, 376, 382, 805, 806, 939-941, 1828, 1875, 1876
Landewednack 1990
Land's End 265, 353, 381, 383, 2029
Launceston 25, 26, 31, 1336, 1337, 1875, 1876, 1878
Levant Mine 600
Liskeard 941, 945, 986
Lizard 382, 383, 939, 1874, 1876
Looe 945, 946, 1156, 1876
Mawnan 939, 1828
Merther 806
Mevagissey 939, 944
Newquay 1875, 1876
Padstow 1335, 1337, 1874-1876
Par 1156
Pelynt 2029
Pencarrow 946
Pendennis 382, 383
Penryn 517, 519, 520, 851, 852, 939, 1224, 2151
Pentewan 940, 941, 943, 944, 1875
Penzance 265, 668, 678, 852, 939, 942, 1156, 1162, 1681, 1875, 1990, 2029, 2033
Petherwin 1331, 1332, 1335, 1336, 1346, 1584, 1612, 1875
Polcreek 939
Polgooth 941, 942, 944
Polgrain 939, 940
Polkerris 1156
Polperro 946, 1161, 2254
Polruan 1156, 1157, 1346
Pordenack Point 382
Rame Head 946
Redruth 519, 604, 658, 1681
St Austell 340, 377, 378, 939-946
St Blazey 941, 1875
St Buryan 381, 383
St Columb 658, 1346
St John 1828
St Ives 381-383, 1990
St Just 942, 945
St Martin 1161
St Michael Caerhays 939
St Michael's Mount 381-383, 1161
St Teath 381, 383
Sennen 381, 383
Tincroft Mine 604
Tintagel 1875, 1876
Tregony 939, 943, 944
Treruffe 604
Tresillian 519
Trevenen 940
Truro 111, 389, 513, 806, 892, 900, 940, 1337, 1874, 2012, 2267
Tywarnhayle Mine 605
Veryan 939, 940, 1876
Wheal Rose 2029

copper deposits 15, 373, 623
fossils localities in 1156, 1157, 1163
geological map of 623
Geological Survey of 208, 210, 261, 263, 265, 266, 268-272, 276, 277, 279-281, 376, 379, 384, 529, 844, 939-946
Geological Transactions of 1163
geology of 264, 378, 390, 939-946, 1583, 1828, 1875, 1876
granite, comparison with Rouvray, France 343
granite exports 517
magnetic bearings in 381-383
mines of 397, 857, 941-945
Ordnance map of 265, 266, 278, 625
proposal for a Mining College in 385, 386, 389, 1652
rocks of 442
Royal Geological Society of 379, 599, 949, 2003, 2029, 2268
Royal Polytechnic Society of 397, 1682, 1990, 2029, 2268
sea level changes 1875
Sedgwick's visits to 1874, 1875
stream tin 604
Taylor's plan for Mining School in 302
temperature in mines in 160
tin mines 941-943, 2221-2228
Cornwall, Cape, tidal range at 2079
Corsica, lichens from 148
Corvo, Capo, Italy
    contorted beds (illn.) 363
    section (illn.) 363
Corwen, Merionethshire 1782
Cosheston, Pembrokeshire
    geology of 351
Côte d'Or, France
    geology of 343, 346
Cotentin, France
    visit of R.A.C. Austen 32
Cothi, Afon 1394, 1397
    lead in 1666
Cotswold Hills 1452
    drift deposits of 310
Courmayeur, Italy 344, 346
Courseulles, France
    geology of 350
Court Henry, Carmarthenshire 1426
Courtown, Ireland 1083, 1086
    Harbour 1084, 1085
    comparison of geology with Bala 557, 558, 1079
    geology of 1085
Coutances, France
    geology of 350
Cowarch, Merionethshire
    lead mine 1947
Cowbridge, Glamorgan 147, 715
    Lias of 176
Cowes, Isle of Wight 585, 587, 589, 590
Coxwell, Oxfordshire
    extraction of iron ore at 153
Coygan, Carmarthenshire 1381, 1382, 1431
Cracow, Poland see Krakow
Craig Twrch, Carmarthenshire 1719, 1724
Crawley Rocks, Gower, Glamorgan
    discovery of Pleistocene vertebrates at 162
Creigiau Ladis, Carmarthenshire 1718, 1719
Crimea 149
Crinow, Pembrokeshire 1485
Croghan, Ireland 762, 766, 769, 1932
    geological section of (illn.) 1072
    geology of 1096, 1933
Cromhall, Gloucestershire 1861, 1863, 1865
Crowan, Cornwall 939, 943
Crowdicote, Derbyshire
    fossils from 1830
Croyde, Devon 1336
    fossils from 1337, 1346
Crug, Carmarthenshire 1400
    geology of 1439
Crunwear, Pembrokeshire 1378
Ctesiphon, Iraq 865

Cuba 352
    Gibara 2015
    Havana 984
    Holguin 2015
    copper mines 1016
    gold 2016, 2017
    publications on, by R.C. Taylor 2015
    sugar 984
Cuckfield, Sussex 105
Cudrefin, Switzerland 347
Cuerdale, Lancashire
    discovery of Viking silver hoard (1840) 1041
Cuffern Mountain, Pembrokeshire
    geology of 351
Culver Cliff, Isle of Wight
    geology of 585
Cumberland
    Alston Moor 1965, 1966
    Buttermere 1870
    Carlisle 595, 1870
    Derwent Water 154
    Eden valley 1871
    Ennerdale 1870
    High Stile 1870
    Keswick 154
    Skiddaw Slate 1025, 1886, 1887, 1889
Cwellyn, Llyn, Caernarvonshire 1777, 1778
Cwm Aman, Carmarthenshire
    Colliery section 1994
Cwmavon, Glamorgan
    works 2010
Cwmfelin Boeth, Carmarthenshire 1991
Cwmgloyne, Pembrokeshire 958
Cwmsymlog, Cardiganshire 1942
Cwmtillery, Monmouthshire
    Coal Measures at 2107
Cwmystwyth, Cardiganshire 1723
    mines 1941, 1942, 1952
Cyffig, Pembrokeshire 1382
Cynwyl, Carmarthenshire 1383
Cynwyl Elfed, Carmarthenshire 1394, 1400, 1426
    geology of 1419
Cystanog, Carmarthenshire
    lead mine at 1435, 1666
Cywarch, Merionethshire see Cowarch
Czechoslovakia
    Giant Mountains 146
    Prague 1022

Dale, Pembrokeshire 948, 1362, 1366, 1367, 1368, 1486, 1706
Dalmatia, offer of specimens from 84
Dalwhinnie, Inverness-shire 1858
Damodar River, India
    coalfield 2137-2139, 2144-2147
    coalfield section (illn.) 2146
Dan, Lough, Ireland 1096
Danube, River 1928
Dargle River, Ireland 1087
Darjeeling, India 709
Darlington, Durham 1588
Darmstadt, Germany 349
    fossil collection at Palace at 349
Dartford, Kent
    geology of 343
Dartington, Devon
    fossils from 1338
Dartmoor, Devon 26, 147, 341, 622, 1006, 1875, 1876, 1879
    Prison, inspection by Capt. Ross 373
    Prison, state of 373
Dartmouth, Devon 179, 1877, 1904, 1906
Davenham Hall, Cheshire 750
Dawlish, Devon
    geology of 341
Deal, Kent 343
    tides 2079

Dean, Forest of, Gloucestershire 152, 310, 420, 700, 747, 779, 1275, 1276, 1278, 1285, 1287, 1410, 2110, 2128
    Coalfield 2111, 2112, 2114, 2115, 2129
    coal of 1465, 1502
    fossil plants of 2115
    Pennant Sandstone in 2115
    railways of 1275, 1277, 1281, 1282
    Sopwith's map of 1280
    Sopwith's model of 2115
Dee, River, glacier 1564
Deeside, geological map of 251
Delhi, India 2138
Delft, Netherlands 349
    Royal Academy 1907
Denbighshire 1663
    Abergele 725, 1942
    Cefn Cave 2042
    Llangollen 561, 562, 813, 1564, 1782, 2133
    Llanrwst 1789
    Pentrefoelas 811
    Ruabon 1956
    Wrexham 1955, 1966, 2132
Denmark
    Chalk fossils of 891
    Copenhagen 291, 597, 891
Dent du Midi, Switzerland 344-346
Derby, Derbyshire 795, 1296, 1600, 1639, 1646
    marble works at 648
    sketch section near (illn.) 1595
Derbyshire 18, 541, 1445, 1583, 1594, 1647
    Alport 1972
    Ashbourne 1595-1600, 1609, 1640, 1831, 1832, 1971
    Ashford 648, 982, 1587, 1624, 1625, 1830
    Bakewell 62, 63, 159, 982, 1582, 1587, 1648, 1832, 1833, 1972
    Barlow 411
    Baslow 1585, 1606, 1613, 1619-1625, 1641, 1654
    Belper 1600
    Bolsover 1246, 1255, 1624
    Bonsall 1655
    Brassington 608, 1832
    Buxton 1165, 1599-1603, 1625, 1830, 1832
    Castleton 608, 1605
    Chelmorton 1624
    Chesterfield 1613, 1614, 1620, 1632, 1643, 1800, 1972
    Crowdicote 1830
    Derby 648, 795, 1296, 1595, 1600, 1639, 1646
    Glossop 1637
    Hartington 1605, 1971
    Hopton 1655
    Kedleston 1646
    Kinder Scout 1627, 1637
    Matlock 159, 1633, 1655
    Matlock Bath 1196, 1197, 1833
    Monyash 648
    Peak 1601, 1604
    Priestcliffe 1625
    Ricklow Dale 648
    Thorpe 1832
    Tideswell 1605, 1633
    Wirksworth 170, 608, 1655
    fossils from 1833
    geological map of 62
    Geological Survey in 1587, 1595-1603, 1605, 1606, 1613, 1632, 1633, 1646, 1830
    Geological Survey map of 1648
    geology of 159, 309, 310, 1585, 1586, 1602, 1603, 1655, 1830, 1832
    goniatites 1584
    Hopkins' geological map of 278, 309, 380, 716, 717, 1878
    Hopkins' proposal for a geological survey of 716, 717
    minerals from 109
    mining 1971, 1972
    model of 228
    proposal for a Mining College in 385, 386, 1652
    rocks from 442

sketch section (illn.) 1606, 1613
underground drainage 1597
visit of W. Buckland 150, 159
Derwent, River (Derbyshire) 1583, 1584, 1627
Derwent Water, Cumberland
floating islands of 154
Devil's Bridge, Cardiganshire 1723, 1941, 1948
Devizes, Wiltshire 2157, 2159
canal 1980
Devon 2, 273, 663, 830, 1020
Abbotskerswell 341
Alverdiscott 624
Ashburton 29-31, 1332, 1336-1338, 1681
Axminster 43, 151, 163, 164, 179, 188, 311, 312, 341
Axmouth 284, 341, 410
Babbacombe 341, 360, 621, 1338, 2086
Baggy Point 31, 1336, 1337, 1340, 1346
Bampton 1340, 1876
Barnstaple 25, 657, 1332, 1336, 1337, 1681, 1875-1877
Beer 341, 433
Berry Head 341
Betsy, Wheal 373
Bickington 1338, 1346
Bideford 34, 378, 621, 832, 1013, 1331, 1336, 1337, 1681, 1859, 1877
Bishopsteignton 36, 341
Blackdown Hills 28, 458, 542-544, 1331
Bolt Head 1876
Bovey Tracey 147
Branscombe 341, 433
Braunton 1336
Brentor 25, 1875
Brushford 1346
Buckfastleigh 29
Budleigh Salterton 341
Carton Bay 179
Chudleigh 29-31, 36, 1335-1338, 1681, 1879
Combe Martin 1336, 1346, 1408, 1875
Croyde 1336, 1337, 1346
Dartington 1338
Dartmoor 26, 147, 341, 373, 622, 1006, 1875, 1876, 1879
Dartmouth 179, 1877, 1904, 1906
Dawlish 341
Devonport 659, 1904
Dowlands 341, 433
Drewsteignton 1876
Exeter 27, 32, 147, 151, 174, 183, 341, 483, 675, 1681, 1873, 1875
Exmoor 31, 622, 2098
Exmouth 341, 2039
Forde Abbey 341
Foreland Point 1336
Franco, Wheal 373
Fremington 31, 1336
Friendship, Wheal 373
Frogmore 694
Goodrington 32
Haldon, Great 28, 29, 174
Haldon Hill 341
Hartland 947
Hatherleigh 1337
Heddon's Mouth 1336
Hembury Hill 29, 30
Holcombe Rogus 1336, 1337, 1340, 1876, 1877
Honiton 341
Hope's Nose 341, 1338
Ideford 36
Ilfracombe 34, 150, 1332, 1336, 1408, 1680, 1881, 1877
Ilsington 33, 844
Ivybridge 844
Kingsbridge Estuary 694
Kingskerwell 341
Kingsteignton 341
Lynton 1336, 1346
Marwood 1336, 1855
Membury 179
Milber Down 27, 28
Newton Abbot 33, 34, 36, 620
Newton Bushell 2086
Oakhampton 1337
Ogwell 25, 28-36, 341, 1337
Paignton 341
Peter Tavy 373
Pilton 1337, 1340, 1346, 1612
Plymouth 12, 37, 39, 115, 167, 294, 301, 302, 373, 657-662, 665-669, 804, 844, 920, 945, 946, 953, 989, 991, 1015, 1139, 1156, 1158, 1335, 1337, 1338, 1346, 1355, 1370, 1371, 1507, 1583, 1584, 1681, 1697, 1876, 2079, 2086-2088
St Marychurch 341
Salcombe 694
Sandy Gate 28
Saunton 1336
Seaton 341
Shaldon 341
Sidmouth 151, 163, 341, 433
Start Point 341, 1876, 1877
Swimbridge 1336, 1337
Tavistock 25, 26, 31, 373, 522, 600, 832, 946, 1875, 1876, 2012, 2097
Tawstock 1346
Teignmouth 24, 33, 34, 36, 341, 1987
Tiverton 2, 24, 209, 257, 658, 834, 835, 898, 1335, 1338, 1873
Tor Bay 24, 360, 620, 1876
Torquay 36, 171, 174, 341, 360, 621, 1051, 1335, 1338, 1346, 1480, 1681
Totnes 31, 372, 1346
Valley of Rocks 1336, 2097
Walkhampton 373
Watcombe 341
Wheal Betsy 373
Wheal Franco 373
Wheal Friendship 373
Woody Bay 1336, 1346
Yarnscombe 624
Yealmbridge 1346
Yealmpton 1346
& Cornwall Natural History Society 1158
W. Buckland travelling to 149, 151, 152, 155, 157, 159, 160, 163, 171
caves of 171
Controversy 26, 29-31, 33-37, 41, 266, 270, 271, 273, 378, 621-624, 837-839, 897, 902, 1012-1014, 1226, 1873, 1875-1877, 1879, 2097
copper deposits 15
coral limestone in inlaid table 1051
faults 24
fossil plants 1873
fossils 1369, 1371, 1480, 1612
fossils, age of 1583
geological map, Exeter to Start Point (illn.) 360
geological map of 147, 188, 497-499, 502, 504
Geological Survey of 208, 209, 259, 260, 268, 270-272, 276, 279-281, 370, 378, 498
geology of 24-31, 34, 36, 141, 341, 360, 620, 804, 892, 1013, 1014, 1481, 1875-1877, 1879
geology of, compared with Ireland 652
geology of, compared with S. Wales 1854
geology of, compared with Egypt 141
head 27, 28
Ordnance map of 278
raised beaches 24
rocks of 442
slate of 694
Devonport, Devon 1904
meteorological observations at 659
Diablerets, Switzerland 345, 346
Dijon, France 146, 343, 346
description of Mass at Cathedral of 343, 346
Dinas Fawr, Carmarthenshire
geology of 351
Dinas Head, Pembrokeshire 957, 1791
geology of 947, 1710
Dinas Mawddwy, Merionethshire 1727, 1852, 1945

Dingle Peninsula, Ireland
    comparison with rocks of Marloes 1523
    geology of 1522, 1523
Dingwall, Ross and Cromarty 921
Dinorwic, Caernarvonshire 1781
Dives, France
    geology of 350
    section (illn.) 350
Diyarbakir, Turkey 864
Dodman Point, Cornwall 939, 941
Dolaucothi, Carmarthenshire 1724, 1725
Dolcoath Mine, Cornwall
    increase in temperature with depth in 160
    sections and plans of 264
Dolgellau, Merionethshire 1656, 1727, 1730-1732, 1738, 1740, 1741, 1743-1749, 1751, 1776, 1798-1800, 1802-1804, 1807, 1852, 1945, 1947-1949, 1968
    geology of 1729, 1733
    geology of, compared with Treffgarne 1733
    National & Provincial Bank 1737, 1739, 1750
Dolwyddelan, Caernarvonshire 1796
Dominican Republic 352, 353
Doncaster, South Yorkshire 18
Donegal, Ireland
    Ordnance maps of 1218
Dorchester, Dorset 124-127, 153, 341, 342, 556, 1770, 1774
Dormington, Herefordshire 1496
Dorset 146, 151, 153, 830
    Abbotsbury 151-153, 155, 157, 341
    Beaminster 179
    Bincombe 1764
    Black Ven 341
    Bournemouth 561
    Bridport 123, 156, 341, 1331
    Cerne Abbas 122, 1757, 1758
    Charmouth 188, 341, 410, 1330
    Chesil Beach 313, 341
    Chideock 123, 1774
    Dorchester 124-127, 153, 341, 342, 556, 1770, 1774
    Down Cliff 169
    Durdle Door 341
    Golden Cap 169, 341, 342, 433
    Holywell 1764
    Lulworth 186, 187, 341, 572-576, 2093
    Lyme Regis 11, 79-81, 130, 157, 161, 163, 169, 177, 179, 183, 188, 296-299, 302, 310, 317, 339, 341-343, 349, 350, 363, 364, 372, 373, 410, 531, 620, 655, 686, 696, 699, 849, 896, 897, 988, 1020, 1222, 1223, 1225, 1330, 1339, 1875, 1980, 2047, 2278
    Melbury 153
    Melcombe Regis 342
    Osmington 151, 341, 575
    Piddletown 128
    Poole 691
    Portesham 2093
    Portland, Isle of 341, 342, 1770, 1774
    Purbeck, Isle of 574, 575, 581, 1800
    Ringstead Bay 341
    Ryme Intrinseca 1759
    Shaftesbury 2092
    Sherborne 47, 342
    Stanley Green 691
    Sturminster Newton 122, 1758
    Swanage 571, 575, 577, 581, 1020, 1806, 2093
    Upwey 124, 125, 572, 575, 2093
    Wareham 572-575
    West Lulworth 572-576
    Weymouth 151, 156, 157, 179, 180, 184, 186, 341, 342, 350, 573, 1331, 1409, 1774, 2093
    Worbarrow 575, 581, 2093
    Yetminster 1759
    Buckland's geological map of 156, 188
    Geological Survey in 122, 123, 128, 208, 572, 574, 575, 1411, 1757, 1758, 1774
    geology of 169, 179, 188, 341, 575, 2093
    watercolour, *Duria Antiquior* 368

Dosso d'Albido, Italy
    section (illn.) 363
Dove, River 1597
Dovedale 1597, 1832, 1971
    caves 1597
Dover, Kent 146, 343
    Castle 343
    sailings to Calais 343, 346
    (illn.) 343
    staircase from town to barracks (illn.) 343
Dowlands, Devon
    geology of 341
    landslip (illn.) 433
Down Cliff, Dorset
    geology of 169
Downham Market, Norfolk 92
Drance, River, Switzerland
    floods of 1818. 344, 346
Drayton Manor, Staffordshire 192, 194, 195, 1168-1171, 1174-1176
Dresden, Germany 146
    geology of 843
Drewsteignton, Devon 1876
Drogheda, Ireland
    railway cuttings 1545
Dropmore House, Buckinghamshire 161, 172
Druidston Haven, Pembrokeshire
    geology of 351
Dryslwyn Castle, Carmarthenshire 1431
Dublin, Ireland 35, 50-52, 89, 260, 557, 558, 564-567, 579, 593, 650, 756-778, 780-785, 787, 804, 814, 815, 818, 819, 823, 825, 843, 903, 970, 1045, 1064, 1071, 1092, 1110, 1115-1118, 1120, 1129, 1540, 1544-1546, 1549, 1551, 1552, 1558, 1559, 1574, 1697, 1938, 1939, 2171
    Archbishops Palace 2091
    British Association meeting (1835) 659, 1567, 2042
    Conybeare in 306
    elevation of coast near 622
    Exchange 758, 760
    geological map of County 1117
    Geological Society of 268, 564, 652, 1022, 1103, 1105, 1541, 1543, 2042
    Geological Survey Office, Custom House 406, 407, 1066-1069, 1073, 1074, 1544, 1545
    Geological Survey Office, St Stephen's Green 558, 760, 779, 801, 817, 821, 828, 831, 1066, 1069, 1071-1083, 1088-1091, 1093-1096, 1099, 1101, 1103-1106, 1114, 1119, 1126, 1130, 1131, 1133, 1135, 1316
    Halls Bridge 1095
    Mechanics Institute 758
    Mountjoy Barracks 1084, 1094
    Mountjoy House 786
    Museum of Economic Geology 402, 403, 593, 758, 760-762, 821, 1078, 1199, 1314, 1538, 1545, 1548
    Museum of Irish Industry 826, 830, 831
    Ordnance map of 140
    Ordnance Survey Office 259, 751-753, 818, 1097, 1127
    Phoenix Park 754, 922, 1544
    Trinity College 579, 593, 1065, 1075, 1091, 1095, 1098, 1109, 1110, 1113, 1125, 1129, 1535, 1541-1543, 1546, 1548, 1554-1556, 1561, 1569, 1847, 2008, 2009
    University Museum 51, 1541
Dudley, Worcestershire 566, 609, 1331, 1354, 1767, 1944
    geology of 2088
    mining disasters at 87-89, 1943
    proposal for School of Mines at 1050
    Wenlock Limestone at 157
Dulas, Anglesey 1960
Dumfries-shire 652
Duncannon, Ireland 755
Dundalk, Ireland 652
    railway cutting near 1105
Dundee, Angus,
    model of 228
Dundry, Somerset
    comparison with fossils of Jura Mountains 343
    comparison with fossils of northern France 350

Hill   794, 1866, 1867, 2122
    Hill, geology of   342
Dungannon, Ireland   1080
    comparison of fossils of, with Bala   557
Dunloe, Gap of, Ireland   781
Dunraven, Glamorgan   150
    Castle   6, 7, 1376
    Lias of   176, 314
    silicified fossils at   176, 314
Dunster, Somerset   1819
Durdle Door, Dorset   341
    (illn.)   341
Durham, County Durham
    coalfield   514
    proposal for Mining College in   385
    Magnesian Limestone   1349
    scale of Ordnance Survey map of   792
Durham, County   1651
    Darlington   1588
    Durham   385, 514, 792, 1349, 1649
    Hartlepool   430
    Jarrow   1586
    Monkwearmouth   1973
Dursley, Gloucestershire   1469, 1503
Dynevor, Carmarthenshire   1431, 1664

Eastbourne, Sussex   511, 512
    geology of   511
Ebbw Fach River, Monmouthshire
    Coal Measures   2107
Ebbw Vale, Monmouthshire
    Coal Measures at   2106
Ecton, Staffordshire   1832
Ecton Mine, Staffordshire   1971
    goniatite from   1585
Eden Valley, Cumberland   1871
Edinburgh   330, 544, 598, 703, 796, 799, 811, 848, 876, 1045, 1163, 1684, 1824
    British Association meeting (1834)   2, 309, 897, 1826
    British Association meeting (1850)   798, 1046, 1061, 1124
    collection of Irish fossils for Survey Museum at   754
    College Museum (illn.)   798
    need for geology museum at   1548
    Royal Society of   595, 596, 1022
    University   796-798
    University Museum of Natural History   798
Edmonton, Canada
    coal at   885
Eglisau, Switzerland   348
    bridge over Rhine at (illn.)   348
Eglwys Fair, Carmarthenshire   1377
Eglwysilan, Mynydd, Glamorgan   2102, 2105
Eglwyswrw, Pembrokeshire   954, 958
Egypt
    Cairo   707
    Nile Delta   82
    Suez   707
    geology of, compared with Devon   141
    gods of Ancient,   86
    pyramids   707
    report on limestone from   548
Eichstätt, Germany
    fossil fish from   343
Eifel, Germany
    fossils   1351
Eiger, Switzerland   348
Elan Valley, Radnorshire   1725
    copper lode at   1724
Elba, Italy
    catalogue of specimens from, for Museum of Economic Geology   1928
    specimens from, for Museum of Economic Geology   1927
Elegug Stacks, Pembrokeshire
    birds of   351
    origin of name   351
Elford, Staffordshire   673

Elgin, Moray
    Mesozoic rocks discovered near   1014
Elmley Lovett, Worcestershire
    trilobite from   624
Ely, Cambridgeshire   1885
Ems, Germany   250
English Channel
    Plesiosaur from Lyme Regis shipped to London via   302
    shape of   492
Ennerdale, Cumberland   1870
Enniscorthy, Ireland   772, 1108, 1110, 1112
Enniskerry, Ireland   1088
Enniskillen, Ireland   1070, 1071, 1109
Enys, Cornwall   513-516, 519, 520, 522, 672
Erfurt, Germany   146
Erie, Lake   885
Esgairmwyn, Cardiganshire
    advertisement in Mining Journal   1949
    lead mine   1948
Essex
    Bures   471, 472
    Chelmsford   472
    Chesterford   676
    Colchester   135, 136, 471, 472, 2068
    Stanway   135, 136
    Ordnance map of   265
Étretat, France
    geology of   350
    section (illn.)   350
Etna, Mount, Italy   897
Etroubles, Italy   344, 346
Exeter, Devon   151, 174, 183, 341, 675, 1681, 1873, 1875
    Cathedral   341
    fossil plants from   32, 33
    geological map of   27
    Institution   147
    Protectionist meeting at   483
Exford, Somerset   1
Exmoor, Somerset/Devon   31, 622, 2098
Exmouth, Devon   341, 2039
    geology of   341
Euphrates, River   865, 866
Europe
    fauna and flora   8
    geological research in   93
    geology of   146
Evian-les-Bains, France   348

Falaise, France
    geology of   350
Falkland Islands
    Darwin on geology of   31
Falmouth, Cornwall   378-380, 383, 513, 521-523, 599, 600, 623, 717, 779, 804, 837, 838, 939, 940, 942-944, 946, 1161, 1225, 1828, 1877, 1990, 2012, 2034
    "Falmouth Foolscap"   340
    Pearce's Hotel   1990
Faringdon, Oxfordshire
    extraction of iron ore near   153
Farleigh, Somerset   1515
    fossils from Oolite at   458
Farnborough, London
    Down House   330, 331
Farnham, Surrey   206
Fécamp, France   350
    geology of   350
    section (illn.)   350
Felin nant yr Eglwys, Carmarthenshire   1426
Fethard, Ireland   407
Ffestiniog, Merionethshire   1796, 1968
Ffosddu, Carmarthenshire   1394, 1400, 1426
Ffynnon, Carmarthenshire   1377
Fichtelgebirge, Germany   141, 146
Fife
    St Andrews   104
    coalfield   848

Place Names *Finsteraarhorn - France*

Finsteraarhorn, Switzerland 348
Fishguard, Pembrokeshire 109, 947-952, 954, 957, 1371, 1383, 1384,
    1386, 1387, 1389, 1400, 1404, 1435, 1707-1709, 1993, 2125
Flaxley, Gloucestershire 1478
Flintshire
    Holywell 739, 1969, 1970
    Mostyn 191
    Prestatyn 1970
    St Asaph 2042
Florence, Italy 168, 256, 347, 348, 363, 620, 1818
    abolition of Natural History Society of 363
    anatomical collection in Museum at 348
    farming in district around 363
    geology of 363
    mineral collection in Museum at 348, 363
    Pleistocene vertebrates in Museum at 363
Flushing, Cornwall
    stone from 520
Foel Cywyn, Carmarthenshire 1426
Foligno, Italy 363
Folkestone, Kent 810
Fondi, Italy 363
Fontainebleau, France
    Palace of 343, 346
Ford Abbey, Devon
    (illn.) 341
Foreland Point, Devon
    geology of 1336
Forest of Dean, Gloucestershire see Dean, Forest of
Formigny, France
    quarries at 350
Foss, River 1566
Foulksmill, Ireland 1121, 1124
Four Mile Bridge, Anglesey 1963
Fowey, Cornwall 198, 668, 1160-1163, 1876
    Consols 434, 513, 515, 941, 942, 2029
    fossils 1156, 1346
Frampton Cotterell, Gloucestershire
    paving stone from 1134
France
    Abbeville 343, 346
    Aiguebelle 347, 348
    Aix les Bains 347
    Annecy 344, 346, 347
    [?]Anse 343, 346
    Argenteuil 343, 346
    Arromanches 350
    Auvergne 309
    Auxerre 146, 343, 346
    Avallon 343, 346
    Balleroy 350
    Bayeux 169, 350
    Beaufort 344, 346
    Beaune 343, 346
    Beauvais 343, 346
    Belegarde 344, 346
    Biarritz 8
    Bondy 343, 346
    Bordeaux 176
    Boulogne-sur-Mer 38, 343, 346, 845, 1334
    Bricquebec 350
    Britanny 302, 1682, 1683
    Bonneville 344-346
    Buet, Mont 344-346
    Burgundy 343, 346
    Caen 163, 350, 610
    Calais 146, 343, 346
    Carentan 350
    Cassel 250
    Cauville 350
    [?]Cerdon [geology of] 344, 346
    Chagny 343, 346
    Chalon sur Saône 343, 346
    Chambery 347
    Chamonix 344, 346, 348
    Cherbourg 32, 350
    Cluses 348
    Collonges 344, 346
    Colmar 348
    Condé sur Noireau 350
    Côte d'Or 343, 346
    Cotentin 32
    Courseulles 350
    Coutances 350
    Dijon 146, 343, 346
    Dives 350
    Étretat 350
    Evian-les-Bains 348
    Falaise 350
    Fécamp 350
    Fontainebleau 343, 346
    Formigny 350
    Frangy 347
    [?]Fromenteau 343, 346
    Grandscamp 350
    Grandvilliers 343, 346
    Grenoble 344, 346
    Harfleur 350
    Hennequeville 350
    Honfleur 161, 343, 346, 350
    Isère, River 344, 346
    Joigny 343, 346
    Joly, Mont 344, 346. 348
    [?]la Barraque 343, 346
    [?] La Chaleur 343, 346
    La Chambre 347
    Lanslebourg 347
    Le Havre 350, 1037, 1328, 1329
    la Môle 344-346
    Les Houches 344, 346
    Limonest 343, 346
    Lion sur Mer 350
    Lisieux 350
    Lison 350
    Littry 350
    Luc 350
    Lucy le Bois 343, 346
    Lyon 343, 344, 346, 363, 850
    Macon 343, 346
    [?]Maillac 344, 346
    [?]Maison Blanche 343, 346
    [?]Maison Neuve 343, 346
    Marseille en Beauvaisis 343, 346
    Marseilles 342, 2071
    Marquise 343, 346
    Meillerie 348
    Mer de Glace 344, 346
    Messery 345, 346
    Meximieux 344, 346
    Miribel 344, 346
    Modane 347
    Mont Cenis 155, 347
    Mont d'Or 343, 346, 2046
    Montebourg 350
    Montivilliers 350
    Montluel 344, 346
    Montmelian 347
    Montpellier 343, 346
    Montreuil 343, 346
    Nantes 1682, 1683
    Nantua 344, 346
    Neufchâtel 146
    Nice 177, 256, 361, 362, 375, 619
    Normandy 299, 532, 1980
    Nuits-St Georges 343, 346
    [?]Ossonne 343, 346
    Paris 12, 93, 133, 146, 157, 161, 168, 217, 317, 322, 325, 326,
        343, 346, 364, 386, 413, 473-476, 484, 486, 489, 490, 492,
        493, 495-510, 534, 566, 601, 640, 679, 854, 855, 966, 1181,
        1182, 1198, 1548, 1683, 1687, 1702, 1848, 2053, 2072, 2073,
        2075, 2085, 2256
    Periers 350

Perrier 32
Plombières 343, 346
Poix 343, 346
Pont Audemer 350
Pont d'Ain 344, 346
[?]Pont de Pary 343, 346
[?]Ponthierry 343, 346
Pont sur Yonne 343, 346
Port en Bessin-Huppain 350
Potigny 350
[?]Publanne 344, 346
Puy de Dôme 155, 343, 346
Quettehou 350
Quiege 344, 346
Quillebeuf 350
Ravenoville 350
Reville 350
Roselend 344, 346
Rouen 350, 491
Rouvray 343, 346
Rumilly 347
St Albain 343, 346
St Bris le Vineux 343, 346
St Georges de Reneins 343, 346
St Germain de Joux 344, 346
St Gervais 344, 346
St Jean de Maurienne 347
St Jouin 350
St Laurent-sur-Mer 350
St Lo 350
St Louis 348
St Martin 344, 346, 348
[?]St Mayence 343, 346
St Michel 347, 348
St Sauveur 32
St Vaast 350
Salève, Mont 191, 344, 346, 347, 354, 355
Sallanches 344, 346, 348
Savoie 344-346
Sélestat 348
Sens 343, 346
Servoz 344, 346, 348
Sèvres 17, 133, 343, 346, 447, 2081
Strasbourg 348, 349
Termignon 347
Thury-Harcourt 350
Thonon-les-Bains 345, 346
Tilly-sur-Seulles 350
Toulouse 343, 1195
Touques 350
Tournus 343, 346
Troarn 350
Trouville 350
Tsigny 350
Ugine 344, 346
Valognes 32, 350
Vermenton 343, 346
Versailles 343, 346
Vierville-sur-Mer 350
Villefranche 177
Villejuif 343, 346
Villeneuve la Guyard 343, 346
Villeneuve sur Yonne 343, 346
Villers-sur-Mer 161, 350
Villerville 350
Vitteaux 343, 346
Vosges 348
Yport 350
Yvoire 345, 346
Academie Royale des Sciences 12, 306, 326, 475, 602
Cretaceous of 27
geological map of 372, 378, 473, 492, 502, 505
geology of Northern 350
Hydrographic Department 1180
ichthyosaurs from Northern 350
Institut de France 12, 306, 326, 475, 602

Institut de France, De la Beche elected corresponding member 602
nummulitic limestone 8
reports on the geology of 34
Société géologique de 93, 493, 497, 566, 583, 588, 1025, 1180, 1182, 1371, 2054
Société Meteorologique de, De la Beche elected member 603
1848 revolution 1192
Franconia
  bone caves 146
Frangy, France
Frankfurt, Germany 146, 349, 1321
  Cathedral 349
  description of 349
Freiberg, Germany 843
Freiburg, Germany 146
  mining school at 389
Fremington, Devon 31, 1336
Freshwater, Isle of Wight 583, 587
Freshwater East & West, Pembrokeshire 1437
  geology of 1406
Fretherne, Gloucestershire 1474, 1513
Fribourg, Switzerland
  Cathedral 348
  description of 348
Friendship, Wheal, Devon 373
Frogmore 1294
Frogmore, Devon 694
Frome, Somerset 961, 1525, 1527, 1530, 1531, 1548, 1555, 1561, 2122, 2130, 2131
Frome, River 342
  fever cases in vicinity of 628
Ffrwd, Carmarthenshire 1426
Fundy, Bay of, Canada 903
Fusaro, Lago di, Italy 363

Gaeta, Italy 363
Galway, Ireland
  Queen's College 1111
Ganges, River 709, 1678
Garron Pill, Pembrokeshire
  quarries at 351
Garth Hill, Glamorgan
  height of 671
Garthmaelwg, Glamorgan
  geological section to S of 715
Gaspé, Canada
  geology of 882, 883, 885, 888
Gateholm, Pembrokeshire
  geology of 351, 1343, 1413
Gateshead, Durham 74
Geneva, Switzerland 146, 217, 344, 346-348, 499, 1200, 1703
  Botanic Garden 344, 346
  De la Beche's lodgings in, comparison with Scotland 344, 346
  Lake of 344-348, 492, 499, 849
  Registrar of 443
  Société de Lecture 344, 346
  Société de Physique et d'Histoire Naturelle, De la Beche elected Honorary Member 606
Genoa, Italy 490
  description of 347
  geology of 363
  Museum 347
Gent, Belgium 349
Germany 141
  Aachen 349, 2046
  Andernach 349
  Auerbach 960
  Bacharach 349
  Bad Kissingen 905
  Baruth 146
  Bavaria 488
  Berlin 141, 146, 201, 1869
  Bingen 349
  Bohemia 146
  Bonn 146, 250, 337, 349, 678, 924, 932
  Boppard 349

Bruchsal 349
Coburg 146
Cologne 146, 250, 343, 349
Darmstadt 349
Dresden 146, 843
[?Durlach] 349
Eichstätt 343
Eifel 1351
Ems 250
Erfurt 146
Fichtelgebirge 141, 146
Frankfurt 146, 349, 1321
Freiberg 843
Freiburg 146, 389
Gotha 146
Gottingen 134
Grillenburg 843
Hamburg 232
Harz 141
Hattersheim 349
Heidelberg 134, 146, 349
Hochheim 349
Hof 141
Ingelheim 349
Julich 349
Karlsruhe 349
Kehl 349
Kelheim 883
Koblenz 349
Konigswinter 349
Leipzig 146
Luneburg 146
Mainz 349
Meissen 843
Melibocus 960
Munich 488, 2077
Neckar, River 349
Neuwied 349
Oberwesel 349
Pappenheim 343, 2077
Rastatt 349
Rheinfelden 348
Rudesheim 349
St Goar 349
Sackingen 348
Saxony 250, 309
Schwarzwald 348
Siebengebirge 349
Solnhofen 2077
[?Thal Ehrenbreitstein] 349
Tharand 843
Thuringer Wald 146
[?Unkelstein] 349
Waldshut 348
Weimar 146
Weinbohla 843
Weinheim 349
[?Welmich] 349
Westerwald 141
Wurzburg 146
  caves of 165, 166, 343
  diluvium of North 306
  English translation of Boué's Memoir on the Geology of 423
  geology of 423, 843
Gerran Foundry 672
Gerrans Bay, Cornwall 940
  fossils from 1157
  geology of 1161
Giant Mountains, Czechoslovakia 146
Giant's Causeway, Ireland 1680
  comparison with St Mary's, Jamaica 354
Gibara, Cuba 2015
Gibraltar
  caves of 167
  Straits of 1975
Giessbach Falls, Switzerland 348

Giessen, Germany 655, 1825
Gileston, Glamorgan 1717
Gilfach, Carmarthenshire
  lead mine at 1435
Girvan, Ayrshire
  fossil collector of 1032
Glamorgan
  Aberdare 98, 2101
  Abergarw 2117
  Aberpergwm 2101
  Bridgend 7, 80, 102, 176, 872, 952, 965, 2010, 2011
  Briton Ferry 879
  Caerphilly 342
  Cardiff 108, 176, 189, 306-309, 396, 671, 715, 958, 1348, 1357, 1358, 1492, 1681, 1922, 1924, 2102, 2126
  Cefn Cribwr 7
  Cowbridge 147, 176, 715
  Cwmavon 2010
  Dunraven 6, 7, 150, 176, 314, 1376
  Eglwysilan, Mynydd 2102, 2105
  Garth Hill 671
  Garthmaelwg 715
  Gileston 1717
  Gower 145-147, 150-154, 158-160, 162, 164-168, 171, 172, 449, 869, 882, 1336, 2010
  Hirwaun 2101
  Llandaff 148, 258, 313
  Llanfabon 2102
  Llangeinor 2116, 2117, 2119
  Llanharan 715
  Llanharry 7, 715, 2181
  Llansamlet 2273
  Maesteg 5, 2117, 2118
  Margam 2010, 2011
  Merthyr Mawr 149
  Merthyr Tydfil 67, 342, 524, 538, 1679, 2034, 2127
  Mumbles, Glamorgan 449, 1336
  Neath 625, 869, 879, 965, 1957-1959
  Ogmore 149, 327
  Penclawdd 880
  Pendoylan 176
  Pentyrch 2103
  Pontypridd 342
  Porthkerry 314
  Pyle 2010
  Quakers' Yard 2101
  Rudry 2241-2243
  Southerndown 176
  Sully 306, 307, 309, 977
  Swansea, Glamorgan 26-31, 33, 34, 36, 146, 147, 150-152, 158, 159, 191, 246, 342, 388, 389, 411, 448-450, 452-465, 513-517, 559, 560, 626, 696, 800, 830, 832, 851, 868-870, 874-879, 919, 947-949, 965, 985, 1016, 1020, 1099-1101, 1250, 1252, 1263, 1398, 1436, 1443, 1479, 1481, 1527, 1585, 1618, 1639, 1640, 1670, 1679, 1879, 1999, 2000, 2029, 2035, 2079, 2096, 2010, 2100, 2126, 2163, 2257, 2277
  Taff's Well 2240
  Taibach 965, 2010
  Canal 342, 1924
  geological map of 145, 146
  map of, by W. Buckland 200
  proposed visit by W. Buckland 158
  silver bowl found in 147
  state of geological knowlege of 283
Glandore, Ireland
  manganese ore from 780
Glasgow 686, 1713, 1721, 1754
  British Association meeting (1840) 1015, 1342, 1680, 1755, 1756, 1858
  Coal Measures fossils of 703
  plan for Museum of Economic Geology in 1755, 1756, 1760
  University, Lord John Russell installed as Rector 1755
Glastonbury, Somerset 341, 1866
Glendalough, Ireland 1938
  lead mines 1936, 1937
Glenmalure, Ireland 1938

lead mines 1935, 1936
Glen Roy, Inverness-shire 1680
    W. Buckland in 1858
Glossop, Derbyshire 1637
Gloucester, Gloucestershire 147, 182, 342, 742, 747, 1275, 1281, 1284,
    1450, 1451, 1462, 1463, 1469-1475, 1477, 1498, 1500, 1501,
    1516, 1527, 1540, 1545, 1550, 1552, 1572, 1814, 2113, 2126
    Bristol to, railway 742, 747, 748
    Bull Inn 1527
    Cathedral 342
    coal supply to 1281
    Dean of 147
    geology of 747
    Great Western railway to 1284
    Robins Wood Hill 1472
    sketch geological sections near (illn.) 1470, 1472
Gloucestershire
    Almondsbury 1867
    Alveston 1860
    Apperley 1470
    Ashleworth 1500
    Aston Ingham 1575, 1576
    Awre 1502
    Badminton 1515, 1516
    Berkeley 1469, 1512, 1513, 1515, 1550
    Bulley 1470, 1471, 1478, 1500
    Cam 747
    Cheltenham 1281, 1445, 1450
    Chipping Camden 731
    Cirencester 342, 806
    Coalpit Heath 2127
    Coleford 1282, 2110-2115
    Cromhall 1861, 1863, 1865
    Dean, Forest of 152, 310, 420, 700, 779, 1275-1278, 1280-1282,
        1285, 1287, 1410, 1465, 1502, 2110-2113, 2115, 2128, 2129
    Dursley 1469, 1503
    Flaxley 1478
    Fretherne 1474, 1513
    Gloucester 147, 182, 342, 742, 747, 1275, 1281, 1284, 1450, 1451,
        1462, 1463, 1469-1475, 1477, 1498, 1500, 1501, 1516, 1527,
        1540, 1545, 1550, 1552, 1572, 1814, 2113, 2126
    Huntley 1471, 1478, 1498, 1500
    Iron Acton 1861
    Longhope 157
    Lydney 2113
    May Hill 1445, 1451, 1454, 1456, 1460, 1499, 1572, 1575-1577,
        1579, 1581, 1611, 1617, 1662, 2266
    Mickleton 731
    Minsterworth 1500
    Mitcheldean 1481, 1502, 1814, 2114
    Newent 1426, 1441, 1443, 1444, 1454-1456, 1462, 1467, 1468,
        1498, 1609
    Newnham 1469, 1476, 1500-1502, 2113
    Newport 342
    Oldbury-on-Severn 1861
    Old Passage 341, 342, 977
    Purton 1512, 1513, 1515,
    Randwick 1474
    Rudford 1478
    Sodbury 176, 1504, 1516, 1863, 1865
    Standish 1473
    Staunton 1450, 1500
    Stroud 342, 1284, 1469, 1503, 1506, 1509, 1515
    Taynton 1471
    Tetbury 342
    Tewkesbury 731
    Thornbury 1863
    Tortworth 747, 1475, 1500, 1503, 1512, 1571, 1579, 1598, 1611,
        1662, 1663
    Westbury on Severn 179, 1474-1476, 1500, 1510, 1609, 1610
    Wickwar 749, 1515, 1863, 1865
    Winchcombe 731
    Wotton-under-Edge 191, 747-749, 751, 1475, 1498, 1500, 1501,
        1503-1516, 1540, 1550, 1685
    Geological Survey in 1470-1478
    Geological Survey map 1545, 1552

Glyder Fawr, Caernarvonshire 1777, 1784
Godor, Carmarthenshire 1401
Goetre-uchaf, Carmarthenshire 1400
Goginan, Cardiganshire
    lead mine 1940, 1942, 1952
Golden Cap, Dorest 341, 342
    section to Sidmouth from (illn.) 433
Golden Grove, Carmarthenshire 1431
    geology of 1432, 1438-1440, 1623
Good Hope, Cape of South Africa
    currents off of 71
    geological survey of 585
Goodrich, Herefordshire 307
Goodrington, Devon
    geology of 32
Goole, Yorkshire 744
Gorey, Ireland 761, 762, 1084, 1085
Gorinchem, Netherlands 349
Gorran Haven, Cornwall 946, 1156-1159, 1337
    geology of 1161, 1163
Gotha, Germany 146
Gottingen, Germany
    University Library 134
Goultrop Roads, Pembrokeshire
    geology of 351
Gower, Glamorgan 145, 152
    Caswell Bay 882
    Cefn Bryn 162
    Crawley Rocks 162
    Mumbles 449, 1336
    Nicholaston 162
    caves of 162, 164-167
    geological map of 145, 151
    geology of 153, 869
Graaff-Reinet, South Africa 19
Grahamstown, South Africa 19
Grandscamp, France
    geology of 350
    section (illn.) 350
Grandson, Switzerland
    chateau at 347
Grandvilliers, France 343, 346
Grassholm, Pembrokeshire
    geology of 351
Graz, Austria 1022
Great Lakes, fossils from 157
Greeb Point, Cornwall 939
Greece 153, 363, 490, 622
    Corfu 363
    Ionian Islands 363
    Tertiary limestones of 548
Greenan, Ireland 1097
Greenland 148, 921
Green Mountains, United States
    geology of 888
Greenwich, London
    boring into London Clay 295
    Hospital 1966
Grenoble, France 344, 346
Gribbin Head, Cornwall 946
Grillenburg, Germany 843
Grindelwald, Switzerland 348
Grindon, Staffordshire 1832
Grisons, Switzerland 153, 347, 348
Gross Monch, Switzerland 348
Guadeloupe 352, 353
Guernsey 836
    granite for roadstone 1269
    sailings from Weymouth 341
Guildford, Surrey 37, 38, 40, 1335, 1358
Gumfreston, Pembrokeshire
    geology of 351
Gurnard Bay, Isle of Wight 588, 590
Gurnards Head, Cornwall 382
Gwaun, Cwm, Pembrokeshire 954
Gwili, Afon 1397

Haarlem, Netherlands 349
    Cathedral 349
    Lake, draining of 1907
    Teylerian Museum 349
Hackness, Yorkshire 1643
Hadleigh, Suffolk 676, 677
Hague, The, Netherlands 349
    W. Strangeways travelling to 171
Haileybury College, Hertfordshire
    geology of area of 809
Haldon, Great, Devon 28, 29
    section in well at 174
Haldon Hill, Devon 341
    geology of Little 341
Halesowen, Worcestershire 812
Halifax, Canada 874-876, 880, 882
Hallstätt, Austria 1022
Hamburg, Germany
    drainage of 232
Hamps, River 1597
Hampshire 146, 151
    Aldershot 206
    Basingstoke 744
    Brockenhurst 1151
    Hordle 1143
    Lymington 561, 1151
    New Forest 418, 1783
    Petersfield 1011, 1023, 1026
    Portsmouth 788, 789
    Ringwood 2042
    Southampton 237, 313, 397, 399, 552, 649, 650, 695, 723, 738,
        756, 790, 792, 863, 1071, 1072, 1604, 1606, 1660, 1743, 1762,
        1775, 1776, 1790, 1819, 1927, 1928, 2167-2172
    Spithead 2034
    Winchester 114
Hamrin Mountains, Iraq 865
Hamstead, Isle of Wight
    geology of 590-591
Harfleur, France
    geology of 350
Harlech, Merionethshire 1776
Hartington, Derbyshire 1605, 1971
Hartland, Devon
    compared with Dinas Head, Pembrokeshire 947
Hartlepool, Durham
    New Colliery 430
Harz, Germany 141
Hastings, Sussex 3, 4, 701, 897
    geology of 293, 534
Hatherleigh, Devon 1337
Hattersheim, Germany 349
Havana, Cuba 984
Haverfordwest, Pembrokeshire 351, 793, 950, 954, 1343, 1362, 1364-
    1366, 1376, 1383, 1384, 1389, 1707-1709, 1815, 2002, 2109
    fossils from 793, 2133
Hawkhurst, Kent 679
Haydon Bridge, Northumberland 1977, 1978
Headon Hill, Isle of Wight 582, 587, 588
Hebrides
    Lias of 176
    specimens for Museum of Economic Geology 2055
    Tertiary volcanic rocks of 921
Heddon's Mouth, Devon 1336
Heidelberg, Germany 134, 146, 349
    Castle 349
    University 349
Helmsley, Yorkshire
    fossils 1328
Helston, Cornwall 43, 376, 382, 939-941, 1828, 1875, 1876
    Grammar School 805, 806
Hembury Hill, Devon
    geological structure of 29, 30
Hemel Hempstead, Hertfordshire
    chalcedony from, for Museum of Economic Geology 1225

Henley-on-Thames, Oxfordshire 342
    Chalk at 308
Henllan Amgoed, Carmarthenshire 1991
Hennequeville, France
    geology of 350
    sections (illn.) 350
Herculaneum, Italy 363
Hereford, Herefordshire 1376, 1482
Herefordshire
    Aymestrey 119, 1354, 1468
    Brampton Bryan 118
    Colwall 1442, 1452, 1479
    Dormington 1496
    Goodrich 307
    Hereford 1376, 1482
    Kington 560, 1726, 1738, 1740, 1742, 1743, 1764, 1765
    Ledbury 1410-1412, 1414, 1426, 1438, 1441-1443, 1445, 1446,
        1449-1451, 1460, 1479, 1498-1500
    Leintwardine 1738, 1740
    Limebrook 119
    Lingen 119
    Marcle, Little 1447, 1451
    Ross-on-Wye 157, 342, 1281, 1287, 1411, 1451, 1455, 1460, 1464,
        1468, 1469, 1483, 1506, 1715, 1717, 1821
    Stoke Edith 1447
    Symonds Yat 307
    Tarrington 1446, 1451, 1453, 1457-1459, 1464, 1467, 1478
    Welland 1461
    Wigmore 1496
    Woolhope 1426, 1441, 1443, 1445-1449, 1451, 1452, 1454, 1460,
        1461, 1463, 1464, 1468, 1469, 1478, 1479, 1482, 1499 1540,
        1611, 1662, 2266
Hermance, Switzerland 345, 346
Hertford, Hertfordshire
    geology of 809
Hertfordshire
    Haileybury College 809
    Hemel Hempstead 1225
    Hertford 809
    Hitchin 129
    Royston 809
    St Albans 92
    Sandridge 92
    Ware 809
Herzogenbuchsee, Switzerland 348
High Stile, Cumbria
    geological section through (illn.) 1870
Hillah, Al, Iraq 865, 866
Himalayas 709, 713
    geological map of 1678
    geological survey of 1678
Hindelbank, Switzerland 348
Hirwaun, Glamorgan 2101
Hitchin, Hertfordshire 129
Hitcham, Suffolk 676, 677
Hochheim, Germany 349
Hodgeston, Pembrokeshire
    geology of 351
Hof, Germany
    occurrence of goniatites 141
Holcombe, Somerset
    Coal Measures at 2132
Holcombe Rogus, Devon 1336, 1337, 1876, 1877
    fossils 1340
Holguin, Cuba 2015
Holnicote, Somerset 1, 2
Holyhead, Anglesey 1128, 1522, 1523, 1939, 1963
    road 1794
Holywell, Flintshire 739, 1969, 1970
Holywell, Dorset 1764
Honddu, River 342
Honfleur, France 161, 350
    fossils from 343, 346
    section (illn.) 350
Honiton, Devon geology of 341

Hook, Ireland 554, 558, 579, 1120
    geology of 1109
Hope's Nose, Devon
    geology of 341, 1338
Hopton, Derbyshire 1655
Hordle, Hampshire
    fossil mammals from 1143
Houches Les, France 344, 346
Houghton, Pembrokeshire
    geology of 351
Howth, Ireland 1127
Hudson Bay, Canada 885, 887
    Company 885
Hudson River, United States 883
Hungary 146, 645, 1928
    Budapest 1927
    criticism of post office in 1927
    geology of 1927
    mining 1927
Huntley, Gloucestershire 1471, 1478, 1497, 1500
Huron, Lake 885-887

Iceland 921
    hot springs 2046
Ideford, Devon
    geology of 36
Ilam, Staffordshire
    Isaac Walton Hotel 1596-1600
Ilfracombe, Devon 34, 150, 1332, 1408, 1680, 1681, 1877
    Carboniferous Limestone fossils found at 1336
    limestone from Mumbles burnt at 1336
Illinois, United States
    geology of 695
Ilsington, Devon 844
    iron workings 33
India 190, 495, 703, 705, 706, 713, 833, 865
    Allahabad 2138
    Aurangabad 97
    Bengal 1678, 2138, 2143, 2146
    Bombay 97, 865
    Calcutta 164, 709, 710, 963, 1678, 1836, 2025, 2032, 2034, 2035, 2137, 2138, 2140, 2141, 2146
    Damodar River 2137-2139, 2143-2147
    Darjeeling 709
    Delhi 2138
    Ganges 709
    Lucknow 1678
    Madras 736
    Meerut 735, 736
    Punjab 2138
    Raniganj 2139-2142
    Roorkee 962-964
    Sundarbans, The 1678
    cholera in 2140, 2141
    D.H. Williams surveying for coal in 416, 445, 446, 479, 2025, 2136-2141, 2143-2148
    Geological Survey of 1115, 1116, 1119, 1121-1123, 1125, 1133, 2006-2007
    geology of 97, 709
    gold in 2146
    plan for a Geological museum in 2028-2035
    plants from 164
    pottery from, for Great Exhibition 482
    Protectionist meeting on 483
    railways planned 2138
Ingelheim, Germany 349
Innsbruck, Austria 1022
Interlaken, Switzerland 348
Inveraray, Strathclyde 1183
    cobalt deposits 15
    nickel deposits 14, 15
Inverness-shire
    Dalwhinnie 1858
    Glen Roy 1680, 1858
Ipswich, Suffolk 1829
    Museum, De la Beche invited to anniversary of 1809

    Museum of Economic Geology at 677, 1985
    Museum, portrait of Bishop of Norwich 1811
    Museum, portrait of Professor Henslow 1810
Iran
    Bakhtiari 867
    Baneh 867
    Kermanshah 867
    Kordestan 865
    Sar Dasht 867
Iraq
    Al Hillah 865, 866
    Babylon 865-867
    Baghdad 865-867
    Ctesiphon 865
    Hamrin Mountains 865
    Karbala 866
    Kirkuk 865
    Mandali 867
    Mosul 864, 867
    Nineveh 865, 867
    Sulaymaniyah 867
    Wasit 865, 866
    archaeology of 865, 867
    geology of 867
Ireland 306, 554, 575, 592, 622, 650, 832, 903, 1285, 1604
    Antrim 2042
    Arklow 762, 1096, 1128, 1931-1934, 2042
    Armagh 558, 1079, 1080
    Arthurstown 755, 1124
    Ashford 1071, 1097
    Avoca 1077, 1096, 1932, 1934-1937, 1939
    Balbriggan 1105
    Ballinagar 1097
    Ballycotton Bay 779
    Ballymanus Bridge 1097
    Baltinglass 1066
    Barristown 772
    Belfast 139, 306, 1111, 1696, 2027
    Birr 1522, 1523
    Bray 1087
    Cahore Point 1084
    Carlow 558, 757, 771, 772, 791, 1101, 1102, 1105, 1109, 1119, 1931, 1933
    Castlecomer 1930, 1931
    [?]Castletown [Bearhaven] 780, 782
    Cleenish 558
    Clonakilty 780
    Clonmel 1125
    Comeragh Mountains 1126
    Connemara 652
    Cork 652, 776, 777, 779, 814, 816, 829, 903, 1045, 1046, 1111, 1114, 1520-1522, 1583, 2026
    Courtown 557, 558, 1079, 1083, 1084, 1086
    Croghan 762, 766, 769, 1072, 1096, 1932, 1933
    Dan, Lough 1096
    Dingle peninsula 1522, 1523
    Donegal 1218
    Drogheda 1545
    Dublin 35, 50-52, 89, 140, 259, 260, 268, 306, 402, 403, 406, 407, 557, 558, 564-567, 579, 593, 659, 751-754, 756-786, 801, 804, 814, 815, 817-819, 821, 823, 825, 826, 828, 830, 843, 922, 970, 1045, 1064,1065,1067, 1068, 1071-1083, 1085, 1088-1099, 1135, 1314, 1316, 1522, 1535, 1538, 1540 -1546, 1548, 1549, 1551, 1552, 1554-1556, 1558, 1559, 1561, 1567, 1569, 1574, 1680, 1697, 1847, 1938, 1939, 2008, 2009, 2042, 2091, 2171
    Duncannon 755
    Dundalk 652, 1105
    Dunloe, Gap of 781
    Enniscorthy 772, 1108, 1110, 1112
    Enniskerry 1088
    Enniskillen 1070, 1071, 1109
    Fethard 407
    Foulksmill 1121, 1124
    Galway 1111
    Giant's Causeway 354, 1680

Place Names   *Ireland - Italy*

Glandore 780
Glendalough 1936-1938
Glenmalure 1935, 1936, 1938
Gorey 761, 762, 1084
Greenan 1097
Hook 554, 558, 579, 771, 1109, 1120
Howth 1127
Kenmare 780, 781
Kerry 319, 1114, 1522
Kildare 554, 565, 567, 568, 770, 814, 821, 1065, 1081, 1086, 1090, 1097, 1098, 1102, 1105, 1106, 1119, 1132, 1266
Kilkenny 757, 771, 772, 791, 815, 1120, 1934
Killaloe 1118
Killarney 780, 781
Killiney 1939
Knockmahon 759, 1123, 1199
Lambay Island 1097, 1963
Limerick 1246, 1537, 1625
Lugnaquillia Mountain 1096, 1936
Malahide 579
Maryborough 761
Monkstown 816
New Ross 755, 772, 1102, 1121
Newry 1127
Newton Barry 772
Newton Head 755
Newtownbarry 1072
Ow River 1097
Rathdrum 772, 1096, 1934-1937, 1939
Ross Carbery 776
Taghmon 759
Tara 761, 762, 766, 1096
Tinahely 1072, 1931, 1932, 1935
Tipperary 815, 1127
Tramore 407, 1085, 1123
Swords 2042
Waterford 553, 751, 757, 771, 772, 776, 777, 791, 815, 817, 819, 1085, 1114, 1120, 1123, 1127, 1128
Wexford 420, 550, 751, 755-757, 759, 772, 775, 779, 791, 818, 1072, 1084, 1096, 1102, 1107, 1109-1111, 1117, 1120-1123, 1133, 2042
Wicklow 420, 757, 770, 772, 791, 814, 1071, 1072, 1077, 1084, 1096, 1108, 1122, 1936, 1939, 1950
Woodenbridge 1096, 1128, 1932-1934
Youghal 418, 776, 777, 779
basalt columns from 343, 346
Board of National Education 2239
Catholic Emancipation 1170
Conybeare in 306
conditions in 558
copper in 319
crime (1847) 1089
disturbances (1848) 1092-1095
early copper mining 2026
famine 756, 761, 1027, 1076, 1077
fisheries 50-52, 2239
Fishery Commissioners 50
fossil fish of 558
geological map of 246, 649, 2172
geological sketches (illn.) 652
Geological Society of 51
Geological Survey of 128, 223, 400, 402, 403, 406, 407, 418, 552, 558, 592, 593, 650, 751-787, 791, 792, 801, 808, 813-818, 820, 821, 825, 828, 831, 833, 1065-1069, 1071-1083, 1085, 1087-1091, 1093-1110, 1114, 1115, 1117, 1119, 1120, 1122-1124, 1126-1128, 1130-1133, 1135, 1313, 1316, 1318, 1407, 1484, 1493, 1538, 1543, 1546, 1547, 1553, 1554, 1556, 1559, 1562, 1730, 1930-1939, 2009
geology of 1046
geology of, comparison with Devon 652, 1583
geology of, comparison with S W Wales 1583
iron ore in 319
limestone in 319
Lord Lieutenant 1078, 1106
Mining Company of 825
Office of Public Works 808

one inch map 1082, 1114
Ordnance Map of 856
Pleistocene fossils of 550
rising (1848) 1089, 1092-1095
Silurian of 557, 558, 563, 565, 567, 568, 762, 770, 1079, 1080, 1081
Iron Acton, Gloucestershire 1861
Ironbridge, Shropshire 1950
Ischia, I. d', Italy 363
Isère, River, France 344, 346, 347
Isis, River (see Thames, River)
Islip, Oxfordshire 142, 1152
Istanbul, Turkey see Constantinople
Italy
    Ancona 363
    Acquapendente 363
    Alessandria 347
    Ameglia 363
    Aosta 344, 346
    Appennines 332, 347, 348, 363, 1025
    Arquata Scrivia 363
    Asti 347
    Bellagio 363
    Bellano 363
    Bolca, Monte 146
    Bologna 82-84, 348
    Bolsena, Lago di 363
    Bolzano 1022
    Bonassola 363
    Borghetto di Vara 363
    Bormio 1022
    Bracciano, Lago di 363
    Capri 363
    Capua 363
    Carrara 363
    Castellammare 363
    Chiavari 363
    Chiavenna 363
    Civita Castellana 363
    Como 363
    Courmayeur 344, 346
    Elba 1927
    Etna, Mount 897
    Etroubles 344, 346
    Florence 168, 256, 347, 348, 363, 620, 1818
    Foligno 363
    Fondi 363
    Fusaro, Lago di 363
    Gaeta 363
    Genoa 347, 348, 363, 490
    Herculaneum 363
    Ischia, I d' 363
    La Salle 344, 346
    La Spezia 317, 347, 363, 1027
    Lecco 363
    [?]Lerici 348
    Levanto 363
    Lierna 363
    Lodi 348
    Loreto 363
    Lucca 347, 363
    Macerata 363
    Mandello 363, 491
    Mantova 146
    Massa 347, 363
    Merano 1022
    Milan 146, 155, 348, 363
    Modena 348
    [?]Mont Carelli 348
    Montefiascone 363
    Monterosi 363
    Naples 171, 363, 1025, 1922
    Narni 363
    Nepi 363
    Novara 348
    Novi 347, 363

Osimo 363
Paestum 363
Parma 146, 348
Pavia 363
Perugia 363
Pescia 363
Piacenza 348
[?]Pietra 348
Pisa 347, 363, 1025
Pistoia 363
Poggibonsi 363
Pompeii 363
Portici 363
Portovenere 363
Pozzuoli 363
Prato 363
Procida, I di 363
Radicofani 363
Rapallo 347, 363
Recanati 363
Recco 363
Riomaggiore 363
Rivoli 347
Romagnano Sesia 343, 346
Rome 256, 343, 346, 363, 1025, 1818
Ronco 363
St Rhemy 344, 346
San Casciano 363
San Primo, Monte 363
San Quirico d'Orcia 363
S Ambrogio di Torino 347
Sarzana 347, 363
Seravezza 363
Sestri 347, 363
Sicily 332, 1025
Siena 363
Somma, Monte 897
Sorrento 363
Spoleto 363
Susa 347, 348
Tavarnelle 363
Terni 363
Terracina 363
Ticino, River 348
Tivoli 363
Torre Annunziata 363
Torre del Greco 363
Turin 347, 348, 363, 974, 1025, 1909-1918, 2003, 2045
Varenna 363
Velletri 363
Venice 146, 148, 151, 1022
Verona 146, 1184
Vesuvius 141, 146, 363, 897, 1921
Vicenza 146
Viso, Monte 347
Viterbo 363
Voghera 363
geology of 332, 347, 363, 1028
offer of specimens from 84
unrest 1025
political situation in 83, 1909-1911
Ivybridge, Devon 844

Jamaica 311, 352-355, 373, 714, 984, 1154, 1921, 2245
Abbey Green 352
Agualta Pen Estate 352
Agua Alta River 352
Agualta Vale Estate 352, 870
Albion 352
Annotto Bay 352, 357, 870
Back River 352
Balcarres 352
Bath 352
Belmont Hill 352
Black Hill 352
Black River 352

Blue Mountains 352, 353, 356-358, 920, 984
Boston Estate 352
Botanic Gardens, curator sought for 686
Brumalia Estate 352
Buff Bay 352, 356, 357
Bull Head 352
Burlington Estate 352
Burnt Savanna 352
Bushy Park Estate 352
Canewood Estate 352
Caratoe Hill 352
Carlton Wood Estate 352
Carpenter Mountains 352, 355
Carrion Crow Hill 352
Cave River 352
Chapleton 352, 358
Charles Town 352, 357
Clarendon 352, 354, 358, 983, 984, 1905
Clarendon Gully 352
Clarendon Mountains 352
Clarendon Park Estate 352
Clifton 352
Cobre, Rio 352
Cold Ridge 352
Collier's River 352
Crawle, The 352, 354
Eolus Valley 352
Essex Valley 352
Fairy Hill Bay 352
Farm Hill 352
Flamstead 352
Folly Point 352
Forster's Cove 352
Fort Haldane 352
Four Paths Estate 352
Golden Valley Estate 352
Green Castle Estate 352
Green Valley 352
Guatamala Estate 352
Halse Hall 301-305, 352, 354, 355, 357-359, 688, 689, 959, 1905, 2023, 2024
Hanbury Pen 352
Harris's Savanna 352
Hay's Savanna 352
Healthshire Hills 352
Hill Side Estate 352
Hope 920
Hope, River 352, 356, 357
Islington Estate 352
Jacks Bay 352
Juan de Bolas Great River 352
Kemp's Hill 352
Kingston 352-354, 356, 357, 920
Konigsburg Estate 352
Lenox Estate 352
Leyton Estate 352
Liguanea Mountains 352
Lime Savanna 352
Llandewey 352
Lodge Estate 352
Long Mountain 352
Luidas Vale 352, 355
Manchester Mountains 352
Manchester Parish 352, 354
Manchineal Harbour 352
Mandeville 352
Marlbro' 352
Martin's Hill 352
May Day Hills 352
Middleton 352
Mile Gully 352
Milk River 352
Minho, Rio 352, 983
Mocho Mountains 352, 354, 983
Moore Town 352
Morant River 352

Mountain River 352
Mount Pleasant 352
Mount Vernon 1016
Navy Island 352
New Forest Estate 352, 355
Newry Mount 352
Norris Estate 352
Nutfield 352
Old Harbour 352, 355, 983
Orange Hill Estate 352
Palmetto Gully 352
Parnassus 352
Petersfield River 352
Pillew Island 352
Pimento Grove 358
Plowden Hill 352
Plantain Garden River 352
Pleasant Hill 352
Port Antonio 352, 357
Portland Cave (illn.) 358
Portland Gap 352, 870
Portland Point 358, 984
Port Maria 352
Port Morant 352
Port Royal 352, 353, 355, 896, 976
Rhymesbury 352
Rio Cobre 352
Rio Grande Valley 352
Rio Minho 352, 983
River Hope 352, 356, 357
Rock Road 352
Round Hill 352
St Andrew's 920
St Andrews Mountains 352
St Ann's Gully 352
St Ann's Parish 358
St Catherines Peak 352, 356, 357
St David's Parish 352, 357
St Dorothy's 983
St Elizabeth's Parish 352
St Elizabeth Savanna 352
St Georges Gap 352, 357, 870
St George's Parish 352, 356
St Jago Savanna 352
St John's Mountains 352, 355
St Mary's Parish 352, 354-356
St Thomas 984
Salt Island 352
Salt River 352
Sandy Gully Estate 358
Santa Cruz Mountains 352
Savanna Point 352
Serge Island 352
Sheerness Estate 352
Sheldon 352
Sixteen Mile Gully 352
Skibo Estate 352
Somerset Estate 352, 357
Spanish River 352
Spanish Town 352, 354
Spring Estate 352
Spring Garden 352
Stony Hill 352, 357
Sunning Hill 352
Swansea Estate 352
Swift River 352, 870
Temple Hall Estate 352
Thomas's River 352
Trelawney 984
Vere 352, 358, 983
Vere Savanna 352
Warwick 352
Williamsfield 352, 358
Wood Hall Estate 352
Woodstock Estate 352
Worthy Park Estate 352

Yallah's Hill 352, 353
Yallah's River 352, 357, 358
agates from 352
analysis of limestone from 332
animal breeds of 329
botanic garden 352, 686
caves of 173, 352, 355, 358, 984
claystone with feldspar crystals from (illn.) 358
coinage of 687
Commissioners of Public Accounts 687
conditions of slaves in 352, 354
copper 1016
De la Beche's journey to 352, 353
De la Beche's paper on geology of 173, 332, 531, 1672, 2020
De la Beche's property in 260, 341, 352, 688, 689, 959, 1905
earthquakes in 352, 356, 357, 896, 976
food prices in 352
geological map of 355
geology of 354-358, 870, 2020
geology of, comparison with Europe 356, 357
land crabs of 130, 352, 355
malaria in 352
mineral springs 352
missionary for De la Beche's estate on 1905
Negro disturbances in 352, 354, 355, 959
plants of 131, 352, 357, 840, 841
Scheme 114, 115
section, Blue Mountains (illn.) 356
section, Hope Valley to St Catherines Peak (illn.) 356
section, Kingston to Buff Bay (illn.) 357
section, Kingston to North Coast (illn.) 356
section, Luidas Vale to Old Harbour (illn.) 355
shells from 653
slavery 113, 114, 352, 354, 355, 357, 359, 1154, 2023
sugar mill in 116
sugar production from De la Beche's estate 373, 959
water supply 983
wood ants of 352
Jameston, Pembrokeshire
    geology of 351
Jarrow, Durham 1586
Jeffreyston, Durham
    geology of 351
Jersey 836
    sailings from Weymouth 341
Joggins, Canada 886
    coalfield 883, 884
    fossil trees at 882
Johnston, Pembrokeshire
    geology of 351, 1815
Joigny, France 343, 346
Joly, Mont, France 344, 346, 348
Jorat, Mont, Switzerland 348
    geology of 348
Julich, Germany 349
Jungfrau, Switzerland 348
    section (illn.) 428
Jura, Argyllshire
    raised beaches on 70
Jura Mountains 344-348, 364
    comparison of Oolite fossils with England 343, 344, 346
    geology of 341
    granite erratics (illn.) (illn.) 347
    specimens from, in Neuchâtel Museum 347

Kanchenjunga, Mount, Nepal 709
Karachi, Pakistan 1038
Karbala, Iraq 866
Karlsruhe, Germany
    description of 349
Katwijkaan Zee, Netherlands 349
    dams at 349
Keban, Turkey
    mines of 864
Kedleston, Derbyshire 1646
Kehl, Germany 349

Kelheim, Germany
    lithographic stone from 883
Kendal, Westmorland 1354
Kenmare, Ireland 780, 781
Kent
    Brasted 533
    Broadstairs 343
    Canterbury 343
    Collingwood 679-684, 1058
    Dartford 343
    Deal 343, 2079
    Dover 146, 343, 346
    Folkestone 810
    Hawkhurst 679
    Margate 343
    Pegwell Bay 343
    Ramsgate 343
    Rochester 343
    Sandgate 580
    Sandwich 343
    Sevenoaks 532
    Sheerness 1293
    Sittingbourne 343
    Tonbridge 1819
    sewers 934
Kermanshah, Iran 867
Kerry, Ireland
    copper and iron ore in 319
    geological map 114
Keswick, Cumberland 154
Kew, British Association observatory at 1279
Keynsham, Somerset
    geology of 307, 342
Kidderminster, Worcestershire
    Lion Inn 1609
Kidwelly, Carmarthenshire 625
    Carboniferous Limestone of 869
    geology of 878
Kildare, Ireland 821, 1065, 1100, 1132
    geological map of 814, 1098, 1101, 1102, 1105, 1106, 1119
    geology of 565
    limestone fossils of 554, 567
    Ordnance map of 1266
    Silurian of 770, 1081, 1086, 1090
Kilkenny, Ireland 771, 772, 1120, 1934
    1" map of 791
    maps of 757, 815
Killaloe, Ireland 1118
Killarney, Ireland 780, 781
Killiney, Ireland 1939
Kilmacthomas, Ireland 1123
Kinder Scout, Derbyshire 1627
    sketch section (illn.) 1637
Kingsbridge Estuary, Devon 694
Kingskerwell, Devon
    geology of 341
Kingsteignton, Devon
    geology of 341
Kingston, Jamaica 356
    description of 352, 353
    geology of 354
    Mr Miller's museum at 352
Kingstown, Ireland 1565-1569
Kington, Herefordshire 560, 1726, 1738, 1740, 1742, 1743, 1764, 1765
Kirkbymoorside, Yorkshire
    fossils 1328
    Kirkdale Cave near 162, 165, 171
Kirkcudbright, Kirkcudbrightshire
    Silurian of 541
Kirkcudbrightshire
    Kirkcudbright 541
    St Mary's Isle 1900, 1901
Kirkdale Cave, Yorkshire 171
    as hyaena den 162, 165
    discovery of 162
    fossils from 1849
    Pleistocene vertebrates of 162
Kirkuk, Iraq 865
Knighton, Radnorshire 119-121
Knockmahon, Ireland 1123, 1199
    tracing of Old Red Sandstone at (illn.) 759
Koblenz, Germany 146, 349
    description of 349
Konigswinter, Germany 349
Kordestan, Iran 865
Krakow, Poland
    geology of 146
Kremnitz
    gold mines 146

Labuan, Malaysia 992, 993
    coal 992, 993
    geology of 992, 993
    map of (illn.) 992
    Taujong Kubong 992, 993
La Chambre, France 347
Lake District
    geology of 1870
Lambay Island, Ireland 1097, 1963
Lampeter, Cardiganshire 1394, 1395, 1428, 1433, 1435, 1486, 1664, 1719
    geology of 1438
Lampeter Velfrey, Pembrokeshire 1380, 1382, 1394
    geology of 1392, 1401, 1429, 1432, 1439
Lamphey, Pembrokeshire
    geology of 351
Lancashire 1190
    Accrington 953-958
    Blackburn 922
    Bolton 890, 1612
    Chorley 328, 1944
    Cuerdale 1041
    Coppull 1944
    Lancaster 327
    Liverpool 136, 600, 611, 706, 739, 791, 879, 880, 903, 1036, 1067, 1986, 2018
    Newgate 109, 110
    Oldham 693
    Ormskirk 1991, 1992
    Patricroft 1039
    Preston 1354, 1356
    Stockport 1637, 1638
    coal 328
    6in Ordnance Survey map of 792
    plan for mining school in 1652
Lancaster, Lancashire
    Duchy of 327, 2175
Landewednack, Cornwall
    coloured section near 1990
Land's End, Cornwall 353, 381, 383, 2029
    Ordnance map of 265
Landshipping, Pembrokeshire
    geology of 2111
Lanslebourg, France 347
La Salle, Italy
    geology of 344, 346
la Sarraz, Switzerland 347
La Spezia, Italy 347, 1027
    analysis of limestone from 317
    people of 363
    section from, to Coregna (illn.) 363
    vertical section, rocks of (illn.) 363
Laufenburg, Switzerland
    geology of 348
Laugharne, Carmarthenshire 1378, 1381, 1432
    geology of 351
Launceston, Cornwall 1336, 1337, 1875, 1876, 1878
    geology of 25, 26, 31
Lausanne, Switzerland 344-348
    Cathedral 348
    geology of 347
Lauterbrunnen, Switzerland 348
    Valley, geology of 348

Lavenza, Italy 363
  section from Carrara Marble Quarries to (illn.) 363
Lawrenny, Pembrokeshire
  geology of 351
Lea, River 809
Leamington, Warwickshire 730, 2084
Lecco, Italy 363
  between Onno and (illn.) 363
  construction of road to Splügen, Switzerland 363
  geology of 363
Ledbury, Herefordshire 1410, 1411, 1412, 1414, 1426, 1438, 1441,
    1442, 1443, 1445, 1446, 1449, 1450, 1451, 1460, 1479,
    1498-1500
Leeds, Yorkshire 1691
  Philosophical Institution 91
  Pottery 1658
  sanitary conditions of 232
Leek, Staffordshire 1830
  sketch section at (illn.) 1639
Le Havre, France 350, 1037
  fossil crocodile from 350
  fossils from 1328, 1329
  Phillips' paper on geology of 1328, 1329
  section (illn.) 350
Leicestershire 1020
  Ashby de la Zouch 1635, 1636, 1639
  Loughborough 1636
  Quorndon 1646
Leiden, Netherlands 349
  description of 349
  Siege of 349
  University of 349
Leintwardine, Herefordshire 1738, 1740
Leipzig, Germany 146
  booksellers in 640
Leissigen, Switzerland 348
Lemberg 146
Lerici, Italy
  from Santerenzo (illn.) 363
  geology of 363
Lerwick, Shetland 549
Les Houches, France 344, 346
Leuven, Belgium 349
Levant Mine, Cornwall 600
Levanto, Italy
  geology of 363
  panorama, Monte Rosso and (illn.) 363
  sections, Capo Mesco and (illn.) 363
Lewes, Sussex 925, 1817
Libertia
  fossil elephant from 343
Liège, Belgium 146
  School of Mines of 244
Lierna, Italy
  geology of 363
Lilleshall, Shropshire 2088
Lima, Peru
  earthquake 896
Limebrook, Herefordshire 119
Limerick, Ireland 1625
  building stone from 1246
  coalfield near 1537
Limonest, France
  copper mine near 343, 346
Lincoln, Lincolnshire
  model of 228
Lincolnshire
  Caistor 676
  Lincoln 228
  Market Rasen 169, 170
  Spalding 1159
Lingen, Herefordshire 119
Linz, Austria 1927
Lion sur Mer, France
  geology of 350
Lisbon, Portugal 1904, 2079

Lisieux, France
  geology of 350
Liskeard, Cornwall 941, 945, 946
Lison, France
  geology of 350
Little Haven, Pembrokeshire
  geology of 351
Littry, France
  geology of 350
Liverpool, Lancashire 611, 706, 739, 879, 880, 903, 1036, 1067, 1536,
    1834, 2018
  and Metropolitan Sanitary Commission 1685
  British Association meeting (1837) 136, 600
  British Association meeting (1854) 791
  sailing time to Ireland 1541
  water supply 1986
Livorno, Italy
  panorama, Alpi Apuane to (illn.) 363
Lizard, Cornwall 382, 383, 939, 1874, 1876
Llanarthney, Carmarthenshire 1383, 1388, 1395, 1431
  geology of 1432
Llanbabo, Anglesey 1960, 1961
Llanbadoc, Monmouthshire 1358
Llanberis, Caernarvonshire 562, 1775, 1776, 1778-1784, 1795-1797
  Dolbadarn Inn 1777
  Pass 1792
Llanboidy, Carmarthenshire 954, 1392, 1397, 1664, 1993
  lead at 1666
Llanbrynmair, Montgomeryshire 1732, 1733, 1736, 1746
  Wynnstay Arms 1728, 1730, 1740
Llancayo, Monmouthshire 1851
  geology of 1357, 1358
Llandaff, Glamorgan 148, 258, 313
Llanddewi Brefi, Cardiganshire 1724
Llanddewi Velfrey, Pembrokeshire 1485
  geology of 1440
  sketch section near (illn.) 1485
Llanddowror, Carmarthenshire 1378, 1382, 1383
  geology of 793, 1431
  section (illn.) 1380
Llandegley, Radnorshire 1770
  mineral spring at 342
  sketch section near (illn.) 1771
Llandeilo, Carmarthenshire 157, 626, 1360, 1382-1385, 1389, 1390,
    1392, 1394-1400, 1431, 1433, 1437, 1666, 1714, 1852
  comparison of geology with North Devon 1336
  Dynevor Park, geology of 1403, 1439
  geology of 1386, 1388, 1403, 1429, 1438, 1486, 1711
  geology of (illn.) 1401
  Geological Survey map of 1623
  section (illn.) 1385, 1388, 1389, 1439
Llandeilo Abercywyn, Carmarthenshire 1431
  geology of 1432
Llandovery, Carmarthenshire 1360, 1388, 1389, 1396-1398, 1401,
    1402, 1404, 1405, 1408 - 1410, 1416, 1426, 1431, 1442, 1445,
    1446, 1448, 1536, 1663, 1664, 1714, 1716, 1718 - 1722
  bank 1405, 1448, 1760
  Castle Inn 1399, 1400
  geology of 1399, 1400, 1407, 1852
  mines near 1424, 1435
Llandrindod Wells, Radnorshire 1770
  geology of 342
  lead mined at 342
  mineral springs at 342
Llandybie, Carmarthenshire
  geology of 1387
Llanelli, Carmarthenshire 878
  coal exports 2126
Llanelwedd, Radnorshire 342
  boulders in rocks at 342
  geology of 342
  quarries at 342
Llanerchymedd, Anglesey 1805, 1961, 1962
Llanfabon, Glamorgan 2102
Llanfachraeth, Anglesey
  Holland Arms 1964

Mona Brewery 1964
Llanfair Clydogau, Cardiganshire 1719, 1940, 1947
Llanfairynghornwy, Anglesey
    geology of 1962
Llanfechell, Anglesey 1963, 1964
    geological section (illn.) 1965
Llanfihangel-uwch-Gwili, Carmarthenshire 1397, 1426
Llanfrechfa, Monmouthshire 1359
    fossils of 1360
Llanfyllin, Montgomeryshire 1744
Llanfynydd, Carmarthenshire
    geology of 1426
Llanfyrnach, Pembrokeshire 954
Llangadog, Carmarthenshire 1388, 1389, 1402, 1408, 1433, 1437, 1852
    coach from Carmarthen 1399
    De la Beche at 336, 1441
    geology of 1431, 1438
    geology of (illn.) 1401
    lead mines to E of 1435
Llangathen, Carmarthenshire 1426
Llangefni, Anglesey 1964, 1965
Llangeinor, Glamorgan 2116, 2117, 2119
Llanglydwen, Carmarthenshire 954, 1397
    Castle 1392
Llangollen, Denbighshire 561, 562, 813, 1782
    Dinas Bran 1564
    geology of 2133
    perched block near (illn.) 1564
Llangunnor, Carmarthenshire 1395
    geology of 1425
Llangurig, Montgomeryshire 1946
Llangwm, Pembrokeshire
    geology of 351
Llangybi, Monmouthshire 1851
Llangynog, Carmarthenshire 1431
    lead mines near 1424
Llanharan, Glamorgan
    geological section near (illn.) 715
Llanharry, Glamorgan 7
    geological section near (illn.) 715
    Trecastle Colliery section (illn.) 2181
Llanidloes, Montgomeryshire 1852, 1945, 1946
Llanllyfni, Caernarvonshire 1968
Llannewydd, Carmarthenshire see Newchurch
Llan Mill, Pembrokeshire 1382, 1392]
Llanover, Monmouthshire 646
Llanrhaeadr-ym-Mochnant, Montgomeryshire 45
Llanrwst, Denbighshire 1789
Llansamlet, Glamorgan 2273
Llansawel, Carmarthenshire 1388, 1397, 1426, 1433
Llansteffan, Carmarthenshire 1380, 1432
    geology of 1381, 1431
Llanteg, Pembrokeshire
    geology of 351
Llanwnda, Pembrokeshire
    jasper at 953
Llanwrda, Carmarthenshire 1426
Llanwrtyd Wells, Breconshire 1852
    geology of 1724, 1725
    mineral springs at 342, 457
Llŷn, Caernarvonshire
    geology of 1791
    marble 1792
Lodi, Italy 348
Loire, River
    canal to Sâonne River 343
London 26, 53-56, 58, 74, 87, 88, 113-115, 117, 148, 149, 152, 154, 161, 164, 176, 178, 180, 191, 201, 244, 248, 326, 399, 400, 686-689, 786, 787, 823, 824, 835, 870-873, 876, 877, 965, 1067, 1578, 1878
    Acton 1835
    Admiralty 20-23, 69, 70, 72, 659, 662, 663, 665, 1845, 2177
    Athenaeum Club 3, 4, 35, 38, 93, 129, 182, 493, 496-499, 535, 536, 538, 797, 850, 869, 895, 1011, 1409, 1872, 1897
    Battersea 106, 107
    Belgrave Square 653, 1014, 1016-1020, 1029-1032
    Berkeley Square 651, 1920
    Blackfriars 684
    Blackheath 295, 305
    Bloomsbury 899, 902
    Board of Trade 60, 221-223, 2179, 2180
    Bridge building stone 172
    British Museum 11, 148, 162, 301, 536, 849, 1166, 1920
    Brixton 715
    Bromley 295
    Brompton 610, 1005
    Brookes Museum 302
    Buckingham Palace 629, 632, 666, 968, 969, 972, 973, 1675
    Bullock's Museum 11, 342
    Burlington Arcade 342
    Camden Town 613
    Carshalton School 1179
    Cavendish Square 656
    Charing Cross 315, 331, 466, 640, 642, 680, 827, 1337
    Chelsea 649, 934, 1925
    City 729
    Colonial Office 637, 674, 1671
    Covent Garden 228, 229
    Craig's Court 1202, 1207, 1211, 1215, 1217, 1222, 1223, 1227-1229, 1236, 1238, 1239, 1250, 1324, 1325, 1337, 1375, 1687, 1827, 1891
    Customs Office 195
    Dock 2067
    Downing Street 636, 1838, 1839
    Egyptian Hall 1297, 1298
    Euston Square 744
    Farnborough 330, 331
    Foreign Office 618, 654
    Geological Society of 19, 26, 29, 32, 37, 40, 41, 47, 54, 109, 130, 141, 145, 150, 152, 156, 160, 162, 163, 173, 175, 178, 181, 189-191, 194, 196, 208, 245, 249, 259, 295, 302, 304, 307, 308, 342, 343, 346, 364, 415, 473, 489, 505, 506, 531, 539, 542-544, 547, 551, 554, 568, 584, 589, 620, 622, 624, 690-692, 701, 721-725, 727, 788, 811, 898, 900, 901, 904, 910, 912, 915, 916, 969, 972, 974, 994, 1009, 1019, 1026, 1042-1045, 1103, 1145, 1148, 1452, 1469, 1531, 1568, 1609, 1873, 1888, 1913, 1974, 2002, 2017-2021, 2050, 2150, 2176, 2274
    Geological Survey Office 402-405, 410, 412-414, 416, 640, 642, 643
    Grays Inn 896
    Greenwich 295, 1966
    Grosvenor Square 655
    Guy's Hospital 1166
    Hampton Court 1251
    Hatton Garden 1671
    Heralds College 2162
    Highgate 1179
    Home Office 59-61, 98-100, 634, 635
    Hounslow 342
    Houses of Parliament 32, 65, 66, 239, 397, 624, 1049, 1214, 2175
    Hoxton 849
    Hunterian Museum 691, 692
    Hyde Park 241
    India House 2006-2008, 2028, 2029
    Kew Gardens 613, 702-705, 711-714
    Kew Green 313
    Kings College 389, 1333, 1337, 1353
    Knightsbridge 989, 2084
    Lambeth 248
    Leicester Square 1908
    Lincoln's Inn 467, 468, 840, 1149
    Maida Hill 2164
    Museum of Practical Geology 14, 19, 41, 62, 73, 92, 103, 109, 110, 131, 192-194, 196, 202, 203, 220, 225, 227, 230, 231, 233, 251, 253, 258, 268, 270, 295, 315, 331, 397-399, 406, 411, 417, 420, 421, 454, 458, 462, 506, 507, 511, 518, 524, 525, 554, 562, 572, 573, 579, 580, 597, 598, 604, 610, 617, 642, 645, 680, 702, 706, 710, 712, 714, 733, 737, 738, 740, 742, 789, 798, 825, 870, 874, 876,-878, 887, 888, 894, 925, 936, 938, 961, 962, 963, 982, 986, 987, 992, 996, 1004, 1005, 1016, 1021, 1029, 1033, 1039, 1051, 1062, 1064, 1073, 1155, 1156, 1178, 1180, 1182, 1185, 1187, 1189, 1192, 1195, 1202,

1203-1242, 1246, 1249-1251, 1253, 1254, 1256-1258, 1260-1262, 1264-1266, 1270-1275, 1278, 1285, 1288, 1289, 1295, 1297-1300, 1301, 1303, 1304, 1306-1309, 1311-1313, 1317, 1322-1325, 1337, 1351, 1357, 1359, 1376, 1391, 1392, 1395, 1405, 1406, 1408, 1412, 1413, 1469, 1490, 1496, 1501, 1507, 1536, 1538, 1563, 1571, 1584, 1598, 1667-1670, 1678, 1685, 1689, 1690, 1692, 1698, 1700, 1703, 1707, 1713, 1747, 1749, 1773, 1782, 1792, 1806, 1824, 1827, 1829-1833, 1865, 1891, 1902, 1909-1911, 1917, 1926-1928, 1966, 1970, 1979, 1999, 2015-2021, 2026, 2028, 2029, 2031-2033, 2035, 2036, 2039-2041, 2054, 2055, 2081-2084, 2090, 2092, 21132132, 2137, 2165, 2272, 2277

Nelson's Column 421

Ordnance Map Office 135, 136, 208, 210, 214, 215, 260-263, 269, 270, 275, 276, 278, 286, 288, 290, 374, 379, 384, 387, 397, 501-503, 505, 511, 512, 529, 622, 691, 692, 717, 839, 855, 901, 902, 951, 1014, 1680, 1682, 1683

Ordnance Office 205-215, 294, 379

Ordnance Survey Office 272, 273, 278, 280, 282-285

Pall Mall 206, 1021, 1693, 2172

Piccadilly 1054, 1057-1059, 1297-1299, 2081, 2082

Pimlico 926, 927, 929-931, 2090

Privy Council Office 1050

Public Record Office 292

Putney 21, 81, 328, 537, 740, 1692

Regent's Park 186, 188, 339, 594, 620, 622-625, 843

Royal Arsenal 294

Royal Institution 64, 103, 526, 527, 545, 1353, 2272

Royal Mint 101, 102, 152, 2164, 2165

Russell Square 1000, 1974

St James's 694, 1328

St Stephen's Crypt 1903

Salopian Coffee House 302, 308, 311, 1408

School of Mines 37, 63, 385, 386, 389, 544, 2180

Scotland Yard 1273

Soho 137, 138, 796, 847

Soho Square 153, 932, 1676, 1677

Somerset House 130, 249, 967, 974, 977, 987, 1042, 1060, 1469, 2154

Surgeons Hall 302

Temple 116, 524

Thurloe Square 1005

Tottenham 1672

Tower 135, 136, 210, 214, 260, 261, 263, 269, 270, 276, 278, 284, 286, 288, 342, 374, 379, 384, 387, 397, 511, 512, 622, 691, 692, 839, 901, 902, 951, 1014, 1156, 1157, 1335, 1823, 2098, 2099

Trafalgar Square 1180, 1687

Treasury 12, 207, 215, 222, 236, 387, 554, 665, 856, 858, 2172

United Services Club 167

University College 103, 389, 741, 1763, 2013, 2051

University of 2049-2051

Wandsworth 636, 1004

Waterloo Station 1151

Westbourne Green 2078

West India Docks 1285

Westminster 239, 256, 454, 649, 1255, 1291

Whitehall 615, 1000, 1168, 1170, 1172-1174, 1176, 1177, 1694

Whitehall Yard 1299, 1301, 1303, 1306

Wimbledon 525

Woods, Office of 65, 73, 219

Woolwich 294, 1179, 1842-1846

Zoo 339

basin 308

cholera in 572

Ordnance Survey of 239

sewerage 1686

Survey of 827, 828

water supply 2063

Longhope, Gloucestershire
quarries at 157

Long Mynd, Shropshire 1573, 1762
Silurian palaeogeography of 1735

Longnor, Staffordshire 1604, 1605, 1830, 1832

Looe, Cornwall 1876
East 945, 946, 1156

Island 946

Loreto, Italy 363

Loughborough, Leicestershire 1636

Luc, France
geology of 350

Lucca, Italy 347, 363
geology of 363

Lucelle, Switzerland
fossil from 343, 346

Lucerne, Switzerland 146, 348
description of 348
geology of 348
Meyer the Bookseller at 348

Lucknow, India 1678

Lucy le Bois, France
Lias near 343, 346

Ludchurch (Yr Eglwyslwyd), Pembrokeshire
geology of 351, 1380
quarries at 351

Ludlow, Shropshire 1434, 1549, 1554, 1556, 1663, 1726, 1740, 1764, 1766, 1953-1955
Silurian of 119, 1406

Lugano, Switzerland 1866

Lugano, Lago di
geology of 363
(illn.) 363
section from Como to (illn.) 363
von Buch's map of 977

Lugnaquillia Mountain, Ireland 1936
geology of 1096

Lulworth, Dorset 186, 341, 572-576
Cove, Dirt Bed at 187
geology of 341, 2093
section of (illn.) 341
vertical strata of Chalk at (illn.) 341

Lundy Island 352, 353

Luneburg, Germany
Chalk pits 146

Luxborough, Somerset
ancient iron mines at 2041

Lydney, Gloucestershire 2113

Lydstep, Pembrokeshire
geology of 351, 1370, 1486
sketch map of geology of (illn.) 1343

Lyme Regis, Dorset 80, 81, 130, 161, 163, 169, 177, 179, 183, 188, 296-299, 302, 310, 317, 339, 364, 373, 531, 620, 686, 699, 849, 896, 897, 1020, 1222, 1223, 1225, 1339, 1875, 1980, 2278
Trinity Hill 341
Whitelands (illn.) 341
Chalk near 350
comparison with fossils of La Spezia, Italy 363
coprolites from 655
crinoid from, in Teyler Museum, Haarlem 349
De la Beche's jetty at (illn.) 341
De la Beche's report on foreshore at 79, 410, 2278
fishing boat (illn.) 341
fossils of 11, 179, 341, 342, 1330, 2047
geology of 341
hydraulic cement from 341, 343, 372
ichthyosaurs from 157, 342, 988
jetties of 341
landslip (illn.) 1339
landslips at 80, 81, 1339
meteorological observations at 349, 696
plesiosaur from 355
repairs to Cobb 341
repairs to Cobb (illn.) 341
rhinoceras tooth from 342

Lymington, Hampshire 561, 1151

Lynton, Devon 1336
fossils from 1346

Lyon, France 343, 344, 346
Cathedral 344, 346
Jardin des Plantes 344, 346
Library 344, 346

Museum 344, 346
manufacture of Florentine silk at 363
silk industry school 850

Maastricht, Netherlands 146, 349
description of 349
fossils from 349
mosasaur from 343, 346, 349
Museum 349
St. Peter's Hill Quarries 349
turtle from 343, 346
Macclesfield, Cheshire 1637, 1638
Macclesfield Forest, Cheshire 1135
Macerata, Italy
description of 363
Machynlleth, Montgomeryshire 1727, 1728, 1852
Macon, France 343, 346
description of 343, 346
wines of 343, 346
Madeira
Conybeare travelling to 313
Madras, India 736
Madrid, Spain 1698, 2054
British Legation 1979
Maesgwrda, Carmarthenshire 1381, 1382
Maesteg, Glamorgan 2117, 2118
visit of De la Beche 5
Maestricht, Netherlands see Maastricht
Maiden Bradley, Wiltshire 49
Maidenhead, Berkshire 308
Bridge 150
geology of 342
Main, River, Germany 349
Mainz, Germany 349
Cathedral 349
description of 349
Library at 349
Majdanpek, Yugoslavia
mines at 1928
Malaga, Spain 1698
Malahide, Ireland 579
Malaysia
Labuan 992, 993
Maldive Islands 331
Mallaen, Mynydd, Carmerthenshire 1718, 1719
Mallwyd, Merionethshire 1728, 1740, 1744, 1945-1947, 1949
Malta 363, 1845
dockyard 657
fossil fish teeth from 707, 708
limestone vase from 707
Tertiary limestones of 548
Malton, Yorkshire 1583
Malvern, Worcestershire 851, 1415, 1418-1426,1429-1434, 1436-1438,
1441-1449, 1452, 1455, 1457, 1458, 1460-1464, 1469, 1475,
1477-1479, 1482, 1483, 1498-1501, 1515-1517, 1525, 1532,
1540, 1542, 1548-1550, 1552-1555, 1557-1563, 1569-1571,
1580, 1592, 1628, 1783, 1865
fossils from 1577, 1583
Memoir 1578-1581, 1584, 1593, 1594, 1611, 1628
Ordnance map of 1416
Welsh placenames near 1617
Malvern Hills, Herefordshire / Worcestershire 110, 1356, 1414, 1416,
1421, 1422, 1441, 1450, 1454, 1455, 1461, 1465, 1466, 1563
Cowleigh Park 1460
Herefordshire Beacon 1452, 1454, 1455, 1462
North Hill 1461
Worcestershire Beacon 1452, 1461, 1462
geology of 112, 1418-1420, 1540, 1549, 1592
geology of compared to Shopshire 1421
sketch section (illn.) 1592, 1663
sketch section (illn.) Wych Road cutting 1449
section (illn.) 2266
Malvern Wells, Worcestershire 114, 1453
Man, Isle of 550, 551, 564, 1102
Castletown 321
Carboniferous Limestone fossils of 573

geology of 321
Manchester 110, 216, 693, 701, 724-727, 903, 955, 1005, 1135, 1349,
1353, 1410, 1444, 1446, 1637, 1652, 1685, 1713, 1725, 2094
Bridgewater Foundry 1039
Deansgate 647
Geological Society 109
Guardian 693
Literary and Philosophical Society 693
Museum 701
Newton Colliery explosion 1944
Royal Hotel 1439, 1440
Royal Institution 78, 1686
British Association Meeting (1842) 536, 1039, 1139, 1415, 1426,
1432-1434, 1436, 1438-1440
lectures by J. Phillips 1342, 1349, 1415, 1649
model of 228
Mandali, Iraq 867
Mandello, Italy 491
geology of 363
Manifold, River 1597
Manorbier, Pembrokeshire
geology of 351
Mansfield, Nottinghamshire 266
Mantova, Italy 146
Marcle, Little, Herefordshire 1447, 1451
Margam, Glamorgan 2010, 2011
Margate, Kent 343
packet to London from 343
Market Rasen, Lincolnshire
fossil crocodile from 169, 170
Marloes, Pembrokeshire 1366, 1409, 1410, 1413, 1437, 1460
brachiopods from 1610
geology of 351, 948, 1343, 1362, 1406, 1432
geology of, compared with Dingle Peninsula 1523
Marlow, Great, Buckinghamshire
Military College 1819
Military College, comparison with Oxford University 342
Marquise, France
geology of 343, 346
Marros, Pembrokeshire
Mountain, geology of 351
Marseille en Beauvaisis, France 343, 346
church at 343, 346
Marseilles, France 324, 2071
Martigny-Ville, Switzerland 344, 346
Martin's Haven, Pembrokeshire
geology of 351
Marwood, Devon 1336
geology of 1855
Maryborough, Ireland 761
Maryland, United States
geology of 695
State Geological Survey 1826
Massa, Italy 347, 363
section ridge between Lucca and, (illn.) 363
section, top of pass between Carrara and (illn.) 363
Massachusetts, United States 1826
Amherst 690-692
Geological Report to Government of 690
State Geological Survey 2015, 2017
Matlock, Derbyshire 1633, 1655
geology of 159
Matlock Bath, Derbyshire 1196, 1197
Museum, sale of 1833
Maulamyaing, Burma 2035
Mauthausen, Austria
granite quarry at 1927
Mawnan, Cornwall 939, 1828
May Hill, Gloucestershire 1445, 1451, 1454, 1456, 1460, 1499, 1579,
1617, 1662
geology 1572, 1575, 1577
Memoir 1578-1581, 1611
sketch map of geology of Acton Ingham near (illn.) 1575, 1576
Mechelen, Belgium 349
Cathedral 349

Mediterranean Sea
    changes in level of 82, 83
    currents 2079
    formation of 1975
    magnetic observations 1845
Meerut, India 735, 736
Meidrim, Carmarthenshire 1382, 1393, 1397, 1400, 1426, 1433, 1437
    geology of 1429, 1431, 1439, 1440, 1663
    section near (illn.) 1394
    trilobites from 1432
Meillerie, France 348
Meiringen, Switzerland 348
Meissen, Germany 843
Melbury, Dorset 153
Melcombe Regis, Dorset
    De la Beche's comments on 342
Melibocus, Germany
    geology of 960
Melin Ricket, Carmarthenshire 1394, 1400, 1426, 1431
Membury, Devon
    Lias quarries of 179
Menai Strait 1492
    J. Trimmer's paper on Diluvium of 181, 496
    S. Haughton's paper on Carboniferous Limestone of 593
Mendip Hills 315, 1503, 1552, 1863
    Carboniferous Limestone of 246, 341
    faults of 496, 2130
    geology 2131
    lead in 179
    mines 852, 2131
    Mining laws and customs of (MS) 110, 111, 112
Mendrisio, Switzerland 363
Merano, Italy 1022
Mer de Glace, France 344, 346
Mergui, Burma
    coal in 2035
Merionethshire
    Arenig 1752
    Bala 554, 555, 557, 563, 609, 1371, 1375, 1376, 1432, 1449, 1461,
        1471, 1492, 1522, 1524, 1525, 1739, 1743, 1745, 1746, 1748,
        1750, 1752, 1753
    Barmouth 1775, 1949
    Cader Idris 1727, 1731
    Corwen 1782
    Cowarch 1947
    Dinas Mawddwy 1727, 1852, 1945
    Dolgellau 1656, 1727, 1729-1733, 1737-1740, 1741, 1743-1751,
        1776, 1798-1800, 1802-1804, 1807, 1852, 1945, 1947-1949,
        1968
    Ffestiniog 1796, 1968
    Harlech 1776
    Mallwyd 1728, 1740, 1744, 1945-1947, 1949
    Mynydd Mawr 1791
    Pennant 1947
Merther, Cornwall 806
Merthyr, Carmarthenshire 1394
    trilobite from 1425
Merthyr Mawr, Glamorgan 149
Merthyr Tydfil, Glamorgan 1679
    colliery explosion at 538
    cost of coal at 2127
    Cyfarthfa Ironworks, description of 342
    De la Beche's comments on 342
    geology of 342
    Metropolitan Sanitary Commission Report on, 67, 524
    Pen y Darren Ironworks 2034
Mesco, Capo, Italy
    panorama from, to Monte Rosso (illn.) 363
    sections (illn.) 363
Mesopotomia 866
Messery, France 345, 346
Mevagissey, Cornwall 944
    Bay 939, 941
Mexico 714
    mines of 302
    specimens for Museum of Economic Geology 2055, 2056

Meximieux, France 344, 346
Michigan, United States
    coalfield 885
    geology of 695
Mickleton, Gloucestershire 731
Middleton Hall, Carmarthenshire 1431, 1437
Middlewich, Cheshire 1135
Midford, Somerset
    geology of 1552
Midlothian
    artesian wells in Coal Measures 246, 247
Milan, Italy 146, 155, 348, 363
    Cathedral 348
    comparison with Turin 363
    description of 348, 363
    J.B. Pentland at 363
    mint at 348
    Natural History Museum 348
    woman (illn.) 363
Milber Down, Devon
    fossils from 27, 28
Milford Haven, Pembrokeshire 158, 948, 1466, 1815
Mill Haven, Pembrokeshire
    geology of 351
Milton, Nottinghamshire 1636
Minehead, Somerset
    visit of H.T. De la Beche 1
Minsterworth, Gloucestershire 1500
Miribel, France 344, 346
Missouri River, United States 1826
Mitcheldean, Gloucestershire 1481, 1502, 1814, 2114
Modane, France 347
Modena, Italy 348
Moel Siabod, Caernarvonshire 1796
Moel Tryfan, Caernarvonshire
    erratics on 2042
    J. Trimmer's paper on marine shells from 181, 496
la Môle, France 344-346
Monkstown, Ireland 816
Monkton Combe, Somerset
    sketch section at (illn.) 1550
Monkwearmouth, Durham
    colliery 1973
Monmouth, Monmouthshire 1610
    election of mayor 342
    geology of 342
Monmouthshire 153
    Abercarn 87, 201
    Abergavenny 342, 1721, 1722
    Abersychan 2107, 2241
    Bedwas 2102
    Blaina 2108, 2127
    Caerleon 1358, 1389
    Chepstow 173, 1388, 2079, 2113-2120
    Clytha 1358, 1359
    Cwmtillery 2107
    Ebbw Fach River 2107, 2241
    Ebbw Vale 2106
    Llanbadoc 1358
    Llancayo 1357, 1358, 1851
    Llanfrechfa 1359, 1360
    Llangybi 1851
    Llanover 646
    Monmouth 342, 1610
    Mynyddislwyn 985, 2105
    Nantyglo 2127
    Newbridge 1679, 2101, 2102
    Newport 87, 342, 1388, 1821, 2126
    Pontypool 948, 1389, 1707, 2101, 2112, 2183
    Risca 1821, 2103, 2105, 2107, 2115, 2242
    St Arvans 2116
    Shirenewton 2116, 2118, 2120
    Trelleck 794
    Trostre 1359
    Twmbarlwm 2105

Usk 1356-1360, 1389, 1406, 1445, 1491, 1579, 1598, 1611, 1662, 1851
    lead ore from 985
Montebourg, France
    geology of 350
Montefiascone, Italy 363
Montenvers, France
    ascent of by De la Beche 344, 346
    geology of 344, 346
Monterosi, Italy 363
Montgomeryshire
    Breidden Hills 1950
    Llanbrynmair 1728, 1730, 1732, 1733, 1736, 1740, 1746
    Llanfyllin 1744
    Llangurig 1946
    Llanidloes 1852, 1945, 1946
    Llanrhaeadr-ym-Mochnant 45
    Machynlleth 1727, 1728, 1852
    Newtown 1746
    Welshpool 45, 607, 1738, 1746, 1768, 1769, 1775, 1803
Montivilliers, France 350
Montluel, France 344, 346
Montmelian, France
    geology of 347
Montpellier, France
    fossils from 343, 346
Montreal, Canada 874, 883-888, 917
    Logan, Cringan & Co. of 880, 882
    Natural History Society of 880
Montreuil, France
    description of 343, 346
Montserrat 352, 353
Monyash, Derbyshire 648
Moray
    Elgin 1014
Moravia
    Devonian limestones of 1022
Morges, Switzerland 345-347
Moscow, Russia 146, 149, 151
    Imperial Society of Naturalists 1002, 1003
Mosel, River [Moselle] 349
Mostyn, Flintshire
    trace fossils from 191
Mosul, Iraq 864, 867
Moudon, Switzerland
    description of 348
Mumbles, Glamorgan 449
    limestone sent to Ilfracombe 1336
Munich, Germany 488, 2077
Musselwick, Pembrokeshire 1367
    Bay, geology of 1432
Mynydd Eglwysilan, Glamorgan 2102, 2105
Mynyddislwyn, Monmouthshire 2105
    Woodfield Pits, lead ore in 985
Mynydd Mallaen, Carmarthenshire 1718, 1719
Mynydd Mawr, Caernarvonshire 1791

Naas, Ireland 1100
Nailsea, Somerset 2122, 2124, 2126, 2127
Namur, Belgium 146
Nantes, France
    geology of, similar to Devon & Cornwall 1682, 1683
Nant Ffrancon, Caernarvonshire 1779, 1780
Nantua, France
    geology of 344, 346
Nant y Car, Radnorshire
    copper mine 1946
Nantycaws, Carmarthenshire
    geology of 1428, 1431
Nant y Creiau, Cardiganshire
    lead mine 1941, 1952
Nantyglo, Monmouthshire
    cost of coal at 2127
Nant y Mwyn, Carmarthenshire
    mines 1666

Naples, Italy 171, 363, 1025, 1922
    antiquities in Museum at 363
    boats with Capri in the distance (illn.) 363
    book censorship in 363
    British fleet at 363
    criticism of royal family of 363
    description of 363
    people of 363
    Section of Grotto del Cane (illn.) 363
    University of 1025
Narberth, Pembrokeshire 1382, 1383, 1389, 1393, 1397, 1406, 1432, 1466, 1485, 1486, 1664, 1991
    fossils 1392
    geology of 1431, 1440
Narni, Italy 363
Neath, Glamorgan 965
    explosion at Eskyn (Eaglesbush) colliery (1849) 1957-1959
    River 625, 879
    Vale of 869
Neckar, River, Germany 349
Needles, The, Isle of Wight 1031
Nepal
    Kanchenjunga, Mount 709
    letter of introduction of Prime Minister of 710
Nepi, Italy 363
Netheravon, Wiltshire 2159
Netherlands
    Amsterdam 349
    Breda 349
    Delft 349, 1907
    Gorinchem 349
    Hague 171, 349
    Haarlem 349, 1907
    Katwijk aan Zee 349
    Leiden 349
    Maastricht 146, 343, 346, 349
    Oosterhout 349
    Ouderkerk 349
    Rotterdam 349
    Scheveningen 349
    Utrecht 349
    Vianen 349
    coal from 349
Nettlecombe, Somerset 1819, 2039, 2275
Netton, Wiltshire 583
Neuchâtel, Switzerland 1548, 1866
    geology of 347
    Lake of 347
    Museum 347
Neufchâtel, France 146
Neuwied, Germany 349
    flying bridge on Rhine at (illn.) 349
Newbiggin, Northumberland
    Bog Hall 144
    fossil trees at 144, 145
Newbridge, Monmouthshire 1679, 2101, 2102
New Brunswick, Canada
    geology of 883
Newcastle Emlyn, Carmarthenshire 958, 1378, 1388, 1394, 1397, 1401, 1719
    geology of 1711
Newcastle upon Tyne 48, 62, 76, 322, 652, 706, 938, 1534, 1648, 1649, 1688
    British Association meeting (1838) 27, 28, 30-32, 322, 513, 516, 652
    coalfield 172, 182, 1834
    College of Practical Science 60, 61, 938
    Museum in 95, 938
    Natural History Society of 322
    Proposal for Mining College in 386, 1652
    temperature in mines at 160
    Queen's Head 1651
Newchurch (Llannewydd), Carmarthenshire 1394, 1397
Newent, Gloucestershire 1426, 1441, 1443, 1444, 1454-1456, 1462, 1468, 1498, 1609
    sulphur spring at 1455

New Forest, Hampshire
    J. Trimmer's work on 418, 1783
Newgale, Pembrokeshire
    geology of 351, 2111
Newgate, Lancashire
    coal mining at 109, 110
New Holland
    bone caves in 796, 797
New Inn, Carmarthenshire 954
New Jersey, United States 1826
    report on geology of 695
Newnham, Gloucestershire 754, 469, 1476, 1500-1502, 2113
    Victoria Hotel 1476
Newport, Pembrokeshire 954, 957, 1709
Newport, Gloucestershire 342
Newport, Monmouthshire 87, 1388, 1821
    coal exports 2126
    De la Beche's comments on 342
    section to Brecon (illn.) 342
Newquay, Cornwall 1875, 1876
New Ross, Ireland 755, 772, 1102, 1121
Newry, Ireland 1127
New South Wales, Australia
    allowances on Geological Survey of, 127
    H.W. Bristow recommended for Geological Survey of, 124
    H.W. Bristow's application to Geological Survey of, 125
    H.W. Bristow's appointment to Geological Survey of, 126, 787-1789
    seeds from 157, 158
    shell from as font in Paris church 343
Newton Abbot, Devon
    geology of 33, 34, 36, 620
Newton Bushell, Devon
    trilobites from 2086
Newton Head, Ireland 755
Newton St Loe, Somerset
    coalfield at 342
Newtown, Isle of Wight 590
Newtown, Montgomeryshire 1746
Newtownbarry, Ireland 772, 1072
New York, United States 875, 1826
    geological collection 883
    geology of 883, 884
    State Geological Survey 2015-2020
    trilobites 1517
New Zealand 229, 930
    coal 612
    Company 612
    fossil birds from 926, 928
    maps of 228
    Otago 612
    seeds from 157, 158
    maritime survey of 612
    Dinornis 1142
    Maoris 1142
Nice, France 177, 256
    abstract of De la Beche's paper on geology of 619
    De la Beche's paper on geology of 361, 362, 364, 375
Niger, expedition 585
Nile, River
    Delta 82
Nineveh, Iraq 865
    excavations at 867
Nolton, Pembrokeshire
    geology of 351, 2112
Norfolk 735
    Estuary project 1988, 2067
    J. Trimmer's work in 1789
    Downham Market 92
    Norwich 623, 624, 1010, 1062, 1873, 1874, 1877, 1885
    Scole 960
    Yarmouth, Great 1885
Normandy, France
    De la Beche's geological map of, 532
    fossils of, compared with British fossils 1980
    section by De la Beche 299

North America
    Wenlock Limestone at Great Lakes 157
Northampton, Northamptonshire 160, 1580
Northamptonshire
    Castle Ashby 1052, 1053, 1055, 1056, 1183
    Nene, River 1183
    Northampton 160, 1580
    Oundle 1183
    drift deposits of North 310
    ironstone 746, 1183
North Carolina, United States
    geology of 695
Northern Ireland see also Ireland
    goniatites from 1583
    Ordnance Survey of 792
North Sea
    sediments of, near Shetland 549
Northumberland 159, 1651
    Allenheads 1977, 1978
    Alnwick 150, 1061
    Haydon Bridge 1977, 1978
    Newbiggin 144, 145
    Ovingham 74
    Ratcheugh Crag 1061
    proposal for Mining College in 385
Norton, Shropshire
    erratics 2042
Norton, Somerset 1530
Norway 456, 549
    Arendal 17
    Heilhau's paper on gneiss of 727
    trilobite from 343
Norwich, Norfolk 623, 624, 1874, 1877, 1885
    Palace 1062
    Prebendaries Close 1873
    proposed British Association Meeting at 1020
Nottingham, Nottinghamshire 291, 1600, 1636, 1643
Nottinghamshire 1644
    Bingham 1636
    Bunny 1636
    Mansfield 266
    Milton 1636
    Nottingham 291, 1600, 1636, 1643
    Worksop 1589
    Quarter Sessions 1303
Novara, Italy 348
Nova Scotia, Canada
    coalfield 875, 876, 885
    geology of 695, 883
Novi, Italy
    geology of 347, 363
Nuits-St Georges, France 343, 346
Nyon, Switzerland 345-347
    chateau of Joseph Bonaparte near 345, 346

Oakhampton, Devon 1337
Oberwesel, Germany 349
Odessa, Russia
    W. Strangeways travelling to 158, 160
Ogmore, Glamorgan
    conglomerate used as ornamental marble 149
    geology of 149
    minerals of 327
Ogofau, Carmarthenshire 1424, 1440, 1666
    gold at 1435
    graptolite from 1403
    Roman working of 1435
    section (illn.) 1403, 1435
    sketch, exposure at 1435
Ogwell, Devon 25, 28-36, 1337
    geology of 341
Ohio, United States
    Carboniferous Limestone fossils 1345
    geology of 695
    plan to establish Museum of Economic Geology in 887
    trilobite from 343

Oldbury, Shropshire 2088
Oldbury, Worcestershire 1943, 1944
Oldbury-on-Severn, Gloucestershire 1861
Oldham, Lancashire
    collapse of mill at 693
Old Passage, Gloucestershire
    fossils in Lias of 341, 977
    ichthyosaurs from 342
Oosterhout, Netherlands 349
d'Or, Mont, France
    copper mine near 343, 346
    mineral springs 2046
Oravita, Romania 1928, 1929
Orbe, Switzerland
    geology of 347
Oregon, United States
    coal in 885
Orkney Islands 456, 549
    specimens for Museum of Economic Geology 2055, 2056
Ormskirk, Lancashire 1991, 1992
Ortegal, Cabo, Spain 1698
Osborne, Isle of Wight 591, 611, 631, 970, 971, 973, 1888
Osmington, Dorset 151, 575
    geology of 341
Ostend, Belgium 488
    sailing from Ramsgate 343
Oswestry, Shropshire 1769, 1955, 2134
    copper in New Red Sandstone near 2134
Otago, New Zealand
    sailing of first settlers for 612
Ottawa, Canada 887
    River 885
Ouchy, Switzerland 348
Ouderkerk, Netherlands 349
Oundle, Northamptonshire 1183
Ovingham, Northumberland 74
Ow River, Ireland 1097
Oxford, Oxfordshire 150-158, 160-162, 164, 165, 167, 169-173, 177, 178, 180, 181, 183, 192, 195, 204, 301, 302, 305, 332, 333, 342, 731, 743, 1623, 1659, 1709, 1926
    Bishop of 148
    British Association meeting (1832) 183
    British Association meeting (1847) 1077, 1086
    comparison with Military College, Great Marlow 342
    Conybeare preaching at 302, 305
    De la Beche on role of divinity in 342
    University Chair of Geology, establishment of 148, 151
    All Souls College 342
    Ashmolean Museum 152, 1022
    Bucklandian Museum 1658
    Christ Church College 152, 306, 307, 309, 342, 364
    Corpus Christi College 144, 146, 147, 167, 342, 1009
    New College 148
    Oriel College 153, 155, 342
    Radcliffe Library 146, 342
    St John's College 342
    University Museum 147, 151, 152, 169, 170, 342
Oxfordshire
    Ardington 2276
    Benson 342
    Blenheim Palace 342
    Burford 342
    Coxwell 153
    Faringdon 153
    Henley-on-Thames 308, 342
    Islip, 142, 1152
    Oxford 144, 146-148, 150-158, 160-162, 164, 165, 167, 169-173, 177, 178, 180, 181, 183, 192, 195, 204, 301, 302, 305-307, 309, 332, 333, 342, 364, 731, 744, 1009, 1022, 1077, 1086, 1623, 1658, 1659, 1709, 1926
    Shotover 1020, 1328
    Stonesfield 302
    Woodstock 342
    drift deposits of 310
    iron ore workings in 153

Pacific Ocean
    Belcher to survey coast of 622
    Denham sailing to 656
    expansion of steam navigation in 612
Padstow, Cornwall 1335, 1337, 1874-1876
Paestum, Italy
    Basilica (illn.) 363
    description of 363
    peasant (illn.) 363
    Temple of Neptune (illn.) 363
Paignton, Devon
    geology of 341
Pakistan
    Karachi 1038
    Rawalpindi 1699
Panama 496, 622
Panteg, Cardiganshire 1381, 1382
Pantyffynnon, Carmarthenshire
    colliery section 1995
Pant-y-Glien Quarry, Carmarthenshire
    copper ore from 626
Pappenheim, Germany 2077
    fossil crabs from 343
Par, Cornwall
    fossils from 1156
Paris, France 12, 93, 133, 146, 154, 161, 168, 326, 343, 346, 364, 413, 442, 473, 475, 476, 486, 489, 490, 492, 493, 495-510, 534, 566, 640, 679, 845, 855, 966, 1181, 1182, 1198, 1548, 1683, 2053, 2072, 2073, 2075, 2085
    Catacombs 343, 346
    Champs-Elysées, Av. des 343, 346
    Ecole polytechnique 386, 507
    Hotel des Invalides 343, 346
    Institut d'Afrique 1848
    Jardin des Plantes 157, 343, 346, 484, 2075
    Jardin du Roi 317, 323, 325, 343, 346
    Louvre 343, 346
    Montmartre 343, 346
    Museum d' Histoire naturelle 317, 325, 343, 346
    School of Mines (Ecole Nationale des Mines) 473, 474, 476, 510, 854, 1181, 1182
    Notre Dame 343, 346
    St Denis 343, 346
    archives 343, 346
    artesian well 1687, 1702
    Commission de Statistique de l'Industrie Minerale 854
    De la Beche in 217
    Director of Museums of 343
    geology of 343, 346
    temperature in artesian well near 601
    temperature at depth below 2256
Parma, Italy 146, 348
    museum of antiquities at 348
    Priapi (illn.) 348
Parret, River 3
Parys Mountain, Anglesey
    copper mine 1960
Patricroft, Lancashire 1039
Paulton, Somerset 2122
Pavia, Italy
    Natural History Cabinet of University of 363
Paviland Cave, Glamorgan
    deposits of 166, 167
    discovery of 165, 166
    Red Lady of 167, 168, 171
    W. Strangways at 172
    visit of Conybeare and Sedgwick 164
Payerne, Switerzland 348
Peak, Derbyshire 1601, 1604
Pegwell Bay, Kent
    geology of 343
Pelynt, Cornwall 2029
Pembrey, Carmarthenshire 878
Pembroke, Pembrokeshire 185, 456, 947, 948, 958, 1342, 1352, 1362, 1364, 1367-1369, 1379, 1466, 1815, 1854, 1991
    Castle 351
    geology of 351, 1385

Place Names   *Pembrokeshire - Pendennis*

Pembrokeshire 152, 153, 258, 873, 1352
   Abereiddi Bay 351, 948-951, 1707
   Abermawr 1709
   Begelly 351
   Benton 351
   Bishops and Clerks 948
   Boulston 351
   Brawdy 351
   Broad Haven 351, 2109
   Brynberian 954
   Bullslaughter Bay 351
   Burton 351
   Caer Bwdy Bay 351
   Caldey Island 351, 1343, 1344, 1358, 1363, 1372, 1583, 1854-1856
   Canaston 1485
   Canaston Bridge 351
   Carew 351, 456
   Castlebythe 954
   Castle Malgwyn 873
   Castlemartin 351
   Cilciffeth, Mynydd 957
   Clarbeston 1991, 1993
   Cosheston 351
   Crinow 1485
   Crunwear 1378
   Cuffern Mountain 351
   Cwmgloyne 958
   Cyffig 1382
   Dale 948, 1362, 1366-1368, 1486, 1706
   Dinas Head 957, 1791
   Druidstone Haven 351
   Elegug Stacks 351
   Eglwyswrw 954, 958
   Fishguard 109, 947-952, 954, 957, 1371, 1383, 1384, 1386, 1387, 1389, 1400, 1404, 1435, 1707-1709, 1993, 2125
   Freshwater East 1406, 1437
   Freshwater West 1406, 1437
   Garron Pill 351
   Gateholm 351, 1343, 1413
   Goultrop Road 351
   Gumfreston 351
   Gwaun, Cwm 954
   Haverfordwest 351, 793, 950, 954, 1343, 1362, 1364-1366, 1376, 1383, 1384, 1389, 1707-1709, 1815, 2002, 2109, 2133
   Hodgeston 351
   Houghton 351
   Jameston 351
   Johnston 351, 1815
   Lampeter Velfrey 1380, 1382, 1392, 1394, 1401, 1429, 1432, 1439
   Lamphey 351
   Landshipping 2111
   Lawrenny 351
   Little Haven 351
   Llanddewi Velfrey 1440, 1485
   Llanfyrnach 954
   Llangwm 351
   Llan Mill 1382, 1392
   Llanteg 351
   Llanwnda 953
   Ludchurch 351, 1380
   Lydstep 351, 1343, 1370, 1486
   Manorbier 351
   Marloes 351, 948, 1343, 1362, 1366, 1406, 1409, 1410, 1413, 1432, 1437, 1460, 1523, 1610
   Marros 351
   Martin's Haven 351
   Milford Haven 158, 948, 1466, 1815
   Mill Haven 351
   Musselwick 1367, 1432
   Narberth 1382, 1383, 1389, 1392, 1393, 1397, 1406, 1431, 1432, 1440, 1466, 1485, 1486, 1664, 1991
   Newgale 351, 2111
   Newport 954, 957, 1709
   Nolton, 351, 2112
   Pembroke 185, 351, 456, 947, 948, 958, 1342, 1352, 1362, 1364, 1367-1369, 1379, 1385, 1466, 1815, 1854, 1991
   Penally 351
   Penberry 351
   Picton Castle 1437
   Plumstone Mountain 351, 949, 957
   Pontfaen 954
   Pontgynon 954
   Porth Clais 351
   Porthllisky 1706, 1707
   Preseli, Mynydd 950-954, 957, 1382, 1383, 1401, 1993
   Puncheston 954
   Pwllcrochan 351, 1486
   Ramsey Island 351, 948, 949, 1706
   Rhoscrowther 1486
   Rhosmaen 1426
   Ridgeway, The 351
   Robeston 351
   Robeston Wathen 1378, 1379, 1432, 1437, 1440
   Roch 351
   Rosemarket 351
   Rosepool 351
   St Ann's Head 351
   St Bride's bay 948, 949, 953, 1410, 1536
   St Davids 204, 351, 948, 949, 957, 958, 1382, 1401, 1706-1708, 1723
   St Florence 351
   Saundersfoot 2111
   Skomer Island 351, 1367, 1368
   Skrinkle 351, 1343, 1362, 1363, 1486, 1854-1856
   Slebech 351
   Smalls, The 351, 948
   Solva 351, 948
   Stackpole 1343, 1362, 1364, 1365, 1368, 1486
   Strumble Head 947, 949, 1815
   Swanlake Bay 351
   Talbenny 351
   Templeton 351
   Tenby 351, 396, 456, 463, 621, 1343, 1344, 1358-1366, 1368-1370, 1461, 1466, 1486, 1584, 1585, 1815, 1854-1856, 1862, 2114, 2126
   Tiers Cross 351
   Treffgarne 351, 949, 1400, 1708, 1733
   Trevine 1706
   Upton 351
   Uzmaston 351
   Walwyn's Castle 351
   West Angle 351, 1343, 1362-1364, 1368, 1371, 1583, 1854, 1855
   Whitechurch 952
   Whitesands bay 351
   Williamston 351
   Wiston 1993
   Wooltack 1343
   accommodation in 954, 1706
   coalmining 1992
   coal miners pay in 2127
   Colby's plan to commence Geological Survey in 283
   De la Beche's notes on geology of 351
   drovers 954
   erratics 2042
   geology of 951
   limeburning in South 351
   potatoes 1706
   A. Thomas' map of 1009
Penally, Pembrokeshire
   geology of 351
Penberry, Pembrokeshire
   geology of 351
   slate quarry near 351
Pencarrow, Cornwall
   fossils from 946
Penclawdd, Glamorgan
   geology of 880
Penddaulwyn, Carmarthenshire
   geology of 1431
Pendennis, Cornwall
   Castle 382
   Point 382, 383

Pendine, Carmarthenshire 1378, 1485
    caves at 351
    geology of 351
Pendoylan, Glamorgan
    chert in Lias at 176
Pennant, Merionethshire
    lead mine 1947
Pennar, Cardiganshire
    geology of 351
Pennsylvania, United States
    Susquehanna 875
    Wilkes-Barre 875
    Wyoming 875
    coalfield 875, 1826
    geological map of 695
    Lyell's fossils from 1398
    State Geological Survey 2015, 2017, 2018
    Stigmaria in fireclays in 875, 876
Penmachno, Caernarvonshire 1968
Penrice Castle, Glamorgan 146, 147, 150-154, 158-160, 162, 164, 167, 168, 171, 2010
    breccia near 149
Penrhyn du, Caernarvonshire
    geology of 1791
Penryn, Cornwall 517, 519, 520, 851, 852, 939, 2151
    granite 1224
Pensford, Somerset 2122
Pentewan, Cornwall 940, 941, 943, 944, 1875
Pentir, Caernarvonshire 1781
Pentrefoelas, Denbighshire 811
Pentyrch, Glamorgan 2103
Penybont, Powys 1744-1746, 1764, 1769-1772
Pen y Gaer, Carmarthenshire 1426
Pen y Mulfre, Carmarthenshire 1382
Pen y parc, Carmarthenshire 1426
Pen y raen, Carmarthenshire 1426
Penzance, Cornwall 668, 678, 939, 942, 1681, 1875, 1990, 2033
    geological meeting at 852, 1162
    Geological Society of 1156, 2029
    Ordnance Map of 265, 939
Periers, France
    geology of 350
    quarries at 350
Peris, Llyn, Caernarvonshire 1791
Perrier, France
    granite 32
Persia 154
    /Turkey Boundary Commission 2149
Perthshire
    Trossachs 352
Peru
    Lima 896
Perugia, Italy 363
Pescia, Italy
    geology of 363
Pesth, Hungary see Budapest
Peterborough, Cambridgeshire 18, 172, 1536
Peterhead, Aberdeenshire 1163, 1164
Petersfield, Hampshire 1011, 1023, 1026
Peter Tavy, Devon 373
Petherwin, Cornwall 1335, 1336, 1875
    fossils from 1331, 1332, 1346, 1612
    goniatites from 1584
Philadelphia, United States 695, 1826, 2015-2021
    Academy of Natural Sciences 1001
    Society 1980
Piacenza, Italy 348
Picton Castle, Pembrokeshire 1437
Piddletown, Dorset 128
Pignone, Italy
    peasant girl of (illn.) 363
Pigyn Shon-Nicholas, Carmarthenshire 1720
Pill, Somerset
    geology of 342
Pilton, Devon
    fossils 1340, 1346, 1612

    quarry at 1337
Pisa, Italy 347, 363
    Leaning Tower of 347
    University of 1025
Pistoia, Italy 363
Pistyllgwyn, Carmarthenshire 1400, 1426
Plas y Golomen, Carmarthenshire 1394, 1426
Platte River, United States 1826
Plombières, France
    geology of 343, 346
    steeple at 343, 346
Plumstone Mountain, Pembrokeshire 957
    geology of 351, 949
Plymouth, Devon 12, 115, 294, 302, 657-662, 665-669, 804, 844, 920, 945, 946, 953, 1156, 1158, 1335, 1337, 1338, 1681, 1876, 2086-2088
    Dockyard 373, 2079
    breakwater 373
    Mechanics Institute 1158
    Plympton 844, 989
    British Association meeting (1841) 37, 39, 1015, 1139, 1355, 1370, 1371, 1507, 1583, 1697
    caves at 301
    China 991
    fossils from 1346
    goniatites from 1584
    Pleistocene vertebrates from 167
    tidal range 2079
Po, River 347, 363
Poggibonsi, Italy
    fair at 363
    fair scene at (illn.) 363
    geology of 363
Poix, France 343, 346
Poland 158
    Krakow 146
    Wieliczka 146
    Wroclaw 146
    W. Strangeways in 158
Polcreek, Cornwall 939
Polgooth, Cornwall 941
    tin mines at 942, 944
    tin veins at (illn.) 944
Polgrain, Cornwall 939, 940
Polkerris, Cornwall
    fossils from 1156
Polperro, Cornwall 946, 2254
    fish beds 1161
Polruan, Cornwall
    fossils from 1156, 1157, 1346
Pomeroy, Tyrone 558, 1079, 1080
Pompeii, Italy
    corn mill, view and section (illn.) 363
    description of 363
    extract from Benucci's guidebook to 363
Pontarllechau, Carmarthenshire
    lead mines near 1435
Pont Audemer, France
    geology of 350
Pont d'Ain, France 344, 346
Pontfaen, Pembrokeshire 954
    Kilkiffeth Farm 954
Pontefract, Yorkshire 744
Ponterwyd, Cardiganshire 1941, 1942, 1948
Pontesbury, Shropshire 1951
Pont Gwladys, Carmarthenshire 1439, 1440
Pontgynon, Pembrokeshire 954
Pont sur Yonne, France 343, 346
Pontypool, Monmouthshire 2183
    fossil plants from 948, 1389, 1707, 2101, 2112
Pontypridd, Glamorgan
    bridge at 342
    bridge at (illn.) 342
Poole, Dorset 691
Pordenack Point, Cornwall 382
Porlock, Somerset 2097

Portalban, Switzerland 347
Port Daniel, Canada
    trilobites from 883
Port en Bessin-Huppain, France
    geology of 350
    section (illn.) 350
Portesham, Dorset
    geology of 2093
Porth Clais, Pembrokeshire
    geology of 351
Porthkerry, Glamorgan
    Lias of 314
Porthllisky, Pembrokeshire
    geology of 1706, 1707
Porthmadog, Caernarvonshire 1775, 1776
Porth Swtan, Anglesey
    geology of 1964
Portici, Italy 363
Portishead, Somerset 2127
    geology of 342
Portland, Isle of, Dorset 341, 342, 1770, 1774
    agriculture of 341
    Bill (illn.) 341
    Bill, lighthouses 341, 342
    fossils of 342
    geology of 341
    sections of (illn.) 341
    Stone, used at Ramsgate harbour 343
    quarries of 341
Port Natal, South Africa
    Geological survey of, proposed by Dr Stanger 585
Portofino, Italy
    panorama to Chiavari (illn.) 363
Portovenere, Italy
    geology of 363
Portrush 1697
Portsmouth, Hampshire 788, 789
Portugal 840
    Lisbon 1904, 2079
Potigny, France
    geology of 350
Potsdam, United States
    fossil footprints 1152
Pozzuoli, Italy
    description of 363
Prague, Czechoslovakia
    Hawle & Corda on trilobites of 1022
    Silurian of 1022
Prato, Italy
    Monte Ferrato, geology of 363
Pravolta, Alpi di, Italy
    granite erratic (illn.) 363
Presburg see Bratislava
Preseli, Mynydd, Pembrokeshire 954, 957, 1382, 1383
    geology of 950-953, 1401
    sketch section, to Wiston (illn.) 1993
    trial for coal near 952
Prestatyn, Flintshire 1970
Presteigne, Radnorshire 118, 119
Preston, Lancashire 1354, 1356
Priestcliffe, Derbyshire 1625
Procida, I di, Italy 363
Puerto Rico 202
Pumsaint (Pumpsaint), Carmarthenshire 1402, 1426, 1433, 1712, 1714-1721
    fossils from 1449, 1713
    section (illn.) 1403
Puncheston, Pembrokeshire 954
Punjab, India
    war in 2138
Purbeck, Isle of, Dorset
    geology of 574, 575, 581, 1800
Puriton, Somerset
    boring for coal at 315
Purton, Gloucestershire 1512, 1515
    geology of 1513

Putney, London 328, 537, 919
    De la Beche at 21
    College of Civil Engineers 81, 981, 1692
Puy de Dôme, France 155
    fossil elephant from 343, 346
Pwllcrochan, Pembrokeshire
    geology of 351, 1486
Pwllheli, Caernarvonshire 1767, 1768, 1786, 1791
Pyle, Glamorgan 2010
Pyrenees 2053
    W.E. Logan's visit to 868
Pysgotwr, Afon 1724

Quakers' Yard, Glamorgan 2101
Quebec, Canada 456
    geology of 874
    Literary and Historical Society of 2001
Quettehou, France
    geology of 350
Quiege, France 344, 346
Quillebeuf, France
    geology of 350
Quorndon, Leicestershire 1646

Radicofani, Italy
    geology of 363
Radnor, Radnorshire 342
Radnorshire
    Builth Wells 121, 336, 342, 401, 431, 1413, 1449, 1562, 1662, 1726, 1729, 1744, 1758-1763, 1768, 1770, 1771, 1851, 1852, 1945, 1946
    Elan Valley 1724, 1725
    Knighton 119-121
    Llandegley 342, 1770, 1771
    Llandrindod Wells 342, 1770
    Llanelwedd 342
    Nant y Car 1946
    Presteigne 118, 119
    Radnor 342
    Rhayader 342, 1723, 1946
Radstock, Somerset 700, 1530, 2130
    Norton Down 2004
    Wellsway Pit 2004
    Coal Measures section 2132
    collieries 2004, 2032
    section of Coal Measures at (illn.) 2004
Rame Head, Cornwall 946
[?Ram Hill]
    geology of 2121
Ramsgate, Kent 343
    description of harbour 343
    geology of 343
    history of 343
    lifeboat 343
    lighthouse 343
    sailing to Ostend 343
Ramsey Island, Pembrokeshire 1706
    geology of 351, 948, 949
Randwick, Gloucestershire 1474
Raniganj, India 2139-2142
Rapallo, Italy 347, 363
Rastatt, Germany 349
Ratcheugh Crag, Northumberland
    glacial striae on 1061
Rathdrum, Ireland 772, 1096, 1934-1937, 1939
    River 1097
Ravenoville, France
    geology of 350
Rawalpindi, Pakistan
    fossil footprints from (illn.) 1699
Reading, Berkshire 744
    Valley of Thames at 308
Racanati, Italy 363
Recco, Italy
    geology of 363
Redruth, Cornwall 519, 604, 658, 1681

Red Sea 1041
    coral reefs of 331
Reuss, River, Switzerland 348
Reville, France
    geology of 350
Rhayader, Radnorshire 342, 1723
    Nant y Car Copper Mine 1946
Rheinfall, Switzerland
    geology of 348
Rheinfelden, Germany
    geology of 348
    gypsum quarries at 348
    section at (illn.) 348
Rhine, River 348, 349, 363, 429
    flying bridge on (illn.) 349
    Gorge 349
    Gorge, geology of 349
    Gorge lead and silver mines near 349
Rhone, River 344, 346, 2079
Rhoscrowther, Pembrokeshire
    geology of 1486
Rhosmaen, Pembrokeshire 1426
Rhymney, River 342
    Valley 2101
Ri[?ck]low Dale, Derbyshire
    marble from 648
Ridge Way 151
Ridgeway, The, Pembrokeshire
    geology of 351
Rigi, Switzerland 348
Ringstead Bay, Dorset
    geology of 341
Ringwood, Hampshire 2042
Riomaggiore, Italy
    geology of 363
Risca, Monmouthshire 1821, 2103, 2105, 2115, 2242
    Coal Measures at 2107
    Waunfawr Colliery 2242
Rivoli, Italy 347
Robeston, Pembrokeshire
    geology of 351
Robeston Wathen, Pembrokeshire 1378, 1379, 1437
    geology of 1432, 1440
Roch, Pembrokeshire
    Castle 351
    geology of 351
Rochester, Kent 343
Rockhampton, Gloucestershire 1861
Rocky Mountains, United States 1826
Rolle, Switzerland 347
Romagnano Sesia, Italy
    fossil elephant from 343
Romania 1929
    Oravita 1928, 1929
    Transylvania 1928, 1929
Rome, Italy 256, 346, 363, 1025, 1818
    Coliseum 363
    Forum 363
    fossil elephant from 343
Ronco, Italy
    geology of 363
Roorkee, India
    Thomason College of Engineering 962-964
Rosa, Monte 347, 348
Rose, Wheal, Cornwall 2029
Roseland, France 344, 346
Rosemarket, Pembrokeshire
    geology of 351
Rosepool, Pembrokeshire
    geology of 351
Ross and Cromarty
    Dingwall 921
Rosscarbery, Ireland 776
Ross-on-Wye, Herefordshire 157, 342, 1281, 1287, 1451, 1455, 1460, 1464, 1468, 1469, 1506, 1715, 1717, 1821
    Geological Survey sheet 1411, 1483

Rotterdam, Netherlands 349
Rouen, France 350, 491
Rouvray, France
    granite at 343, 346
    Lias near 343, 346
Royston, Hertfordshire
    geology of 809
Ruabon, Denbighshire 1956
Rudesheim, Germany 349
Rudford, Gloucestershire 1478
Rudry, Glamorgan 2241-2243
Rugby, Warwickshire 744
    fossil bones from 157, 170
Rumilly, France
    geology of 347
Runcorn, Cheshire
    erratics 2042
Russia 152
    Moscow 146, 149, 151, 1002, 1003
    Odessa 158, 160
    Siberia 342
    St Petersburg 149, 150, 401, 430, 651, 966, 967, 1841
    Consulate General 1841
    Emperor of 651
    Imperial Mining Department 1841
    Old Red Sandstone 1348
    Silurian of 1018
    W. Strangways in 149, 151, 157, 158
Ryde, Isle of Wight 587
    Scientific Society 591
Ryme Intrinseca, Dorset
    geology of 1759

Saane, River, Switzerland 348
Sackingen, Germany 348
Saginaw Bay, United States
    coalfield 885
St Albain, France 343, 346
St Albans, Hertfordshire
    Architectural Society 92
San Ambrogio di Torino, Italy
    geology of 347
St Andrews, Fife
    St Leonard's College 104
St Ann's Head, Pembrokeshire
    geology of 351
St Arvans, Monmouthshire 2116
St Asaph, Flintshire
    erratics 2042
St Audries, Somerset 284
St Austell, Cornwall 939-946
    Bay 940
    De la Beche at 340, 377, 378
St Bernard, Col du Grand, Switzerland
    red snow 148, 344, 346
St Blazey, Cornwall 1875
    tin mines at 941
St Brides Bay, Pembrokeshire 1410, 1536
    geology of 948, 949, 953
St Bris le Vineux, France 343, 346
St Buryan, Cornwall 381, 383
San Casciano, Italy
    geology of 363
St Clears, Carmarthenshire 793, 1377, 1378, 1381, 1392, 1394. 1397, 1400, 1426, 1431, 1432, 1486, 1663, 1664, 1666, 1712, 1993
    Blue Boar Inn 1377, 1378
    geology of 1383, 1393, 1407, 1429, 1437, 1440
St Columb, Cornwall 658
    fossils of 1346
St Davids, Pembrokeshire 204, 948, 957, 958, 1706-1708, 1723
    agriculture around 1706
    Cathedral 351
    copper mine near 949, 1706
    geology of 351, 947, 949, 1382, 1401, 1706
    Head, geology of 948
    J. Kidd's 1814 paper on 351

St Florence, Pembrokeshire
  geology of 351
St Genis-Pouilly, France 344, 346
St Georges de Reneins, France 343, 346
St Germain de Joux, France
  geology of 344, 346
  strata near (illn.) 344, 346
St Gervais, France
  geology of 344, 346
St Goar, Germany 349
St Gotthard Pass, Switzerland 146, 344, 346
  analysis of limestone from 317
  minerals from 348
St Govan's Head, Pembrokeshire 1815
St Helens, Lancashire
  colliery explosion at (1849) 1647
St Helens, Isle of Wight 583
  geology of 584-586, 590
St Ives, Cornwall 381-383, 1990
St Jean de Maurienne, France
  geology of 347
St John, Cornwall 1828
St Jouin, France
  geology of 350
  section (illn.) 350
St Just, Cornwall 942, 945
St Lawrence River, Gulf of 874, 885, 887, 888
  geology of 883
St Laurent-sur-Mer, France
  geology of 350
St Leonards, Sussex 623
St Lo, France
  geology of 350
St Louis, France 348
Ste Marie, France
  geology of 350
St Martin, Cornwall 1161
St Martin, France
  geology of 344, 346, 348
  stratigraphy at (illn.) 348
St Marychurch, Devon
  geology of 341
St Mary's Isle, Kirkcudbrightshire 1900, 1901
St Maurice, Switzerland 344, 346
St Michael Caerhays, Cornwall 939
St Michael's Mount, Cornwall 381-383, 1161
St Michel, France
  geology of 347
St Petersburg, Russia 149, 150, 430, 894, 966, 967, 1841
  Imperial Academy of Science of 401, 651
San Primo, Monte, Italy
  geology of 363
  section, Bellagio Point to (illn.) 363
San Quirico d'Orcia, Italy 363
  meteorite fall near, (1794) 363
St Rhemy, Italy
  customs at 344, 346
St Sauveur, France
  visit of R.A.C. Austen 32
St Teath, Cornwall 381, 383
St Truiden, Belgium 349
St Vaast, France
  geology of 350
Salève, Mont, France 191, 344, 346, 347, 354, 355
  De la Beche's ascent of 345, 346
  geology of 344-347
  geology of, comparison with Jamaica 354, 355
Salcombe, Devon 694
Salford 693
Salisbury, Wiltshire 743, 1770
Salisbury Plain, Wiltshire 151
Sallanches, France 344, 346
Saltcoats, Ayrshire 541
Salthill, Berkshire 342
Salzburg, Austria 156
  models of 1022

Samsun, Turkey 864
Sandgate, Kent 580
Sandhurst, Berkshire 41
Sandown, Isle of Wight 582-586, 588-592, 2093
Sandown 41
Sandridge, Hertfordshire 92
Sandwich, Kent 343
  salt pans at 343
Sandy Gate, Devon 28
Sâonne, River, France
  at Lyon 344, 346
  canal to River Loire 343, 346
Sar Dasht, Iran 867
Sardinia
  Cagliari 2077
  bone breccia 2077
  lichens from 148
Sargasso Sea
  seaweeds 669
Sarzana, Italy 347, 363
  section of the Lignite Formation of Caniparola near (illn.) 363
Saskatchewan, Canada
  coal in 885
Saundersfoot, Pembrokeshire 2111
Saunton, Devon 1336
Savoie
  criticism of government of 344-347
Sawdde, Afon 1388, 1390, 1398, 1399, 1401, 1435, 1437, 1438, 1717, 1764, 1852
  lead in 1666
Saxony, Germany 250
  geology of 309, 843
Scandinavia
  Silurian of 1018
Scarborough, Yorkshire 706, 1037, 1895
  railway sections 1581
Schaffhausen, Switzerland 348, 530
Scheidegg, Grosse, Switzerland 348
Schemnitz
  gold mines 146
Scheveningen, Netherlands 349
Schwarzwald (Black Forest), Germany 348, 349
Scilly Islands 379
Scole, Norfolk
  Brome Hall 960
Scotland 541
  Agricultural Chemistry Association of 980
  coal 848
  geological map of 247
  Geological Survey of 596, 980, 1046
  glaciation of 189
  Macculloch's geological map of 158, 271, 623
  plan for mining school in 1652
  plan for Museum of Economic Geology for 1755, 1756, 1760
  proposed visit by W. Buckland 158, 159
  raised beaches of, compared with R. Seine 980
  Trigonometrical Survey of 596
  value of contours on 1" map of 790
Seaton, Devon
  geology of 341
Seaview, Isle of Wight 585-587
Seigne, Col de la Italy/France 344, 346
Seine, River 343, 350
  comparison with Scotland raised beaches 980
  frozen, winter 1828-9. 363
Sélestat, France 348
Sempacher See, Switzerland 348
Sennen, Cornwall 381, 383
Sens, France
  Cathedral relics 343, 346
  geology of 343, 346
Seravezza, Italy
  marble quarry at 363
Servoz, France 348
  copper and lead mines near 344, 346
  geology of 344, 346

Sestri, Italy 347
    geology of 363
Sevenoaks, Kent 533
Severn, River 150-153, 1410, 1447
    coprolites from rocks along 655
    crossing 342
    Estuary, fossil vertebrates of 179
Sèvres, France 17, 2081
    De la Beche invited to 133
    porcelain factory 343, 346, 447
Shaftesbury, Dorset
    Motcombe House 2092
Shaldon, Devon 341
Shanklin, Isle of Wight
    sands of 1020
Sheerness, Kent 1293
Sheffield, Yorkshire 530, 1587, 1637
Shepton Mallet, Somerset 342, 2131
    Lias of 176
Sherborne, Dorset 47, 342
Shetland Isles 549
    Lerwick 549
    specimens for Museum of Economic Geology 2055, 2056
Shirenewton, Monmouthshire
    boring for coal near 2116, 2118, 2120
Shotover, Oxfordshire 1020
    fossils 1328
Shrewsbury, Shropshire 1952-1954, 2134
    erratics near 2042
Shropshire
    Bishop's Castle 607, 1735, 1736, 1738, 1739, 1769, 1771
    Bridgnorth 2088
    Cardington 45
    Caer Caradoc 1744
    Chirbury 1738, 1740, 1741, 1768
    Church Stretton 44, 45, 563, 1733-1737, 1766, 1767, 1771, 1796
    Clee Hill 310, 1452, 1766, 1952, 1955
    Clun 119, 121
    Coalbrookdale 2088, 2094
    Ironbridge 1950
    Lilleshall 2088
    Long Mynd, The 1573, 1735, 1762
    Ludlow 119, 1406, 1434, 1549, 1554, 1556, 1663, 1726, 1740, 1764, 1766, 1953-1955
    Norton 2042
    Oldbury 2088
    Oswestry 1769, 1955, 2134
    Pontesbury 1951
    Shrewsbury 1952-1954, 2042
    Wellington 1950, 2088
    coalfield 1950, 2088, 2134
    Geological Survey sheet 1951
    geology of, compared with Malverns 1421
Siberia
    mammoth preserved in ice from 342, 343, 346
Sicily, Italy 1025
    paper by C.G.B. Daubeny on 332
Sidmouth, Devon 151, 163, 341
    geology of 341
    Peak Hill 341
    section to Golden Cap (illn.) 433
Siebengebirge, Germany 349
Siena, Italy 363
    Cathedral 363
Sierra Leone
    slavery 113
Sierra Marena, Spain
    coalfield 1979
Siirt, Turkey 867
Simplon Pass, Switzerland 146, 348
Sittingbourne, Kent 343
Skomer Island, Pembrokeshire 1367, 1368
    geology of 351
Skrinkle, Pembrokeshire 1343, 1362
    geology of 351, 1363, 1486, 1854-1856
Skye, Isle of Mesozoic rocks of 1014

Slebech, Pembrokeshire
    geology 351
Slough, Berkshire 744
Smalls, The, Pembrokeshire
    geology of 351, 948
Snowdon, Caernarvonshire 559, 1461, 1778, 1795, 1796, 1968
    Crib Goch 1781
    geology of 1777, 1792
    peak (illn.) 436
    J. Trimmer's paper on Diluvium of 181
Sodbury, Gloucestershire 1504, 1516, 1863, 1865
    Lias of 176
Soho, London 137, 138, 796, 847
    Square 153, 932, 1676, 1677
Solnhofen, Germany 2077
Solva, Pembrokeshire
    geology of 351, 948
Somerset
    Abbots Leigh 342
    Banwell 2128
    Bath 91, 109, 110, 155, 157, 246, 307, 342, 458, 748, 1008, 1020, 1033, 1166, 1180, 1409, 1503, 1512, 1515, 1525, 1528-1530, 1552, 1554, 1556, 1558, 1699, 1751, 1883, 1884, 2093
    Batheaston 246, 655, 1980
    Bathford 1550, 1552, 1566
    Beckington 961
    Bishop's Lydeard 2
    Bleadon 2041, 2097-2100, 2150
    Blue Anchor 179, 1819
    Brendon Hills 2041
    Bridgwater 176, 284, 315, 2041, 2126
    Bridgwater Level 3
    Bruton 1550
    Chard 3, 2126
    Chew Magna 2131
    Chewton Mendip 176
    Claverton 1528-1530
    Clevedon 342, 1863, 1864
    Coleford 2130
    Dundry 342, 343, 350, 794, 1866, 1867, 2122
    Dunster 1819
    Exford 1
    Exmoor 31, 622, 2098
    Farleigh 458, 1515
    Frome 961, 1525, 1527, 1530, 1531, 1548, 1555, 1561, 2122, 2130, 2131
    Glastonbury 341, 1866
    Holcombe 2132
    Holnicote 1, 2
    Keynsham 307, 342
    Luxborough 2041
    Midford 1552
    Minehead 1
    Monkton Combe 1550
    Nailsea 2122, 2124, 2126, 2127
    Nettlecombe 1819, 2039, 2275
    Norton 1530
    Paulton 2122
    Pensford 2122
    Pill 342
    Porlock 2097
    Portishead 342, 2127
    Puriton 315
    St Audries 284
    Shepton Mallet 176, 342, 2131
    Taunton 341, 1340, 1341, 2041, 2275
    Uphill 1609
    Wambrook 390
    Watchet 342, 1331, 1819, 2275
    Wellow 1550, 1552, 1555
    Wedmore 176
    Wells 167, 176, 341, 2122, 2128, 2129
    Weston-super-Mare 1609, 2150
    Whately 961
    Wick 284
    Winsford 1, 2

Wookey Hole 167
Wrington 42, 44, 1759, 2086-2088, 2122, 2125
Yatton 2124-2127
Yeovil 1758, 1759
acreage of 3
boring for coal in 315
British Association Surveying in 284
Buckland travelling to 152, 155
Geological Survey of/in 208, 1411
report on farming in 4
Society 2041
Somma, Monte, Italy 897
Sorrento, Italy 363
   cliffs between Vico & Sorrento (illn.) 363
   part of the Piano di Sorrento (illn.) 363
   part of the Piano di Sorrento, Vesuvius in the distance (illn.) 363
South Africa
   Good Hope, Cape of 71, 585
   Grahamstown 19
   Natal 585
   Port Natal geological survey 585
   Sundays River 19
   fossil molluscs from 19
   fossil vertebrates from 19
   specimens from, presented to Geological Society of London 19
South America
   silver mines 13
   temperature in mines in 160
Southampton, Hampshire 237, 313, 738, 791, 863, 1604, 1605, 1660, 1762, 1775, 1776, 1790, 1819
   British Association meeting (1846) 552, 723, 756, 1071, 1072, 1743
   inconvenience of Ordnance Map Office at 399
   Ordnance Map Office 397, 399, 611, 649, 650, 695, 790, 792, 1927, 1928, 2167-2172
Southerndown, Glamorgan
   lead in Lias of 176
South Stoke, Somerset 91
Spa, Belgium 250
Spain 1009
   Barcelona 1700
   Cordoba 1979
   Madrid 1698, 1979, 2054
   Malaga 1698
   Ortegal Cabo 1698
   Pyrenees 868
   Rio Tinto 1698
   Trafalgar, Cape 2079
   anthracite 868
   caves of 167
   geology of, compared with S W England 1181
   geological survey of 2054
   nickel minerals from, for Museum of Practical Geology 1698
Spalding, Lincolnshire
   Principal Coast Officer, Customs at 1159
Spezia, Gulf of, Italy 363
   De la Beche's paper on geology of 365, 366
   geological map of (illn.) 437
   geology of 363, 365, 366
   Lerici from Santerenzo (illn.) 363
   panorama of (illn.) 363
   sections, Coregna (illn.) 363
   west side of Isola Palmaria (illn.) 363
Spithead, Hampshire 2034
Splügen, Switzerland
   construction of road from Lecco, Italy 363
Sri Lanka see Ceylon
Stackpole, Pembrokeshire 1343, 1362, 1365, 1368
   geology of 1364, 1486
Staffa, Argyllshire 1680
Staffordshire 558, 743, 817, 1135
   Alstonefield 608, 1830-1832
   Burton upon Trent 1634, 1635
   Drayton Manor 192, 194, 195, 1168-1171, 1174-1176, 1597
   Cheadle 1967
   Consall 1967
   Ecton 1585, 1832, 1971
   Elford, Staffordshire 673
   Grindon 1832
   Ilam 1596-1600
   Leek 1639, 1830
   Longnor 1604, 1605, 1830, 1832
   Stoke on Trent 17, 1195
   Walsall 1767
   Wetton 1832
   Wolverhampton 566, 1767, 1773, 1784
   coalfield 1834
   Coal Measures plants from 334
   Geological Survey mapping of coalfield 1782, 1784, 1799, 1801, 1804
   iron smelters 1183
   Lias of 731
   mines of 1191
   mines of Lord Granville in 616
   potteries 1194
Staines, Surrey 1178
Standish, Gloucestershire 1473
Stanley Green, Dorset 691
Stanway, Essex 135, 136
Start Point, Devon 341, 1876, 1877
Staubbach Falls, Switzerland 348
Staunton, Gloucestershire 1450, 1500
   Swan Inn 1450
Steeple Ashton, Wiltshire 2157-2159
Steiermark, Austria 146
Stockholm, Sweden 2005
Stockport, Lancashire 1637
   sketch section near (illn.) 1638
Stoke Edith, Herefordshire
   fossils from 1447
Stoke on Trent, Staffordshire 17, 1195
Stonesfield, Oxfordshire
   Megalosaurus from 302
Stowmarket, Suffolk 95
Strasbourg, France 348, 349
   Library 349
   religious procession at 349
Stratford upon Avon, Warwickshire 112, 731
Stroud, Gloucestershire 342, 1469, 1503, 1509, 1515
   railway 1284
   Roman villa near 342
   sketch section near (illn.) 1506
Strumble Head, Pembrokeshire 1815
   geology of 947, 949
Sturminster Newton, Dorset 122, 1758
Styria 146, 644, 645
Suez, Egypt 707
Suffolk
   Aldeburgh 2041
   Bures 471, 472
   Combs 677
   Hadleigh 676, 677
   Hitcham 676, 677
   Ipswich 677, 1809-1811, 1829, 1985
   Stowmarket 95
   Swefling 18
Sulaymaniyah, Iraq 867
Sully, Glamorgan 306, 307, 309, 977
Sundays River, South Africa 19
Sundarbans, The, India 1678
Sunderland, Durham 172, 903
Superior, Lake 604
   copper near 887
   geology of 887
Surrey
   Bagshot Heath 308
   Farnham 206
   Guildford 37, 38, 40, 1335
   Staines 1178
   sewers of 243, 934
Sursee, Switzerland 348

Susa, Italy
  marble quarries near 347
  Napoleon's road at 347
Sussex
  Beachy Head 511, 512
  Brighton 470, 748, 1026, 1281, 1303, 1770, 1817, 1818
  Buxted 64
  Eastbourne 511, 512
  Hastings 3, 4, 293, 534, 897
  Lewes 925, 1817
  St Leonards 623
  Uckfield 64
  Hastings Sands 516
Swanage, Dorset 571, 577, 581, 1020, 1806
  geology of 575, 2093
Swanlake Bay, Pembrokeshire
  geology of 351
Swansea, Glamorgan 26-31, 33, 34, 36, 146, 147, 150-152, 158, 159,
    342, 388, 389, 453, 513-517, 559, 560, 626, 696, 800, 830,
    832, 868, 869, 875-879, 919, 947, 948, 1263, 1398, 1436,
    1443, 1585, 1618, 1679, 1879, 1999, 2000, 2029, 2035, 2079,
    2096, 2100, 2163
  Burrows Lodge 457, 985
  Lilliput 949, 2257
  Parkwern 449, 450, 452, 454, 455, 458, 462, 465, 1481
  Penllergare 458, 463, 870, 965
  Port Tennant 919
  Rhyddings 870, 2010
  Royal Institution of South Wales 463, 868, 874, 1250, 1252
  Singleton 461
  Sketty 191, 449, 452, 456-460, 462, 464
  Valley 1639, 1640
  British Association meeting (1848) 246, 461-463, 851, 1020,
    1099-1101, 1639, 1640
  coal exports 2126
  copper ore sales at 452
  copper smelting 1016
  copper works at 516, 869
  De Breos charter to 459, 460
  De la Beche at 1479
  Hafod copper works 1679, 2277
  Infirmary 456
  lack of railway to 448, 1020
  Rebecca Riots 1527
  R. Phillip's inspection of copper smelting at 411, 1670
Sweden
  Stockholm 2005
    Academy of Science 2005
    iron ores from 1032
    minerals from in museum in Paris 343, 346
    Wenlock Limestone of 157
Swefling, Suffolk 18
Swimbridge, Devon 1336, 1337
Swindon, Wiltshire
  fossils 1328, 1329
Switzerland 141, 640
  Aarberg 348
  Aare, River 348
  Andeer 363
  Auvernier 347
  Basel 348
  Bern 146, 202, 348, 1028, 1866
  Bex 344, 346
  [?Botzberg (illn.)] 428
  Brienzer See 348
  Carouge 345, 346
  Chevroux 347
  Chiasso 363
  Chur 363
  Coppet 344-347
  Cudrefin 347
  Dent du Midi 344, 346
  Diablerets 345, 346
  Drance, River, floods (1818) 344, 346
  Eglisau 348
  Eiger 348
  [?Engelberg] 348
  Fiescherhorn 348
  Finsteraarhorn 348
  Fribourg 348
  Geneva 146, 217, 344-348, 443, 492, 499, 606, 849, 1200, 1703
  Giessbach Falls 348
  Gletscherhorn 348
  Grandson 347
  Grindelwald 348
  Grisons 153, 347, 348
  Gross Monch 348
  Hermance 345, 346
  Herzogenbuch See 348
  Hindelbank 348
  Interlaken 348
  Jorat, Mont 348
  Jungfrau 348, 428
  la Sarraz 347
  Laufenburg 348
  Lausanne 344-348
  Lauterbrunnen 348
  Leissigen 348
  Lucelle 343, 346
  Lucerne 146, 348
  Lugano 1866
  Martigny-Ville 344, 346
  Meiringen 348
  Mendrisio 363
  Morges 345-347
  Moudon 348
  Neuchâtel 347, 1548, 1866
  Nyon 345-347
  Orbe 347
  Ouchy 348
  Payerne 348
  Portalban 347
  [?Reichenbach, River/Falls] 348
  Reuss, River 348
  Rheinfall 348
  Rigi 348
  Rolle 347
  Saane, River 348
  Schaffhausen 348, 530
  Scheidegg, Grosse 348
  St Gotthard Pass 146, 344, 346, 348
  St Maurice 344, 346
  Sempacher See 348
  Simplon Pass 146
  Splügen 363
  Staubbach Falls 348
  [?Suleck] 348
  Sursee 348
  Thun 348
  Thuner See 348
  Thusis 363
  Unterseen 348
  Valangin 347
  Velan, Mont 344, 346
  Veyrier 345, 346
  [?Wellhorn] 348
  Wetterhorn 348
  Yverdon 347
  Zillis 363
  Zofingen 348
  Zug 348
  Zuger See 348
  Zurich 146, 348, 1866
  Zurich See 348
  Zweilutschinen 348
Swords, Ireland
  erratics 2042
Sydney, Australia 977, 2178
Syfynwy, Afon
  geology of 953
Symonds Yat, Herefordshire 307

Taf, Afon 1377, 1379, 1383, 1430, 1432, 1434
Taff, River 342
    Valley 2101
Taff's Well, Glamorgan
    water analysis (1760) 2240
Taghmon, Ireland 759
Taibach, Glamorgan 965, 2010
Talbenny, Pembrokeshire
    geology of 351
Taliaris, Carmarthenshire 1397, 1401, 1419, 1426, 1429, 1432, 1433, 1439, 1714, 1715, 1717
Tal-Sarn, Carmarthenshire 1724
Tamar, River 622
Tara, Ireland 761, 766
    Hill 762, 1096
Tarporley, Cheshire
    Oulton Park 487
Tarrington, Herefordshire 1446, 1451, 1453, 1457-1459, 1464, 1467, 1478
Tasmania, Australia
    Hobart 444
    Maria Island 444
    convict stations on 444
    fossil plants from 712
    geology of 444
    Governor of 444
    Museum 444
Taunton, Somerset 341, 1340, 1341, 2275
    Museum of Somersetshire Society 2041
Taurus Mountains, Turkey 864
Tavarnelle, Italy
    geology of 363
Tavernspite, Carmarthenshire 1378
Tavistock, Devon 373, 379, 522, 600, 832, 946, 1875, 1876, 2012, 2097
    geology of 25, 26, 31
    Institution & Reading Room 373
Tawstock, Devon
    Corffe Quarry fossils 1346
Taynton, Gloucestershire 1471
Teifi, Afon, 1492
Teignmouth, Devon
    clay exports from 24, 33, 34, 341, 1987
Teme, River 118, 1444
Temiscamingue, Lac, Canada 886
Temple, London 116
Templeton, Pembrokeshire
    geology of 351
    quarries at 351
Tenasserim, Burma
    Capt. Tremenheere's paper on tin mines of 1678
    Tremenheere in 2035
Tenby, Pembrokeshire 396, 456, 463, 1343, 1344, 1358-1360, 1362-1366, 1368-1370, 1461, 1466, 1854-1856, 1862
    Castle Hill 1363, 1365
    Hean Castle 2114
    coal exports 2126
    geology of 351, 621, 1361, 1486, 1815
    goniatites from 1584, 1585
Tennessee, United States
    geology of 695, 1826
Termignon, France 347
Terni, Italy 363
    Cascata de Marmore 363
Terracina, Italy 363
Tetbury, Gloucestershire 342
Tewkesbury, Gloucestershire
    flooding at 731
    progress of Geological Survey at 731
Texas, United States 75
Thames, River 72, 150, 343, 809, 1293
    frozen (winter 1828-9) 363
    geology of estuary of 295
    Isis, River 342
    Valley of 308
Tharand, Germany 843

Thillerton 2
Thonon-les-Bains, France 345, 346
Thornbury, Gloucestershire 1863
Thorpe, Derbyshire 1832
Thun, Switzerland 348
Thuner See, Switzerland 348
Thuringer Wald, Germany 146
Thury-Harcourt, France
    geology of 350
Thusis, Switzerland 363
Tibet 703, 709, 713
Ticino, River, Italy 348
Tideswell, Derbyshire 1605, 1633
Tienen, Belgium 349
Tiers Cross, Pembrokeshire
    geology of 351
Tigris, River 864
Tilly-sur-Seulles, France
    geology of 350
Tinahely, Ireland 1072, 1931, 1932, 1935
Tincroft Mine, Cornwall 604
Tintagel, Cornwall 1875, 1876
Tinto, Rio, Spain
    minerals from, for Museum of Practical Geology 1698
Tipperary, Ireland
    maps of 815, 1127
Tisbury, Wiltshire 1807
Tiverton, Devon 2, 24, 209, 257, 658, 834, 835, 898, 1335, 1338, 1873
Tivoli, Italy 363
Tonbridge, Kent 1819
    Grammar School 1819
Tone, River 3
Tongeren, Belgium 349
Tor Bay, Devon 1876
    De la Beche's Manuscript on geology of 360
    De la Beche's paper on geology of 620
    raised beaches of 24
    section, Tor Abbey Sands (illn.) 360
Torquay, Devon 171, 174, 341, 360, 620, 621, 1051, 1335, 1338, 1681
    Hope's Nose 341
    Kent's Cavern, comparison of fossils from, with those of Italy 363
    fossils from 1346, 1480
    geology of 36
Torre Annunziata, Italy 363
Torre del Greco, Italy 363
Tortworth, Gloucestershire 747, 1475, 1500, 1503, 1512, 1571, 1579, 1598, 1611, 1662, 1863
Totnes, Devon 498
    fossils from 1346
    geology of 31, 33, 372
Toulouse, France 1195
    fossil elephant from 343
Touques, France 350
Tournus, France 343, 346
Tower, London 135, 136, 284, 1823
    armour collection at 342
    Ordnance Map Office 135, 136, 208, 210, 214, 215, 260-263, 269, 270, 275, 276, 278, 286, 288, 290, 374, 379, 384, 387, 397, 501-503, 505, 511, 512, 529, 622, 691, 692, 717, 839, 855, 901, 902, 951, 1014, 1680, 1682, 1683, 2098, 2099
Trafalgar, Cape, Spain 2079
Tramore, Ireland 407, 1085, 1123
Transylvania, Romania 1928, 1929
Trecastle, Breconshire 1400, 1407, 1408, 1426, 1721, 1722
    section (illn.) 1405
Treffgarne, Pembrokeshire
    geology of 351, 949, 1400, 1708
    geology of compared with Dolgellau area 1733
Trefilan, Cardiganshire 1724
Trefriw, Caernarvonshire 1789, 1793
Tregaron, Cardiganshire 1724, 1725
Tregony, Cornwall 939, 943, 944
Trelleck, Monmouthshire 794
Tremadoc, Caernarvonshire 1778
    Lingula from 1610, 1611
Treruffe, Cornwall 604

Tresillian, Cornwall 519
Trevenen, Cornwall 940
Trevine, Pembrokeshire 1706
Trinidad
    sugar 1776
Troarn, France
    geology of 350
Troedrhiw, Carmarthenshire 1426
Trossachs, Perthshire
    comparison with Jamaica 352
Trostre, Monmouthshire
    geology of 1359
Trouville, France
    geology of 350
    section (illn.) 350
Trowbridge, Wiltshire 744
Truro, Cornwall 513, 806, 892, 900, 940, 1337, 1874, 2012
    Institution 111
    Mining College at 389, 2267
Tsigny, France
    geology of 350
Turin, Italy 974, 1912, 1913
    Academy 1909-1911, 1914-1918, 2003, 2045
    comparison with Milan 363
    description of 347
    Natural History Museum 347
    University of 1025
    University Antiquities Museum 347
    University Library 347
Turkey 622
    Ararat Mount 865, 867
    Bitlis 867
    Constantinople (Istanbul) 153, 154, 622, 864, 1922, 2149
    Diyarbakir 864
    Keban 864
    /Persia Boundary Commission 2149
    Samsun 864
    Siirt 867
    Taurus Mountains 864
    Van Golu 867
Twmbarlwm, Monmouthshire
    height of 2105
Tynfron, Cardiganshire
    lead mine 1952
Ty'n y Groes, Caernarvonshire 1775
Typica, Carmarthenshire 1431
Tyrol, Austria 141, 153, 156, 172, 348
Tyrone, Ireland
    Dungannon 557
    Pomeroy 558
Tywarnhayle Mine, Cornwall 605
Tywi, Afon 1375, 1383, 1385, 1388, 1394, 1395, 1397, 1400, 1401,
        1426, 1429-1432, 1434, 1437-1439, 1466, 1492, 1711, 1715
    anticline (illn.) 1440
    section (illn.) 1433, 1438, 1715

Uckfield, East Sussex 64
Ugine, France 344, 346
United States
    Alabama 8
    Albany 883, 2019
    Amherst 690-692
    Appalachian Mountains 883
    Boston 8, 690, 879, 880, 1709, 1826, 1827
    California 13
    Champlain Lake 883
    Cleveland 885
    Connecticut 695
    Green Mountains 888
    Hudson River 883
    Illinois 695
    Maryland 695, 1826
    Massachusetts 690-692, 1826, 2015, 2017
    Michigan 695, 885
    Missouri River 1826
    New Jersey 695, 1826
    New York 875, 883, 1517, 1826, 2015-2020
    North Carolina 695
    Ohio 343, 695, 887, 1345
    Oregon 885
    Pennsylvania 695, 875, 876, 1398, 1826, 2015, 2017, 2018
    Philadelphia 695, 1001, 1826, 1980, 2015-2021
    Platte River 1826
    Potsdam 1152
    Rocky Mountains 1826
    Saginaw Bay 885
    Tennessee 695, 1826
    Texas 75
    Virginia 695, 1827, 2015, 2018
    Washington 55
    Yale College 1906
    biogeographical comparisons with Europe 8
    coalfields 885
    copper mines in 1695
    cost of postage in 2015, 2016
    state geological surveys in 692, 695, 1826, 2015-2019
Unterseen, Switzerland 348
Uphill, Somerset
    igneous rocks at 1609
Upton, Pembrokeshire
    geology of 351
Upwey, Dorset 124-127, 572, 575, 2093
Usk, Monmouthshire 1358-1360, 1389, 1445, 1579, 1598, 1611, 1662
    geology of area of 1357-1360, 1851
    Old Red Sandstone fish from 1491
    Silurian 1406, 1851
    Three Salmons Inn 1356, 1357
Usk, River 153, 342, 1359
Uzmaston, Pembrokeshire
    geology of 351

Valangin, Switzerland
    geology of 347
Valley of Rocks, Devon 1336, 2097
Valognes, France 350
    visit of R.A.C. Austen 32
Van Diemen's Land see Tasmania
Van Golu, Turkey 867
Varenna, Italy
    geology of 363
Vatican, Italy
    Piazza Belvedere 1025
    St Peters 363
    Seeing the Raphaels at the Vatican (illn.) 363
    Visiting the Vatican by torch light (illn.) 363
Velan, Mont, Switzerland 344, 346
Velletri, Italy 363
Venice, Italy 146, 151, 1022
    specimens from 148
Ventnor, Isle of Wight 96
Vermenton, France
    geology of 343, 346
Verona, Italy 146, 1184
Versailles, France
    description of 343, 346
Veryan, Cornwall 939-941, 1876
Vesuvius, Italy 141, 146, 897, 1921
    crater of (illn.) 363
    De la Beche's ascent of 363
    panorama, Vesuvius to Capri (illn.) 363
    section of (illn.) 363
Veyrier, Switzerland 345, 346
Vianen, Netherlands 349
Vicenza, Italy 146
Vico, Italy 363
    cliffs between Vico and Sorrento 363
    (illn.) 363
Vienna, Austria 146, 153, 363, 639-645, 1022, 1282, 1834, 1927, 1928
    Natural History Society 639, 1949
    W. Strangeways in 158
Vierville-sur-Mer, France
    geology of 350

section (illn.) 350
Villefranche, France
    Greensand of 177
Villejuif, France 343, 346
Villeneuve, Italy 344, 346
Villeneuve la Guyard, France 343, 346
Villeneuve sur Yonne, France 343, 346
Villers-sur-Mer, France
    fossil crocodile from 161
    geology of 350
    section (illn.) 350
    slate quarries at 350
Villerville, France
    geology of 350
    section (illn.) 350
Virginia, United States
    geology of 695
    State Geological Survey 2015, 2018
    State Geologist 1827
Viso, Monte, Italy 347
Viterbo, Italy 363
Vitteaux, France
    geology of area around 343, 346
Voghera, Italy 363
Volga, River
    fossil elephant from 343, 346
Vosges, France
    geology of 348

Wakefield, Yorkshire 57, 744, 1808, 2044
Waldshut, Germany
    geology of 348
Wales 541
    coke manufacture 107
    Conybeare and Buckland in South 146
    flora of South 700
    geological map of 1492
    Geological Survey of South Wales 207, 269, 270, 283, 504, 623, 624, 839, 872-874, 878, 947, 948
    geology of 625
    glaciation of 189, 537
    investigation of colliery explosion in South Wales 414
    lead mining & smelting in 524
    Mines Inspector for South Wales 59, 89
    North Wales geology 1019
    Ordnance maps of North Wales 287
    section of S Wales coalfield (illn.) 342, 715
    Sedgwick in 30
    South Wales as source of Lias for hydraulic cement for Eddystone lighthouse construction 372
    South Wales Coalfield 145, 1334
    visit of W. Buckland to South Wales 150
Walkhampton, Devon 373
Walsall, Staffordshire 1767
Walwyn's Castle, Pembrokeshire
    geology of 351
Wambrook, Somerset
    section (illn.) 390
Wardour, Vale of, Wiltshire
    flints and cherts from 581
Ware, Hertfordshire
    geology of 809
Wareham, Dorset 572-575
Warminster, Wiltshire 49, 543
Warrington, Cheshire 701
Warwickshire
    Birmingham 35, 36, 291, 334, 464, 698, 701, 743, 794, 812, 1030, 1109, 1134, 1185-1197, 1261, 1264, 1334, 1356, 1499, 1688, 1691, 1696, 1785, 1943, 2034, 2096
        Leamington 730, 2084
        Rugby 157, 744
        Stratford upon Avon 112, 731
Washington, United States 55
Wasit, Iraq 865, 866
Watchet, Somerset 1819, 2275
    belemnites 1331
    ichthyosaur from 342
Watcombe, Devon
    geology of 341
Waterford, Ireland 552, 751, 771, 772, 776, 777, 815, 817, 1085, 1101, 1120, 1123, 1127, 1128
    1" map of 791
    geological map of 1114
    maps of 757, 819
Waun-gron, Carmarthenshire 1380
Wedmore, Somerset
    Lias of 176
Weilitzka see Wielicza
Weimar, Germany 146
Weinbohla, Germany
    quarry 843
Weinheim, Germany 349
Welland Common, Worcestershire 1461
Wellington, Shropshire 1950, 2088
Wellow, Somerset
    geology of 1550, 1552, 1555
Wells, Somerset 167, 341, 2122, 2128, 2129
    Cathedral 341
    Lias of 176
Welshpool, Montgomeryshire 45, 607, 1738, 1746, 1768, 1769, 1775, 1803
Wentworth, Yorkshire 18, 309, 625
Wernddu, Carmarthenshire 1426
West Angle, Pembrokeshire 1362, 1364, 1368
    geology of 351, 1343, 1363, 1371, 1583, 1854, 1855
West Ashton, Wiltshire 2157-2159
Westbury-on-Severn, Gloucestershire 1475, 1476, 1500
    fossil insect from 1609, 1610
    Garden Cliff 1474, 1510, 1609
    Rhaetic Bone Bed at 179, 1510
Westerwald, Germany 141
West Lavington, Wiltshire 2157-2159
West Lulworth, Dorset 572-576
Westmorland 1853
    Kendal 1354
    Ordnance Survey Map of 791
Weston-super-Mare, Somerset 2150
    igneous rocks at 1609
Wetterhorn, Switzerland 348
Wetton, Staffordshire 1832
Wexford, Ireland 550, 751, 755-757, 759, 772, 779, 1072, 1084, 1096, 1102, 1107, 1109-1111, 1117, 1120-1123
    erratics 2042
    geology of 1109, 1110
    map of 818, 1133
    1" map of 791
    scale of map of 756, 757, 775
    W.W. Smyth on mines of 420
Weymouth, Dorset 151, 153, 156, 157, 179, 180, 184, 186, 341, 573, 891, 1409
    belemnites 1331
    coast near (illn.) 342
    fort 342
    geology of 341
    geology of, comparison with Northern France 350
    Ridgway railway cutting 2093
    sailings to Guernsey and Jersey 341
Whately, Somerset
    crinoid from 961
Wheal Betsy, Devon 373
Wheal Franco, Devon 373
Wheal Friendship, Devon 373
Wheal Rose, Cornwall 2029
Whitby, Yorkshire
    discovery of plesiosaur at 109
    ichthyosaur from 1849
Whitechurch, Pembrokeshire
    metal mine near 952
Whitecliff Bay, Isle of Wight 583, 588
    geology of 582, 585, 586, 589
Whitesands Bay, Pembrokeshire geology of 351

Wick, Somerset 284
Wicklow, Ireland 1071, 1072, 1077, 1086, 1122, 1939, 1950
    geological map of 814, 821, 1096, 1097, 1099, 1108, 1122, 1813
    geology of 1084, 1096
    Head 1084
    map of Avoca Mines 1077
    maps of 757, 1084
    1" map of 791
    Mountains 1936
    Silurian of 770
    W.W. Smyth on mines of 420, 772
Wicklow Head, Ireland
    geology of (illns.) 1084
Wickwar, Gloucestershire 1515, 1863, 1865
    railway tunnel & cuttings at (illn.) 749
Wieliczka, Poland
    salt mines 146
Wight, Isle of 95, 532, 534, 556, 569, 1020, 1072, 1299, 1770, 1980
    Alum Bay 581, 582
    Atherfield 539
    Bembridge 584, 590
    Brading 583, 585, 587
    Brook 581
    Cliff End 582, 586, 587
    Colwell Bay 590
    Cowes 585, 587, 589, 590
    Culver Cliff 585
    Freshwater 583, 587
    Gurnard Bay 588, 590
    Hamstead 590, 591
    Headon Hill 582, 587, 588
    Needles 1031
    Newtown 590
    Osborne 591, 611, 631, 970, 971, 973, 1888
    Ryde 587, 591
    St Helens 583, 584, 586, 590
    Sandown 582-586, 588-592, 2093
    Seaview 585-587
    Shanklin 1020
    Ventnor 96
    Whitecliff Bay 582, 583, 586, 588, 589
    Yarmouth 590
    Eocene crocodile from 1144
    fossils compared with Belgian 591
    geology of 2093
    Ordnance 6" map of 790
    section, Brading Harbour to St Helens (illn.) 587
    section, Brading Harbour to Seaview (illn.) 587
Wigmore, Herefordshire 1496
Wigtown, Wigtownshire 1896
Williamston, Pembrokeshire
    geology of 351
Wiltshire 146, 744
    Ashley 1550, 1552
    Box 1550-1552, 1557, 1558
    Bradford-on-Avon 1166, 1503, 1552
    Chippenham 90
    Devizes 1980, 2157, 2159
    Maiden Bradley 49
    Netheravon 2159
    Netton 583
    Salisbury 743, 1770
    Salisbury Plain 151
    Steeple Ashton 2157-2159
    Swindon 1328, 1329
    Tisbury 1807
    Trowbridge 744
    Wardour, Vale of 581
    Warminster 49, 543
    West Ashton 2157-2159
    West Lavington 2157-2159
Winchcombe, Gloucestershire 731
Winchester, Hampshire 114
Windsor, Berkshire 744, 1294
Windsor Castle, Berkshire 9, 630, 633, 971, 1673, 1674
    sewers 2057-2066
    water analysis from well at 219
Winsford, Somerset 1, 2
Wirksworth, Derbyshire 608, 1655
    cave near 170
Wiston, Pembrokeshire
    sketch section from Preseli to (illn.) 1993
Woburn, Bedfordshire
    Abbey 1840
Wolverhampton, Staffordshire 566, 1767, 1773, 1784
Woodenbrige, Ireland 1096, 1128, 1932-1934
Woodstock, Oxfordshire 342
Woody Bay, Devon 1336
    fossils from 1346
Wookey Hole, Somerset
    Pleistocene vertebrates from 167
Woolhope, Herefordshire 1426, 1441, 1443, 1445-1449, 1451, 1452, 1454, 1460, 1461, 1463, 1464, 1468, 1469, 1478, 1479, 1482, 1499, 1540, 1611, 1662
    section (illn.) 2266
Wooltack, Pembrokeshire
    Park 1343
Worbarrow, Dorset 575, 581
    geology of 2093
Worcester, Worcestershire 1376, 1416, 1516, 1532, 1548, 1551, 1609, 1618, 1784, 2084
    elephant tooth from near 342
Worcestershire
    Abberley Hills, Worcestershire 1452, 1536, 1540, 1548, 1553, 1554, 1557, 1558, 1563, 1571, 1573, 1577-1581, 1611, 1617, 1618, 1631
    Alfrick 1540
    Bredon Hill 731, 1434, 1452
    Dudley 87-89, 157, 566, 609, 1050, 1331, 1354, 1767, 1943, 1944, 2088
    Elmley Lovett 624
    Halesowen 812
    Kidderminster 1609
    Malvern 851, 1415, 1416, 1418-1426, 1429-1434, 1436-1438, 1441-1449, 1452, 1455, 1457, 1458, 1460-1464, 1469, 1475, 1477-1479, 1482, 1483, 1498-1501, 1515-1517, 1525, 1532, 1540, 1542, 1548-1550, 1552-1555, 1557-1563, 1569-1571, 1577-1581, 1583, 1584, 1592-1594, 1611, 1617, 1628, 1783, 1865, 2266
    Malvern Hills, Herefordshire / Worcestershire 110, 112, 1356, 1414, 1416, 1418-1422, 1441, 1449, 1450, 1452, 1454, 1455, 1460-1462, 1465, 1466, 1540, 1549, 1563, 1592, 1663
    Malvern Wells 114, 1453
    Oldbury 1943, 1944
    Worcester 342, 1376, 1416, 1516, 1532, 1548, 1551, 1609, 1618, 1784, 2084
Worksop, Nottinghamshire 1589
Wotton-under-Edge, Gloucestershire 1501, 1503-1516, 1540, 1550, 1685
    Geological Survey at 191, 747-749, 751, 1475, 1498, 1500
Wrexham, Denbighshire 1955, 1966, 2132
Wrington, Somerset 42, 44, 1759, 2086-2088, 2122, 2125
    mines at 2125
    sketch section near (illn.) 2125
Wroclaw, Poland
Wurzburg, Germany 146
Wye, River 308
    at Builth Wells 342
    geology of valley of 307
Wye, River (Derbyshire) 1627

Yale College, United States 1906
Yarmouth, Isle of Wight 590
Yarmouth, Great, Norfolk 1885
Yarnscombe, Devon
    lead mine near 624
Yatton, Somerset 2124-2127
Yealmbridge, Devon
    fossils of 1346
Yealmpton, Devon
    fossils of 1346
Yemen 1041

Yeovil, Somerset 1758, 1759
Yetminster, Dorset
  geology of 1759
Ynys Brechfa, Carmarthenshire 1426
Ynyscedwyn, Breconshire 455
York, Yorkshire 62, 322, 1328, 1331, 1334, 1335, 1340, 1347, 1349, 1362, 1370, 1373, 1374, 1376, 1388, 1389, 1397, 1398, 1401, 1479, 1485-1489, 1491-1493, 1495, 1496, 1505, 1515, 1523, 1525-1527, 1533, 1535, 1536, 1554, 1566, 1569, 1579-1581, 1587, 1594, 1604, 1606-1608, 1610, 1636, 1644, 1651, 1652, 1767, 1849
  British Association meeting (1831) 183
  British Association meeting (1844) 651, 1301, 1303, 1375, 1415, 1501, 1525, 1563
  City and County Bank 1462, 1489, 1526, 1533, 1582, 1589, 1610, 1634
  Museum 95, 1329, 1341, 1849
  Philosophical Society 1349, 1353, 1404, 1408, 1481, 1527, 1589, 1849, 2003
  St Mary's Lodge 1348, 1352, 1353, 1375, 1405-1417, 1480-1484, 1490, 1493, 1517, 1532, 1570-1572, 1578, 1583, 1584, 1586, 1588-1593, 1611, 1612, 1614-1618, 1628, 1631, 1642, 1643, 1645, 1647-1650, 1653-1656, 1662, 1858
Yorkshire 1020, 1433, 1574, 1582, 1878
  Barnsley 76, 2044
  Bolton Abbey 162
  Bradford 528
  Bridlington 1642
  Castle Howard 14
  Doncaster 18
  Goole 744
  Hackness 1643
  Helmsley 1328
  Kirkbymoorside 162, 165, 171, 1328
  Kirkdale Cave 162, 165, 171, 1849
  Leeds 91, 232, 1658, 1691
  Malton 1583
  Pontefract 744
  Scarborough 706, 1037, 1581, 1895
  Sheffield 530, 1587, 1637
  Wakefield 57, 744, 1808, 2044
  Wentworth 18, 309, 625
  Whitby 109, 1849

York 62, 95, 183, 322, 651, 1301, 1303, 1328, 1329, 1331, 1334, 1335, 1340, 1341, 1347-1349, 1352, 1353, 1362, 1370, 1373-1376, 1388, 1389, 1397, 1398, 1401, 1404-1417, 1462, 1479-1493, 1495, 1496, 1501, 1505, 1515, 1517, 1523, 1525, 1526, 1527, 1532, 1533, 1535, 1536, 1554, 1563, 1566, 1569-1572, 1578-1584, 1586-1594, 1604, 1606-1608, 1610-1612, 1614-1618, 1628, 1631, 1634, 1636, 1642-1645, 1647-1656, 1662, 1767, 1849, 1858, 2003
  Coal Measures 528
  building stones of churches in 528
  coalfield 172
  fossils of 1329, 1833
  geological survey of 1536, 1578
  Geological Society of West Riding of 625
  goniatites from 1584
  investigation of colliery explosion in 414
  6" Ordnance Survey map of 792
  pottery manufacture in 1658
Youghal, Ireland 418, 776, 777, 779
Yport, France
  geology of 350
  section (illn.) 350
Ysbyty Ifan, Caernarvonshire 811
Ystradgynlais, Breconshire
  stone for Houses of Parliament 65
  Sutton Quarry, 65
Ystumtuen, Cardiganshire
  lode 1941
Yugoslavia
  Majdanpek 1928
Yverdon, Switzerland 347
Yvoire, France 345, 346

Zillis, Switzerland 363
Zofingen, Switzerland 348
Zug, Switzerland 348
Zuger See, Switzerland 348
Zurich, Switzerland 146, 348, 1866
  criticism of museum at 348
  description of 348
Zurich See, Switzerland
  fossils from coal mine at 348
Zweilutschinen, Switzerland 348

SUBJECT

actualism 179
Académie Royale des Sciences, Institut de France 12, 306, 326
    De la Beche elected Corresponding Member 475, 602
Admiralty Coal Enquiry (1845-1850) 20-23, 328, 488, 540, 634, 674, 733,734, 919, 981,1667, 1668, 1690, 1820, 2156
agate, moss 677
agriculture 40, 194, 195
    agricultural chemistry 398
    Agricultural Chemistry Association of Scotland 980
    agricultural geology 418
    Portland, Isle of 341
    use of sewage manure in 68
    value of geological maps to 372
    value of Museum of Economic Geology to 397
air, analysis of 233
alabaster 2275
Alexander, HMS 148
American Philosophical Society 2015, 2020, 2021
ammonites
    from France 1981
    lappets 2002
    Oxford Clay (illn.) 2002
analysis
    of air 233
    of bronze 196
    of copper ores 1671
    of coprolites 1669
    of lead ores 1671
    of limestone 317, 332, 333
    of slag 1188, 1189
Andalusian Mining Company 614
Annals of Philosophy 849
Antiquarian Society 147, 1060, 1926
Antiquaries, Society of 2081
archaeology 1041
    of Iraq 865, 867
Architectural Society 103
art
    Berlin Museum of 922
    Kaulbach's method of fresco painting 924
artesian wells
    Jamaica 983
    Midlothian 246, 247
    Paris 601, 1687, 1702
    Trafalgar Square 1687
Artillery Institution 294
Astronomical Society
    monthly Notices of 669
    astronomy
    connection with geological phenomena 847
Athenaeum Club, London 3, 4, 35, 38, 93, 129, 413, 493, 496-499, 535, 536, 538, 797, 850, 869, 895, 1011, 1022, 1409, 1499, 1531, 1537, 1872, 1882, 1897, 1926, 1984, 2039, 2072, 2073, 2075, 2088
    J. Phillips elected to 1411, 1412
    presentation of W. Buckland's *Reliquiae Diluvianae* 182

banks
    Barnett, Hoare & Co., London 1489, 1491, 1526, 1533, 1582, 1589, 1610, 1634
    Cocks and Biddulph, London 1321, 1323, 1668, 1730, 1739, 1750, 1760, 1780, 1796, 1800
    David Jones & Co, Llandovery 1405
    Jones, Lloyd & Co, London 1405
    National & Provincial, Dolgellau 1737, 1739, 1750
bear, Gower 162
belemnites 1148, 1331
    nomenclature of 1330,
    Oxford Clay (illn.) 2002

birds, Arctic 148
birds, fossil
    from Montmartre 343
    from New Zealand 926, 928, 1142
bitumen 478
blackjack 1715
bone
    effects of carbonic acid on 135
    caves, Baruth, Germany 146
    caves, New Holland 796, 797
Bone Bed, Ludlow 1513
Bone Bed, Rhaetic 179, 731, 1447, 1510
botanical specimens 218
    for Lady Mary Cole 157, 158, 162, 167
    from Lady Mary Cole to W. Buckland 164
    from Jamaica 840
    from Mr Lambert to W. Buckland 164
    for Jane Talbot 146-151, 155, 162
Bovey Formation, Tertiary 24, 27, 28, 32-34
brachiopods 141
    Lingula 1610, 1611
    Marloes 1610
    Silurian 119
    Tremadoc 1610, 1611
brass rubbings 92
breakwaters 109, 110
bricks, Assyrian 17
Bristol Diamonds 341
Bristol Institution 109, 301-305, 342, 355, 357
    lithograph of crocodile from Le Havre presented by De la Beche 977
    sale of shares in 977
    specimens from Mary Anning 977
    von Buch's map of Lake of Lugano presented by De la Beche 977
British Association for the Advancement of Science 312, 397, 400, 658, 659, 975, 1524, 1527, 1846
    Committees and Reports
        Committee of Illustration of British Fossil Reptiles 1136-1141, 1150
        Railway Section Committee 189, 193, 397, 1375, 1376, 1858, 1859
        Report on Fish 1150
        Report on British Fossil Mammals 1139
        formation of local committees 310
    Kew Observatory 1279
    meetings
        Birmingham (1839) 35, 36, 1261, 1264, 1334, 1696
        Birmingham (1849) 464, 812, 1030, 1109, 1192, 1628, 1785
        Bristol (1836) 35, 266, 378, 622, 658, 902, 1015, 1875
        Cambridge (1833) 1826
        Cork (1843) 903, 1520, 1521
        Dublin (1835) 659, 1567, 2042
        Edinburgh (1834) 2, 309, 897, 1826
        Edinburgh (1850) 798, 1046, 1061, 1124
        Glasgow (1840) 1015, 1342, 1680, 1755, 1756, 1858
        Liverpool (1837) 136, 600
        Liverpool (1854) 791
        Manchester (1842) 536, 1039, 1139, 1415, 1426, 1432-1434, 1436, 1438-1440
        Newcastle (1838) 27, 28, 30-32, 322, 513, 516, 652
        Oxford (1832) 183
        Oxford (1847) 1077, 1086
        Plymouth (1841) 37, 39, 1015, 1139, 1355, 1370, 1371, 1507, 1583, 1697
        Southampton (1846) 552, 723, 756, 1071, 1072, 1743
        Swansea (1848) 246, 461-463, 851, 1020, 1099-1101, 1639, 1640
        York (1831) 183
        York (1844) 651, 1301, 1303, 1375, 1415, 1438, 1501, 1525, 1563

Council meetings 1408, 1468, 1528, 1530, 1531, 1554, 1579, 1612, 1628, 1640
    Geological Survey maps displayed at meetings of 651, 791
Phillip's, J., post with 1349, 1353, 1535, 1547
Phillips' *Catalogue of .... fossils of Great Britain* 1330
presidency of 914
presidency of Section C 1015, 1016, 1030
publication of local floras & faunas 464
recommendation on contouring Irish maps 1532
surveying in Somerset by 284
British Museum, London 11, 148, 301, 572, 578, 699, 849, 1022, 1029, 1041, 1528, 2040
    anti-Royal Society attitude in 570
    Buckland appointed to Committee of 1883
    building stone of 1920
    claim to Geological Survey collection 1217
    excavations in Iraq 867
    ichthyosaurs in 342
    offered J. Chaning Pearce's collection 1166
    proposed appointment of J.S. Miller 162
    proposed position for De la Beche 199
    purhcase of Koch's Missouri vertebrates 536
    Robert Brown at 1058
    specimens for 865, 866
    specimens from Kirkdale Cave for 1849
    W. Smith's collection 1537
bronze manufacturers, Birmingham 1185
Buckingham Marble 193
building stones 728
    British Museum 1920
    Caen Stone 610
    Houses of Parliament 32, 65, 66, 397, 528, 610, 648, 948, 979, 982
    Old Red Sandstone, Brecon Beacons 342
    Pennant Sandstone, Bristol 342
    Ramsgate Harbour 343
    roads 237
    Survey collection of 777, 778, 1266
    value of geological maps in search for 372, 397
    Yorkshire churches 528
Bullock's Museum, London 11
    ichthyosaurs formerly in 342
    painting formerly in 343
    purchase of specimen by De la Beche 342
    sale of 342

Cambrian 84, 330, 1610, 1617
    /Silurian boundary 791, 1018, 1019
Cambrian Pottery, Swansea
    problems of 461
Cameo-Portrait, Taking a (illn.) 363
canals
    Brecon & Monmouthshire 342, 1821, 2183
    Glamorgan 342, 1924
    Netherlands 349
    Stroud 342
capucines (illn.) 363
carbonic acid
    experiments on effects of 135, 136
Carboniferous 141
    age for Culm 26, 30, 36
    climate 32
    plants discovered in Devon grauwacke 1330
    trace fossils in 191
    Limestone 314
        fossils 458
        fossils, Isle of Man 573
        fossils, Ohio 1345
        Ireland 246
        Mendips 246
        Pembrokeshire 351, 1372
        Scotland 247
        South Wales 342
carriages
    Chamonix charabanc (illn.) 344, 346
    description of French 343, 344, 346
    Dutch Waggon (illn.) 349

French gentleman's (illn.) 343, 346
Quarante sous par poste (illn.) 343, 346, 344
Sledge Coach, Amsterdam (illn.) 349
Catholic Emancipation 1170
caves 1006
    Bordeaux 176
    Burrington 42
    deposits, Gower 162, 165-167, 171
    Dovedale 1597
    fossils 2150
    Germany 146, 165, 166, 168, 343
    Gibraltar 167
    Gower 162, 164-167, 171
    Jamaica 173, 352, 355, 358, 984
    Kirkdale 162, 165, 171
    New Holland 796, 797
    Paviland 165-168, 171
    Pendine 351
    Plymouth 301
    Spain 167
    Wirksworth 170
    Wookey Hole 167
cement, hydraulic,
    Lyme Regis 341, 343, 372
    South Wales 372
cemeteries, public 233
central heat, theory of 247, 377, 658, 2144
    experimental verification in mines 160, 599-601
chalcedony
    Devon/Dorset 341
    Hemel Hempstead 1225
Chalk
    Denmark 891
    Devon 28, 341
    Dorset 341, 342, 350
    erratics from Antrim 2042
    fossils of 148, 343
    France 343, 350
    Henley 308
    joints in 511, 512
    Kent 343
    Poland 146
    Saxony 309
    Sussex 148, 511, 512
    swallow holes in 150
Chartists 1368, 1634, 2028
Chemical Club 149
chert
    Dorset/Devon 341
china clay 940, 989, 990
    Lee China Clay Co. 990
"Chippenham Ammonites" 109
cholera 339
    in India 2140, 2141
    in London 572
    in Paris 498
Church
    Established 303
    Methodist 303
    Moravian 303
coach
    Brecon & Bristol 1759
    fares 1587
Coal Measures plants 164, 166, 171, 172, 309, 334, 701, 703-705, 711, 961
    collection of Edmund Tyrell Artis 18
    Devon 32, 33
    Forest of Dean 2115
    identified for De la Beche by J. Lindley 890
    India (illn.) 2144
    J. Lindley's lecture notes on 862
    Stigmaria in seat earths 695, 711, 875, 876, 883, 2139
    trees 144, 145, 334, 882, 1612
Coal Measures
    Abersychan 2107, 2241
    Blaina 2108

Bristol 742
  colliery sections 1994, 1995, 2107, 2108, 2128, 2132
  Cwm Tillery 2107
  Durham 514
  Ebbw Fach 2241
  Ebbw Vale 2106
  Fife 848
  fossils 955, 956, 1956
  France, northern 350
  India 2136-2141, 2143-2147
  Jamaica 356
  Lancashire 955
  Midlothian 246, 247
  molluscs 955, 956
  Pembrokeshire 351
  Risca 2107, 2242
  Rudry 2243
  Shropshire 1950, 2088, 2134
  Somerset 315, 342, 961, 2004, 2122-2124, 2128, 2129, 2132
  South Wales 342, 2117, 2118
  South Wales, section of (illn.) 342, 715
  Staffordshire 334, 1967
  Yorkshire 528
coal 1427, 1976, 2161
  Admiralty Coal Enquiry (1845-50) 20-23, 328, 488, 540, 636, 674, 733, 734, 919, 981, 1194, 1315, 1820, 2156
  Australia 612
  Belgium 488
  Bohemian brown 524
  Borneo 1315
  Bovey Formation 28
  Canada 874, 875, 883
  crop workings 342
  Devon 2
  Pembrokeshire 1992
  Forest of Dean 1465
  Ireland 1930
  Labuan 992, 993
  Lancashire 328
  miners pay 2127
  mining 76, 85, 87, 88, 109-111, 351, 1649, 1653, 1834
  Netherlands 349
  Newcastle 1834
  Newgate, Lancashire 109-111
  New Zealand 612
  Northumberland 144
  Pembrokeshire 351, 952
  price of 2126, 2127, 2183
  root beds below 334
  Scotland 848
  Spain 868, 1979
  Staffordshire 1834
  structure of 584, 585
  structure of (illn.) 1465
  Tasmania 444
  trade 2126
  trial near Banwell, Somerset 2128
  trial near Mynydd Preseli, Pembrokeshire 952
  United States 875, 885
  value of geological maps in search for 372
  Warlich's Patent 22, 919
coastguard service
  conditions of service 1158
  pay and pension in 1158
cobalt deposits, Inverary 15
coin dies 1825
coke manufacture 107
collections
  Adam, William 1833
  Amouroux, J.V.F. 350
  Artis, Edmund Tyrell 18
  Atherstone, W. Guybon 19
  Austen, Robert A.C. 37, 1334, 1342, 1358
  Bain, Andrew Geddes 19
  Baker, John 49
  Bakewell Institute 1832

Benett, Etheldred 91
Bern Museum 348
Birch, Col. Thomas J.B. 169, 170, 300, 341, 699
Blisset, Joseph 302, 304
Boromeo, Count Vitello 363
Braikenridge, George Weare 296, 297, 300
Brieslak, Mr 348, 363
Bright, Benjamin Heywood 1442, 1457, 1458
Bright, Richard 109, 300
Brongniart, Alexandre 343, 346
Buckland, William 301
Bullock, William 296, 298
Bunbury, Charles James Fox 904
Caen Museum 350
Clark, Mr 580
Clark, Mr 800
Conway, Mr 1360, 1389
Cumberland, George, Junior 320
Cumberland, George, Senior 296, 300, 341
Damon, Robert 573
Darmstadt Palace 349
De la Beche, Henry Thomas 297, 300, 341, 431
De Luc, Andre & Jean Andre 344, 346
Dillwyn, Lewis Weston 458
Dixon, Frederick 573
Dree, Marquis de 484
Dyer, Dr 300
Edwards, Frederick Erasmus 569
Enniskillen, Lord 51, 558, 1075
Florence Museum 348
Forbes, Edward 554
Gilbertson, William 1356
Glasgow 1755
Gresswell, Rev. 747, 748
Gulpeth, Mr 51
Harding, Maj. William 1334, 1342, 1352, 1376, 1469
Hastings, Lady 569, 1144, 1151
Hawker, Rev. Peter 342
Hennah, Rev. Richard 373, 1334, 1338
Holl, Harvey Buchanan 1334
Howse, Mr 800
Humphreys, J.D. 800
Hutton, William 322
Jeffreys, John Gwyn 572-574, 800
Jelly, Henry 1503
Johnson, James 296, 297, 299, 300, 320, 342
King, Mr 800
Lee, John Edward 1334, 1352, 1358, 1360, 1389
Loscombe, Mr 800
Maastricht Museum 349
Matlock Bath Museum 1833
Meismer, Prof. 348
Milan Museum 348
Miller, Johann Samuel 300, 1330
Museum d'Histoire naturelle, Paris 343, 346
Neuchâtel Museum 347
Ottley, Mr 1352
Oxford University Museum 147, 151
Paccard, Dr Michel Gabriel 344, 346
Page, Mr 300
Pattison, S.R. 1334, 1352, 1337
Peach, Charles William 1158, 1161
Pearce, J. Chaning 1166, 1503
Purdue, Rev. John 551, 555, 1704
Rashleigh, Philip 941
Sanders, William 1335, 1336
Sedgwick, Adam 1334
Setter, Mr 1830
Shepard, Prof. 1345
Skinner, Mr 301
Smith, William 1537
Somersetshire Society Museum, Taunton 2041
Stanhope, Lady Hester 146
Tankerville, Lord 148
Teyler Museum, Haarlem 349
Walton, William 1503

collections - elephant

  Wilkinson, Charles Hunnings 342
  Williams, Rev. David 1332, 1334, 2041, 2099, 2100, 2150
  Wilson, Lady 159
  York Philosophical Society 1481
colonies, British 113, 114, 151, 152
Colonial Office 635, 674, 1671
comet
  (July 1819) 343
  (c.1843) 2020
contours 956
  value of 790
copper
  Bankhart's patent for 1670
  deposits
    Canada 887
    Carmarthen 626
    Devon & Cornwall 15
    Ireland 319
    Jamaica 1016
    lode, Elan Valley 1724
    near Oswestry 2134
  early mining, Ireland 2026
  mines 1838
    America 1695
    Cuba 1016
    mine near Brecon 342
    Fowey Consols, Cornwall (illn.) 434
    Wheel Franco, Cornwall 373
    Wheel Friendship, Cornwall 373
    near Mont d'Or, France 343
    Nant y Car 1946
    Parys Mountain 1960, 1961
    Pembroke Mine, Cornwall 945
    near Servoz, France 344/346
    St. Davids 949, 1706
    ore sales, Swansea 452
    products collection 2277
  production 1694
  smelting, Swansea 411, 869, 1016, 1670
  smelting with honey 1041, 1951
  Vesuvius 146
coprolites
  anaylsis 655
  analysis, cost of 1669
  in agriculture 196, 655
  Rhaetic Bone Bed 179
Coprolitic Vision (illn.) 432
coral reefs
  Belcher to investigate 622
  C. Darwin on origin of 331
  R.A.C. Austen's theory 28
corals 1997
  Cornwall & Devon 1346
  limestone in inlaid table 1051
  Milne Edwards Monographs 486
  Silurian 119
Cotham Marble
  insect fossils in 1610
Creation, unity of design in 204
Cretaceous
  Chalk 28, 148, 150, 308, 309, 341-343, 346, 350, 511, 512, 891, 1225, 2042
  of France 27, 177, 343, 346
  Gault 1529
  Greensand 27, 28, 34, 49, 169, 174, 177, 341, 350, 458, 585, 1757
  Speeton Clay 1589
  Wealden 532, 534, 929
crinoids
  (illn.) 341
  in collection of P. Hawker 342
  in Paris museum 343, 346
crocodile
  comparison with ichthyosaurs 298
  comparison with plesiosaur 302
  from Honfleur 343, 346
  from Isle of Wight (Lady Hastings' Collection) 1144

  from Le Havre 350
  from Market Rasen 169, 170
crustacea
  from Monte Bolca 343
  from Pappenheim 343
  from Woolhope (illn.) 1446
  Jurassic 341
  Jurassic (illn.) 341, 342
Culm 25, 26, 28-31, 33, 35, 621, 902, 1876, 1879
  mines 832
  Pembrokeshire 351
curling stones
  Ailsa Craig 1032
cystoids 566, 567
  (illn.) 563

datum level for Ordnance Survey 205
Davy Safety Lamp 1649
  improvement of 85
?Dawey's Engine Scheme 511, 512
Deluge, Mosaic 148
  Buckland on 152-154, 162
Devil and Friar (illn.) 363
Devon Controversy 26, 29-31, 33-37, 41, 266, 270, 271, 273, 378, 621-624, 837-839, 897, 902, 1012-1014, 1226, 1332, 1873, 1875,-1877, 1879, 1880, 2049, 2050, 2097, 2098
Devonian 84, 141, 250
  comparison of rocks of Devon & Pembrokeshire 1854
  naming of period 39
  diachronism 652
Diluvium 173, 177
  France 344, 346
  Germany 306
  Ireland 306
  J. Trimmer on North Wales 181
  Yates on 797
Dinornis 1142
drift deposits
  mapping of 1743, 1789
  N.E. Wales 2134
  shells in 554, 1746
  Southern England 310
  Stratford upon Avon 112
drovers, in Pembrokeshire 954
Duria Antiquior 180, 182
  watercolour by De la Beche (illn.) 368
  German parody of 183

earthquakes
  Chile 2051
  East Indies 896
  Jamaica 352, 356, 357, 896
  Peru 896
  De la Beche's theories on origin of 352, 357
East India Company 47, 331, 445, 446, 479, 896, 962, 963, 1125, 1706, 1836, 2030, 2031
  Directors' visit to Museum of Economic Geology 2028, 2030
  request to, for assistance for D.H. William's widow 416
  request for successor to Williams as geologist 446
East Middlesex Water Works 513
echinoids
  (illn.) 341
  in Chalk 343
eclipse (8 July 1842) 1442
Edinburgh Philosphical Journal
  criticism of 849
Electrical Society
  Fox's clay specimens sent to 599, 601
electricity
  experiments with 657, 658
  related to geology 318
electrotyping 1037, 2078
elephant, fossil
  Kirkdale Cave 162
  Romagno Sesia 343
  Rome 343

elevation
  craters of 501
  theory of 491, 493-495
Elk, Irish 306, 1088, 1874
engines, pumping 511, 513-515
English Agricultural Society 397, 398
engraving
  new process for banknotes 152
  on copper 342
erratics
  J. Trimmer on 2042
  Runcorn, Cheshire 2042
  St Asaph, Flintshire 2042
Exeter and Bristol Railway 25

facies change 1434
facts
  need for collection of 883
"Falmouth Foolscap" 340
faults
  Derbyshire 309
  Devon 24
  South Wales 2101
feldspar
  analysis of 17
Female Aid Society 1000
fish 946
  analysis of scales of 2047
  casts from Russia 1029
  Cornwall 1163
  from Speeton Clay (illn.) 1589
  Lias 341
  Lias (illn.) 341
  Monte Bolca, 146, 343
  naming of *Dapedium politum* 342
  of Ireland 558
  oldest fossils 8
  Old Red Sandstone, Usk 1491
  paper by T.R. Egerton 487
  scales (illn.) 1491
  spines 341
  spines (illn.) 341
  teeth (illn.) 341, 707, 1446
  teeth, Malta 707, 708
  tracks 191
fisheries
  Ireland 50-52, 2239

Gaelic 1617
galena see lead
Gardener's Chronicle
  article on phosphate in 40
gas, collection of 2050
gasworks
  Birmingham 291
  Nottingham 291
gastropods
  Bellerophon (illn.) 1496
  from Eastnor Park, sketch (illn.) 1446
Gault 1529
geological dictionary
  need for 470
Geological Society of Ireland
  De la Beche elected Vice President of 51, 756
Geological Society of London 26, 29, 32, 37, 40, 41, 47, 109, 141, 178,
    181, 189, 208, 249, 295, 302, 307, 308, 311, 317, 318, 342,
    364, 489, 539, 545-547, 620, 622, 690, 701, 788, 797, 876,
    877, 890-892, 897-899, 902, 905, 909, 912, 925, 926, 928, 929,
    953, 966, 967, 977, 987, 989, 999, 1011, 1018-1020,
    1022-1024, 1026, 1038, 1103, 1114, 1144, 1146, 1166, 1402,
    1612, 1826, 1827, 1870, 1871, 1875, 1881-1884, 1886, 1892,
    1898, 1904, 1906, 1975, 1982, 2051, 2075, 2086-2089, 2149,
    2176
  Anniversary Addresses
    De la Beche 969, 972, 1145, 1148, 1913
    Horner 722
    Murchison 1452
  charter of 304
  Council 245, 1873
  design of seal 173
  exhibition of Gower cave material at 162
  Greenough's geological map 145
  index of colours on geological maps 208, 259
  membership, officers, staff
    appointment of successor to Edward Forbes 544
    De la Beche elected President 554, 723-725, 904, 910, 994,
      1019
    De la Beche nominated as Secretary 175
    duties of Foreign Secretary 898
    Edward Turner resigning as Secretary 2050
    election of Dufrenoy as Honorary Member 473
    election of Elie de Beaumont as Honorary Member 473
    membership of J. Michelotti 974
    membership of S. Stutchbury 2002
    post of Assistant Secretary 915, 916, 1469
    Presidency of 584, 589, 622, 901, 1609
    resignation of Edward Forbes as Curator & Librarian 542, 543,
      547
    resignation of James Nicol 1045
    staff holidays 1042
    W. Strangeways proposed as Vice President 152
  meetings 189, 191, 194, 302, 568, 1009, 1531
    discussion at meetings of 531
    order of publication of papers read at 620
    dining club 54, 1026
  Museum 912
    collections of 551, 624, 1974, 2274
    comparison with Paris 343, 346
    possible transfer to Museum of Economic Geology 543
    casts of USA fossil footprints sent to 691, 692
    ichthyosaurs in Museum of 342
    invited to purchase Rev. D. Williams' collection 2150
    presentation of American specimens to 2017-2021
    presentation of specimens from South Africa 19
    specimens from India 190
  portrait of De la Beche 415
  publications 130, 150, 156, 160, 163, 196, 539, 727, 811, 1043,
    1044
    suppression of T. Weaver's paper 1873
    Geological Survey papers 1103
  Wollaston Medal award 141, 505, 506, 721, 900, 1568, 1888, 2259
Geological Society of West Riding of Yorkshire 625
Geological Survey
  see Survey, Ordnance Geological
Geological Survey Bill 1582
Geological Survey of Canada 883-885
  aims of 881
  allocation of funds to commence 875, 885
  annual reports 885, 886
  application of W.E. Logan for post of Provincial Geologist 876-878
  appointment of W.E. Logan as Provincial Geologist 881
  appointment of E.S. De Rottermond 885
  budget of 885
Geological Survey of India 2006, 2007
  need for Assistants 1115, 1122
  Thomas Oldham appointed 1115, 1116, 1119, 1122, 1123, 1125,
    1133, 2008
Geological Survey of Ireland 128, 223, 400, 402, 403, 406, 407, 418,
    552, 592, 751-787, 808, 816, 817, 821, 825, 833, 1067, 1123,
    1313, 1407, 1547, 1554, 1556, 1730, 1930-1939, 2009
  accounts and grants 1072, 1081, 1090, 1095, 1100, 1108, 1109,
    1117, 1120, 1124, 1126, 1130
    inadequate grant of 767-770, 772, 773, 776, 1071
  staff
    appointment of G.V. Du Noyer 1085
    appointment of H. James 1562
    appointment of J.B. Jukes 813
    appointment of T. Oldham 784, 1065
    De la Beche's dispute with Henry James 402, 403
    J. Phillips 1538, 1543, 1546, 1553, 1554, 1556, 1559
    proposal to appoint Agricultural Geologist 418

resignation of H. James 780, 781
resignation of T. Oldham 1131
training 767
collection of building stones 774, 778
collections for colleges 767, 768, 770, 776, 779
communication of results of 1103
cost of 1" geological map 1122
costs of fieldwork 1115
drift maps 815
fossil collections of 771
maps for 1127, 1128
Museum 402, 403, 593, 758, 760-762, 769, 770, 1078, 2009
Office, Custom House, Dublin 406, 407, 1066-1069, 1073, 1074
Office, St. Stephens Green 558, 760, 799, 801, 817, 821, 828, 831, 1066, l069, 1071-1083, 1088-1091, 1093-1096, 1099, 1101, 1103-1106, 1114, 1119, 1124, 1126, 1130, 1133, 1135, 1316
plan to amalgamate with Geological Survey of Great Britain 1484
postponement of 1493
progress of 791, 792, 1090, 1093, 1094, 1096-1102, 1105, 1107, 1109, 1110, 1117, 1122, 1123, 1126, 1132
railway sections 1081, 1083, 1085, 1105
right of entry 1087
scale of maps 650, 756, 772, 775, 815, 818
slowness of 814
soil collecting by 765, 767, 768, 770, 772, 776, 779, 814, 820, 1318
glacial striae
Ratcheugh Crag, Northumberland 1061
glacial theory 809, 1016, 1507, 1564
controversy concerning 8, 37, 189, 310, 311, 395
Darwin convinced by 537
De la Beche cartoon on 396
De la Beche cartoon on (illn.) 395
glaciers
Grindelwald 348
on Mont Blanc 344, 346
on Mont Blanc (illn.) 344, 346
Wales 537
glass manufacture, Birmingham 1186, 1188
gold
Australia 2178
Cuba 2016, 2017
India 2146
lectures on, for emigrants to Australia 647
mine, Wales 1435
mines, Kremnitz and Schemnitz 146
need for volume on 647
supply 13
goniatites 141
occurrence of 1584, 1585, 1859
gradual changes 1031
Grand Junction Canal Company
request for copies of railway sections 417
granite 512
Alps 344, 346
as paving stone for London 2151
at Rouvry, France 343, 346
Cornwall 343, 346
Dartmoor 26
direction of veins in Alps 348
effects of carbonic acid on 135
erratic, Alpi di Pravolta (illn.) 363
erratics, Jura (illn.) (illn.) 347
exports, Cornwall 517
Guernsey 1269
in Museum of Practical Geology 192
in Chalk in Saxony 309
Italian Lakes 363
near Laufenburg 348
near Lyon 344, 346
quarries (Aberdeen) 192
quarrying in Devon 24
Spain 1700
St Vaast, France 350
weights of 521
graptolites
Aberystwyth area 1727

Cornwall 1163
discovered in Skiddaw Slates 1025
Llandeilo - Carmarthen area 1388, 1393
(illn.) 1394, 1405
grauwacke 26, 31-33, 341
Carboniferous plants in Devon 1330
Great Exhibition (1851) 1691
award of medal to H.T. De la Beche 9
exhibition catalogue 530
Indian pottery for 482
international jury of 244, 476
lecture by H.T. De la Beche 10
manufacturers exhibiting 78
L. Playfair invited to be Travelling Commissioner for 617, 1692
specimens from, for Museum of Practical Geology 1178
Great Merioneth Mining Company 1949
Greensand
Bovey Tracey 27, 28, 34
Devon 174, 341
Dorset 169, 341, 1757
fossils of 49, 341, 458
France 177, 350
Isle of Wight 585
Gryphaea
in Kimmeridge Clay of Le Havre 1329
guano
from West Africa 194
from Peru 194
gypsum
Iraq 866
mines at Bex, Switzerland 344, 346
quarries at Rheinfelden, Germany 348

Hastings Sand, Dorset 2093
head, Devon 27, 28, 34
Health of Towns Commission
see Sanitary Commission, Metropolitan
Herald's College 2162
Hercynian as stratigraphic name for Cambrian 185
Herschel see Uranus
hippopotamus
Kirkdale Cave 162
Home Office 59-61, 98-100, 414, 1647, 1648, 1656, 1657
horse
Kirkdale Cave 162
hospitality
offer of, to H.T. De la Beche 1, 6, 79, 129
Hunterian Museum, London 691, 692
American fossil footprints sent to 692
Huttonian Theory 179, 921
hyaena
den 162
Kirkdale Cave 162
Gower 162
Paviland 164

ichthyosaurs 1514
analysis of bone of 2047
anatomy of 296-300, 302
collections
Annings 297
G.W. Braikenridge 296, 297
W. Bullock 296, 298
G. Cumberland 296
De la Beche 297, 300
Rev. P. Hawker 342
J. Johnson 296, 342
William Moore 988
J. Chaning Pearce 1166
C. Wilkinson 342
comparison with iguana 298
comparison with monitor lizard 298
De la Beche on 341
discoveries of 341
localities
Bath 342

    Northern France 350
    Lyme Regis 157, 341, 342, 988
    Old Passage 342
    Watchet 342
    Whitby 1849
  Home's reaction to Conybeare & De la Beche's paper on 161
  in Muschelkalk 489
  lack of specimens in Paris 343, 346
  lithograph, *Awful Changes* (illn.) 367
  lithograph of skull of *I. vulgaris* 342
  named *Ichthyosaurus* 342
  named *Proteosaurus* 342
  paper by Cuvier on 303
  skull donated by De la Beche to Cuvier 325
  species named *I. tenuirostris*, *I. communis*, *I. platyodon* 297, 342
  species named *I. vulgaris*, *I. longirostris*, *I. platyodon* 342
  species of 341
  specimen sent by De la Beche to Royal College of Surgeons 699
  teeth 341, 342
  teeth (illn.) 341
  viviparity 1166
  watercolour, *Duria Antiquior* (illn.) 368
iguana
  comparison with ichthyosaurs 298
Iguanodon
  Mantell and Owen on 568
  specimen exhibited by Mantell 929
illustrations
  buildings
    Basilica, Paestum 363
    Caerphilly Castle 342
    Forde Abbey, Devon 341
    Edinburgh College Museum 798
    factory 1637
    flying bridge, on Rhine at Neuwied 349
    Lyme Cobb repairs 341
    Lyme, De la Beche's jetty at 341
    Eglisau bridge 348
    Museum of Practical Geology, Piccadilly front 2230
    Museum of Practical Geology, The Great Hall 2231
    Pontypridd Bridge 342
    plan, New Forest House, Jamaica 352
    Temple of Neptune, Paestum 363
    tomb of M.L. Watson 2080
    villa 363
    wall 728
    Pompeii Corn Mill (view & section) 363
  by Elizabeth De la Beche 340
  cartoons and lithographs
    Char a Banc Chamonix 344, 346
    Devil and Friar 363
    Henri and self 363
    A Coprolitic Vision (lithograph) 432
    Awful Changes (lithograph) 367
    The Irregularities of Sol visited upon his System, 395
    Quarante souse par poste 343, 346
    Roman girl and Franciscan Monk 363
    Roman peasant and Antiquary 363
    'Royal skeleton' 363
    seeing Mont Blanc & the Valley of Chamonix 346
    seeing the Raphaels at the Vatican 363
    Taking a Cameo-Portrait 363
    Visiting the Vatican by torchlight 363
    watercolour, *Duria Antiquior* 368
  drinking vessel at Clarendon Races, Jamaica 352
  fossils
    ammonite 2002
    belemnite 2002
    Bellerophon 1496
    ?*Calamites* 2144
    Canadian trilobites 883
    crab 341
    crinoids 341
    crustacean, Woolhope 1446
    cystoid 563
    Devon fossils 804

    echinoderms 341
    fish teeth 341
    fish from Speeton Clay 1589
    fish scales 1491
    fish tooth 1446
    fish 341
    fish spines 341
    fossil footprints, Rawalpindi 1699
    gastropod 341
    gastropod, Eastnor Park 1446
    graptolite 1394, 1405
    ichthyosaur skulls 297, 298
    ichthyosaur teeth 297, 299, 341
    ichthyosaur jaw 296, 299
    insect fossil, Westbury on Severn 1609
    Lingula 1610
    lithograph, trilobite from Builth Wells 431
    ?lobster claw 342
    Nucula 1405
    ostracod 1372
    oyster, Blue Marl, Weymouth 341
    plesiosaur (complete) 302
    plesiosaur rib 298
    rostroconch 1459
    sharks tooth 707
    starfish, Bala 609
    trees, Coal Measures 144
    trilobite pygidium, Wenlock 1333
    trilobite, distorted 336
    trilobites, Bala 609
  geological features and structures
    block of granite upon the Jura, Neuchâtel 347
    block of granite upon the Jura, Neuchâtel (2) 347
    claystone with feldspar crystals, Jamaica 358
    contorted beds, Capo Corvo 363
    geological sketches, Ireland (x 6) 652
    granite erratic, Alpi di Pravolta 363
    jointing, Tasmania 444
    Portland Dirt Bed (*Trans Geol Soc*, 2nd Ser, **4**, (1836) p.13) 184
    Portland Dirt Bed (*Trans Geol Soc*, 2nd Ser, **4**, (1836) p.14) 187
    river courses and differential erosion 307
    fold, Savoie Alps 347
    coal structure 1465
    exposure, Ogofau 1435
    vertical section, rocks of La Spezia 363
    vertical strata of Chalk, Lulworth Cove 341
    interior, Fowey Consols Copper Mine 434
    landslip, Lyme Regis 1339
    lunar crater 1523
    perched block near Llangollen 1564
    fault in limestone near Baslow 1585
    strata near St Germain 344, 346
    tin veins, Polgooth, Cornwall 944
    Old Red Sandstone at Knockmahon, Ireland 759
    strata at St Martin 348
  Geological Society seal 173
  landscape views and panoramas
    Durdle Door 341
    Head, Beer 433
    crater of Vesuvius 363
    Dover 343
    Dover, staircase from town to Barracks 343
    river scene (?Avon) 342
    road, Arc Valley, Savoie 347
    Lago di Lugano 363
    Lake Como 363
    lake scene 363
    landscape with bridge (?Ponte della Maddalena) 363
    Lerici from Santerenzo, Gulf of Spezia 363
    Glacier de Boisson, Chamonix 344, 346
    Port of Como 363
    Portland Cave, Vere, Jamaica 358
    Alpi Apuane to Livorno (panorama) 363
    Capo Mesco and Monte Rosso (panorama) 363
    Gulf of Spezia (panorama) 363
    Lake Como (panorama) 363

Porto Fino to Chiavari (panorama)   363
Vesuvius to Capri (panorama)   363
Rezzonico, Lake Como   363
mountain scene   363
Naples boats with Capri in the distance   363
cliffs between Vico and Sorrento   363
between Onno and Lecco, Lake Como   363
boat and oar, Lake Como   363
Branscombe Cliffs, Devon   341
Portland Bill   341
mangrove swamp, Salt River, Jamaica   352
Part of the Piano di Sorrento   363
Part of the Piano of Sorrento, Vesuvius in the distance   363
Ashwood Dale, Buxton   1165
Blue Mountain Peak, Jamaica   352
Via Mala   363
Vico   363
view from the Villa Pliniana, Lake of Como   363
?view near Abergavenny   342
view of Dorset coast near Weymouth   342
Snowdon Peak   436
Botzberg   428
Jungfrau   428
Monte Ferrato   363
near Braco   363
Rheinfelden   348
west side of Isola Palmania, Gulf of Spezia   363
Whitelands near Lyme   341
Tuscan countryside   363
near Camerton   155
leaf rubbings   439
lithograph of an island   371
maps
    coloured, geology SE of Carmarthen   1425
    coloured, Marlstone   1552
    coloured, May Hill   1575, 1576
    Glamorgan by W. Buckland   200
    Labuan   992
    geology, Exeter to Start Point, Devon   360
    geology, Gulf of Spezia   437
    geology, Ashbourne   1640
    geology, Llangadog   1401
    geology, Llandeilo   1401
    Blue Mountains, Jamaica   352
    Brecon area   1410
    geology, Lydstep Haven   1344
    Gloucester - Malverns levelling   1462
mines and mining
    mine ventilation   975
    Mining Chronicle   377
    water pump   672
museums
    storage cabinets for Geological Survey collections   1494
    suggested display, Museum of Practical Geology   1039
    table case, Museum of Economic Geology   1217
    fossil cabinet   751
Parma, priapi   348
people and costume
    back view of a peasant girl's head, Milan & Como   363
    Capucines   363
    De la Beche family photographs of   422
    De la Beche portrait   415
    De la Beche, Rosalie Torre, photographs of   422
    De la Beche, photographs of   422
    Dutch woman's hat   349
    fair scene, Poggibonsi   363
    Florentine woman   363
    French peasant woman   343, 346
    French postillion   343, 346
    Frenchwoman's hat   344, 346
    hats   344
    Italian women's hats, Turin   347
    Italian women's hats, Sarzana   347
    Lazzaroni   363
    man's hat, Strasbourg   349
    method of wolf hunting   363
    Milanese woman   363
    Napoleon, Pedigree of   348
    ostlers   343, 346
    Paestum peasant   363
    peasant girl of Pignone   363
    peasant woman with spindle   363
    photographs, De la Beche and family   422
    Portrait a la Rembrandt   363
    portrait, De la Beche (1848)   415
    reclining woman (unfinished)   363
    shoes, Italian   348
    Swiss woman's hat   348
    Three French peasants   343, 346
plan for horse engine   438
Sargans   363
section, toadstone   1625
section, ?Carpathian Mountains   1936
sections, England
    Abberley Hills   1577
    Bristol to Gloucester railway, Charfield   748
    Bristol to Gloucester railway, Wickwar   749
    Castleton   1605
    Clee Hill   1766
    Cooks Folly to Abbot's Leigh, Bristol   342
    Derbyshire   1606, 1613
    Devon/Dorset Coast   433
    Dorset/Devon coast   341
    Gloucester area   1470, 1472
    High Stile   1870
    Inferior Oolite   1552
    Isle of Wight   587
    Isle of Portland sections   341
    Jurassic beds thinning towards Mendips   1552
    Kinder Scout   1637
    Derby   1595
    Landewednack to Cadgwith   1990
    Leek   1639
    Lulworth Cove   341
    Malvern   2266
    Malvern Hills   1592, 1663
    May Hill   2266
    Mesozoic rocks, Severn Vale   1453
    Monkton Combe   1550
    Pennant Quarry at Fishponds, Bristol   742
    Radstock Coalfield   2004
    Stockport   1638
    Stroud   1506
    Wrington   2125
    Tor Abbey Sands   360
    Wambrook, Somerset (De la Beche, 1839 *Report on the geology of Cornwall*, Figures 34 & 35)   390
    Woolhope   2266
    Wych Road, Malverns (Phillips, 1848, *Memoirs of the Geological Survey*, **2**(1), p. 64)   1449
sections, France
    Arromanche to Port-en-Bessin coast   350
    Fecamp to Le Havre coast   350
    Hennequeville coast   350
    Honfleur   350
    Port en Bessin to Grandscamp   350
    Villers-sur-Mer to Dires coast   350
    Villerville to Trouville coast   350
    valley of Chamonix at the Chapeau   348
sections, India
    Damodar Coalfield   2146
sections, Ireland
    Croghan,   1072
    SW Ireland   1522
    Wicklow Head   1084
sections, Italy
    Bagni di Lucca   363
    Bellagio Point to Monte San Primo   363
    Capo Corvo   363
    Como to Lake of Lugano   363
    Dosso d' Albido, Lake Como   363
    La Spezia to Coregna   363

of Vesuvius 363
ridge between Lucca and Massa 363
top of pass between Carrara and Massa 363
from the Carrara Marble Quarries to Lavenza 363
from the Marble Quarries westward, Carrara 363
Capo Mesco and Levanto 363
Coregna, Gulf of Spezia 363
near Carrara 363
near Bellano 363
Rochetta 363
Grotto del Cane, near Naples 363
Lignite formation of Caniparola, near Sarzana 363
from near Como to Monte Francesca 363
sections, Jamaica
Blue Mountains, Jamaica 356
Hope Valley to St Catherines Peak, Jamaica 356
Kingston to Buff Bay, Jamaica 357
Kingston to north coast, Jamaica 356
Luidas Vale to Old Harbour, Jamaica 355
sections, Wales
Abernant 1394
Carneddau, Builth Wells 342
Llanddowror 1380
Llandeilo 1385, 1388, 1389, 1439
Meidrim 1394
Newport to Brecon 342
Pumsaint 1403, 1435
Trecastell to Garthmaelwg 715
Trecastle 1405
Tywi valley 1433, 1438, 1715
near Llanddewi Velfrey 1485
Anglesey 1936
Cardiff to Caernarfon 1492
Llandegley 1771
Llanfechell 1965
near Brechfa 1714
Preseli to Wiston 1993
Swiss pipe 348
transport
?De la Beche's boat at Lyme 341
Chamonix Char a Banc 344, 346
coach 344, 346
coach (*Quarante sous par poste*) 343, 346
Dutch waggon 349
French gentleman's carriage 343, 346
Lyme fishing boat 341
sailing boat 433
sledge coach, Amsterdam 349
illustrations
criticism of standard of geological 2015
meteorological 90
insects, fossil
Cotham Marble, Bristol 1610
Westbury on Severn 1609, 1610
Institut d'Afrique
De la Beche proposed as Membre Titulaire 1848
Institute of Civil Engineers 1816
Irish Office 859
iron
beds in Coal Measures 516
beds in Hastings Sands 516
Coalbrookdale 2094
Commission 789, 1322
mines, ancient 2041
works, Cyfarthfa 342
iron ore
blackband ironstone 230
in Ireland 319
Merthyr Tydfil 342
mines, Lake Como, Italy 363
need for analysis of 789
need for study of British 1192
Northamptonshire 746, 1183
smelting of, with turf as fuel 319

Sweden 1032
workings in Oxfordshire 153
Irregularities of Sol visited upon his system
cartoon by De La Beche (illn.) (1841)

Jameson's Journal 189
Jupiter, colour of 1523
Jurassic
Coral Rag, France 350
Kimmeridge Clay 299, 1328, 1329
Lias 176, 341-343, 372, 655
of Jura Mountains 347
Oxford Clay 161, 2002
Portland Stone 184, 187, 575
France 350
Purbeck 574-577

Kenmore Mining Association 319
Kew Gardens
marine glue for glazing Great Palm House 613
killas 341
Kimmeridge Clay 299
Le Havre 1328, 1329
Yorkshire 1329
King Coals Levee (poem) 150, 151, 159

lakes, temperature of Swiss 345-348
landslips
Lyme Regis 80, 81, 1339
on railways 192
lazzaroni (illn.) 363
lead
at Llanboidy 1666
Devil's Bridge area 1940
in Monmouthshire 985
South Devon coast 179
in Lias of South Wales 176
in limestone 179
in New Red Sandstone 179
in R. Cothi 1666
in R. Sawdde 1666
Ireland 1935, 1936
in water supply 1175
lode, Ystumtuen 1941
mines,
Alston Moor 1965, 1966
Bwlch Gwyn 1941
Carmarthen 1435
Cowarch 1947
Cwmsymlog 1942
Cwmystwyth 1941, 1942, 1952
Cystanog 1435, 1666
Esgairmwyn 1948, 1949
Gilfach 1435
Goginan 1940, 1942, 1952
Llandovery 1424, 1435
Llangadog 1435
Llangynog 1424
Llandrindod Wells 342
Nant y Creiau 1941, 1952
Pennant 1947
Pontarllechau 1435
Tynfron 1952
Servoz, France 344, 346
mining & smelting, Wales 524, 739, 1666
on Mr. Lloyd's property, Carmarthenshire 1715
production 1694
& silver mine, Wheal Betsy, Devon 373
& silver mines, Rhine Gorge 349
Lias, Jurassic 176
Bath 342
coprolites in 655
Devon 341
Dorset 341, 372
fossils of 341, 342

ichthyosaurs from Lyme 342
Northern France 343
South Wales 176, 372
lichen
as cause of red snow 148
lighthouses
Eddystone, contruction of 372
Portland Bill 341, 342
Ramsgate 343
lightning
conductors for ships 657, 659, 661-664, 667
Paliamentary Enquiry on shipwreck by 661-663
lime
burning, South Pembrokeshire 351
limestone
analysis of 317, 332, 333
in Ireland 319
Linnean Society 920, 2003
lithographic stone
consumption of, in Britain 883
from Kelheim 883
lithography 342, 396
De la Beche's ability in 533

Magnesian Limestone 95
crinoids of 1349
& Murchison 1409
stratigraphic position of 1417
magnetic observations
in Mediterranean 1844, 1987
mammals
Owen's Report on British Fossil Mammals 1139
Owen's paper on Hordle mammals 1143
mammoths
found by Mikhail Adams 343, 346
preserved in ice 148, 342, 343, 346
teeth from Italy 347
maps, geological
by students 1109
coloured 25, 290
Conybeare's plan for map of Bristol 302
cost of engraving 650
economic advantages of 372
Geological Survey maps displayed at 1844 York BAAS meeting 651
index of colours on Geological Survey maps 208, 259, 1822
of Austria 640, 644, 645
of Derbyshire 278, 309
of France 372, 378, 473, 492, 502, 504, 505
of Himalayas 1678
of Ireland 306, 650, 756, 757, 1105
structural nomenclature of 309
value of, to agriculture 372
colouring of map of France 504
scale of Ireland maps 650, 756, 772, 775, 815, 1114, 1122
second edition of Greenough's map 622, 624, 625
shading of metamorphic aureole 1118
maps, topographical 32, 757
1" map of Ireland 1082, 1128
1" map of Scotland 790
6" map of Isle of Wight 790
6" map of Lancashire 792
6" map of Yorkshire 792
construction of Ordnance base maps 274
cost of engraving 648
for Museum of Economic Geology 1211, 1214-1216, 1218
Ireland 1127, 1128
need for accurate maps 19, 265, 276, 278, 399
of Austrian Empire 363, 643, 645
of Westmorland 791
ordnance, criticised 1470, 1473, 1477, 1502
scale of Durham map 792
value of contours 790
marble
Buckinghamshire 193
Derbyshire 648, 982
Llŷn 1792

Mechanics Institutions 646
Megalosaurus
from Stonesfield 302
mercury
mines, California 13
supply 13
metals
origin of ores 524
meteorites
used for tools by Eskimos 148
1794 fall near San Quirico d'Orcia, Italy 363
meteorological observations 341-344, 346-349, 352, 353, 659, 660, 696
millstones
from Andernach, Germany 349
from Monte Ferrato, near Prato, Italy 363
mineral deposits
in Devon & Cornwall Memoir 273
veins 1198
minerals
collection in Paris 343, 346
Derbyshire 109
Germany 337
Ogmore 327
Sweden 343, 346
mines and mining
coal 76, 85, 87, 88, 109-111, 351, 1649, 1653, 1808, 1834
disaster, Aberdare 98
disaster, Ardsley Main Colliery 85
disaster, Barnsley (1849) 76
disaster, Merthyr Tydfil 538
disasters, Dudley 87-89, 1943
explosions 2014
explosion, Neath 1957-1959
explosion, St Helens 1647
explosions investigations 414, 1308, 1657, 2044
Trecastle Colliery, Llanharry 2181
cobalt, Inveraray 15
copper, Devon & Cornwall 15, 1834, 1876
Brecon 342
Fowey Consols, Cornwall 434
Mont d'Or, France 343
Nant y Car, Rhayader 1946
Parys Mountain, Anglesey 1960, 1961
Servoz, France 344, 346
St. Davids, Pembrokeshire 949
Wheel Franco, Cornwall 373
Wheel Friendship, Cornwall 373
Culm, Bideford 832
Devon & Cornwall 397, 1834
gold, Ogofau, Carmarthenshire 1435
Hungary 1928
implements for Museum of Economic Geology 1970
Inspection Act 89
Inspectors of 1591
Inspector for South Wales 59, 89
lead 524
Alston Moor 1965, 1966
Bwlch Gwyn 1941
Carmarthen 1435
Cowarch 1947
Cwmsymlog 1942
Cwmystwyth 1941, 1942, 1952
Cystanog 1435
Esgairmwyn 1948, 1949
Gilfach 1435
Glendalough 1936, 1937
Glenmalure 1935, 1936
Goginan 1940, 1942, 1952
Llangadog 1435
Llangynog 1424
Nant y Creiau 1941, 1952
Nant y Mwyn 1666
Pennant 1947
Pontarllechau 1435
Tynfron 1952
Servoz, France 344, 346

    &silver, Rhine Gorge 319
    &silver, Wheel Betsy, Cornwall 373
    leases 15, 16
    lectures 225
    Llandovery 1424, 1435
    manpower 16, 74
    mercury (quicksilver), California 13
    Mexico 302
    need for engineers 1191
    need for instruction in 1653
    need for legislation to regulate 524, 1191
    nickel, Inveraray 14, 15
    Northern England 322
    overseas 74
    plan for registration of 48
    safety 1191
    school, Cornwall 302, 389, 1652
        Derbyshire & Yorkshire 1652
        Dudley 1050
        Lancashire 1652
        Newcastle 1652
        Scotland 1652
        South Wales 1652
    schools 61, 385, 386, 389, 1652
    silver, South America 13
    Staffordshire 1191
    surveying 76
    temperature increase with depth in 377, 599, 601, 658, 2144
    terms 351
    tin, Cornwall 941-943, 945, 2221-2228
    training of engineers in 938
    ventilation 975, 1834, 2000
    ventilation (illn.) 975
    ventilation by use of steam 62
    ventilation (Government enquiry) 57, 1650, 1652, 1655, 1656
    Wrington 2125
    Yugoslavia 1928
Mining Commission 2111
Mining Engineer, Government 16
Mining Records Office 76, 399, 1250, 1970, 2120
    in Jermyn Street building 130
    Keeper of 871, 872, 1249, 1250, 1254, 1999
    role of British Association in establishment of 397
    salary of Keeper of 1259
Miocene flora of Europe 8
models
    of Dundee by F.A. Carrington 228
    British Isles 228
    Derbyshire 228
    Manchester to Lincoln 228
    mining, in Museum of Economic Geology 397
    offered for purchase 229
    of Salzburg 1022
    scales of 228
molasse
    Alps 347
molluscs
    ammonites 1981, 2002
    belemnites 1148, 1330, 1331, 2002
    Coal Measures 955, 956
    collection of R.A.C. Austen 28
    fossil, from South Africa 19
    freshwater, of Purbeck 574
    from Blackdown 458
    from France 489
    gastropods 1446, 1496
    gastropods (illn.) 341
    Gryphaea 343, 489
    Nucula (illn.) 1405
    orthocone cephalopods 342
    Ostrea 1328
    oyster (illn.) 341
Moon
    model of 679
    crater (illn.) 1523

monitor lizard,
    comparison with ichthyosaurs 298
Monmouthshire Canal Company
    charges on train road 1821
Mosaic chronology 204
mosaic, Roman
    from South of France 1923
mosasaur
    from Maastricht 343, 346, 349
Museum of Economic Botany
    plants donated by De la Beche 714
Museum of Economic Geology, Dublin 402, 403, 593, 758, 760-762,
    821, 1078,1199, 1314, 1538, 1545, 1548
Museum of Economic Geology
    see Museum of Practical Geology
Museum of Irish Industry 826
    acquisitions of 830, 831
Museum of Practical Geology, London 14, 19, 41, 62, 92, 103, 109, 110,
    131, 196, 202, 203, 220, 225, 230, 231, 251, 253, 258, 270,
    295, 315, 331, 399, 406, 411, 417, 420, 454, 458, 462, 506,
    507, 511, 518, 524, 525, 562, 597, 598, 604, 610, 617, 642,
    645, 680, 702, 706, 710, 712, 714, 737, 738, 740, 742, 789,
    798, 825, 874, 878, 887, 888, 938, 962, 963, 986, 992, 996,
    1004, 1005, 1016, 1021, 1033, 1064, 1185, 1187, 1189, 1206,
    1238, 1242, 1337, 1351, 1359, 1376, 1392, 1395, 1405, 1406,
    1408, 1412, 1413, 1490, 1496, 1507, 1536, 1571, 1584, 1598,
    1667-1670, 1678, 1685, 1689, 1692, 1700, 1703, 1713, 1747,
    1749, 1773, 1782, 1792, 1824, 1831, 1832, 1865, 1891, 1902,
    1926, 1999, 2015-2021, 2026, 2031, 2035, 2039-2041, 2054,
    2132, 2137, 2165
    account of progress of 268
    agricultural chemistry course 1303, 1304
    budget & accounts of 73, 1205-1213, 1216, 1219, 1220, 1222,
        1225-1230, 1234, 1235, 1237, 1239-1241, 1271, 1273-1275,
        1288, 1289, 1309, 1311-1313, 1323, 1324, 2036
    buildings and facilities
        building of 554
        chemical laboratory at 870, 1690
        cost of Jermyn Street building 1306, 1307
        expansion into No. 5 Craig's Court 1254, 1257, 1260, 1261,
            1264-1266
        foundation of [Museum of Economic Geology] 1202, 1203,
            1224
        granite hall in 192
        laboratory 1246, 1249, 1250, 1253, 1260-1262, 1272, 1273,
            1300, 1301
        lecture room 1254
        move from Craig's Court 1325
        naming of 1203
        new building 733
        notice to quit Craig's Court 1073
        opening of new museum 1806
        plans for Jermyn Street building 1300, 1301, 1303, 1317
        proposed purchase of Egyptian Hall, Piccadilly 1297, 1298
        proposed purchase of Jermyn Street site 1298, 1299
    collections and displays
        collection of mining implements 1970
        collection of copper products 2277
        display cases for 1217-1220, 1225, 1226, 1236, 1238, 1239
        display cases for (illn.) 1217
        donation of Gault fossils by Clark 580
        donation of model dredge by Bull 579
        drawers required 2272
        duplicate material for India 2029, 2031-2033
        metal castings for display in 421
        mining dial 1220, 1221
        model of Arran for 1707
        models for 1214
        presentation by T. Weaver of early geological book 2090
        soils collection 1258, 1270, 1285
        specimens for 193, 194, 572, 573, 604, 874, 876, 877, 894, 925,
            936, 950, 961, 982, 987, 1029, 1039, 1051, 1062, 1155,
            1156, 1178, 1185, 1192, 1195, 1203-1205, 1207, 1222,
            1223, 1225, 1228, 1229,1231,1232, 1236, 1256, 1262,
            1266, 1295, 1357, 1391, 1469, 1496, 1501, 1563, 1698,

1827, 1830, 1833, 1909-1911, 1917, 1927, 1928, 1966, 1979, 2039, 2055, 2056, 2081-2084, 2092, 2113
    suggestions for iron ore display (illn.) 1039
    United States Geological Survey reports for 2015-2020
    vases for 1221, 1236
    Ordnance maps for 1211, 1214-1216, 1218, 1237, 1266
  days of opening 1829
  exchange of publications 1180, 1182, 2016
  guide book 740
  & mining explosions enquiry 1308
  purpose of 397
  staffing
    appointment of chemist 233, 398, 1232, 1233
    De la Beche nominated Director 1209, 1214, 1215, 1217
    duties of chemist in 398
    E. Forbes proposed for post at 1538
    salary of curator 1278
    porter 1251, 1261
  visit by Directors of East India Company 2028
  visitors 1227
  visitors to 227
  work on Commissions 1322

New Red Sandstone
  Bristol 342
  Devon 36, 174
  footprints in 306
  lead in 179
  Leek 1639
  Liverpool 1986
nickel,
  deposits, Cornwall 945
  deposits, Inveraray 14, 15
  minerals from Spain 1698
Nucula
  sketch (illn.) 1405

ocean currents 71, 669, 670, 2079
Old Red Sandstone 141
  Brecon Beacons 342, 1385, 1390, 1400, 1407
  Bristol 341
  Devon 28, 31, 36
  West Wales 351, 1431, 1432
  fish 1491
  fossils 917
  Murchison on 1011, 1348
  Pembrokeshire 351
  Russia 1348
  Shropshire 119
  thickness 1481
  trace fossils in 191
  Usk 1357, 1491
Oolite
  France 343
  France, comparison with England 343
  Jura, comparison with England 344, 346
orders, foreign 654
Ordnance, Board of 207-215, 260, 261, 269, 272, 273, 276, 277, 280-282, 285, 286, 288-290, 379, 397, 401, 774, 775, 856, 1561, 1822, 1823
  estimates 623
Oriental Club 2029
ostlers
  (illn.) 343, 346
ostracods
  from R. Avon (illn.) 1372
Ostrea 1328
ox
  Kirkdale Cave 162
Oxford Clay, Jurassic
  ammonites 2002

Palaeontographical Society 461, 471, 1150, 1919
  application to Treasury 2038
  Monographs 94-96, 486

Parliament
  Bill for mines registration 48
  Building Stone Commission for new Houses of 32, 65, 66, 397, 528, 610, 648, 948, 979, 982, 1197, 1242-1244, 1246, 1255, 1266, 1291, 1293
  drainage problems of 239
  Enquiry on shipwreck by lightning 661-663
  House of Commons acoustics & ventilation 1214
  plot against 238
  security of 240
  sewers of 238, 240
  visit of Queen Victoria 240
peasants
  back view of girl's head (illn.) 363
  French (illn.) 343, 346
  Frenchwoman (illn.) 343, 346
  Paestum (illn.) 363
  Pignone (illn.) 363
  Roman and antiquary (illn.) 363
  woman with spindle (illn.) 363
pencil manufacture of 999
Pennant Sandstone
  building uses, Bristol 342
  in Forest of Dean 2115
  occurrence 625
  section of Quarry at Fishponds Bristol (illn.) 742
Permian
  Magnesian Limestone 95, 1349
Philosophical Club
  meeting of 570
phosphate deposits 40
photography 1197, 1897
  cost of 1469
  proposed views of N. Wales mountains by Calvert Jones 807
  use of, in geology 1196, 1469
  value to palaeontology 506
planets, colours of 1523
plants
  donated by De la Beche to Museum of Economic Botany 714
  rare, in Britain 464
plants, fossil 303
  Baffin Bay 862
  Bovey clay 24
  Coal Measures 18, 164, 166, 171, 172, 309, 334, 701, 703-705, 711
  Exeter area 32, 33
  Forest of Dean 2115
  J. Lindley's lecture notes 862
  Lias 131
  Mr Skinner's donation to Bristol Institution of 301
  neglect of, by botanists 302
  Portland/Purbeck 187
  South Wales 700, 948
  Staffordshire 334
  Van Diemen's Land 712
Pleistocene
  elephants, Essex/Suffolk 471
  fossils, Ireland 550
  vertebrates 42, 157, 162, 164-168, 170, 172, 250
  vertebrates in Arno Valley, Italy 363
  vertebrates in Palace at Darmstadt 349
plesiosaurs 180, 298
  anatomy of 300
  bones found by De la Beche in France 350
  description of Anning specimen 302
  discovery at Caen 163
  discovery at Lyme Regis 163, 169, 170, 302, 355, 357
  discovery at Whitby 109
  discovery of jaw 297
  in Muschelkalk 489
  in possession of Mrs Mary Anning 299
  in possession of De la Beche 163, 169, 170
  in possession of T. Birch 300
  sketches of 302, 307
Polar Dinner 137
postillions 343, 346
  French (illn.) 343, 346

potatoes, Pembrokeshire 1706
Potato Famine 756, 761, 1027, 1076, 1077
potteries
    Copeland, Aderman 461
    Indian 482
    problems of 461
    Staffordshire 1194
    Stoke 1195
    Wood, Enoch 461
pottery
    from Suffolk 677
    Wedgwood, for Museum of Practical Geology 936
Precambrian 330
printing techniques 183
    use of transparent paper 180
Privy Council Office 805
Protectionists meeting at Exeter 483
Proteosaurus 157, 300
    naming of 342
    see also ichthyosaurs
pterodactyl 179
publications, international exchange of 55

quarrying
    manpower 24
Quarterly Review 955-957
quartz, Bristol Diamonds 341

railways
    Board 718, 719
    Bristol to Bath 1859
    Bristol to Gloucester 742, 747, 748
    Exeter and Bristol 25, 1859
    Exeter Railway Co. 1865
    expansion of 57, 189
    fares 1587
    Forest of Dean 1275, 1277, 1281,1282
    gauge of 1588
    Great Western 1284, 1308, 1310, 1588, 2126, 2127, 2150
    India 2138
    landslips on 192
    London to Birmingham 744
    Midland 1627
    role of Geological Survey in providing advice to 1594
    Section Committee, British Association 189, 193, 397, 1375, 1376, 1858, 1859
    sections 406, 417, 743-745, 747-749, 1581, 1583, 1615, 1764
    sections, Ireland 1081, 1083, 1085, 1105
    South Eastern 468
    South Wales 189, 1924
    South Wales, opening of 807
    speed of 1588
    Swansea 448
    Taff Vale Railway Act 2244
    tolls on iron and coal 2244
raised beaches
    Devon 24
    Scotland 70, 980
Rebecca Riots
    Swansea 1527
Reform Bill 895
reindeer
    Ireland 1088
religion
    De la Beche's attitudes to 341-344, 346, 363
    education for workers on De la Beche's Jamaican estate 1905
    Mass at Dijon 343, 346
    procession at Aosta 344, 346
    procession at Nantua 344, 346
    procession at Strasbourg 349
    role of in Oxford University 342
reptiles
    BAAS British Fossil Reptiles report 1136-1141, 1150
Rhaetic Bone Bed 179
rhinoceros
    Kirkdale Cave 162
    teeth from Lyme 342
    Wirksworth 170
river valleys
    formation of 307, 308, 429
roads
    Arc Valley, Savoie (illn.) 347
    condition of French 343
    condition of German 349
    repairs to 342
    stone from Cornwall 519
    tolls on Dutch 349
    value of geological maps in construction of 372
Roman girl and Franciscan monk (illn.) 363
Roman peasant and antiquary (illn.) 363
rotation of earth 893
Royal Agricultural College 981
Royal Agricultural Society 40, 629, 1274, 1789, 2156
Royal College of Surgeons 302, 405, 911, 1137-1150, 1152, 1153
    De la Beche's visit to 342
    ichthyosaur sent by De la Beche to 699
    Library 1147, 1149
    Museum of, comparison with Paris 343, 346
Royal Dublin Society 823, 1107, 1555
    decision to arm 1095
    exhibition of Manufactures 1123
Royal Geological Society of Cornwall 379, 599, 2003
Royal Hammerers 818, 1956
Royal Institution, London 64, 103, 526, 527, 1031
    invitation to Edward Forbes to lecture at 545
    J. Phillips as Lecturer at 1353, 2272
    specimens from Kirkdale Cave for 1849
Royal Institution, Manchester 78, 1686
Royal Institution of South Wales, Swansea 463, 868, 874, 1250, 1252
Royal Irish Academy 757, 764, 1068, 1554
Royal Military College 108, 513
Royal Mint, London 101, 102, 152, 1824, 1825, 2164, 2165
Royal Society 189, 564, 658, 664, 911, 914, 1104, 1105, 1194, 1531, 1876,1896
    W.G. Armstrong proposed as Fellow 743
    articles in Athenaeum about 568
    Charter Committee 1052
    Committee to select building stone for London Bridge 172, 301
    Conybeare showing Plesiosaur drawing to 302
    Copley Medal 309, 1006
        to W. Buckland 165
        to W.S. Harris 658, 1006, 2050
    De la Beche's offer to chair meeting 412
    De la Beche proposed as Fellow 342
    meetings 570, 1060
    Officers and Council
        Council 1052, 1054-1056, 1059, 1104, 1844, 2050
        resignation of Lubbock as Treasurer 2051
        T. Bell proposed as Secretary 913
        W.R. Grove proposed as Secretary 570
        limiting term of officers 912, 1842
    Presidency 638, 682, 906-908, 1056-1058, 1842-1844, 2051
        appointment of D.G. Gilbert as President 153
        J.F.W. Herschel proposed as President 1842-1844
        Peacock suggested as President 1843
        refusal of Presidency by J.F.W. Herschel 682, 906, 907, 1058, 1843
        need for scientific President 1844
        resignation of Sir Joseph Banks as President 153
        resignation of Lord Northampton as President 1842
    rooms of 1060
    state of 570
    Wintringham Bequest/Medal 1053, 1055, 1056
Royal Society of Edinburgh 595, 596, 1022
Russell Institution 103

salt
    Cheshire 1135
    mines, Poland 146
Sanitary Commission, Metropolitan 56, 67, 68, 226, 231, 232, 234-239, 242, 248, 467, 656, 729, 828, 933, 934, 1295, 1322, 1684-1686, 1919, 2044, 2068

De la Beche invited to join Commission 615
sanitation 106, 233, 729, 933, 934
 London 2063
Sanitary Bill 2044
School of Mines, London 547, 584, 630, 635, 646, 1195, 1650-1652, 1773, 2180
 Duke of Cornwall's Exhibitions 1674, 1675
 funding of 850
 Inaugural lecture by De la Beche 63, 685, 856, 986
 Museum of Practical Geology as Mining School 740, 1250
 plan for 1333
 proposal by D.T. Evans 524
 proposal for a College of Mines 385, 386, 389, 850
 proposal for Iron & Coal Masters Prize 1050
 R.A.C. Austen's collection 37
science
 practical applications of 10, 40, 322, 397, 398, 630, 710
 training of teachers in 1693
Science and Art, Department of 1693
sea level changes 24, 284
 Cornwall 1875
 Mediterranean Sea 82
sedimentation at river mouths 2079
sewage 934
 Prince Albert on use of 629
sewers 729
 Metropolitan Sewers Bill 239, 242
 cleaning 1978
 Commission 226
 Kent 934
 of Parliament 238, 240, 243
 Surrey 243, 934
 Windsor 2057-2066
sheep, Tibetan, for Britain 1836
shells
 collections 800
 from Jamaica 653
Sidon, H.M.S. 707, 708
Silliman's Journal 8
Siluria 39
Silurian 8, 84, 109, 118, 141, 250, 330, 1331
 Baltic 1018
 /Cambrian boundary 791, 1018, 1019
 comparison of, in Wales and Ireland 557, 558, 568, 1079
 Dingle compared with Marloes 1523
 West Wales 1406, 1433, 1437, 1523
 Ludlow 119, 1406
 May Hill 1572
 naming of 39, 185
 N.E. Wales 2133
 of Bala 555
 of Constantinople 622
 of Gloucestershire 747
 of Ireland 557, 558, 563, 565, 567, 770, 1080, 1081
 of Kirkcudbright 541
 palaeogeography, Welsh Borders 1735
 Prague 1022
 Russia 1018
 Scandinavia 1018
 Silurian System 29-31, 34-36, 38, 39
 trace fossils of 543
 Usk 1357, 1406, 1851
 Wenlock 119, 1771
 Wenlock Limestone
  analysis of 1814
  North America 157
  Russia 157
  Sweden 157
 silver
  and lead mines, Rhine Gorge 349
  and lead mine, Wheel Betsy, Cornwall 373
  mines, California 13
  supply 13
Skiddaw Slates
 discovery of graptolites in 1025, 1886, 1887, 1889
 Sedgwick's paper on fossils in 1890-1893

J. Percy on 1188, 1189, 1193
slate
 Abereiddy 1707
 Pembrokeshire coast 947
 quarry near Penberry, Pembrokeshire 351
slavery 113, 114, 302, 352, 354, 355, 373, 2233
 De la Beche's attitude to 352, 354, 355, 359, 369, 2023
 De la Beche's "plan for emancipating the negroes" 369
 emancipation of slaves 301, 357, 369, 1154, 2023
 religious instruction for slaves 359
smelting 63
smoke emissions, control of 998
smoke enquiry 1308, 1309, 1311
snow
 origin of red snow 148, 149, 343
Snowdonian
 as stratigraphic name for Cambrian 185
Society of Arts, Manufactures and Commerce
 lecture by De la Beche on 1851 Great Exhibition (2 Dec 1851) 10, 999
Society for the Diffusion of Useful Knowledge 622
 invitation to De la Beche to publish on geology 257
soils
 analysis of 4, 1274, 1705, 2028, 2031
 Chalk soils for vines 343
 collection of 1258
 samples, Ireland 765, 767, 768, 770, 772, 776, 779, 814, 1318
solar system 847
Somersetshire Society
 Museum 2041
Southern's rule 521, 522
Speeton Clay, Cretaceous
 fish from (illn.) 1589
sphalerite 1715
springs, mineral 2046
 Aachen 349
 Aix les Bains, France 347
 Bath, Jamaica 352
 Buxton 1601
 Llandegley 342
 Llandrindod Wells 342
 Llanwrtyd Wells 342, 457
 Park Wells 342
 plants of 457
 St Gervais, France 344, 346
 Taff's Well 2240
springs
 artesian 246, 247
 thermal 246, 2046
stalagmite, Burrington cave 42
Stationery Office 935, 936, 1127, 1128, 1204, 1223, 1233, 1812, 1813
steam, velocity of 522, 523
steam engines 522
 for sugar mill 116
 water consumption of 116
steel manufacture 530
stratigraphy
 correlation based on lithology 36
 nomenclature 39, 185, 532, 1011, 1014, 1350, 1397, 1439
 use of fossils in 24-27, 33, 36, 38, 119, 350, 1013, 1018, 1331, 1388, 1393, 1850
submerged forest 788
sugar manufacture 1776
Survey, Ordnance Geological 1, 2, 62, 79, 83, 84, 103, 119, 120, 270, 278, 312, 330, 388, 396, 399, 401, 410, 412-414, 416, 595, 651, 696, 748, 889, 939-941, 1022, 1049, 1113, 1417, 1420, 1635, 1703, 1814, 1902, 2079, 2168, 2169, 2174
 account of progress 268
 accounts of surveyors 121, 744, 949, 1404, 1405, 1587, 1604, 1634, 1648, 1654, 1730, 1737, 1742, 1743, 1795, 1797, 1802, 2113, 2127
 budget/estimates of 207, 208, 213, 222, 267, 397, 1323
 collections
  fossil collections from Ireland 751-754
  fossils collected on Cornwall & Devon Survey 387, 1984
  proposal for Museum of, in Scotland 596

proposals for storage of fossil collections of 1484, 1493, 1494, 1521, 1665
    need to publicise collection of 1488
colouring of maps 208, 259, 290
correspondence 1300
engraving of maps 2170-2172
equipment 2167
esprit de corps 126
expenses of Devon and Cornwall mapping 209-212, 214, 261, 263, 279, 376, 379
gift of maps and sections 201, 224
inconvenience of Trigonometrical Survey's move to Southampton to 399
instruments of 288
litigation over access to land 123
mapping
    commencement of Cornwall Survey 261
    Cornwall 210, 263, 265, 269, 270, 276, 277, 280, 281, 379, 384, 387, 844, 939-946
    Derbyshire Survey by Hopkins 380
    Devon Survey, conditions for 208
    Devon 208, 209, 270, 276, 280, 281, 370, 378, 379, 387, 1872
    extension into South Wales 207, 269, 283, 379, 384, 623, 624, 839, 872-874, 878, 879, 947-952, 1381, 1388, 1394, 1397-1400, 1409, 1410, 1436, 1466, 1706-1715, 1991, 2010, 2011, 2101-2109, 2112, 2113, 2115-2120
    in Bristol area 2121-2131
    in Dorset 122
    in Forest of Dean 2110-2112, 2114
    in Gloucestershire 191, 1470-1478, 1498, 1503, 1504
    in Shropshire 607
    in North Wales 2133, 2134
    in Welsh Borders 1441-1469, 1478, 1479, 1482, 1498
    Scotland 596, 980
    Yorkshire & Derbyshire 1536, 1578, 1587, 1595-1603, 1605, 1606
    section through Wales 1413
    Shropshire sheet 1951
    state of (1838) 384
    property damage during 44, 45
map of Wales 1492
original proposals and costs 259, 260
plan to amalgamate with Irish Geological Survey 1484
presentation of maps to St Petersburg Imperial Academy of Science 401
progress of 379, 744, 1409, 1411, 1770, 1782, 1784, 1793, 1799
proposal for continuation beyond Devon and Cornwall 269
proposed transfer to Department of Woods 397, 1299, 1300, 1302, 1303, 1306
publications 259, 272, 273, 275, 276, 281, 283, 284, 337, 1539
    Figures and descriptions of British Organic Remains 1153, 1162, 1489, 1490, 1492, 1495, 1534, 1559
    Memoirs 937, 973, 1181, 1184
    Memoirs, Volume 1 MSS 408, 409
    Memoirs, Volume 1 (1846) 726, 737, 860, 1590, 1607, 1734, 1747
    Memoirs, Volume 2, 861, 1578-1581, 1593, 1594, 1607, 1614, 1618-1622, 1626, 1628, 1629, 1631, 1658, 1661, 1662, 1752, 1910
request for maps 244
role of, in providing advice to railways 1594
staff
    appointment of assistants 215, 262
    appointment of Edward Forbes 544
    appointment of H.H. Howell 730
    employment of J. Phillips 387, 397, 400, 1337, 1341, 1347, 1353-1355, 1373, 1384, 1389, 1390, 1518, 1534, 1535, 1559, 1560, 1574
    proposed appointment of R. Owen as Honorary Palaeontologist 405
    rank of survey staff under Department of Science and Art 593
    rates of pay on 125, 128, 1787, 2135
    request for employment on 47, 99, 105, 258, 541, 732, 922, 1154, 1157, 1164, 1896, 1900, 1901
    R.M. Murchison attached to 1008
    salaries of Jukes & Selwyn 997

salary of Ibbetson as surveyor of railway sections 745
salary of W.W. Smyth as Mining Geologist 414
staff matters 289
termination of employment of T.E. James 404, 406, 795, 1739
Transfer Bill of 221, 222
visit of Hornes and Haner to 640-643
Survey, Ordnance Trigonometrical 2, 212, 388, 397, 399, 649, 716, 827, 1306, 2105
    abolition of office of Treasurer of 942
    appointment of H. James as Director-General 791
    appointment of H.J. Maguire as leveller 922
    budget of 236
    Certificate of employment of De la Beche on 370
    datum level 205
    display of maps in public institutions 2040
    in Cardiff 189
    in Cornwall 265, 939-
    in Ireland 259, 306, 1068
    march route of De la Beche as Geologist to 2074
    Master General of 260, 397
    of London 239
    of North Wales 287
    of Scotland 596
    of towns 232, 235
    progress of 792, 948
    state of (1838) 384
surveying 1467
    instruments 108
swallow holes
    in Chalk 150
    plants in 152

Tertiary 93, 1770
    Barton Beds 582
    Bovey Formation 24, 27, 28
    faults 295
    fossils 458, 548, 1143, 1148, 1151
    Isle of Wight 582-592
    limestone from Egypt 548
    limestone of Malta 548
    London Clay 295
    of Italy 332
    volcanics 921
theology 204
tides 2079
    guages 2067
till, Cardigan 1746
tin mines, Cornwall 941-943, 945, 2221-2228
toadstone 309, 1585, 1586, 1624
    sketch section in (illn.) 1625
tortoise
    comparison with plesiosaur 302
    from Aarberg, Switzerland 348
tourmaline
    from Bovey Tracey 147
trace fossils 878
    Buckland on 191
    in New Red Sandstone 306
    naming of 191
    Mostyn 191
    value of studying modern organisms 543
Trade, Board of 60, 221-223, 617, 918, 1694, 1695
Treasury, London 12, 207, 215, 222, 236, 387, 414, 554, 649, 665, 856, 858, 1205, 1206, 1208, 1209, 1211, 1212, 1217, 1222, 1227, 1229, 1234, 1235, 1250, 1253, 1256, 1265, 1273, 1292, 1295, 1311-1313, 1323, 1327, 1347, 1984, 2036-2038, 2172
    Report of Treasury Commission 221
Triassic 141
    conglomerate at Ogmore 149
    Keuper 1470, 1471
    New Red Sandstone 36, 174, 179
    Rhaetic Bone Bed 179
    vertebrates, Bristol 903, 918
trees, fossil 2163
    in situ 144, 1612
trilobites 851, 1567, 2054, 2088

Bala (illn.) 609
Builth Wells 336, 342, 431
Canada (illn.) 883
Carmarthen area 1393, 1432
comparison of New York and Welsh 1517
Cornwall 1157
Devon 27, 1338, 2086
distorted 336
eye structure 110
Elmley Lovett 624
in De la Beche's collection (illn.) 431
Ireland 1097
Malverns 1499, 1577
Norway 343
Ohio 343
Prague 1022
Snowdonia 1779
Wenlock (illn.) 1333
Turko-Persian Frontier Commission 864-867, 2149
turquoise from Russia 149
turtle fossils from Maastricht 343, 346

University College, London
    Chair of Mineralogy at 103
Uranus [as Herschel]
    colour of 1523

vertebrates
    crocodile from Market Rasen 169, 170
    De la Beche on appearance of air breathing vertebrates 341
    discovery of fossil 11, 109
    fossil crocodile purchased by De la Beche 161
    fossil, from South Africa 19
    from Missouri 536
    ichthyosaurs 296
    Pleistocene 42, 162, 164-168, 170, 172, 349
    Pleistocene from Rugby 157, 170
    plesiosaurs 163, 169, 170, 180
    Rhaetic Bone Bed 179
    Triassic, from Bristol 903, 918
volcanoes 141
    craters 5101

water
    analysis
        rain 233
        Jamaica 116
        Taff's Well 2240
        Trafalgar Square artesian well 1687
        Windsor Castle 219
    pump (illn.) 672
    divining 150
    supply
        Bath 246
        Jamaica 983
        Liverpool 1986
        London 2063
waterworks
    Copenhagen 291
    East Middlesex 513
    Nottingham 291
Wealden, Cretaceous 532, 534
    Mantell on fossils of 929
    Mantell on freshwater origin of 534
Wedgwood pottery
    for Museum of Practical Geology 936
Welsh
    costume 342
    language 342
    placenames near Malvern 1617
    woman and baby, photograph (illn.) 422
Wenlock Limestone
    analysis of 1814
    Baltic & North America 157
    Dudley 157
    Russia 157
Wernerian school 348
windmills 511
wood
    analysis of fossil 2047
    fossil 19
    fossil, Lias 341, 2047
    preservation of fossil 862
Woods, Office of 65, 73, 219, 234, 238, 250, 407, 418, 421, 666, 668, 695, 697, 766, 857, 978, 979, 1201-1272, 1274-1298, 1300, 1303-1305, 1307, 1308, 1310-1327, 1336, 1689, 1696, 1902, 1903, 1999
    Chief Commissioner of 398, 407, 410, 420, 421, 1248, 1251
    Museum of Economic Geology in care of 397
    proposed transfer of Geological Survey to 397, 1299-1303, 1306
Yoredale Series 1584

Zoological Society 926
zoology, economic 50

Fig. 1 Fossil trees in Northumberland by William Buckland (**144**)

Fig. 2 Ichthyosaur skull structure: William Daniel Conybeare to De la Beche 1821 (**297**)

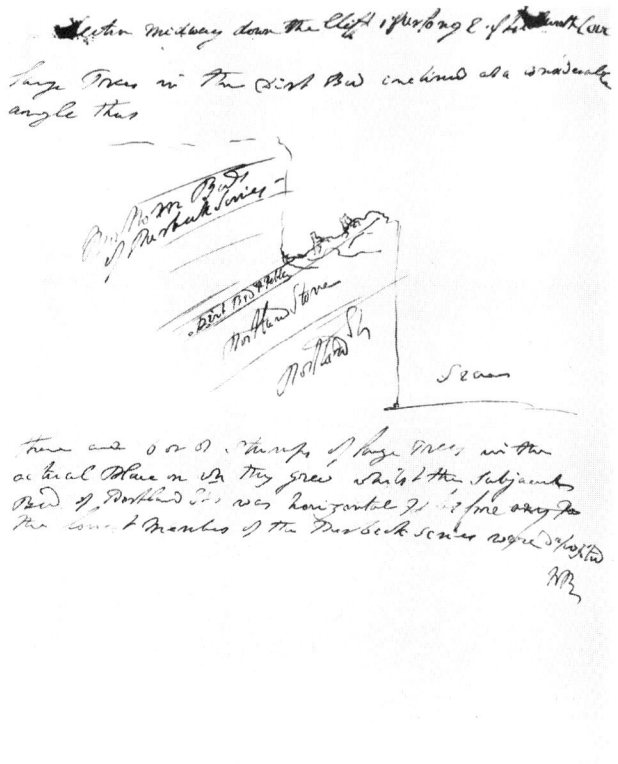

Fig. 3 Section of Portland Dirt Bed near Lulworth, Dorset by William Buckland (**187**)

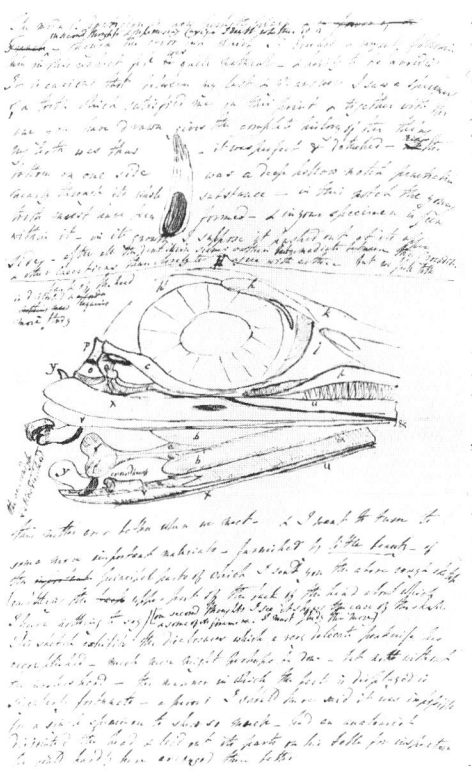

Fig. 4 Ichthyosaur skull structure: William Daniel Conybeare to De la Beche 1821 (**297**)

Fig. 5 Durdle Door, Dorset (**341**)

Fig. 6 Branscombe Cliffs, Devon (**341**)

Fig. 7 Vertical strata of Chalk, Dorset (**341**)

Fig. 8 The Bill of Portland, Dorset (**341**)

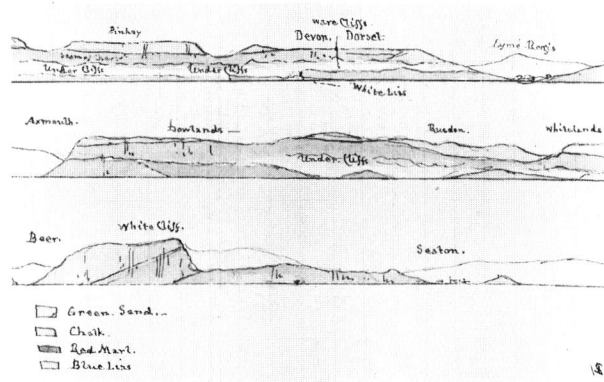

Fig. 9 Geology of the coast from Beer, Devon to Lyme Regis, Dorset (**341**)

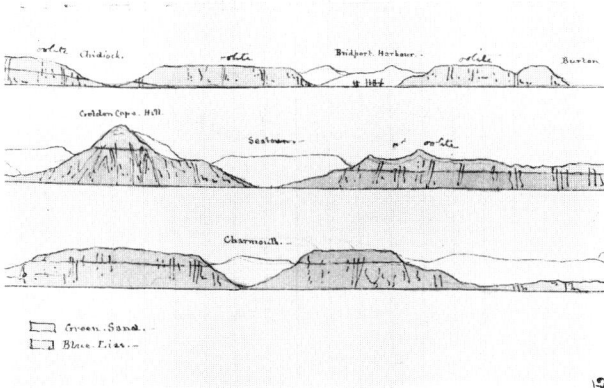

Fig. 10 Geology of the coast from Charmouth to Burton Bradstock, Dorset (**341**)

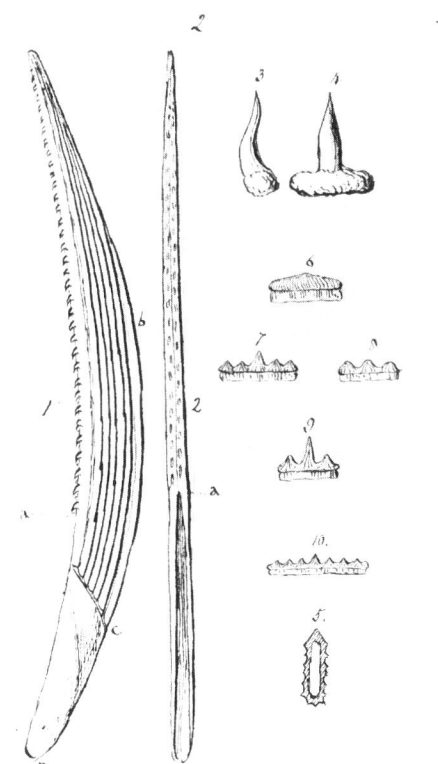

Fig. 11 Fish teeth and spines, Lias, Dorset (**341**)

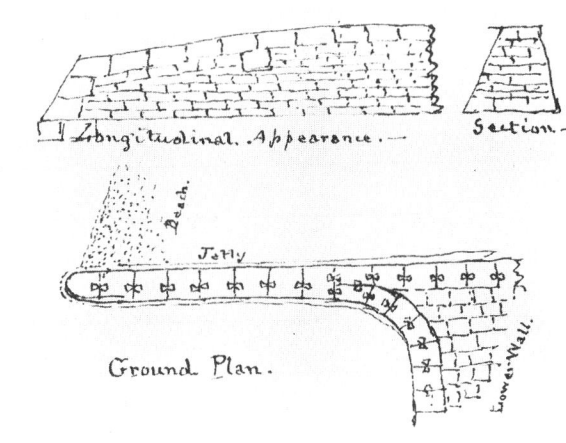

Fig. 12 De la Beche's jetty at Lyme (**341**)

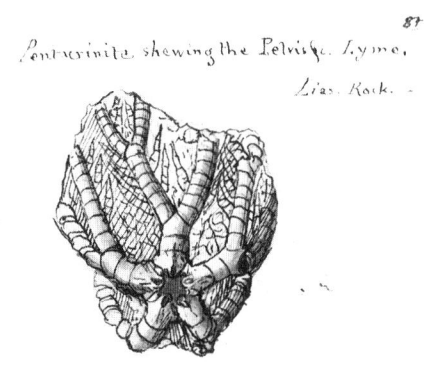

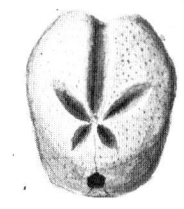

Fig. 13 Pentacrinite, Lias, and Echinite, Greensand, Lyme (**341**)

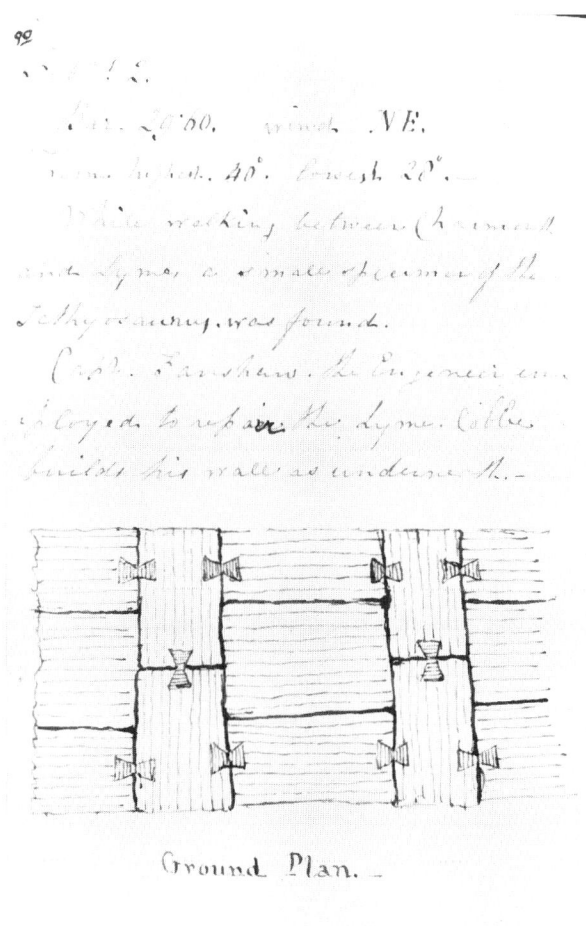

Fig. 14 Plan of Cobb walls, Lyme (**341**)

Fig. 15 Caerphilly Castle (**342**)

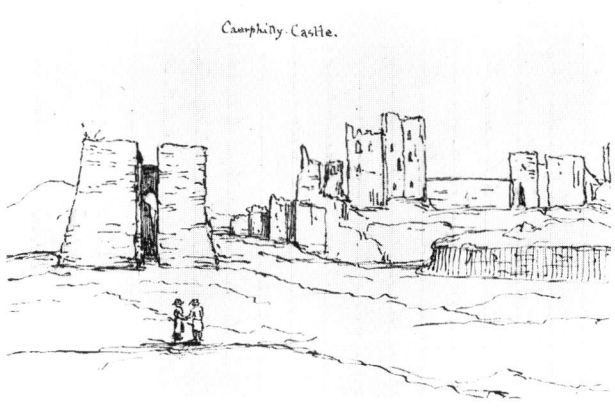

Fig. 16 Caerphilly Castle (**342**)

Fig. 17 Section through the Carneddau, Builth, South Wales (**342**)

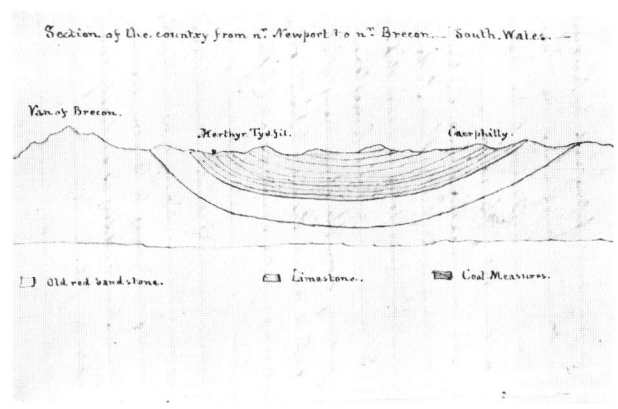

Fig. 18 Section from Newport to Brecon, South Wales (**342**)

Fig. 19 *Seeing Mont Blanc and the Valley of Chamonix* (**346**)

Fig. 20 *A Chamonix Char-a-Banc* (**346**)

Fig. 21 Section of Jamaica from Kingston to Buff Bay (**357**)

Fig. 22 Section from Lluidas Vale to the sea near Old Harbour, Jamaica (**355**)

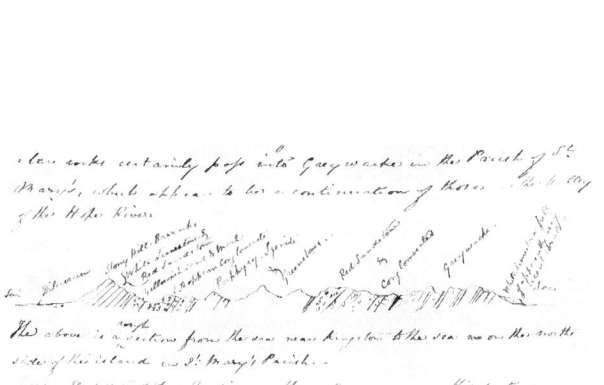

Fig. 23 Section north across Jamaica from Kingston (**356**)

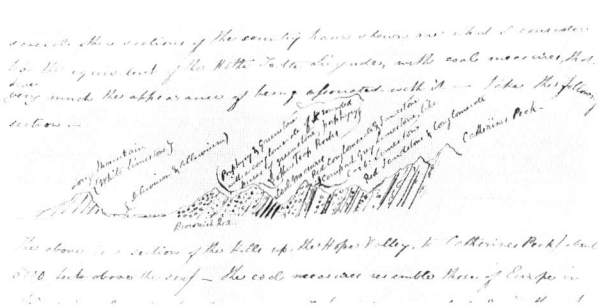

Fig. 24 Section along the Hope valley to Catherine's Peak, Jamaica (**356**)

Fig. 25 Geological map of Tor Bay by De la Beche (**360**)

Fig. 26 *Seeing the Raphaels at the Vatican* (**363**)

Fig. 27 *Milanese* (**363**)

Fig. 28 *Taking a Cameo-portrait* (**363**)

Fig. 29 *Visiting the Vatican by torchlight* (**363**)

Fig. 30 *Peasant girl of Pignone* (**363**)

Fig. 31 *Capucines* (**363**)

Fig. 32 *Crater of Vesuvius* (**363**)

Fig. 33 *Lazzaroni* (**363**)

Fig. 34 Transported [erratic] block. Alpi di Pravolta, Lake of Como (**363**)

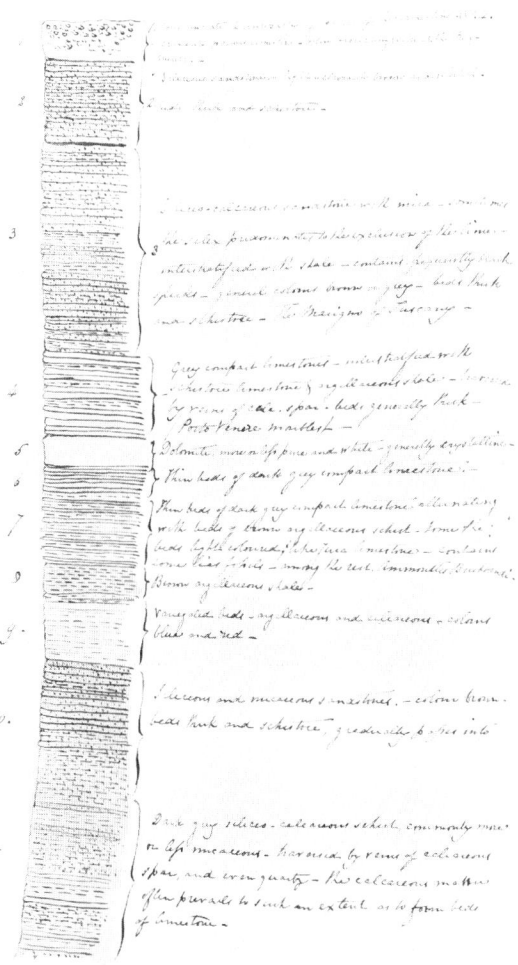

Fig. 35 *General section of the stratified rocks in the neighbourhood of La Spezia* (**363**)

Fig. 36 Section from Bagni di Lucca (**363**)

Fig. 37 *Portrait à la Rembrandt* (**363**)

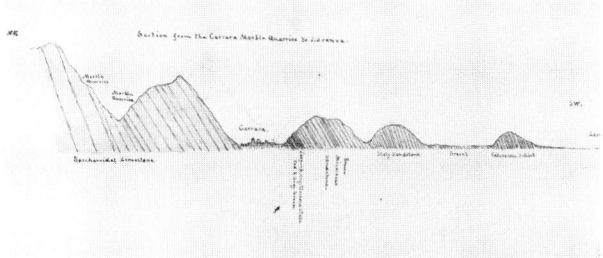

Fig. 38 Section from the Carrara Marble Quarries to Lavenza (**363**)

Fig. 40 *Henri and self* (**363**)

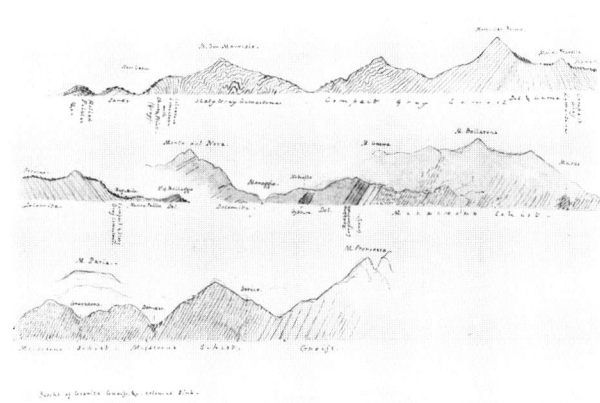

Fig. 39 Sections from near Como to Monte Francesca (**363**)

Fig. 41 *Lerici from Santerenzo, Gulf of Spezia* (**363**)

Fig. 42 *Roman peasant and antiquary* (**363**)

Fig. 43 Monk and the Devil (**363**)

Fig. 44 Vico (above) and cliffs between Vico and Sorrento (**363**)

Fig. 45 Coast from Vesuvius to Capri (**363**)

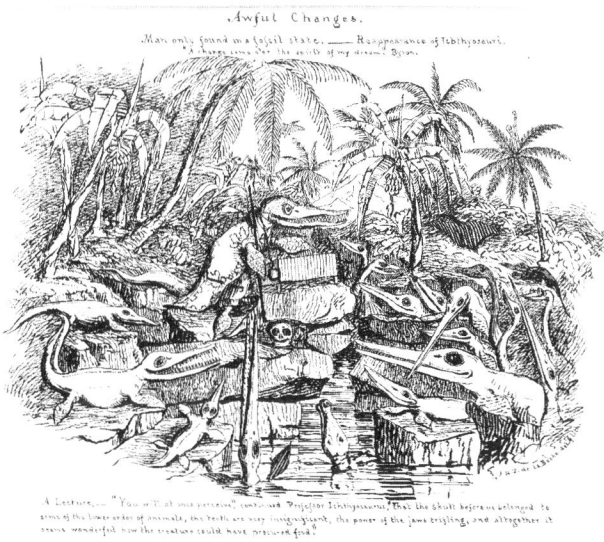

Fig. 46 *Awful Changes* (**367**)

Fig. 47 Pencil changes on reverse of *Duria Antiquior* watercolour (**368**)

Fig. 48 *Duria Antiquior:* original watercolour (**368**)

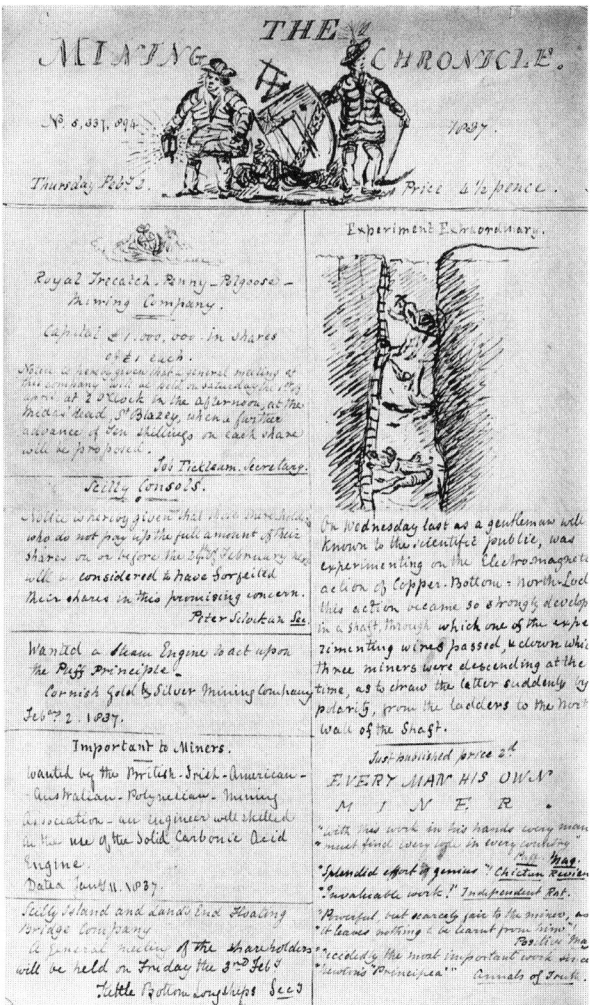

Fig. 49 *The Mining Chronicle* (**377**)

Fig. 50 *Heat in Mines* (**377**)

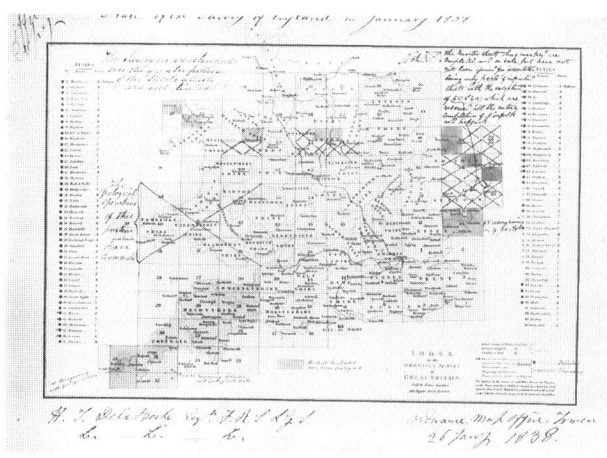

Fig. 51 *State of the Survey of England in January 1838* (**384**)

Fig. 52 *The Irregularities of Sol* (**395**)

Fig. 53 Henry De la Beche, 1848 (**415**)

Fig. 54 De la Beche and his daughters Rosie (standing) and Bessie (seated) (**422.1**)

Fig. 55 De la Beche and his three eldest grandchildren, Mary, Henry, and Elizabeth Amy Dillwyn (**422.2**)

Fig. 56 Rosie De la Beche and Bessie Dillwyn (**422.5**)

Fig. 57 De la Beche, Rosie, and Tartar (**422.3**)

Fig. 58 Lewis Llewelyn Dillwyn and Badger (**422.7**)

Fig. 59 Bessie Dillwyn (**422.9**)

Fig. 60 Lewis Llewelyn Dillwyn and his daughter Amy (**422.6**)

Fig. 61 Bessie Dillwyn and her daughter Amy (**422.10**)

Fig. 62 *A Coprolitic Vision* (**432**)

Fig. 63 Gaff-rigged schooner off Dorset coast (**433**)

Fig. 64 Hooken Undercliff, Beer Head, Devon (**433**)

Fig. 65 Beer Head and Whitecliff from Dowlands Cliff, Devon (**433**)

Fig. 66 On N. of Snowdon Peak (**436**)

Fig. 67 Fowey Consols Copper Mine, Cornwall (**434**)

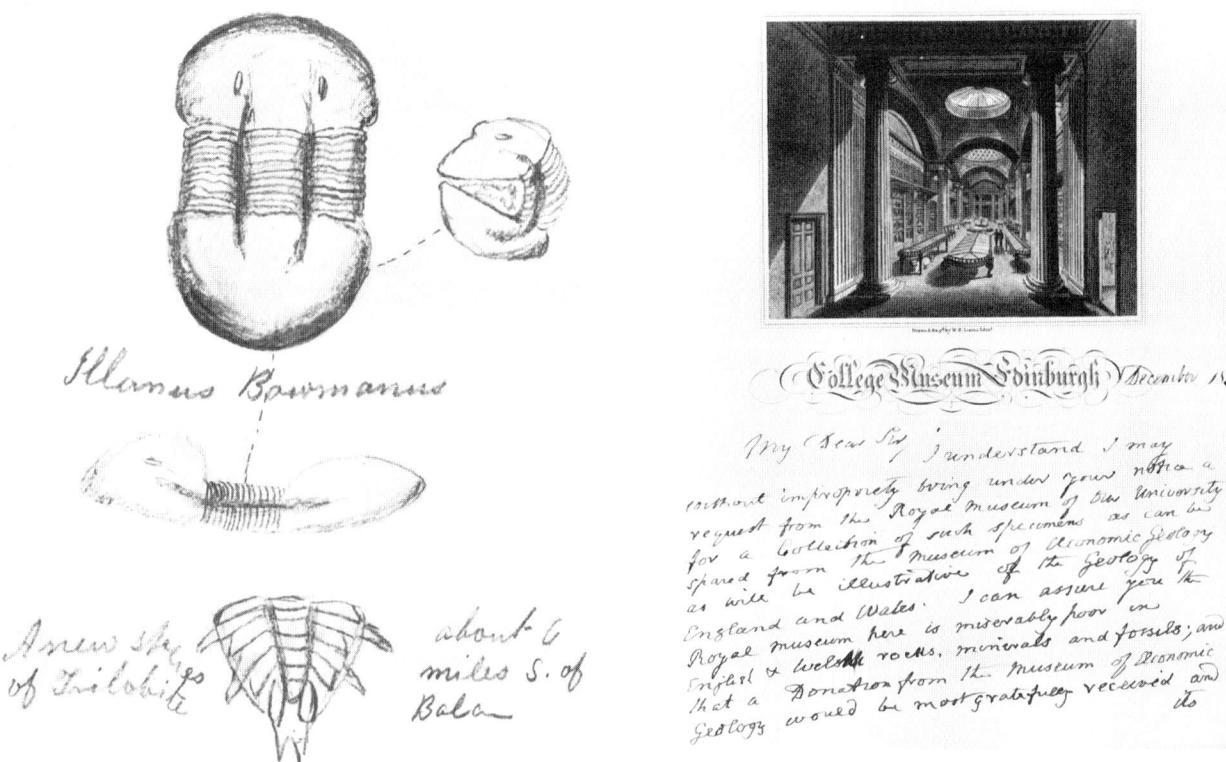

Fig. 68 Trilobites from Bala by Richard Gibbs (**609**)

Fig. 69 Letter from Robert Jameson on Edinburgh College Museum letterhead (**789**)

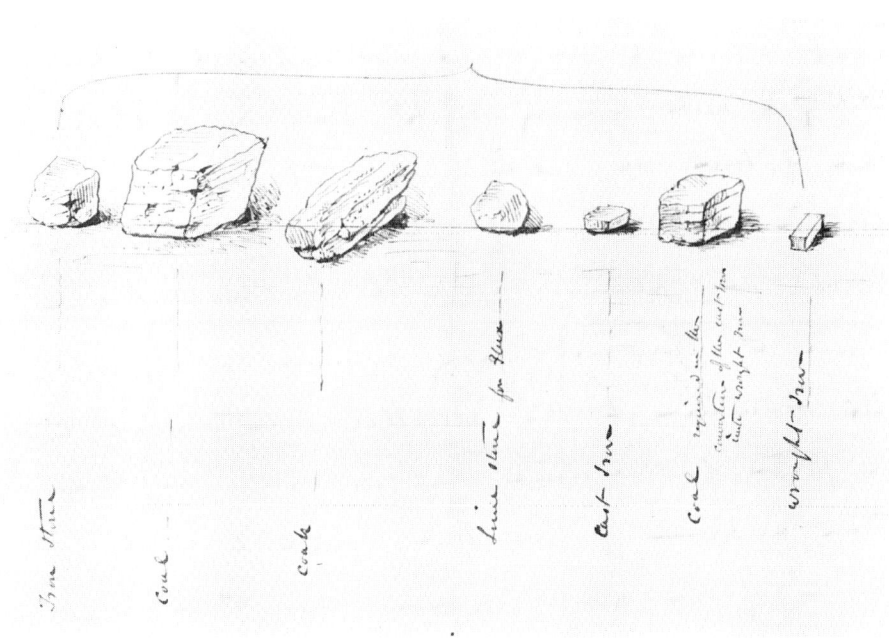

Fig. 70 Arrangement of specimens of materials used in iron-making: James Nasmyth to De la Beche, 1842 (**1039**)

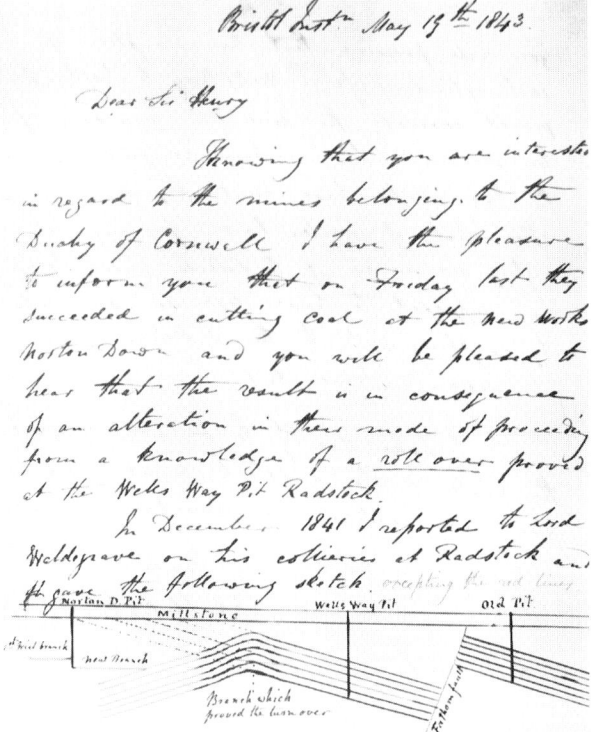

Fig. 71 Section of the Dowlands landslip, Devon: John Phillips to De la Beche, 1840 (**1339**)

Fig. 72 Section through the Radstock coalfield: Samuel Stutchbury to De la Beche, 1843 (**2004**)

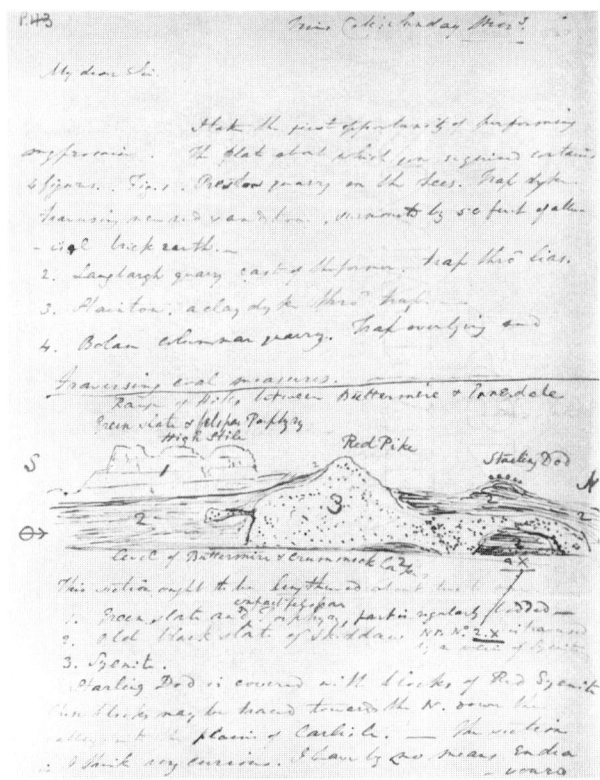

Fig. 73 Section through part of the Lake District: Adam Sedgwick to De la Beche, 1830 (**1870**)

Fig. 74 Piccadilly front of the new Museum of Economic [Practical] Geology (**2230**)

Fig. 75 The Great Hall, Museum of Practical Geology (**2231**)

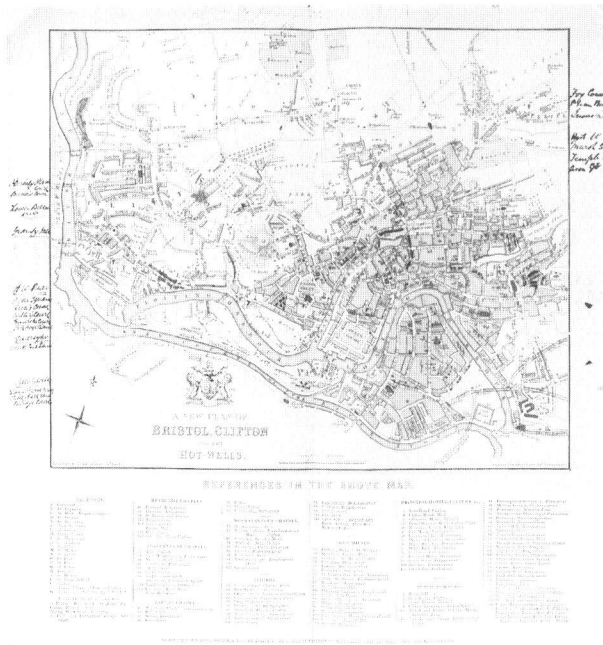

Fig. 76 Map of Bristol annotated to show outbreaks of cholera and fever, c.1843 (**2232**)

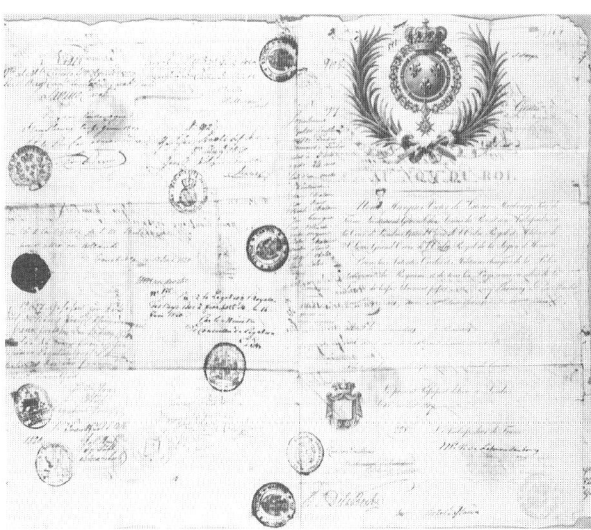

Fig. 77 Passport used by De la Beche on his 1819-1820 tour of the continent (**2246**)

Fig. 78 Medal struck for De la Beche and presented to James Holmes (**2280**)

Fig. 79 Miniature bust of De la Beche c. 1845 (**2283**)

REFERENCES

[* = Publications in which individual items from the archive are cited]

*Andrews, J.H., 1975. *A paper landscape. The Ordnance Survey in nineteenth-century Ireland.* Oxford: Clarendon Press, xxiv+350pp.

Bailey, E.B., 1952. *Geological Survey of Great Britain.* London: Thomas Murby & Co., xii + 278pp.

*Bassett, D.A., 1977. Foreword. *In* McCartney, P.J., *Henry De la Beche: observations on an observer.* Cardiff: Friends of the National Museum of Wales. xiii + 77pp.

*Bassett, M.G., 1985. 'Transition rocks and grauwacke' - The Silurian and Cambrian Systems through 150 years. *Episodes,* **8** (4), 231-235.

Bassett, M.G., 1990. [Letter to the editor]. *The Linnean. Newsletter and Proceedings of the Linnean Society of London,* **6** (1), p.12.

*Boylan, P.J., 1967. Dean William Buckland, 1784-1856. A pioneer in cave science. *Studies in Speleology,* **1**(5), 237-253.

Boylan, P.J., 1981. The role of William Buckland (1784-1856) in the recognition of glaciation in Great Britain. *In* Neale, J. and Flenley, J. (eds) *The Quaternary in Britain. Essays, reviews and original work on the Quaternary published in honour of Lewis Penny on his retirement.* Oxford: Pergamon Press, pp.1-8.

Boylan, P.J., 1984. *William Buckland (1784-1856): scientific institutions, vertebrate palaeontology and Quaternary geology.* Unpublished PhD thesis, University of Leicester.

Buckland, W., 1823. *Reliquiae Diluvianae; or, observations on the organic remains contained in caves, fissures, and diluvial gravel, and other geological phenomena, attesting the action of an universal deluge.* London: John Murray, vii + 303pp.

*Burkhardt, F. & Smith, S., 1986. *The correspondence of Charles Darwin. Volume 2. 1837-1843.* Cambridge: Cambridge University Press, xxxiii + 603pp.

*Burkhardt, F. & Smith, S., 1988. *The correspondence of Charles Darwin. Volume 4. 1847-1850.* Cambridge: Cambridge University Press, xxxiii + 711pp.

Butcher, N., 1996. Henry De la Beche and the geological map of southwest England. *In* Donovan, S.K. (ed.) De la Beche Meeting. An international conference to commemorate the bicentennial of the birth of Sir Henry De la Beche and to celebrate the 40th anniversary of the Geological Society of Jamaica. Articles, field guide, programme and abstracts, June 7th-8th, 1996, University of the West Indies, Mona, Kingston 7, Jamaica. *Contributions to Geology, UWI, Mona* No. 2, pp.9-10.

Cannon, W.F., 1970. William Buckland. *In* Gillispie, C.C. (ed.) *Dictionary of Scientific Biography* Vol II. New York: Charles Scribner's Sons, pp.566-572.

*Cartmill, M., 1992. Dragons in Eden. *Natural History,* **12/92**, pp.14-18.

Challinor, J., 1969. De la Beche, Sir Henry Thomas. Reprinted in Williams, T. (ed.) 1994. *Collins biographical dictionary of scientists.* Glasgow: Harper Collins, p.132.

*Chang, S.-K., 1997. *Fossil.* Seoul: Daewon Publishing Co., 128pp.

*Chubb, L.J., 1958. Sir Henry Thomas De la Beche. *Geonotes. Quarterly Newsletter of the Jamaica Group of the Geologists' Association. De la Beche Memorial Number,* December 1958, pp.1-26.

Colvin, H.M. (ed.), 1973. *The history of the King's Works. Volume VI 1782-1851.* London: HMSO, pp.460-461.

Cook, K.S., 1987. *Design sources of the first Ordnance Geological Survey map: Henry De la Beche and the Geological Society of London.* Paper read and distributed at the 12th International Conference on the History of Cartography, Paris 1987.

*Crowther, P.R., 1984. Edmund Artis: the obituary of 1849. *Durobrivae: a review of Nene Valley archaeology,* **9**, pp.14-15.

*Davies, G.L., 1969. The University of Dublin and two pioneers of English geology. William Smith and John Phillips. *Hermathena. A Dublin University Review,* **109**, 24-36.

Davies, G.L., 1974. First official geological survey in the British Isles. *Nature, London,* **249**, p.407.

*de Beer, G.R. and North, F.J., 1950. Sir Henry De la Beche's attempt on Mont Blanc in 1819. *Alpine Journal,* **57**, pp.493-500.

Donovan, S.K., 1996. De la Beche, C.A. Matley and the Jamaican 'Palaeozoic'. *In* Donovan, S.K. (ed.) De la Beche Meeting. An international conference to commemorate the bicentennial of the Birth of Sir Henry De la Beche and to celebrate the 40th anniversary of the Geological Society of Jamaica. Articles, field guide, programme and abstracts, June 7th-8th, 1996, University of the West Indies, Mona, Kingston 7, Jamaica. *Contributions to Geology, UWI, Mona* No. 2, pp.15-19.

*Dott, R.H., 1996. Lyell in America - his lectures, fieldwork and mutual influences 1841-1853. *Earth Sciences History,* **15**(2), pp.101-140.

Draper, G., 1996. De la Beche's "Remarks on the geology of Jamaica": context and content. *In* Donovan, S.K. (ed.) De la Beche Meeting. An international conference to commemorate the bicentennial of the Birth of Sir Henry De la Beche and to celebrate the 40th anniversary of the Geological Society of Jamaica. Articles, field guide, programme and abstracts, June 7th-8th, 1996, University of the West Indies, Mona, Kingston 7, Jamaica. *Contributions to Geology, UWI, Mona* No. 2, pp.2-8.

Draper, G. and Dengo, G., 1990. History of geological investigation in the Caribbean region. *In* Dengo, G. and Case, J.E. (eds) *The geology of North America, Volume H, the Caribbean Region*. Boulder: Geological Society of America, pp.1-14.

*Edmonds, J.M., 1956. William Buckland (1784-1856). *Nature, London*, **178**, pp.290-291.

*Edmonds, J.M. & Douglas, J.A., 1976. William Buckland F.R.S. (1784-1856) and an Oxford geological lecture 1823. *Notes and Records of the Royal Society of London*, **30**(2), pp.141-167.

Eyles, V.A., 1971. Henry Thomas De la Beche. *In* Gillispie, C.C. (ed.) *Dictionary of Scientific Biography* Vol IV. New York: Charles Scribner's Sons, pp.9-11.

Eyles, V.A. and Eyles, J.M., 1955. Two geological centenaries: G.B. Greenough, F.R.S. (1778-1855) and Sir Henry De la Beche, F.R.S. (1796-1855). *Nature, London*, **175**, pp.658-660.

Flett J.S., 1937. *The first hundred years of the Geological Survey of Great Britain*. London: HMSO, 280pp.

Fowles, J., 1982. *A short history of Lyme Regis*. Wimborne: Dovecote Press, 53pp.

*Gardiner, B.G., 1993. Editorial. *The Linnean. Newsletter and Proceedings of the Linnean Society of London*, **9**(1), pp.1-3.

Geikie, A., 1895. *Memoir of Sir Andrew Crombie Ramsay*. London: Macmillan, xi + 397pp.

*Gordon, [E.O. (A.B.)], 1894. *The life and correspondence of William Buckland, D.D., F.R.S., sometime Dean of Westminster, twice President of the Geological Society, and first President of the British Association*. London: John Murray, xvi + 288pp.

Hamilton, W.J., 1856. Obituary notice of H.T. De la Beche. *Quarterly Journal of the Geological Society of London*, **12**, p.xxxiv.

Harley, J.B., 1969. Introduction to the 1969 edition. *In* C. Close, *The early years of the Ordnance Survey*. [1st ed. 1926] Newton Abbot: David & Charles Reprints, pp.xi - xxxi.

Harley, J. B., 1971a. Place-names on the early Ordnance Survey maps of England and Wales. *Cartographic Journal*, **8**, pp.91-104.

*Harley, J.B., 1971b. The Ordnance Survey and the origins of official geological mapping in Devon and Cornwall. *In* K.J. Gregory & W.L.D. Ravenhill, *Exeter essays in geography in honour of Arthur Davies*. University of Exeter, pp.105-123.

*Harper, D.A.T. & Parkes, M.A., 1996. Geological Survey donations to the Geological Museum in Queen's College Galway: 19th Century inter-institutional collaboration in Ireland. *The Geological Curator*, **6**(6), pp.233-236.

*Harrington, B.J., 1883. *Life of Sir William E. Logan, Kt*. London: Sampson Low, Marston, Searle & Rivington, xv+432pp.

Harrison, W.J., 1908. Henry Thomas De la Beche. *Dictionary of National Biography*, Vol II. London: Smith, Elder and Co., pp.73-74.

*Herries Davies, G.L., 1982. Mary Anning. *In* G.Y. Craig & E.J. Jones, *A geological miscellany*. Oxford: Orbital Press, pp.104-105.

*Herries Davies, G.L., 1983. *Sheets of Many Colours. The mapping of Ireland's rocks 1750-1890*. Royal Dublin Society Historical Studies in Irish Science and Technology No.4, xiv + 242pp.

*Herries Davies, G.L., 1995. *North from the Hook. 150 years of the Geological Survey of Ireland*. Dublin: Geological Survey of Ireland, xi + 342pp.

*Holbrook, M., 1992. *Science preserved. A directory of scientific instruments in collections in the United Kingdom and Eire*. London: HMSO, 271pp.

*Howe, S.R., Sharpe T. & Torrens, H.S., 1981. *Ichthyosaurs: a history of fossil 'sea-dragons'*. Cardiff: National Museum of Wales, 32pp.

*Howes, C.J., 1988. The Dillwyn diaries 1817-1852, Buckland and caves of Gower (South Wales). *Proceedings of the University of Bristol Spelaeological Society*, **18**(2), pp.298-305.

Hunt, R., 1886. William Buckland. *In* Stephen, L. and Lee, S. *Dictionary of National Biography*, volume III, pp.206-208. Oxford University Press.

*James, F.A.J.L. (ed.), 1996. *The correspondence of Michael Faraday. Volume 3, January 1841-December 1848, Letters 1334 to 2145*. London: Institution of Electrical Engineers.

*Jones, I.M., 1990. Scientific visions: the photographic art of William Henry Fox Talbot, John Dillwyn Llewelyn and Calvert Richard Jones. *Transactions of the Honourable Society of Cymmrodorion*, 1990, pp.117-172.

*Lang, W.D., 1939. Mary Anning (1799-1847) and the pioneer geologists of Lyme. *Proceedings of the Dorset Natural History and Archaeological Society*, **60**, pp.142-164.

Lang, W.D., 1941. Early days of natural history at Charmouth. *Proceedings of the Dorset Natural History and Archaeological Society*, **62**, pp.97-113.

*Lang, W.D., 1958. Mary Anning's escape from lightning. *Proceedings of the Dorset Natural History and Archaeological Society*, **80**, pp.91-93.

Matley, C.A., 1951. *Geology and physiography of the Kingston District, Jamaica.* Kingston: Institute of Jamaica, vii + 141pp.

*McCartney, P.J., 1975. Henry De la Beche - a new kind of geologist. *Amgueddfa. Bulletin of the National Museum of Wales*, **21**, pp.13-28.

*McCartney, P.J., 1977. *Henry De la Beche: observations on an observer.* Cardiff: Friends of the National Museum of Wales. xiii + 77pp.

Miller, D.J., 1996. The contribution of Sir H.T. De la Beche to the geology of southwest England and to the identification of 'head' deposits. *In* Donovan, S.K. (ed.) De la Beche Meeting. An international conference to commemorate the bicentennial of the Birth of Sir Henry De la Beche and to celebrate the 40th anniversary of the Geological Society of Jamaica. Articles, field guide, programme and abstracts, June 7th-8th, 1996, University of the West Indies, Mona, Kingston 7, Jamaica. *Contributions to Geology, UWI, Mona* No. 2, pp.11-14.

*Morrell, J. 1988. Science and government: John Phillips (1800-74) and the early Ordnance Geological Survey of Britain. *In* N.A. Rupke, (ed.) *Science, Politics and the Public Good.* London: Macmillan, pp.7-35.

*Morrell, J. & Thackray, A., 1981. *Gentlemen of Science. Early years of the British Association for the Advancement of Science.* Oxford: Clarendon Press. xxi & 592pp.

*North, F.J., 1928. *Geological maps. Their history and development with special reference to Wales.* Cardiff: National Museum of Wales and the Press Board of the University of Wales, vi + 133pp.

North, F.J., 1933a. *Maps. Their history and uses, with special reference to Wales. A handbook to a temporary exhibition July to October, 1933.* Cardiff: National Museum of Wales and the Press Board of the University of Wales, 45pp.

North, F.J., 1933b. From Giraldus Cambrensis to the geological map. *Transactions of the Cardiff Naturalists' Society*, **64** (1931), pp.20-97.

*North, F.J., 1934a. From the geological map to the Geological Survey. Glamorgan and the pioneers of geology. *Transactions of the Cardiff Naturalists' Society*, **65** (1932), pp.41-115.

North, F.J., 1934b. The Museum of Practical Geology [letter to the Editor]. *Nature, London*, **134**, p.419.

*North, F.J., 1935. Dean Conybeare, Geologist. *Transactions of the Cardiff Naturalists' Society*, **66** (1933), pp.15-68.

*North, F.J., 1936a. Further chapters in the history of geology in South Wales; Sir H.T. De la Beche and the Geological Survey. *Transactions of the Cardiff Naturalists' Society*, **67** (1934), pp.31-103.

North, F.J., 1936b. De la Beche and his activities, as revealed by his diaries and correspondence.[abs.] *Proceedings of the Geological Society of London*, **92**, no.367, cxxviii-cxxx; *Abstracts of the Proceedings of the Geological Society of London*, No.1314, June 19th 1936, 1935-36, pp.104-106.

*North, F.J., 1938. Verses about Buckland. *Nature, London*, **142**, pp.1040-1041.

North, F.J., 1939a. H.T. De la Beche: geologist and businessman. *Nature, London*, **143**, pp.254, 255.

*North, F.J., 1939b. The Ordnance Geological Survey: its first Memoir, 1839. *Nature, London*, **143**, pp.1052-1054.

*North, F.J., 1942. Paviland Cave, the "Red Lady", the Deluge, and William Buckland. *Annals of Science*, **5**(2), pp.91-128.

*North, F.J., 1943. Centenary of the glacial theory. (Notes on manuscripts and publications relating to its origin, development and introduction into Britain). *Proceedings of the Geologists' Association*, **54**(1), pp.1-28.

North, F.J., 1944. Geology's debt to Henry Thomas De la Beche. *Endeavour*, **3**, (9) (Jan1944), pp.15-19.

North, F.J., 1951. Sir H.T. De la Beche: his contributions to the advancement of science, and the circumstances in which they were made. *Bulletin of the British Society for the History of Science*, **1** (5), p.111.

*North, F.J., 1956. W.D. Conybeare, his geological contemporaries and Bristol associations. *Proceedings of the Bristol Naturalists' Society*, **29**(2), pp.133-146.

*North, F.J., 1959. Sir H.T. De la Beche and the Geological Survey. *Glamorgan College of Technology, Treforest, Mining Society Magazine*, **2**, pp.59-64.

*North, F.J., 1966. The 'Red Lady' of Paviland. *In* Williams, S. (ed.) *Glamorgan Historian*, **3**, pp.123-137. Cowbridge: S.Williams.

*Oldroyd, D.R., 1991. The Archaean Controversy in Britain: Part I - The rocks of St David's. *Annals of Science*, **48**(5), pp.407-452.

*Oldroyd, D.R., 1992. The Archaean Controversy in Britain: Part II - The Malverns and Shropshire. *Annals of Science*, **49**(5), pp.401-460.

*Oliver, R., 1986. The origins of the Ordnance Survey quarter-inch mapping in Great Britain 1837-72. *Sheetlines. The Newsletter of the Charles Close Society*, **15**, pp.9-14.

Painting, D., 1987. *Amy Dillwyn.* University of Wales Press, Cardiff, x + 110pp.

Porter, A.R.D., 1990. *Jamaica. A geological portrait.* Kingston: Institute of Jamaica Publications Ltd, xi + 152pp.

Pym, H.N., 1882. *Memories of old friends. Being extracts from the journals and letters of Caroline Fox of Penjerrick, Cornwall from 1835 to 1871.* 2nd edition, Vol 1, London: Smith, Elder & Co., 333pp.

Randall, H.J. and Rees, W., (eds) 1963. Diary of Lewis Weston Dillwyn. *South Wales and Monmouthshire Record Society, Publication No.5,* iii +110pp.

Reeks, M., 1920. *Register of the Associates and Old Students of the Royal School of Mines and History of the Royal School of Mines.* London: Published under the auspices of the Royal School of Mines (Old Students') Association, 233pp + 212pp.

*Robinson, E., 1996. Field guide to the natural bridge at Riversdale, Lluidas Vale, and the eastern end of the Central Inlier, Jamaica. *In* Donovan, S.K. (ed.) De la Beche Meeting. An international conference to commemorate the bicentennial of the Birth of Sir Henry De la Beche and to celebrate the 40th anniversary of the Geological Society of Jamaica. Articles, field guide, programme and abstracts, June 7th-8th, 1996, University of the West Indies, Mona, Kingston 7, Jamaica. *Contributions to Geology, UWI, Mona* No. 2, pp.27-33.

Rose, E.P.F., 1996. Geologists and the army in nineteenth century Britain: a scientific and educational symbiosis? *Proceedings of the Geologists' Association,* **107**, pp.129-141.

Rudwick, M.J.S., 1972. *The meaning of fossils. Episodes in the history of palaeontology.* London: MacDonald, vi + 287pp.

*Rudwick, M.J.S., 1975. Caricature as a source for the history of science: De la Beche's anti-Lyellian sketches of 1831. *Isis,* **66**, pp.534-560.

Rudwick, M.J.S., 1976. The emergence of a visual language for geological science 1760-1840. *History of Science,* **14** (3), pp.149-195.

*Rudwick, M.J.S., 1985. *The Great Devonian Controversy. The shaping of scientific knowledge among gentlemanly specialists.* Chicago & London: University of Chicago Press, xxxiii + 494pp.

*Rudwick, M.J.S., 1989. Encounters with Adam, or at least the hyaenas: nineteenth century visual representations of the deep past. *In* Moore, J.R. (ed.), *History, humanity and evolution: essays for John C. Greene.* Cambridge: Cambridge University Press, pp.231-251.

*Rudwick, M.J.S., 1992. *Scenes from deep time. Early pictorial representations of the prehistoric world.* London: University of Chicago Press, xiii +280pp.

*Rupke, N.A., 1983. *The great chain of history. William Buckland and the English School of Geology (1814-1849).* Oxford: Clarendon Press, xii + 322pp.

*Rupke, N.A., 1994. *Richard Owen. Victorian naturalist.* New Haven & London: Yale University Press, xviii + 462pp.

*Secord, J.A., 1985. John W. Salter: the rise and fall of a Victorian palaeontological career. *In* A. Wheeler & J.H. Price (eds) From Linnaeus to Darwin: commentaries on the history of biology and geology. Papers from the Fifth Easter Meeting of the Society for the History of Natural History 28-31 March 1983 'Natural History in the early nineteenth century'. *Society for the History of Natural History Special Publication* No.3, pp.61-75.

*Secord, J.A., 1986a. *Controversy in Victorian geology. The Cambrian-Silurian Dispute.* Princeton: Princeton University Press, xvii + 363pp.

*Secord, J.A., 1986b. The Geological Survey of Great Britain as a research school, 1839-1855. *History of Science,* **24,** pp.223-275.

*Sharpe, T., 1985. Henry De la Beche and the Geological Survey in Swansea. *Gower,* **36**, pp.5-12.

*Sharpe, T., 1997. The archive of H.T. De la Beche (1796-1855) in the National Museum of Wales, Cardiff, UK. *Journal of the Geological Society of Jamaica,* **37**, pp.29-35.

*Stafford, R.A., 1984. Geological surveys, mineral discoveries and British expansion 1835-71. *Journal of Imperial and Commonwealth History,* **12**(3), pp.5-32.

*Stafford, R.A., 1989. *Scientist of Empire. Sir Roderick Murchison, scientific exploration and Victorian imperialism.* Cambridge: Cambridge University Press, xii + 293pp.

*Taylor, M.A., 1994. The plesiosaur's birthplace: the Bristol Institution and its contribution to vertebrate palaeontology. *Zoological Journal of the Linnean Society,* **112**, pp.179-196.

*Taylor, M.A., 1997. Before the dinosaur: the historical significance of the fossil marine reptiles. *In* J.M. Callaway & E.L. Nicholls (eds) *Ancient marine reptiles.* San Diego & London: Academic Press, xix-xlvi.

*Taylor, M.A. & Torrens, H.S., 1987. Saleswoman to a new science: Mary Anning and the fossil fish *Squaloraja* from the Lias of Lyme Regis. *Proceedings of the Dorset Natural History and Archaeological Society,* **108**, pp.135-148.

*Thackray, J.C., 1985. Separately-published prints of fossils in nineteenth century Britain. *Archives of Natural History,* **12** (2), pp.175-199.

Thomas, H.M., 1979. *Grandmother extraordinary. Mary De la Beche Nicholl 1839-1922*. Barry: Stewart Williams, 205pp.

*Tickell, C., 1996. *Mary Anning of Lyme Regis*. Lyme Regis: Philpot Museum, 32pp.

*Torrens, H.S., 1980 Collections and collectors of note No.28 Colonel Birch (c.1768-1829). *The Geological Curator*, **2**(9/10), pp.561-562.

*Torrens, H.S., 1995a. Mary Anning (1799-1847) of Lyme: 'the greatest fossilist the world ever knew'. *British Journal for the History of Science*, **28**, pp.257-284.

*Torrens, H.S., 1995b. The dinosaurs and dinomania over 150 years. *In* Sarjeant, W.A.S. (ed.), *Vertebrate fossils and the evolution of scientific concepts. Writings in tribute to Beverly Halstead by some of his many friends*. Gordon and Breach, pp.255-284.

*Torrens, H.S. & Taylor, M.A., 1990. Collections, collectors and museums of note, No. 50. Geological collectors and museums in Cheltenham 1810-1988. A case history and its lessons. *Geological Curator*, **5** (5) (for 1988), pp.175-213.

Wilson, H.E., 1985. *Down to Earth. One hundred and fifty years of the British Geological Survey*. Edinburgh: Scottish Academic Press, 189pp.

*Winder, C.G., 1965. Logan and South Wales. *Proceedings of the Geological Association of Canada*, **14**, pp.103-120.

*Zaslow, M., 1975. *Reading the rocks. The story of the Geological Survey of Canada 1842-1972*. Ottawa: Macmillan, 599pp.